0008812903

D1465462

STANDARD LOAN

Renew Books on PHONE-it: 01443 654456

Books are to be returned on or before the last date below

HERBICIDES

PHYSIOLOGY, BIOCHEMISTRY, ECOLOGY

2nd Edition

Volume 1

Edited by
L. J. AUDUS
Department of Botany, Bedford College
London University

1976

ACADEMIC PRESS
LONDON · NEW YORK · SAN FRANCISCO
A Subsidiary of Harcourt Brace Jovanovich, Publishers

ACADEMIC PRESS INC. (LONDON) LTD.
24/28 Oval Road
London NW1

United States Edition Published by
ACADEMIC PRESS INC.
111 Fifth Avenue
New York, New York 10003

Library of Congress Catalog Card Number: 75 21774
ISBN: 0 12 067701 6

Printed in Great Britain at The Whitefriars Press Ltd.
London and Tonbridge

LIST OF CONTRIBUTORS

F. T. ADDICOTT, *Department of Botany, University of California, Davis, California 95616, U.S.A.* (p. 191)

FLOYD M. ASHTON, *Department of Botany, University of California, Davis, California 95616, U.S.A.* (p. 220)

D. E. BAYER, *Department of Botany, University of California, Davis, California 95616, U.S.A.* (p. 220)

R. C. BRIAN, *Imperial Chemical Industries, Plant Protection Limited, Jealott's Hill Research Station, Bracknell, Berkshire, England* (p. 1)

MARTIN J. BUKOVAC, *Department of Horticulture, Michigan State University, East Lansing, Michigan 48823, U.S.A.* (p. 335)

P. M. CARTWRIGHT, *Department of Agricultural Botany, University of Reading, Whiteknights, Reading, Berkshire, England* (p. 55)

JOE H. CHERRY, *Department of Horticulture, Agricultural Experiment Station, Purdue University, West Lafayette, Indiana 47907, U.S.A.* (p. 525)

CHR. J. GORTER, *Englaan 22, Wageningen, The Netherlands* (p. 127)

J. R. HAY, *Canada Department of Agriculture, Regina Research Station, Regina, Saskatchewan, Canada* (p. 365)

J. L. HILTON, *U.S. Department of Agriculture, Agricultural Research Service, Northeastern Region, Agricultural Research Center, Beltsville, Maryland 20705, U.S.A.* (p. 483)

N. P. KEFFORD, *Department of Botany, University of Hawaii, Honolulu, Hawaii 96822, U.S.A.* (p. 427)

R. C. KIRKWOOD, *Department of Biology, University of Strathclyde, Glasgow, Scotland* (p. 444)

A. J. LINCK, *College of Agriculture, University of Minnesota, St Paul, Minnesota 55101, U.S.A.* (p. 83)

D. E. MORELAND, *U.S. Department of Agriculture, Agricultural Research Service, Southern Region, Crop Science Department, N. Carolina State University, Raleigh, N. Carolina 27607, U.S.A.* (p. 493)

P. W. MORGAN, *Department of Plant Sciences, College of Agriculture, Texas A & M University, College Station, Texas 77843, U.S.A.* (p. 255)

R. S. MORROD, *Imperial Chemical Industries, Biochemistry Section, Plant Protection Division, Jealott's Hill Research Station, Bracknell, Berkshire, England* (p. 281)

AUBREY W. NAYLOR, *Department of Botany, Duke University, Durham, N. Carolina 27706, U.S.A.* (p. 397)

C. PARKER, *ARC Weed Research Organization, Begbroke Hill, Yarnton, Oxford, England* (p. 165)

O. M. VAN ANDEL, *Botanical Laboratory, University of Utrecht, The Netherlands* (p. 127)

W. VAN DER ZWEEP, *Institute for Biological and Chemical Research on Field Crops and Herbage, Wageningen, The Netherlands* (p. 127)

J. L. P. VAN OORSCHOT, *Institute for Biological and Chemical Research on Field Crops and Herbage, Wageningen, The Netherlands* (p. 305)

PREFACE

The first edition of this book was prompted by the dramatic growth of herbicide research which followed the wartime discoveries of the auxin-type herbicides and the need to take stock of the plethora of complex data relating to weed killer action which had been the outcome of those new activities. In the intervening twelve years, this upward trend has continued, although perhaps it has slackened somewhat. I think it can truthfully be said that over this period there have emerged no new principles in herbicide science which were not described or at least adumbrated in the first edition. However, in a subject having ramifications into almost every branch of biology and roots in an actively progressive substratum of chemical industry, there is little wonder that a tremendous accumulation of detailed information, the development of more and more sophisticated techniques, and the continuing flow of new types of compound with new spectra of selectivities demand another stocktaking.

From the outset, it was obvious that if the level of treatment in the first edition was to be maintained, then the second edition would have to be much larger. Many branches of the subject covered by one chapter in the first edition had themselves so ramified that separate chapters in the second edition were called for, to give a logical treatment to the subject. The growing impact of herbicides use on modern society had created new interests and opened up new facets of research, and these demanded the introduction of completely new chapters. Even for the first edition, some critics had complained that the title of the book did not adequately describe its breadth of cover. During the writing of the second, several contributors voiced similar criticisms, now undoubtedly with much greater justification. The result has been that, on a consensus of contributors, the second edition has been retitled, to cater for the ecological aspects of herbicide science which have been so considerably amplified and augmented in it. From the practical point of view, this has necessitated the production of a two-volumed edition. Although the two volumes are not separately sub-titled, their contents fall reasonably logically into two distinct categories. The first volume, after an introductory chapter on history and classification, deals with the effects of herbicides on every aspect of the growth, development, structure and functioning of higher plants, crop as well as weed species. It comprises seventeen chapters in all, fourteen of which are completely new and written by experts in those branches of the subject into which the major topics of the first edition have been split. The second volume of fifteen chapters, covers what might be called the environmental aspects of herbicide science. It includes the physical and chemical behaviours of herbicides in the environment, the mutual interactions of herbicides with soil organisms of all kinds, the effect of environmental factors on herbicide effectiveness, and the impact of herbicides on natural and man-made higher plant communities. Again, eleven of these chapters are completely new. An indication of the thoroughness with which the changes since the first edition have been reviewed, is that of the total of about 5,000 references to original research articles, nearly three-quarters were published after the completion of the first edition, leaving a residue only slightly smaller than the total references of the first edition. On the other hand, this residue of pre-1963 papers is evidence of the proper regard which has been given to a full and balanced cover of each subject.

In spite of the escalating costs of finding and developing herbicides, a steady stream of new, economically viable molecules has continued over the twelve years. In his first edition chapter, Dr. R. C. Brian, who continues his review of the early history and chemical classification of herbicides, catalogued the names and structures of 95 separate compounds. In this edition he lists 156. The treatment of his new chapter follows the lines of the first, but additional technical details such as solubility in solvents, type of action etc. are now supplied for each molecule listed.

In the last decade, two aspects of plant growth and its regulation by chemical "growth regulants" have been receiving increased attention. The first is the fine structure of the cell and its component organelles as revealed by the electron microscope, a technique which has revolutionized the study of cell cytology. Its use to identify the organelles disrupted by the toxic action of herbicides, to pin-point the sites of operation in the cell, and to provide clues to their underlying molecular mechanisms, is now well established and producing valuable results. The subject has now grown to a stature that has warranted the inclusion of a completely new chapter by Professor A. J. Linck who has described and comprehensively catalogued the structural disturbances observed in a wide number of plants and a wide variety of herbicide treatments. The second aspect of plant growth which has enjoyed increasing attention from plant physiologists in recent years is dormancy. Those interested in weed control have been increasingly conscious of the implications of dormancy, its induction and removal, for the effective action of herbicides, and it was felt that this too deserved the separate chapter written by Dr. C. Parker. The three remaining chapters dealing with growth aspects of herbicide action, i.e. general growth responses by Dr. Phyllis Cartwright, morphogenic responses by Drs. W. van der Zweep, Christina J. Gorter and D. van Andel, and abscission responses by Professor F. T. Addicott, follow patterns similar to those of the first edition, although their updating has involved an almost complete re-writing.

In the first edition, a single chapter on plant composition and metabolism was, of necessity, a pot-pourri, with no particular theme or coherence. Since then research has tied up many loose ends, and from this mixture has crystallized a number of more clearly defined components, which can now be treated more logically in separate chapters. Thus the several metabolic activities are now included in separate chapters on biochemical mechanisms. Effects on plant composition, particularly those changes brought about by herbicides in the structural and storage components of the plant (i.e. carbohydrates, lipids and proteins) and the supply to and distribution in the plant of the raw materials for their construction (i.e. mineral nutrients), have been dealt with by Professor Floyd M. Ashton and Dr. David E. Bayer. This has left the complementary subject of the supply of carbon and water to the plant as a whole. Another new chapter, by Dr. J. L. P. van Oorschot, surveys the impact of herbicide application on those aspects of the physiology of weed and crop which concern the carbon dioxide interchange with the atmosphere, the uptake of water from the soil and its loss by transpiration from the leaves, all vital considerations, not only for the effectiveness of the herbicide on weeds, but also for the well being of the associated crop plant. These aspects of herbicide action had been covered in the first edition in another inescapable miscellany of more or less disparate phenomena. The residue of the topics left after the removal of the gas and water-exchange aspects, contained material concerned mainly with protoplasmic membrane properties and function. A chapter updating and co-ordinating these features of herbicide action, has been written by Dr. R. S. Morrod.

The first edition contained a chapter on the effects of herbicides on endogenous regulator systems, auxins, inhibitors, etc. In the new edition there is no such chapter, but aspects of it have been more appropriately integrated into the relevant chapters on mechanisms. The exception is ethylene, which in spite of its long-established and extremely dramatic modifying actions on the growth, development and tropic movements of plants, had not until recently, been regarded as a normal component of their natural regulatory systems. Now ethylene is virtually established as a natural hormone, and it is clear that many growth responses, natural and man-induced, including some which are the result of herbicide application, may be mediated via an ethylene-controlled system. Professor Page W. Morgan has reviewed this subject in a separate chapter.

Mechanisms of phytotoxicity are now expanded to four separate chapters, as demanded by the increasing emphasis which is being given to this research area. Professor N. P. Kefford has made a novel assessment of the thorny, and as yet unresolved problems of toxic action via the dislocation of normal growth processes; it is here that disturbances of natural growth regulant systems is reviewed. Respiratory and intermediary metabolism and the associated phenomena of the supply of high-energy molecules is surveyed by Dr. R. C. Kirkwood. A complementary chapter on the disruption of the biochemical components of the photosynthetic systems of plants and the associated flow of high-energy carriers is surveyed by Drs. James L. Hilton and Donald E. Moreland. Both these last chapters make a point of introducing their topics by outline descriptions of the biochemical systems involved, so that the complexities of herbicide action therein should be capable of appreciation by the reader not closely conversant with plant biochemistry. For a long time it has been apparent that the control of many aspects of plant growth, and development by regulator substances is via the modulation of nucleic acid and enzyme protein metabolism. Certainly the mechanisms of some of the actions of the auxin-type herbicides are linked to such metabolisms. Recent advances in this area are covered by a chapter by Professor Joe H. Cherry.

In the first edition only one chapter was devoted to herbicide behaviour in the plant, and this covered uptake from the environment, transport about the plant body and ultimate degradative or sequestering metabolism. Here, as for aspects of mechanisms already noted, there has been a particularly massive increase in research effort, and the information and understanding thus acquired has justified three separate chapters, on herbicide entry via leaf, bark and root by Professor Martin J. Bukovac, on herbicide transport and accumulation by Dr. J. R. Hay, and on herbicide metabolism by Professor Aubrey W. Naylor.

The second volume starts with five chapters on aspects of herbicide behaviour in the soil, a topic needing only two chapters in the first edition. Dr. G. S. Hartley has reconstructed his chapter on the physical aspects of herbicide behaviour. Except for updating and an improved analytical approach, the cover is unchanged. Transformations of herbicide molecules now occupy two chapters. The first, by Drs Phillip C. Kearney and D. D. Kaufman, shows what a transformation has taken place in the last decade or so, in our understanding of the molecular details of herbicide degradation by soil micro-organisms. On the other hand, soil scientists have realized that a multitude of non-biological transformations can also take place in the soil environment, some of which can result in the rapid loss of the active weed killer molecule; the full complexities and extent of such processes are comprehensively reviewed by Professor D. G. Crosby. Over the years there has been a small, but steady interest shown by soil microbiologists on the impact of applied herbicides on various component species of the soil microflora, their biochemical activities and, to some extent, the ecological equilibria between various populations. Thus a complex, not to say a rather bewildering picture has built up, a situation which has been carefully and systematically sifted by Dr. Erna Grossbard in her chapter. It is to be regretted that until recently, scant attention has been given to the effects of pesticides, and in particular herbicides, on the multifarious populations in the soil fauna, which also have an important role in maintaining soil fertility. Happily this deficiency is being rectified, and it is gratifying that enough information has now accrued to allow Drs. J. van der Drift and H. Eijsackers to contribute a separate chapter on this subject.

Inevitably continuing studies on phytotoxicity reveal, either incidentally or by directly orientated research, the nature of the complex of factors, both internal and environmental, which modify in one direction or another the lethality of the applied molecule. The original chapter by Dr. Ewert Åberg on factors intrinsic to plants which determine their susceptibilities, such as maturity, disease, genetic make-up etc. has been brought up-to-date in collaboration with Dr. Vilmos Steckó. So much is now known about the environmental factors which may modify the susceptibility of the plant to herbicides, that this too has been given a separate chapter. In it, Professor T. J. Muzik has analysed the effect of weather conditions, before, during and after treatment on plant responses, and has looked, not only at those actions which directly modify the herbicide behaviour in the plant, but also at those which arise from modifications of its morphology.

The important subject of herbicide selectivity has been given three chapters instead of two. There is an updated chapter by Dr. K. Holly on the effects of formulation and application methods and one largely rewritten by Professor R. L. Wain and Dr. M. S. Smith on the relationship of metabolism of the applied compound within the target plant to its suceptibility. A third aspect, not covered in the first volume, is the extent to which selectivity may depend on specific uptake and movement parameters within the plant. Although this is still an area of study which has not received the attention it deserves, yet there is sufficient data of importance and interest to warrant the short chapter by Dr. J. A. Sargent.

The remainder of the new edition is devoted to a set of topics which, although they have important contacts with and implications for what has gone before, nevertheless are sufficiently important and circumscribed to need separate chapters. The stimulatory effects of sublethal concentrations of herbicides have now been firmly established by an intensified programme of work in recent years, and Professor S. K. Ries, who has played a considerable part in this programme, has contributed a chapter on them.

Although the usefulness of herbicides in such ecological matters as range management has long been recognized, it took the widespread military use of defoliant herbicides in the Vietnam war to underline the extensive and perhaps irreversible changes that excessive herbicide application can bring about in plant communities. In a new chapter on herbicides in higher plant ecology, Drs. J. M. Way and R. J. Chancellor have made a balanced and completely objective assessment of the changes which have been induced, intentionally or unintentionally, in natural, seminatural and agricultural plant communities in terrestrial and in aquatic habitats.

The toxicity of herbicides to animals, particularly man, is of obvious importance to users, and quite apart from the tragic, yet happily rare cases of accidental poisoning, is a subject on which manufacturers and consumers alike need to be unequivocally informed. Such information is provided in a chapter by the late Dr. J. M. Barnes.

Since the first edition, the sophistication, sensitivity and specificity of methods used for the assay of herbicide traces in soil, and in plant and animal material, have been vastly improved by extensive refinements, both of techniques and equipment. A much extended chapter on the extraction, purification and quantification of such traces has been provided by Drs. R. J. Hance and C. E. McKone with a most comprehensive documentation of published material, which will be invaluable to research workers looking for techniques appropriate to their particular problem. The edition concludes with a chapter on the way in which modern chemical industry looks for new herbicides to meet the ever present demand for enhanced toxicity and improved selectivity. It is written by Dr. D. T. Saggars, a scientist of wide experience in screening techniques, and traces the whole complicated process from the synthetic chemist's flask to the final emergence of a commercially viable molecule.

One of the most difficult and potentially harrowing tasks of an editor in organizing a treatise such as this, is the avoidance of gross overlaps or wide lacunae between contiguous components of the overall mosaic. In this instance, my task has been made easy by the sympathetic and readily cooperative approach of all the contributors, for which I tender them my sincere thanks.

Bedford College,   L. J. AUDUS
London University,
January 1976

CONTENTS

1. THE HISTORY AND CLASSIFICATION OF HERBICIDES

R. C. BRIAN

2. GENERAL GROWTH RESPONSES OF PLANTS

P. M. CARTWRIGHT

8. Effects on Ethylene Physiology

P. W. MORGAN

9. EFFECTS ON PLANT CELL MEMBRANE STRUCTURE AND FUNCTION

R. S. MORROD

10. EFFECTS IN RELATION TO WATER AND CARBON DIOXIDE EXCHANGE OF PLANTS

J. L. P. VAN OORSCHOT

11. HERBICIDE ENTRY INTO PLANTS

MARTIN J. BUKOVAC

12. HERBICIDE TRANSPORT IN PLANTS

J. R. HAY

13. HERBICIDE METABOLISM IN PLANTS

AUBREY W. NAYLOR

14. DISLOCATION OF DEVELOPMENTAL PROCESSES

N. P. KEFFORD

15. ACTION ON RESPIRATION AND INTERMEDIARY METABOLISM

R. C. KIRKWOOD

16. Actions on Photosynthetic Systems

D. E. MORELAND AND J. L. HILTON

17. ACTIONS ON NUCLEIC ACID AND PROTEIN METABOLISM

JOE H. CHERRY

CONTENTS OF VOLUME 2

CHAPTER 1

THE HISTORY AND CLASSIFICATION OF HERBICIDES

R. C. BRIAN

*Imperial Chemical Industries, Jealott's Hill Research Station
Bracknell, Berkshire, England*

I. INTRODUCTION AND HISTORY

Weeds are not a problem only of recent years. They were alluded to in the parables of Jesus Christ (St. Matthew, Chap. 13) and hence, even 2,000 years ago they must have been very much in people's minds. He referred to the deleterious effect of weeds in two ways. First, in the parable of the sower they choke and reduce the yield of the crop. Second, in the parable of the tares, sown by an enemy, the crop itself is disturbed and its growth impaired when weeds are removed from the crop. The removal of weeds also without disturbing the crop are two of the important advantages inherent in the techniques of chemical weedkilling.

In A.D. 1971, LeRoy Holm (1971) reports that more energy is still expended on the weeding of man's crops than on any other single human task! Although weeds have challenged man's effort to survive ever since he tilled the soil for food, and much of the world's crop production lies in the tropical zones, it is there we find that advanced weed control methods are only practised on a limited scale. Hand weeding is still carried out widely in the tropics. Before the introduction of chemical weedkilling, four measures were adopted to eradicate or limit the spread of weeds. These were manual weeding, crop rotation, ploughing and various methods of preventing weed seeds being dispersed in crop seed. The methods suffer from a basic weakness, namely that they are aids to control but they cannot prevent weeds growing with the crop. The hoe was necessary to select and remove weeds growing beside a crop plant. The cost of weeding has increased greatly during this century as the cost of labour has increased and selective weed control by hand today would be economically prohibitive—if indeed the labour required for the operation was available.

Chemicals such as salt and various industrial by-products such as smelters' wastes, have been applied to roadsides and paths to rid them of vegetation, for some hundreds of years. Chemical weedkilling however, may be considered to have been born in 1896 when Bonnet, a French grape grower, observed that the Bordeaux mixture he applied to his vines as a protection against downy mildew, turned the leaves of yellow charlock *(Sinapis arvensis)* black. The weedkilling properties of sulphates of ammonia, zinc, iron and other metals were then soon observed.

Later milestones in weed control were the introduction of the first organic chemical, 2-methyl-4,6-dinitrophenol (DNOC) in 1932 and, a few years later, the important discovery that chemicals related structurally to plant hormones could be used as selective weedkillers. Here at last, was the first chemical substitute for the hoe. The use of non-selective and, later, selective residual chemicals such as the substituted phenylureas and triazines, and the non-residual chemicals diquat and paraquat may be considered as more recent milestones. Paraquat has been developed as a chemical substitute for the plough.

Chemicals are solving many problems but they are also creating new ones. As groups of weeds are brought under control, others which are resistant to the chemicals thrive with reduced competition. An increase in grassy weeds occurred after the introduction of selective hormone weedkillers. The new problems are being solved by the use of herbicide rotations or by mixtures, in order to broaden the spectrum of weeds that may be controlled. The astronomic growth in the control of weeds by chemicals which has occurred in the 70 years since 1896 can be seen in Table I (Furtick, 1970). The chemical weed-control arsenal available to the farmer, horticulturist and indeed to anyone with land where weeds may impair efficiency, contains chemicals of

TABLE I

ESTIMATED WORLD CONSUMPTION OF HERBICIDES AT CONSUMER LEVEL IN 1968

Area	Consumption (Millions of $)
North America	550
Latin America	80
Near East, Southeast Asia, Oceania	80
Japan	70
Western Europe	60
Africa	40
TOTAL	880

widely differing physical and biological properties. Their discovery and properties relating to weed control will now be considered in more detail.

Historically the inorganic chemicals were the first to be exploited.

A. INORGANIC CHEMICALS

Observations of the phytocidal effects of Bordeaux mixture by Bonnet and of the effectiveness of 6% copper sulphate were reported to the Agricultural Association at Rheims on 12th December, 1896. Copper sulphate was thus instituted as a selective weedkiller for the destruction of charlock, *Sinapis arvensis* L., in cereal crops. It was further developed simultaneously but independently by Schultz (1909) in Germany and by Bolley (1908) in America (see Rademacher, 1940). This was undoubtedly a physical selectivity depending on the preferential wetting of charlock leaves compared with cereal leaves. Other sulphates were tested in quick succession so that in 1897 a French farmer, Martin, of Pont d'Ardres, and Duclos in the same year, used sulphuric acid, ferrous sulphate and copper nitrate with good results, as quoted by Bissey and Butler (1930).

The rate of development increased progressively in the following three years and by 1900 it was known that many commercial fertilizers also acted as weedkillers. Independently, Heinrich in 1899 and Stender in 1900 (Rademacher, 1940), found that Chile saltpetre ($NaNO_3$), ammonium sulphate and certain potassium salts were just as effective as ferrous sulphate. Ten years later, Remy carried out more extensive work and kainit and calcium cyanamide resulted (Remy and Vasters, 1914).

The high nitrogen content of calcium cyanamide and of potassium in kainit required the exercise of care in their use in order to maintain a correct nutrient balance in the soil. Therefore it was often inadvisable to use large doses of either compound alone. Moreover, Remy found as early as 1914 that a 10% addition of calcium cyanamide to kainit was more phytocidal to various plants than a double quantity of either by itself.

Sodium arsenite was also studied at an early date. Bolley (1901) compared copper sulphate, sodium arsenite and arsenate and concluded that although sodium arsenate was ineffective, "copper sulphate and sodium arsenite may be effectively used against all weeds which have a surface upon which the solution will spread". Although other arsenic derivatives (e.g. the sulphoarsenites, sulphoarsenates and sulphoxyarsenates) were covered by patent (Raymond, 1939) sodium arsenite has proved to be superior and has been used in considerable quantities in a number of countries. Other herbicides are now replacing it mainly because of its high mammalian toxicity.

Rabaté (1911) must be given the credit for much of the basic work on sulphuric acid for weed destruction in cereals even though its use had been advocated by agricultural societies since about 1900. He demonstrated clearly the real virtues of the acid and subsequently it rapidly replaced copper sulphate in France. The work was taken on in Germany by Gelpke in 1913, an account of his work appearing in a 1918 report of the German Agricultural Society. A review by McDowall (1933) covers the use of sulphuric acid at home and abroad up to 1933. A section on the selective action of sulphuric acid and a number of sulphates would be incomplete without reference to the extensive work of Morettini (1915) in Italy and of Korsmo (1932) in Norway.

The toxicity of boron compounds had been observed in 1876 by Peligot (1876) but further interest in boron compounds awaited the investigations of Thompson and Robbins (1926) involving the use of borax and boric acid for the eradication of barberry *(Berberis)*. They were unsuccessful. Offord (1931) made further tests but could not commend boron compounds for weed control. Borax has however, found a place for weed control on non-crop sites at rates equivalent to 10–24 cwt B_2O_3/acre (1,250–3,000 kg/hectare) and for the eradication of St. John's Wort and poison ivy (Stodard, 1944). After heavy applications, it remains in the soil for up to two years depending on climatic conditions and soil properties. The situation is summed up by Crafts and Raynor (1936) following extensive glasshouse and field experiments on boron compounds. Their criticisms include the toxicity to economic crops and the speed with which they are leached from the soil after all but heavy doses. For effective soil sterilization, dosages must be heavy.

The early interest in chlorates no doubt stemmed from the occasional injury to crops resulting from the use of sodium nitrate as fertilizer. Sjollema (1896) suggested this to be due to potassium perchlorate known to be present as impurity and de Caluwe (1900) conducted experiments with rye (*Secale cereale*) using perchlorate and chlorate. Guthrie and Helms (1904) found toxic effects resulting from the inclusion of 0·001% sodium chlorate in soil in which wheat and maize plants were growing. The first attempt to use chlorate for the destruction of weeds appears to have been made in Australia about 1901 where it was tested against prickly pear (*Opuntia*) without much success. Later in Switzerland and France under the name "mort-herbes" it was used for the control of weeds on roadside verges. The later development of sodium chlorate as a herbicide was contemporary with that of the borates. Åslander in 1926 was perhaps the first to demonstrate that sodium chlorate was effective for killing deep-rooted perennial weeds by soil applications (Åslander, 1926, 1928). A number of American workers, notably Loomis et al. (1931), Muenscher (1932) and Crafts (1935) confirmed later that it was most effective when applied to the soil. A full account of the use of both borates and chlorates up to 1942 is given by Robbins et al. (1942).

Yet another inorganic salt was found to possess herbicidal properties when Harvey (1931b) found that 15 cwt/acre (1,900 kg/hectare) of ammonium thiocyanate rendered soil sterile for up to four months. Smaller doses were also effective but for shorter periods of time. Ammonium thiocyanate sprays destroyed annual weeds in cereal crops in later experiments by Singh and Das (1939).

One of the most recent innovations in the field of inorganic chemical herbicides is the use of ammonium sulphamate as a contact weedkiller or as a soil sterilant. This was not patented as a herbicide until 1942 (Cupery and Tanberg, 1942). The reader is referred to a comprehensive review on inorganic chemicals as herbicides by Long (1934).

B. ORGANIC CHEMICALS

1. *Nitrophenols and Anilines*

The first use of organic chemicals for weed control is contained in a patent by Truffaut and Pastac (1935) covering the use of nitrophenols as selective herbicides. The sodium salt of 2-methyl-4,6-dinitrophenol became widely used in Europe under the common name DNOC and in the USA under the name "Sinox". This chemical was the forerunner of a family of nitrophenols, all of which show marked selective properties. Some are active by contact action, some by absorption from the soil by the roots.

DNOC itself controls weeds in cereals and is used for · pre-harvest desiccation of potatoes. Crafts (1946) improved on the activity of DNOC finding that 4,6-dinitro-2s-butylphenol (dinoseb) was more active on some weed species but less so on economic crops such as peas. Dinoterb, first described by Poignant and Crisinel (1967) is similar to dinoseb but it is more selective on cereals and less dependent for action on climatic conditions such as temperature. It is used mostly for post-emergence weed control in lucerne and pre-emergence weed control in peas and beans. Bromofenoxim (Green, 1969) is one of the most recent introductions to this family of contact herbicides. It has a powerful contact action on dicotyledons and it is used for selective weed control in cereals. A number of nitro-compounds is taken up mainly by the roots. Nitralin introduced by Shell in 1966 and trifluralin both control annual grass and weeds in a number of crops when incorporated into the top 1–2 in. (2·5 to 5 cm) of soil. Nitrofen is similar but this chemical should not be incorporated in soil but placed as a layer on the soil surface.

2. *Hormone-Type Herbicides (phenoxyalkane carboxylic acids)*

In the late 1920s and early 1930s, the seeds were being sown for what is now acknowledged to be the greatest advance in the history of weedkilling—the introduction of selective hormone-type weedkillers. Thus in 1928, Went summed up the conditions for plant growth in the terse statement "Ohne Wuchsstoff kein Wachstum" (Went, 1928). The isolation of this "Wuchsstoff" in the form of auxins A and B was claimed by Kögl *et al.* (1934) who later extracted heteroauxin (auxin C) from urine (Went and Thimann, 1937). The latter was shown to be indol-3yl-acetic acid, known in older literature as heteroauxin and now abbreviated to 1AA. Attempts to synthesize auxins A and B have since failed but 1AA was first prepared by E. Fischer (1886) in Germany. Synthetic compounds related to 1AA were applied to a wide range of plants by Zimmerman and Wilcoxon in 1935 and a number of physiological and morphological effects was observed from the use of naphth-1yl-acetic acid (NAA), naphth-2yl-oxyacetic acid (NOXA) and γ-(indol-3yl)butyric acid (1BA) (Zimmerman and Wilcoxon, 1935). These substances found immediate uses as rooting compounds in horticulture, but the discovery that related chemicals could be used for selective weed control was not made until early in the Second World War when it was shown by Templeman (Slade *et al.* 1945) and by Nutman *et al.* (1945) working independently in England that such plants as charlock (*Sinapis arvensis*) and sugar beet were killed by NAA, but cereals were unharmed by it. Subsequently, tests indicated that 2-methyl-4-chlorophenoxyacetic acid (MCPA) and 2,4-dichlorophenoxyacetic acid (2,4-D) were much more effective. Because of the exigencies of war, these results were not published until 1945. Meanwhile, Marth and Mitchell (1944) in America had also discovered the herbicidal effects of growth regulator chemicals. After 1945, a rapid expansion ensued in the use of MCPA and 2,4-D so that in 1949, about 10,000 tons of 2,4-D were manufactured in the USA alone for the control of broad-leaved weeds.

Specific uses were found for other chemicals in this group. Thus Hamner and Tukey (1944) laid the foundations for the use of 2,4,5-trichlorophenoxyacetic acid (2,4,5-T) as an effective brushkiller when they reported that it was more effective than 2,4-D for controlling bindweed and various woody plants. More recently 2-(2,4,5-trichlorophenoxy)propionic acid (fenoprop or silvex)—the α-methyl derivative of 2,4,5-T, was shown to be superior to 2,4,5-T on certain woody species such as oak. The salts of this acid were introduced in 1953 by the Dow Chemical Company under the trade name "Kuron". Yet another α-methyl derivative was found to have specific properties. Boots Pure Drug Co. filed an application for the use of ($\pm$)-2-(4-chloro-2-methylphenoxy) propionic acid (mecoprop) in 1956 (Lush and Stevenson, 1959) having found it to be particularly effective against cleavers (*Galium aparine*) and chickweed (*Stellaria media*) which are resistant to phenoxyacetic acids. Two further chemicals in this group, namely γ-(4-chloro-2-methylphenoxy)butyric acid (MCPB) and 2,4-dichlorophenoxyethyl hydrogen sulphate (2,4-DES) owe their activity to a biological conversion of the compounds to the respective phenoxyacetic acids (Wain, 1957) (see Chapter 9, vol. 2).

3. *Benzoic and Phenylacetic Acids*

Other aromatic acids have found limited applications as herbicides. For example, 2,3,6-trichlorobenzoic acid (2,3,6-TBA) and its aldehyde were tested in the field as early as 1948 at the Plant Protection Limited research station, Jealott's Hill, as an all-purpose, non-selective herbicide. 2,3,6-TBA was also investigated in the USA and found to be promising for the pre-emergence and early post-emergence control of weeds in maize at doses of 0·5–1·0 kg/ha. It is absorbed by plant leaves and roots and can persist in soil for long periods. Bentley (1950) reported high physiological activity for this chemical and Minarik *et al.* (1951) showed that of 200 derivatives of benzoic acid, only seven were active growth inhibitors. It is reasonable to assume that the more detailed work of Zimmerman and Hitchcock (1951) on 2,3,6-TBA itself in which they observed promotion of cell elongation, proliferation of tissue and the induction of adventitious roots, gave impetus to the development of practical applications for this useful chemical. It is now being used for the control of bindweed and couch grass and is effective at low rates for killing a variety of woody species and evergreens which are resistant to the phenoxy acids. It is also used in combination with hormone weedkillers to improve the control of dicotyledon weeds. A major weakness of 2,3,6-TBA in some circumstances is its long residual toxicity in the soil (see Chapter 2, vol. 2). 2-methoxy-3,6-dichlorobenzoic acid (dicamba), is a translocated post-emergence herbicide. It is used for the control of docks in established grassland and for the control of bracken. It is usually used at low rates of application (130 g/ha) in conjunction with other appropriate herbicides.

One phenylacetic acid has achieved success in commercial application. 2,3,6-trichlorophenylacetic acid (fenac) was introduced in 1957 and experimental work indicated that at 3 kg/ha, it controls annual weeds and quackgrass (*Agropyron repens*) for an entire growing season (Beatty, 1958). At 2 kg/ha, it can be used selectively to control annual weeds and seedling perennials in maize and sugar beet.

4. *Halo-Aliphatic Acids*

The herbicidal action of 1% chloroacetic acid was described in 1951 by Zimmerman *et al.,* when they claimed that several weed species were killed within 24 h. It also had a selective action, no damage being sustained by corn plants, roses and carnations (Zimmerman *et al.* 1951). It is recommended as a contact herbicide for controlling weeds in kale, sprouts, cabbage and leeks at rates of about 20 kg/ha.

Two further derivatives have found a place in agriculture. The grass-killing properties of sodium trichloroacetate (TCA) were observed by McCall and Zahnley (1949). It is absorbed by the roots of plants and it has found commercial outlets in the control of couch grass at 15−30 kg/ha when used in conjunction with some tillage operations; it also controls wild oats (*Avena fatua*) in sugar beet, peas and kale at rates of about 7 kg/ha prior to planting.

By substituting one chlorine atom of TCA by a methyl group, one obtains 2,2-dichloropropionic acid which was given the name "dalapon" by the Dow Chemical Company (1953). Introduced as the sodium salt, it is selectively toxic to grasses in much the same way that 2,4-D is to broad-leaved plants. Unlike TCA, it is absorbed by the plants both through the roots and the leaves. Its principal use is for the control of annual and perennial weed grasses of arable land including couch using rates up to 20 kg/ha.

5. *Amides*

The amide naptalam was known as long ago as 1949 (Hoffman and Smith, 1949). Amides now constitute a moderately-sized family of herbicides which have the important property of inhibiting seed germination and/or growth of seedlings. Thus naptalam (*N*-1-naphth-yl-phthalamic acid) and allidochlor (*NN*-diallylchloroacetamide) are used as pre-emergence herbicideswith a persistence in the soil of up to eight weeks. Some grasses are particularly sensitive to allidochlor (Hamm and Speziale, 1956). Benzadox, however, introduced in 1965, is a contact herbicide, the main use of which is the post-emergence control of weeds in sugar beet. Another group of amides contain the phenyl group and may also be termed anilides since they are related structurally to aniline

 NH₂ .

The members of this sub-group are similar in their biological properties and are used as selective herbicides for controlling annual grasses and broad-leaved weeds in a number of crops such as maize, beans, cotton, radish and brassicas. Alachlor, butachlor and propachlor are taken up through the roots and are used as pre-emergence selective weedkillers. Their soil persistence is up to 12 weeks in the case of alachlor. Butachlor is one of the most recent introductions, its herbicidal properties being first described by Baird and Upchurch (1970). This chemical is similar in herbicidal properties to the other members of the family but it is additionally used as a pre-emergence herbicide to control grass and broad-leaved weeds in seeded and transplanted rice.

Anilides used post-emergence include monalide, pentanochlor and propanil. They have low soil persistence but have found useful outlets for weeding umbelliferous crops such as carrots, celery, parsley. Tomatoes also are tolerant to pentanochlor. Propanil may be used for controlling weeds in rice and potatoes.

Oryzalin, a sulphonamide, is a recent introduction. Like other members of the family, it is also finding uses as a pre-emergence weedkiller particularly in orchards and vineyards.

6. *Nitriles*

Nitriles are among the more recent introductions into the herbicide field. Dichlobenil (2,6-dichlorobenzonitrile), an inhibitor of actively-dividing meristematic cells was the first to be developed; its herbicidal properties were first described by Koopman and Daams (1960). It is used as a selective weedkiller applied pre- or post-emergence, as an aquatic weedkiller and at application rates up to 20 kg/ha, for total herbicide action. More recently, the herbicidal properties of ioxynil and bromoxynil were described by Wain (1963) and independently by Carpenter and Heywood (1963). In the same year, ioxynil was introduced as a selective herbicide for the control of broad-leaved weeds in cereals with little residual activity. Bromoxynil is similar but less active than ioxynil. Bromobenzocarb is another member of this family.

7. *Aryl Carbamates*

The biological activity of compounds of the carbamate series has been known for many years having been demonstrated on animals in 1927 (Frankel, 1927).

Two years later, phenylurethane was shown to retard the germination of oats and wheat with resulting abnormal growth (Friesen, 1929). Other workers within the following 15 years noted the similarity of their action to that of colchicine but it was during the war years 1939–45 that Templeman and Sexton (1945) found that isopropyl *N*-phenylcarbamate (propham) and other aryl carbamic esters were toxic to monocotyledons but not to dicotyledons. The family of compounds (which are absorbed by the roots and not the foliage) are, therefore, used to control grasses in crops such as peas and beet at rates of 2–4 kg/ha. One of the most serious limitations of propham however, is its failure to control crabgrass (*Digitaria* spp.). This was later shown by De Rose to be controlled satisfactorily by the 3-chloro derivative. Thus he found that isopropyl-*N*-(3-chlorophenyl)carbamate (chlorpropham) controlled crabgrass at the rate of 1 mg/kg soil whereas it was unaffected by 10 mg of propham per kg soil (De Rose, 1951).

Two further carbamates, 2,3-dichloroallyl-di-isopropyl-thiolcarbamate (di-allate) and butinol-*N*-(3-chlorophenyl)carbamate (BiPC) were described at the Fifth British Weed Control Conference as potential wild oat (*Avena fatua*) killers. The former is incorporated into the top 4 in. (10 cm) of soil just before sowing the crop, whereas the latter is sprayed on to the foliage at times carefully selected in respect to the stage of growth of the crop and weeds.

With similar attention to the stage of plant growth, barban (4-chlorobut-2-ynyl-*N*-(3-chlorophenyl)carbamate is also a highly selective herbicide. Applications up to 0·75 kg/ha to crops at the 2–3 leaf stage kill wild oats without significant damage to cereals and broad-leaved crops such as broad beans and sugar beet. More recent additions to the range include carbetamide first described by Desmoras *et al.* (1963), phenmedipham (Arndt and Kötter, 1967) and the sulphur analogue asulam (Cottrell and Heywood, 1965).

8. *Substituted Ureas*

An important contribution was made to the history of herbicides when Bucha and Todd (1951) found that N^1-(4-chlorophenyl)-NN-dimethylurea (monuron) was very effective in controlling many plant species especially annual and perennial grasses. In the following year, McCall reported that, after extensive field trials and laboratory tests, monuron had proved to be an effective soil sterilant at rates of 20 kg/ha. It had also shown promise as a selective herbicide since some crops were found to be tolerant to small amounts of the material (McCall, 1952). It is now used for the pre-emergence control of broad leaved and grass weeds in crops such as cotton, sugar cane, pineapple and asparagus at rates of 1–5 kg/ha. It was subsequently found that the chemical was extremely persistent and soil could be sterilized to higher plants for periods of a year or two depending on the amount applied.

Pre-emergence control is possible because the substituted ureas are taken up both by foliage and through the roots. Their persistence and depth of action in the soil depends on physical properties such as water solubility and their degree of binding to soil.

Ureas other than monuron having varying solubilities in water were developed in quick succession. Thus Sharp *et al.* (1953) described *NN*-dimethyl-N^1-phenylurea (fenuron) as a promising chemical for the control of deep rooted perennial weeds, especially field bindweed (*Convolvulus arvensis*). This chemical is less strongly held to soil and hence is effective to greater soil depths than is monuron. On the other hand N^1-(3,4-dichlorophenyl)-*NN*-dimethylurea (diuron) with low water solubility, is extremely persistent in soils. An even less soluble urea, N-butyl-N^1-(3,4-dichlorophenyl)-N-methylurea (neburon) is now accepted for the control of annual weeds in certain evergreens, for the control of common and mouse-ear chickweed (*Stellaria media* and *Cerastium vulgatum* resp.) and shows promise for annual weed control in seedling alfalfa, tomatoes and strawberries.

Although the general physical properties of the urea herbicides are similar, they have greatly varying half-life periods in the soil. By a suitable choice of chemical, weeds may be controlled in crops for extended periods of time. Fluometuron with a half-life period of about 60 days in the soil, is an example. Absorbed mainly through the roots of plants, it was introduced as a herbicide, to which cotton is tolerant. It has been possible to develop this chemical for the control of broad-leaved and grass weeds in cotton at rates as low as 1−2 kg/ha.

The alkoxyurea herbicides are also similar in action, but are more soluble in water and less persistent in soil. Monolinuron and linuron with the more recent introductions such as metobromuron and chlorbromuron (Green *et al.* 1966) are used mainly for weed control in carrots, groundnuts, peas, soyabeans and potatoes at rates of 0·75−2·0 kg/ha.

Other members of this interesting family of herbicides are cycluron, noruron and dichloralurea and the reader is referred to pp. 26−27 for their chemical formulae. Siduron, however, is different. The herbicidal properties of this chemical were first described by Varner *et al.* (1964) and it has the ability to control crabgrass and annual weed grasses in turf grasses or cereals as well as in broad-leaved crops.

9. *Thiocarbamates and Dithiocarbamates*

The thiocarbamate family of herbicides is toxic to germinating seeds following soil incorporation before sowing. Uptake occurs by roots and through parts of the shoot system that are underground. They are mainly volatile and non-persistent. The most widely used are diallate (Hannah, 1959) for the control of wild oats and blackgrass in brassicas and sugar beet and the closely related triallate (Friesen, 1960) for wild oats and blackgrass in cereals. Eptam (Antognini *et al.* 1957) is also particularly useful for controlling *Agropyron repens* and perennial *Cyperus spp.* Other herbicides in this family are butylate (Gray *et al.*, 1962), cycloate and molinate. One of the most recent introductions is benthiocarb, a herbicide specifically for use in controlling weeds in rice.

The volatility of these chemicals is a mixed blessing. They must be incorporated into the soil with a consequent dilution of chemical and the danger of uneven distribution and variable activity. On the other hand, incorporation places the chemical in the zone of germinating weeds thus making the treatment less affected by rainfall.

10. *Bipyridylium Quaternary Ammonium Salts*

Although quaternary ammonium salts in general are known to be phytocidal, the bipyridylium quaternary salts are novel both in chemical and biological properties.

ICI's interest in quaternary ammonium compounds commenced at the Jealott's Hill laboratories as early as 1947, when dodecyltrimethylammonium bromide was tested in the field and found to produce toxic symptoms in plants. Little was done however, until further quaternaries were evaluated in 1955 when the bipyridylium quaternary ammonium salts were found to be unique compared with conventional chemicals in this family. First, they are highly active and plants are killed rapidly. Whilst 5–10 kg/ha of the conventional quaternaries were required to damage plants severely under glasshouse conditions, only $\frac{1}{16}$ kg/ha of what are now known as diquat and paraquat salts were equivalent in activity. Second, they are rapidly and strongly adsorbed to soil particles (Brian *et al.* 1958). Of a large number of related bipyridylium compounds now tested, these two chemicals have proved to be the most potent. Their biological properties in general are very similar but diquat salts are found to be somewhat more toxic to broad-leaved weeds than are the paraquat salts, whereas the reverse is true in grasses. Potato haulm destruction is possible with as little as $\frac{1}{2}$ kg/ha of diquat. Calderbank (1968) reviews the herbicidal properties of diquat and paraquat.

The bipyridylium quaternary salts are rapidly absorbed by leaves and hence their activity is undiminished by rain falling as little as half an hour after spraying.

Important outlets for paraquat depend on its high activity and rapid action and on its ability to adsorb strongly and rapidly to soil particles. Its uses include weed control before crop emergence, directed weed control in row crops, weeding around trees in orchards and forests and stubble cleaning. The killing of old grassland before ploughing is a further outlet for paraquat. Perhaps the most important development is the use of chemical sprays containing paraquat to replace ploughing and subsequent cultivations with considerable advantages to farm management. Thus it is being used increasingly in minimum cultivations and direct drilling techniques for the production of cereals and kale and for grassland reseeding (Halliday, 1973). Diquat may be used with paraquat for weed control but it is more frequently used alone for the desiccation of potato haulm before harvesting and of seed crops such as clover. Both diquat and paraquat are used to control submerged aquatic weeds.

Morfamquat is a highly selective bipyridylium quaternary salt and has been used as a contact post-emergence herbicide to control a number of annual weeds in cereals (Brian, 1972).

H.P.B.E.(1)–2

11. *Pyridines*

The herbicidal properties of picloram (4-amino-3,5,6-trichloropicolinic acid) were first described by Hamaker *et al.* (1963). It was introduced in the same year by the Dow Chemical Company under the trade name "Tordon". Picloram and its salts are absorbed by leaves and roots and it is then translocated to new growth. It is used for controlling broad-leaved weeds (except *Cruciferae*) in grasses. Rates as low as 0·02 kg/ha will control some annual weeds but doses up to 3 kg/ha may be required for deep-rooted perennials. It may persist in the soil for almost 12 months after high application rates.

12. *Pyridazines*

Novel biological properties for maleic hydrazide (MH) were discovered by Schoene and Hoffman (1949). It is translocated in plants and inhibits cell division but not cell extension. They observed that it had a pronounced though temporary inhibiting effect on plant growth, an inhibition which in favourable environmental conditions was not accompanied by visible damage to the plants. It has found applications for retarding the growth of grass, hedges and trees, for the inhibition of the sprouting of potatoes and onions and for the prevention of sucker development in tobacco.

Pyrazon is similar to lenacil (see below) in biological properties though somewhat higher application rates must be used to control weeds in beet crops.

13. *Pyrimidines: Uracils*

Bucha *et al.* (1962) first reported on the biological activity of bromacil, terbacil and lenacil. They are all absorbed mainly through plant roots. Bromacil is non-selective and is recommended for general weed control on non-crop land. Lenacil and terbacil are sclective and control many annual and some perennial weeds with no damage to a variety of crops.

14. *Triazines*

The first preparation of what is now known as simazine was made by Hofmann (1885). He prepared a compound for which he quoted no melting point or analysis but to which he assigned the formula now associated with simazine. Further chemicals in the triazine group were synthesized by Pearlman and Banks (1948) but in the early 1950s certain triazines were introduced as herbicidal chemicals by Geigy Agricultural Chemicals. A series of triazines was synthesized by Gysin and Knüsli in Geigy's laboratories. The first two chemicals to show herbicidal activity were chlorazine, 2-chloro-4,6-bis(diethylamino)-1,3,5-triazine (Geigy 444) (Gysin, 1960) and later 2-chloro-4,6-bis(ethylamino)-1,3,5-triazine (Simazine). Of this family, simazine and atrazine are by far the best known and they are characterized by high toxicity to an extensive range of monocotyledons and dicotyledons when applied to the soil. The lack of movement of simazine and the resulting depth protection to plant roots that this affords, adds to its selectivity in some crops. It may be used for weed control in deep-rooted crops such as citrus, deciduous fruits, asparagus and coffee. Its persistence in the soil makes it valuable for long-term non-selective weed control. Atrazine is more soluble in water and hence is less dependent on soil moisture for effective weed control. (It also has less residual activity than simazine.) It is therefore preferred for example, for the drier conditions in which maize (*Zea mays*) is grown.

An important property of both simazine and atrazine is their metabolism in maize where they are degraded nonenzymically to the inactive hydroxy derivative (Castelfranco *et al.* 1961). Hence they have found extensive use for controlling annual grasses and broad-leaved weeds in maize. Other members of the family, ametryne, prometryne and desmetryne (to mention but a few) are similar in biological properties (Gysin and Knüsli, 1960, Gysin, 1962), though in addition to root uptake, they are also absorbed through foliage. This group of methylthiotriazines find uses in both pre-emergence and post-emergence situations. Yet another triazine, prometon, is used only as a non-selective herbicide for the control of most annual and perennial broad-leaved and grass weeds on non-crop areas. Trietazine, first described by Gysin and Knüsli (1958) was introduced in 1972 as a component of herbicide formulations for weed control in potatoes and peas.

15. *Miscellaneous Heterocyclic Compounds*

Amitrole (3-amino-1,2,4-triazole) was first discovered to have defoliating properties in tests on cotton conducted in 1952 by the Texas Agricultural Experimental Station (Hall *et al.* 1953). Subsequent tests in Texas and elsewhere showed that it had promise not only as a cotton defoliant but as a general plant growth suppressant. Absorbed by both leaves and roots, it has proved to be effective as a pre- and post-emergence spray and as a selective herbicide (Behrens, 1953).

Endothal (7-oxabicyclo-(2,2,1)-heptane-2,3-dicarboxylic acid) is an oxygen heterocyclic compound forming several isomers of which the most active is the "exo-cis". Its biological properties were described as early as 1951 by Tischler *et al.* (1950) and these include pre- and post-emergence control of weeds in sugar beet and spinach. It is now known not to be widely selective but sugar beet is tolerant to concentrations which kill a number of weeds commonly associated with sugar beet such as *Polygonum* spp. It has also defoliant and desiccant properties and can therefore be used as a pre-harvest treatment in a number of crops.

Benazolin, first reported to be biologically active by Leafe (1964), is a sulphur heterocyclic. It is highly specific in its activity and may be used to control chickweed and cleavers in crops such as clover.

16. *Thiocarbonates*

Thiocarbonates comprise a small family of herbicides with the characteristic group —CS.SS.CS—. Proxan (isopropylxanthic acid) is now of little commercial interest but dimexan and EXD are used as contact herbicides for controlling seedling dicotyledonous weeds before the crops emerge. They are non-persistent in soil. Dimexan is also a pre-harvest desiccant for onions and peas.

17. *Glycine Derivative*

MON-0573, (*N*-(phosphonomethyl) glycine), was first disclosed by Baird *et al.* (1971). As inorganic and amine salts, it is claimed to be a broad-spectrum, post-emergence herbicide that is effective for the control of many annual and perennial weeds of both grass and broad-leaved species at rates ranging from 0·56 to 4·5 kg/ha (Monsanto, 1971). Low crop selectivity is a problem to be considered when carrying out weed control by directed spray applications in crops such as cotton and soybeans. However, tree and shrub type crops such as citrus, nut, olive and vines are quite tolerant to trunk-directed sprays. Tolerance to foliage contamination varies from crop to crop and casual contact should be avoided. There is no apparent residual activity in the soil or pre-emergence effect when the chemical is sprayed on to the soil or if it comes into contact with soil indirectly through decomposing weed vegetation.

18. *Organo-Arsenic Compounds*

Cacodylic acid $((CH_3)_2AsO.OH)$ which forms sodium and potassium salts is used as a non-selective, post-emergence herbicide for controlling weeds in non-crop areas and as a desiccant and defoliant for cotton. DSMA $(CH_3AsO(ONa)_2)$ also an arsenic derivative has selective herbicidal properties. It is applied post-emergence and is used for the control of grass weeds in cotton and in rubber plantations.

19. *Oils*

An account of herbicides would be incomplete without allusion to the hydrocarbon oils. These consist of discarded lubricating oils or diesel oils and have been used for many years to clear waste ground of vegetation. The selection of a phytotoxic oil is frequently based mainly on its availability in a given locality and stove oil for example, has been used in California for weeding carrots. (Some umbelliferous crops are tolerant to phytotoxic oils). In the Eastern States of America, Lackman (1945) has carried out a number of experiments with paint thinners. Crafts and Reiber (1948) have found that selective action is greatest in oils of medium boiling range and containing about 25% of unsaturated and aromatic hydrocarbons. A list of solvents giving their phytotoxicity is given by Ivens (1952). Oils, however, are bulky, relatively expensive and insufficiently persistent and where possible, they are being replaced by the herbicides previously described.

II. Chemical Classification

There is today an extensive array of herbicides of widely differing chemical types which influence the metabolism and hence the growth and behaviour of plants in a number of ways. Two closely related chemicals may also behave quite differently because physical factors such as their adsorption to plant components, volatility, acid or base dissociation differ and these factors may to varying degrees, change the apparent activity of the chemicals in a plant. Thus there is a temptation to consider each herbicide as an individual chemical and the reader of literature on herbicides may be excused for not seeing wood for trees. "Pesticide Manual" classifies herbicides in alphabetical order (Martin, 1972) and this presents the discovery, preparation and physical and biological properties of all current herbicides. Indeed integration to enable the reader to embrace the subject as a whole, is not possible. Our knowledge of the biochemistry of herbicides is incomplete and it cannot yet be made the basis of a rigorous classification. Moreover, there is no simple relation between the chemical structure of a herbicide and its biochemical behaviour and little of fundamental value is obtained from a classification based purely on chemical formulae. Nevertheless, in our present state of knowledge, formulae provide the only comprehensive basis on which to group the extensive variety of herbicides, but the limitations involved are clear. A structural formula is at best a code which gives little indication of fundamental properties of a molecule such as bond energies, electron dispositions and steric factors, all of which are essential parameters in the ultimate biological behaviour of a herbicidal molecule.

In the following classification, chemicals are divided broadly into inorganic and organic. The organic chemicals are then sub-divided into families such as aliphatic and aromatic acids and nitriles, amides, ureas and triazines where a chemical group is common to a number of herbicides. Esters and various salts of the same herbicides are not included separately and water of crystallization is not included in the list of organic chemicals. "Weed Control Handbook" (Fryer and Evans, 1968) is amongst the books consulted in compiling this classification.

The common name is quoted with an asterisk(*) which signifies a name approved by the British Standards Institution and/or with a dagger(†), signifying one approved by the Weed Society of America. Apart from the structural formulae, the tables also quote the molecular weight of a herbicide to the nearest 0·5, the best solvents for the herbicide arranged in the order of solubility and its type of action also arranged in the order of priority of application. The following code has been used in the text:

CODE FOR TABLES

Solubility	*Type of Action*

ac = acetone
al = alcohols
b = benzene
dmf = dimethylformamide
eth = ether
org = organic solvents
pyr = pyridine
w = water
xy = xylene

s = uptake by roots
    from the soil

ct = local action following
    uptake by the foliage

ft = translocation following
    uptake by the foliage

Where no entry is made for molecular weight, solubility and type of action, this signifies that the chemical is no longer of commercial interest.

A. INORGANIC CHEMICALS

	Structural formula	Mol. wt.	Solubility	Type of activity
Ammonium sulphamate AMS†	$(NH_4)O.SO_2.NH_2$	114	w	ct, ft, s
Ammonium sulphate	$(NH_4)_2SO_4$	132	w	ct
Ammonium thiocyanate	NH_4CNS	76	w	ct, s
Calcium cyanamide	$CaCN_2$	80		
Cupric sulphate pentahydrate	$CuSO_4.5H_2O$	250	w	ct
Cupric nitrate trihydrate	$Cu(NO_3)_2.3H_2O$	241.5	w	ct
Ferrous sulphate heptahydrate	$FeSO_4.7H_2O$	278	w	ct
Magnesium sulphate/potassium chloride (Kainit)	$MgSO_4/KCl$		w	ct
Potassium cyanate	$KCNO$	81	w	ct
Sodium arsenite	Na_3AsO_3 $(+NaAsO_2+Na_4As_2O_5)$		w	ct
Sodium tetraborate decahydrate	$Na_2B_4O_7.10H_2O$	381	w	s
Sodium chlorate	$NaClO_3$	106.5	w	s, ct, ft
Sodium chloride	$NaCl$	58.5	w	ct
Sodium nitrate	$NaNO_3$	85	w	ct
Sulphuric acid	H_2SO_4	98	w	ct

B. ORGANIC CHEMICALS

(i)   NITROPHENOLS AND ANILINES

Common name	Chemical name	Mol. wt.	Solubility	Type of activity
DNOC*, DNC†	2-methyl-4,6-dinitrophenol	198	al=3·5% w	ct, s
Dinoseb*, DNBP†	4,6-dinitro-2-S-butylphenol	240	w=0·1%	ct, s
Dinoterb acetate*	4,6-dinitro-2-t-butylphenol/acetate	282	xy, ac=V.Sol	s
PCP†	Pentachlorophenol	266	Na-salt w=33%	ct, s
Nitrofen*†	2,4-dichlorophenyl-4'-nitrophenyl ether	284	w=1 ppm	s, ct
Fluorodifen*	2,4'-dinitro-4-trifluoromethyl-diphenyl ether	328	ac=75% b=52%	s, ct

(i) NITROPHENOLS AND ANILINES—continued

Common name	Chemical name	Mol. wt.	Solubility	Type of action
Bromofenoxim*		461	ac=1%	ct, ft
Trifluralin*†	2,6-dinitro-NN-dipropyl-4-trifluoro-methylaniline	335	ac=40%, xy.	s
Nitralin†	4-(methylsulphonyl)-2,6-dinitro-NN-dipropylaniline	345	ac=36%	s
Dinitramine†	2,6-dinitro-3-amino-4-trifluoromethyl-NN-diethylaniline	322	ac=57%, al	s

(ii)   HORMONE TYPE HERBICIDES (PHENOXYALKANE CARBOXYLIC ACIDS)

(a) Phenoxyacetic acids

Common name	Chemical name	Mol. wt.	Solubility	Type of action
4 CPA*†	4-chlorophenoxyacetic acid			
2,4-D*†	2,4-dichlorophenoxyacetic acid	221	w=0·06% (Na-salt w=4·5%)	ft, s
MCPA*†	4-chloro-2-methylphenoxyacetic acid	200·5	w=0·1% (Na-salt w=27%)	ft, s
2,4,5-T*†	2,4,5-trichlorophenoxyacetic acid	255·5	al, ac, eth w=0·025%	ft, s

(b) α-Phenoxypropionic acids

$$\text{CH}_3 \\ \text{O.CH.COOH}$$

(benzene ring, positions 2, 3, 4, 5, 6)

Dichlorprop*† (2,4-DP)	α-(2,4-dichlorophenoxy) propionic acid	235	w = 0.35% (Na-salt = 66%)	ft, s
Mecoprop*† (CMPP)	(±)-2-(4-chloro-2-methylphenoxy) propionic acid	213·5	w = 0.06% (Na-salt w = 42%)	ft, s
Fenoprop* / Silvex† } 2,4,5-TP	2-(2,4,5-trichlorophenoxy) propionic acid	269·5	ac, al; w = 0.014%	ft, s

(c) γ-Phenoxybutyric acids

$$\overset{\gamma}{\text{OCH}_2}\overset{\beta}{\text{CH}_2}\overset{\alpha}{\text{CH}_2}\text{COOH}$$

(benzene ring, positions 2, 3, 4, 5, 6)

2,4-DB*†	γ-(2,4-dichlorophenoxy) butyric acid	249	ac	ft, s
MCPB* 4-(MCPB)†	γ-(4-chloro-2-methylphenoxy) butyric acid	228·5	ac, al	ft, s
2,4,5-TB* 4-(2,4,5-TB)†	γ-(2,4,5-trichlorophenoxy) butyric acid	283·5	ac	ft, s

Several β-phenoxyethyl derivatives were marketed but they are now of little commercial interest.

(iii)   BENZOIC AND PHENYLACETIC ACIDS

COOH structure (benzene ring, positions 2, 3, 4, 5, 6)

CH₂COOH structure (benzene ring, positions 2, 3, 4, 5, 6)

Common name	Chemical name	Mol. wt.	Solubility	Type of action
2,3,6-TBA†	2,3,6-trichlorobenzoic acid	225·5	al, ac, w	ft, s
Dicamba*†	2-methoxy-3,6-dichlorobenzoic acid	221	w=0·8% (Na-salt=38%)	ft, s
Tricamba	2-methoxy-3,5,6-trichlorobenzoic acid			
Chloramben* Amiben†	3-amino-2,5-dichlorobenzoic acid	206	ac=23% al, w	s, ft
Dinoben	3-nitro-2,5-dichlorobenzoic acid	236		
Dacthal Chlorthal-Methyl	2,3,5,6-tetrachloroterephthalic acid dimethyl ester	332	b=25% ac=10%	s
Chlorfenac* Fenac†	2,3,6-trichlorophenylacetic acid	239·5	al, ac, eth	s, ft

Structure: dimethyl ester with COOCH₃, Cl, Cl, COOCH₃ groups on benzene ring

(iv) HALO-ALIPHATIC ACIDS

Common name	Chemical name	Formula	Mol. wt.	Solubility	Type of activity
Monoxone ⎫ SMA ⎭	Na-monochloroacetate	$Cl.CH_2.COO.Na$	116·5	w	ct
TCA	Trichloroacetic acid	$Cl_3.C.COOH$	163·5	w, al, eth	s, ct, ft
Dalapon*†	2,2-dichloropropionic acid	$CH_3.CCl_2.COOH(Na)$	165	w, al	ft, ct, s
Chloropon	2,2,3-trichloropropionic acid	$Cl.CH_2.CCl_2.COOH$			
Mendok	2,3-dichloro-2-methylpropionic acid	$Cl.CH_2\!\!\overset{\displaystyle CH_3}{\underset{\displaystyle Cl}{\rule[-0.5em]{0pt}{1.5em}\!\!C\!\!}}COOH(Na)$			
Erbon	2-(2,4,5-trichlorophenoxy)-ethyl-2,2-dichloropropionate	$CH_3.CCl_2.COO.CH_2.CH_2.O-$			

(v) AMIDES $\begin{matrix} R_1 \\ R_2 \end{matrix} {>} N.\overset{\text{O}}{\overset{\|}{C}} - R_3$

Specific groups

Common name	R_1	R_2	R_3	Mol. wt.	Solubility	Type of activity
Allidochlor*† CDAA†	—CH₂CH:CH₂	—CH₂CH:CH₂	—CH₂Cl	173·5	w = 2%	s
Benzadox*	—OCH₂COOH	H	-phenyl	195	ac, al.	ct
Diphenamid*†	—CH₃	—CH₃	—CH(diphenyl)	293	w = 1·6% ac = 19% xy	s
Pronamide* Propyzamide	CH₃ —C.C≡CH CH₃	H	3,5-dichlorophenyl	255	Org	ct, s
Propachlor*†	-phenyl	—CH(CH₃)₂	—CH₂Cl	211·5	ac, b	s
Monalide*	4-chlorophenyl	H	CH₃ —C.CH₂CH₂CH₃ CH₃	239·5	xy = 10%	ct, s
Propanil*†	3,4-dichlorophenyl	H	—CH₂CH₃	218	al = 54%, w	ct, ft
Dicryl	3,4-dichlorophenyl	H	—C=CH₂ CH₃	230		

Name						
Pentanochlor* Solan†	3-chloro-4-methylphenyl	H	$\begin{array}{c}CH_3\\-CH.CH_2CH_2CH_3\end{array}$	239·5	xy, ac	ct, ft
Karsil	3,4-dichlorophenyl	H	$\begin{array}{c}CH_3\\-CH.CH_2CH_2CH_3\end{array}$	260	xy, ac	ct, ft
Cypromid*†	3,4-dichlorophenyl	H	$\begin{array}{c}CH_2\\-CH\\CH_2\end{array}$	230	xy, ac, al	ct
Alachlor	2,6-diethylphenyl	—CH₂OCH₃	—CH₂Cl	269·5	eth, ac, b, al	s
Naptalam* NPA†	(naphthyl)	H	$-CH_2Cl$ COOH (phenyl)	291	Na-salt w=30%	s
Oryzalin†	$\begin{array}{c}O_2N\\CH_3CH_2CH_2\\CH_3CH_2CH_2\end{array}$N— (benzene ring with SO₂.NH₂ and NO₂)			346	ac, al	
Suffix	—CH(CH₃).COOC₂H₅	3,4-dichlorophenyl	-phenyl	366	ac=70%	ft

(vi) NITRILES

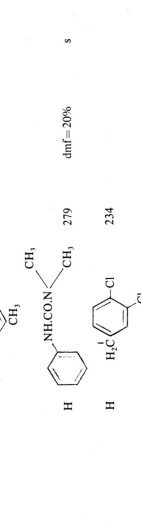

Common name	Chemical name	Mol. wt.	Solubility	Type of activity
Dichlobenil*†	2,6-dichlorobenzonitrile	172	Low in most solvents	s, ct
Bromoxynil*†	4-hydroxy-3,5-dibromobenzonitrile	277	Na-salt w = 4% ac, al	ct, ft
Ioxynil*†	4-hydroxy-3,5-di-iodobenzonitrile	371	Na-salt w = 14% ac, al	ct, ft

(vii) ARYL CARBAMATES

$$\begin{matrix} R_1 \\ R_2 \end{matrix} > N - \overset{\overset{O}{\|}}{C} - OR_3$$

Common name	R₁	R₂	R₃	Mol. wt.	Solubility	Type of action
(a) Alkyl			Specific groups			
Terbutol†	CH₃	H	(CH₃)₃C — C(CH₃)₃, CH₃	277	ac, al	s
Karbutilate	t-butyl	H	NH.CO.N(CH₃)CH₃	279	dmf = 20%	
Dichlormate / Sirmate	CH₃	H	H₂C, Cl, Cl	234		s

(b) Aryl

Name	Aryl		R			
Propham*, IPC†	Phenyl	H	$-CH(CH_3)CH_3$	179	al, eth	s
Chlorpropham*, CIPC†	3-chlorophenyl	H	$-CH(CH_3)CH_3$	213·5	org	s, ct
Chlorbufam*, BiPC	3-chlorophenyl	H	$-CH(CH_3)C\equiv CH$	223·5	ac, al	s
Barban*†	3-chlorophenyl	H	$-CH_2C\equiv C.CH_2Cl$	258	b, ac	ft
Phenmedipham*	H_3C— (phenyl with NH.COOCH$_3$)	H	CH_3	300	ac, al	ft
Asulam*	H_2N—SO$_2$—phenyl	H	CH_3	230	Na-salt w > 60%, ac, al	ft, s
Nisulam	O_2N—SO$_2$—phenyl	H	CH_3	260	Na-salt w = 40%, ac, al	ft, s
SWEP	dichlorophenyl (Cl, Cl)	H	CH_3	220	ac = 46%	ft, s

(viii) SUBSTITUTED UREAS R_1—N.C.N with H, O (C=O), and R_2, R_3

$$R_1-\underset{H}{N}.\underset{\|}{C}.N\underset{R_3}{\overset{R_2}{<}}\quad(O)$$

(a) Alkyl

Common name	Specific groups			Mol. wt.	Solubility	Type of action
	R_1	R_2	R_3			
Fenuron*†	(phenyl)	CH_3	CH_3	164	w = 0.4%	s, ct
Monuron*†	(4-Cl-phenyl)	CH_3	CH_3	198·5	w = 0.02% ac, b	s
Diuron*†	(3,4-diCl-phenyl)	CH_3	CH_3	233	ac, b	s
Chlortoluron*	(Cl-CH₃-phenyl)	CH_3	CH_3	212·5	ac, b	s, ct
Metoxuron*	(Cl-OCH₃-phenyl)	CH_3	CH_3	228·5	w = 0.07% ac, b	ct, s

Name	Structure			MW	Solubility	
Fluometuron*†	CF_3-phenyl	CH_3	CH_3	232	al, ac	s, ct
Chloroxuron*†	Cl-phenyl-O-phenyl	CH_3	CH_3	290·5	ac	s
Cycluron*† (OMU)	$-CH$ $CH_2$$CH_2$$CH_2$ / $CH_2$$CH_2$$CH_2$ ring	CH_3	CH_3	198	al, ac	s
Neburon*†	3,4-dichlorophenyl	CH_3	C_4H_9	275	low in most solvents	s
Benzthiazuron*	benzothiazole -C(N)(S)-	H	CH_3	207	pyr, ac, xy	s
Methabenzthiazuron*	benzothiazole -C(N)(S)-	H	$\left(CH_3 \begin{array}{c} CH_3O \\ -N-C- \end{array}\right)$	221	ac = 12% al = 7%	s
Azolamid	HN ring $N.CONHCH_2CH(CH_3)_2$, $=O$			185		

(viii) SUBSTITUTED UREAS—*continued*

Common name	Specific groups			Mol. wt.	Solubility	Type of action
	R_1	R_2	R_3			
Methazole†	(2,4-dichlorophenyl)	(ring: N—O—CO / CO—N—CH₃)		261	xy = 55%	s, ft
Siduron*†	(phenyl)	H	—CH(CH₂—CH₂—CH₂ / CH—CH₂)—CH₃	232	dmf = 10%	s
(b) Alkoxy						
Monolinuron*†	(4-chlorophenyl)	CH_3	OCH_3	214·5	w = 0·06% al, ac, b, xy	s, ct
Linuron*†	(3,4-dichlorophenyl)	CH_3	OCH_3	249	ac, al, b, xy	s, ct
Metobromuron*†	(4-bromophenyl)	CH_3	OCH_3	259	w = 0·03% ac, al	s, ct
Chlorbromuron*	(3-chloro-4-bromophenyl)	CH_3	OCH_3	293·5	ac, xy	s, ct

(ix)

(a) Thiocarbamates

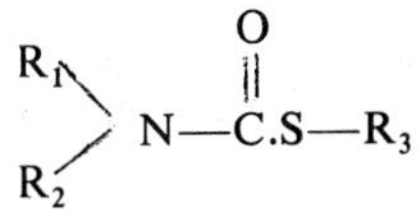

$$R_1 \diagdown \atop R_2 \diagup N-\overset{\overset{\displaystyle O}{\|}}{C}.S-R_3$$

Common name	Specific groups			Mol. wt.	Solubility	Type of action
	R_1	R_2	R_3			For rice
Benthiocarb	C_2H_5	C_2H_5	H_2C—⟨C₆H₄⟩—Cl	257·5	Org	s
EPTC*†	C_3H_7	C_3H_7	C_2H_5	189	al, ac, b, xy	s
Vernolate†	C_3H_7	C_3H_7	C_3H_7	203	ac, xy	s
Di-allate* Diallate† } Avadex	—$CH(CH_3)_2$	—$CH(CH_3)_2$	—$CH_2ClC:CHCl$	270	al, ac, b, xy	s
Tri-allate* Triallate† } Avadex BW	—$CH(CH_3)_2$	—$CH(CH_3)_2$	—$CH_2ClC:CCl_2$	304·5	al, ac, b	s
Butylate*†	—$CH_2CH(CH_3)_2$ C_4H_9	—$CH_2CH(CH_3)_2$	C_2H_5	217	xy, ac	s
Pebulate*†		C_2H_5	C_3H_7	203	al, ac, b, xy	s
Cycloate†	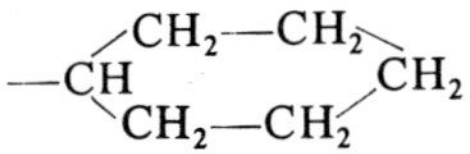 —CH⟨cyclohexyl⟩	C_2H_5	C_2H_5	215	ac, al, b, xy	s
Molinate*†	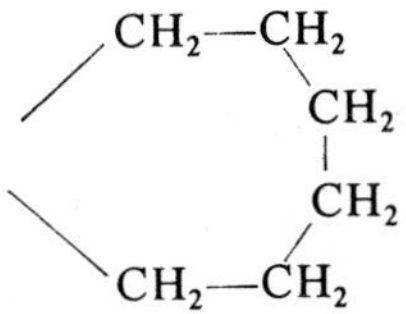 ⟨azepane ring (R₁, R₂)⟩		C_2H_5	187	ac, al, b, xy	s

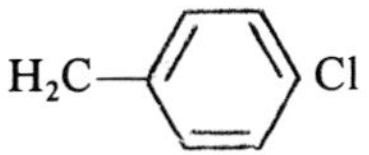

(ix)—continued

(b) Dithiocarbamates

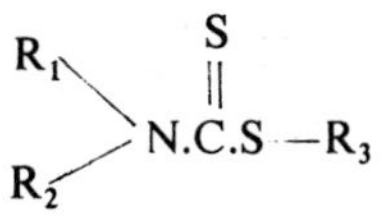

$$R_1\diagdown \atop R_2\diagup N.C.S-R_3 \quad (\overset{S}{\underset{\|}{}})$$

Common name	Specific groups			Mol. wt.	Solubility	Type of action
	R_1	R_2	R_3			
Metham-sodium*⎫ SMDC†	CH_3	H	Na	129	w = 72% al	s
Sulfallate* CDEC†	C_2H_5	C_2H_5	$-CH_2CCl{:}CH_2$	224	ac, b, eth	s

(x)   HETEROCYCLIC COMPOUNDS

(a) Bipyridylium quaternary ammonium salts

Common name	Chemical name	Structural formula	Mol. wt.	Solubility	Type of action
Diquat dibromide*†	1,1¹-ethylene-2,2¹-bipyridylium dibromide (9,10-dihydro-8a, 10a-diazoniaphenanthrene dibromide)	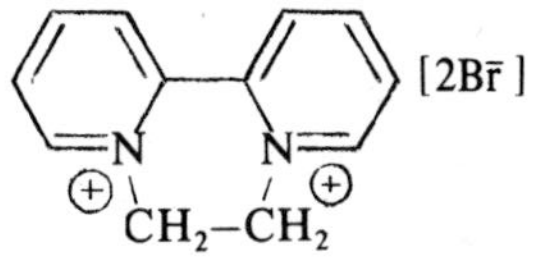 [2Br⁻]	344 (cation = 184)	w	ct, ft
Paraquat dichloride*†	1,1¹-dimethyl-4,4¹-bipyridylium dichloride	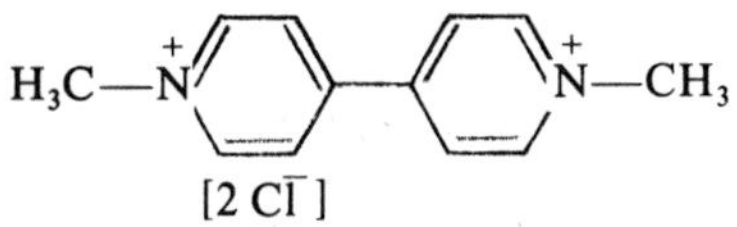 [2Cl⁻]	257 (cation = 186)	w	ct, ft

Name	Structure	M.W.	w	ct, ft
Morfamquat dichloride*	1,1¹-bis-(3,5-dimethylmorpholino carbonylmethyl)4,4¹-bipyridylium dichloride)	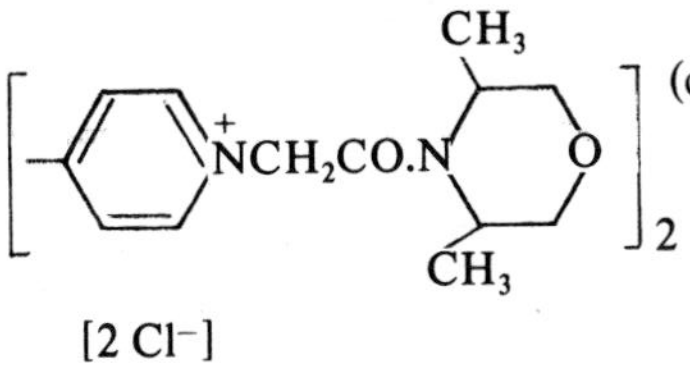 539·5 (cation = 468·5)	w	ct, ft

[2 Cl⁻]

(b) Pyridines

Pyriclor†	2,3,5-trichloro-4-pyridinol	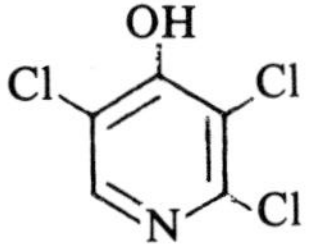 198	xy, al, ac	ft, s, ct
Haloxydine	3,5-dichloro-2,6-difluoro-4-hydroxy-pyridine	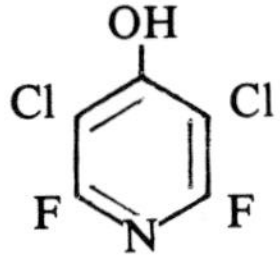 200	xy, al, ac	s
Picloram*†	4-amino-3,5,6-trichloropicolinic acid	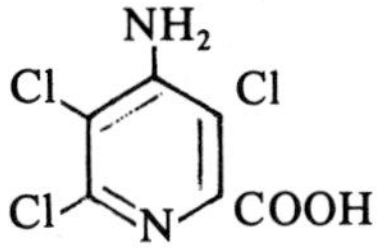 241·5	ac, al	ft, s

(c) Pyridazines

| Credazine | 3-(2-methylphenoxy)-pyridazine | 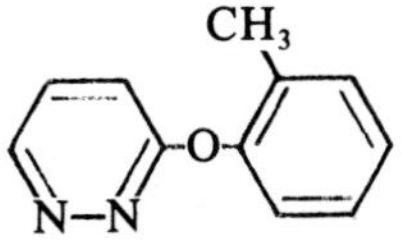 186 | w = 0·2% org | ft |

(x) HETEROCYCLIC COMPOUNDS—*continued*

Pyridazines (*continued*)

Common name	Chemical name	Structural formula	Mol. wt.	Solubility	Type of action
Pyrazon*†	5-amino-4-chloro-2-phenyl-pyridazin-3(2H)-one		221·5	al, ac	ct, s
Maleic hydrazide } MH†	1,2-dihydropyridazine-3,6-dione		112	w=0·6%	ft, s
Norflurazon	4-chloro-5-methylamino-2(3-trifluoromethylphenyl)pyridazin-3(2H)-one		303·5		

(d) Pyrimidines: Uracils

Isocil	5-bromo-3-isopropyl-6-methyluracil		274		s, ct

Bromacil*†	5-bromo-6-methyl-3,5-butyluracil		261	w=0.08%	s, ct
Terbacil†	5-chloro-6-methyl-3-t-butyluracil		216·5	w=0·07% al, ac, xy	s, ct
Lenacil*†	3-cyclohexyl-5,6-trimethylene-uracil		234	pyr	s, ct
Bentazon	3-isopropyl-2,1,3-benzothiadiazin-4-one-2,2-dioxide		240	w=0·05% ac, al	ct

(x) HETEROCYCLIC COMPOUNDS—*continued*

Common name	Chemical name	Structural formula	Mol. wt.	Solubility	Type of action
Flumezin	2-methyl-4-(3-trifluoromethyl-phenyl)-1,2,4-oxadiazin-3,5-dione		274		

(e) Triazines

			Specific groups			
Common name	R_2	R_4	R_6	Mol. wt.	Solubility	Type of action
Simazine*†	Cl	—NH.C$_2$H$_5$	—NH.C$_2$H$_5$	201·5	al = 0·04%	s
Simeton	OCH$_3$	—NH.C$_2$H$_5$	—NH.C$_2$H$_5$			
Simetryne	SCH$_3$	—NH.C$_2$H$_5$	—NH.C$_2$H$_5$			
Atrazine*†	Cl	—NH.C$_2$H$_5$	—NH.CH(CH$_3$)$_2$	215·5	al = 2%	s, ct
Atraton*	OCH$_3$	—NH.C$_2$H$_5$	—NH.CH(CH$_3$)$_2$	211	w = 0·2% org	s, ct
Atratone†						
Cyanazine	Cl	—NH.C$_2$H$_5$	—NH.C(CH$_3$)$_2$.CN	240·5	al = 4·5%	s, ct

Compound	X	R₁	R₂	M.p.	Solubility	
Ametryne*†	SCH₃	—NH.C₂H₅	—NH.CH(CH₃)₂	227	org	ct, s
Propazine*†	Cl	—NH.CH(CH₃)₂	—NH.CH(CH₃)₂	229·5	low solubility in most solvents	s
Prometon*	OCH₃	—NH.CH(CH₃)₂	—NH.CH(CH₃)₂	225	w=0·08%	s, ct
Prometone†					al, ac, b	
Prometryne*†	SCH₃	—NH.CH(CH₃)₂	—NH.CH(CH₃)₂	241	org	ct, s
Trietazine*	Cl	—NH.C₂H₅	—N(C₂H₅)₂	229·5	b, ac	ct, s
Ipazine	Cl	—N(C₂H₅)₂	—NH.CH(CH₃)₂			
Chlorazine	Cl	—N(C₂H₅)₂	—N(C₂H₅)₂			
Desmetryne*†	SCH₃	—NH.CH₃	—NH.CH(CH₃)₂	213	w=0·06 org	ct, s
Terbutryne* / Terbutryn†	SCH₃	—NH.C₂H₅	—NH.C(CH₃)₃	241	Org	s, ct
Cyprazine	Cl	—NH.CH(CH₃)₂	—NH.CH⟨CH₂–CH₂ (cyclopropyl)⟩	227·5	ac, al	ct, s

Cyprazine third substituent:

$$-NH.CH\!\!\begin{array}{c} CH_2 \\ | \\ CH_2 \end{array}$$

Metribuzin

Structure: $t\text{-}C_4H_9$—C(=O)—C(—NH₂)=N—N=C—SCH₃ (triazinone ring)

	M.p.	Solubility
Metribuzin	214	al=45% w=0·1%

(x) HETEROCYCLIC COMPOUNDS—*continued*

(f) Miscellaneous heterocyclic compounds

Common name	Chemical name	Structural formula	Mol. wt.	Solubility	Type of action
Aminotriazole* } Amitrole†	3-amino-1,2,4-triazole		84	w=28% al	ft, s
Chlorflurazole*	4,5-dichloro-2-trifluoromethyl benzimidazole		255	w=0·006%	ct
Endothal* } Endothall†	3,6-endoxohexahydrophthalic acid		186	w=10%	s, ct
Bentranil	2-phenyl-3,1-benzoxazinone		223	b, eth	ct, ft
Benazolin*	4-chloro-2-oxobenzothiazolin-3-ylacetic acid		243·5	Na-salt w=40% ac	ft

		Structure	M.W.	Solubility	
Pyrazon*†	5-amino-4-chloro-2-phenylpyridazin-3(2H)-one		221·5	al, ac	s, ft
Oxadiazon	2-butyl-4-(2,4-dichloro-5-isopropyloxyphenyl)-1,3,4-oxadiazolin-5-one		345	ac=60% al	s, ft
(xi)	THIOCARBONATES				
Proxan Dimexan*	Isopropylxanthic acid Di-(methoxythiocarbonyl) disulphide	$(CH_3)_2CHO.CS.SHNa$ $CH_3O.CS.SS.CS.OCH_3$	214	b, ac, al	ct
EXD†	Di-(ethoxythiocarbonyl) disulphide	$C_2H_5O.CS.SS.CS.OC_2H_5$	242	b, ac, al	ct
(xii)	GLYCINE DERIVATIVE				
Glyphosate MON-0573	N-(phosphonomethyl) glycine	$HOOC.CH_2.NH.CH_2.\overset{O}{\overset{\|}{P}}(OH)_2$	169	w=1%	ft

(xiii)   ORGANO-ARSENIC COMPOUNDS

Common name	Chemical name	Structural formula	Mol. wt.	Solubility	Type of action
Cacodylic acid†	Dimethylarsinic acid	$(CH_3)_2As.OH$ (with =O)	138	w	ct
MSMA†	Monosodium methylarsonate	CH_3, OH / As / O, ONa	162	w	ft
DSMA†	Disodium methylarsonate	$CH_3.As$ with O, ONa, ONa	184	w	ct
(xiv) OILS					
(xv) MISCELLANEOUS					
DMPA ⎫ Zytron ⎭	O-(2,4-dichlorophenyl)-O'-methyl-N-isopropyl phosphoroamidothioate	Cl, Cl (ring), OP with S, OCH₃, NH.CH(CH₃)₂	314		
Bensulide*†	N-[2-(O,O-di-isopropyldithio-phosphoryl)ethyl] benzene sulphonamide	$(iPro)_2PS.SCH_2CH_2NHSO_2$ (phenyl ring)	397	ac, al, xy	s, ft

III. CLASSIFICATION BY PHYTOTOXICITY

In view of the complexity of the system into which herbicides are introduced, it is not surprising to find that their mode of action within plants is clearly understood in very few cases. For the majority, a number of physiological and morphological effects have been observed, but there is a measure of conjecture in deciding what is direct or indirect effect in relation to herbicidal action. It presents the anomaly of studying a system under abnormal conditions when the system under normal conditions is still incompletely understood.

Various authors have attempted to group herbicides by their most important biochemical effects in the plant. Ashton and Crafts (1973) provide the following three examples.

Van Overbeek (1964) finds that just two physiological actions account for the herbicidal activity of about 70 named herbicides—the hormone weedkillers produce growth abnormalities and the triazines and substituted ureas inhibit photosynthesis.

King (1966) in "Weeds of the World" classifies herbicides as contact, inhibitors of cell growth, growth-regulators, inhibitors of growth and tropic responses, inhibitors of chlorophyll formation and of photosynthesis.

The third example is the classification of Moreland (1967) who uses only three biochemical headings, namely modifications of respiration and mitochondrial electron transport, inhibition of photosynthesis and the Hill reaction and interference with nucleic acid metabolism and protein synthesis. Only one biochemical effect—inhibition of photosynthesis—figures in all three classifications. One must conclude that there exists no generally accepted classification of herbicides based on their physiology and the following survey confirms this view. No attempt will be made to group them by biochemical or phytotoxic effect particularly as classification may give undue weight to effects which are not directly concerned in the herbicidal activity.

Information on the inorganic chemicals is very scanty and in general, reliance must be placed on the early literature for work on their mode of action. As present-day biological tools were then unavailable, the conclusions from much of the work in the early part of this century are general in character and frequently based on the slenderest of evidence. Moreover, in view of the high application rates required for adequate toxicity, economics rather than biological activity has been an overriding factor.

A. INORGANIC CHEMICALS

Little is known of the mode of action of inorganic compounds. For the purpose of this account therefore they are discussed in alphabetical order as it is not possible to classify them according either to their chemical or biological properties.

1. *Ammonium Sulphamate* (AMS)

Ammonium sulphamate prolongs the dormant period of plants and, with heavy applications, plants can be kept in the dormant stage until starch and sugar reserves are exhausted, when the plant dies (Robbins *et al.* 1952). It is not known, however, by what process dormancy is established.

2. *Ammonium Sulphate*

Weed control by ammonium salts is largely due to the toxic action of the ammonium ion. Ammonia is absorbed into the plant cell very rapidly and it has been shown by Harvey (1911) that cell sap that is normally heavily buffered at an acid pH, may become alkaline under the influence of the ammonia. Increased alkalinity causes rapid death of the cell and it is suggested that in addition, ammonia has a toxic action on cell protoplasm, possibly by the formation of a complex with the plasma proteins (Bokorny, 1915).

3. *Ammonium Thiocyanate*

Ammonium thiocyanate is extremely toxic to plant cells and very rapid in action but the nature of its action is unknown. According to Harvey (1931*a*) it is considered as a protoplasmic poison inhibiting certain enzymes such as catalase (Landen, 1934) and coagulating proteins. In this respect its affinity for iron may be of importance. It has also been observed to reduce the germination and respiration rates in potato tubers at a concentration of 2% (Ranjau and Kaur, 1954). Moreover, at the optimum temperature for plant growth it has been found that 10^{-4} N ammonium thiocyanate stimulated root formation on the stems of bean and geranium by 500%. At lower temperatures, the same dose killed the plants (Novikov and Barannikova, 1954).

4. *Calcium Cyanamide*

Little work has been carried out on the mode of action of calcium cyanamide. Its action however is not plasmolytic and it has been noted microscopically that, after treatment, the cell contents become granular. This may result from a coagulation of cell proteins.

5. *Copper Sulphate and Nitrate and Ferrous Sulphate*

Salts of heavy metals such as copper inactivate many enzymes at high concentrations and are general protein precipitants. It is not established, however, that copper acts in this way. Nevertheless, copper sulphate has been found to depress the photosynthetic activity of *Chlorella* exposed for 20 min to a 10^{-7} M solution (Greenfield, 1941).

Ferrous sulphate induces what appears to be immediate plasmolysis of cells but on the other hand, Åslander (1927) found that a 5% solution completely destroyed *Sinapis* plants without inducing plasmolysis. The chloroplasts moreover were healthy.

As a general comment, in a system such as a plant cell where iron, copper, magnesium, etc. ions are in competition for certain vital complexing centres in the cell, normal metabolic activity will only result when they are present in the correct relative amounts at these centres. It is reasonable to assume that if the cellular concentration of copper or iron is considerably increased, modification of the normal equilibrium conditions will occur with serious effects on the activity of the cell.

6. *Sodium Arsenite*

In spite of the large scale on which sodium arsenite is used in practice, comparatively little work has been carried out on its mode of action. Nobbe *et al.* (1884) examined the toxic action of potassium arsenite on maize and oats and concluded that it affected the protoplasm of the root, modifying its osmotic properties. The absence of bleeding from the cut stumps was held to support this view. Later Stewart and Smith (1922) suggested a chemical reaction of arsenite with chlorophyll but there is little direct experimental evidence for this view. More recently, using *Avena* coleoptile sections and etiolated pea stems Christiansen *et al.* (1949) reported inhibition of respiration and of growth by sodium arsenite and a conversion of reducing sugars to non-carbohydrates. An action similar to iodoacetate whereby it combines with an enzyme dependent on free—SH groups was also suggested by Thimann and Bonner (1949). Sodium arsenite has also proved to upset normal mitosis in the roots of *Vicia narbonensis* (Mallah and Dawood, 1956). Concentrations down to 0.01 N induced gross cytoplasmic and nuclear derangements in which metaphase and anaphase chromosomes rapidly clumped together and finally became agglutinized. At 0.005 to 0.0005 N, prophase stages ceased and the spindle apparatus ceased to function. Concentrations of 10^{-4} N to 5×10^{-5} N reduced the number of C-mitoses and 2.5×10^{-5} N for 48 h disturbed the spindle polarization and resulted in multipolar distributions of the chromosomes and in the formation of multinucleated cells. Further comments on arsenic biochemistry are in the section "organo-arsenic compounds".

7. *Sodium Tetraborate*

Virtually nothing has been published on the mode of action of borate as a herbicide but a little is known about its action as a micronutrient. In this capacity it is known to be closely related to the movement of sugars, the sugar-starch balance, protein synthesis and respiration. Its observed effect on cell elongation may reflect its influence on sugar movement and on auxin transport.

It has also been shown that sodium tetraborate gives pronounced inhibition of chlorophyll formation in wheat seedlings grown in the dark over a range of concentrations (Brebion and Scuflaire, 1954) but there is no evidence that this is directly responsible for the toxicity of borax. In addition, toxic concentrations of borax have been shown by Leaf (1953) to inhibit the uptake of water by roots.

8. *Sodium Chlorate*

It was concluded by Yamasaki (1929) that the toxicity of sodium chlorate resided in its reduction within the plant to sodium hypochlorite. Susceptible plants showed a higher content of reducing substances than resistant varieties and the latter were rendered susceptible after being allowed to absorb formaldehyde. Moreover, although more chlorate ions were absorbed by susceptible plants, less of the salt could be found in the tissues. Åberg (1948) supports this view and found an antagonism between sodium chlorate and nitrate. It appears therefore that the toxicity of chlorate involves a reduction to hypochlorite in the affected cells by enzymes which are normally involved in the reduction of nitrate, light being essential for such activity in shoots. Goksøyr (1951) however does not accept this view since a number of micro-organisms can reduce nitrate but not chlorate. Instead, he concludes from his experiments with the mould *Aspergillus oryzae* that chlorate inhibits the nitrate-reducing apparatus.

An effect of sodium chlorate on catalase activity has been reported by Neller (1931) using bindweed (*Convolvulus*) roots. In cases of severe toxicity, the activity was reduced to about 50% of that of the untreated roots. In view of the extremely high efficiency of hydrogen peroxide destruction by catalase, it is doubtful if the above reduction in activity would materially affect the equilibrium of peroxide present within the cell.

Laboratory experiments with the filamentous alga *Nitella clavata* (Offord and d'Urbal, 1931) indicate that chlorates can also exert what they describe as a powerful plasmolysing effect at concentrations down to 0·01 M.

9. *Sodium Chloride and Nitrate*

The rapid action of these salts when applied at high concentrations is generally held to be that of plasmolysis of the root cells and consequent interference with the water uptake by the plant.

10. *Sulphuric Acid*

Åslander (1927) concluded that sulphuric acid penetrates leaf tissue rapidly, instantly destroying the protoplasm. The acid can unite with the Mg atom of chlorophyll *in vitro,* destroying the chlorophyll. Chloroplasts also were destroyed in his *in vitro* experiments since none could be detected in affected cells. Cell walls were unaffected. These conclusions are in general agreement with earlier ones of Brenner (1918) and all agree that sulphuric acid does not cause plasmolysis of plant cells. Korsmo (1930) however, regarded the action of the acid as due primarily to its ability to attract or bind the water of plant cells, its action being greatest under dry conditions when the bound water cannot be replaced.

In contrast to the inorganic chemicals, extensive work has been carried out on the mode of action of organic chemicals. As a result, much more is known of their action, but much remains to be done. In most cases, a wealth of facts have yet to be fitted into a co-ordinated theory of action. This is particularly so with the auxin herbicides.

1. Nitrophenols and Anilines

Phenolic herbicides kill plants in two ways. At high rates of application plant membranes are destroyed, water-logging of the leaves occurs and desiccation follows. The action is more rapid on sunny, warm days and in this respect the action resembles that of the bipyridyl quaternary ammonium salts. Dinitrophenols uncouple oxidative phosphorylation (van Overbeek, 1964). Inhibition of energy-requiring processes results and a loss of reservoirs of energy-rich phosphate as well as stimulation of glycolysis (Simon, 1953). This process may predominate on cold, dull days when the action is very slow and plants lose vigour and fail to grow. A further manifestation of the action of substituted phenols on plant tissue is the stimulating effect on respiration. This amounts to as much as 200–300% of that of the controls (Simon, 1953; Gaur and Beevers, 1959).

2. Hormone-type Herbicides

Much is known about the biochemical actions of the hormone weedkillers on plants but which actions are cause and which effect is still unresolved. Chlorophenoxy acids have profound effects on the growth and structure of plants. According to Ashton and Crafts (1973) these chemicals produce epinastic bending, cessation of growth, tumour formation, and secondary root induction. Meristematic cells also cease to divide, and cells which normally would elongate expand only radially. In mature plants, parenchyma cells swell, divide and produce callus tissue and expanding root primordia. Root elongation stops and root tips swell. Young leaves stop expanding, roots fail to absorb water and salts, photosynthesis is inhibited and the phloem becomes plugged.

These chemicals are known to modify nucleic acid metabolism in plants (Hanson and Slife, 1969) and they interact with numerous enzyme systems (Woodford et al. 1958). Van Overbeek (1964) notes that the nature of 2,4-D-induced growth results from an imbalance of hormones, possibly in the auxin-kinin interaction.

Most recently, ethylene production by 2,4-D-treated plants has been studied with particular interest (Hanson and Slife, 1969; Holm and Abeles, 1968). A further factor operates in the case of the phenoxybutyric acids and ethyl hydrogen sulphates. To become active, these chemicals must first be metabolized to the corresponding phenoxyacetic acids.

The reader is referred to Chapters 14, 15 and 17 for full discussions of these various responses and for a critical assessment of the ultimate mechanisms which underlie the action of the phenoxyalkane carboxylic acid herbicides.

3. *Benzoic and Phenylacetic Acids*

The substituted benzoic acids cause many symptoms in plants which are typical of the phenoxyacetic acids. Examples are tissue proliferation (Zimmerman and Hitchcock, 1951) epinasty of young shoots, inhibition of apical meristems of dicotyledons and of geotropic curvature of roots (Keitt, 1960). Dicamba has also been observed by Wuu and Grant (1966) to inhibit mitosis in barley. There is increasing evidence that the benzoic acids have auxin-type properties (Keitt and Baker, 1966) but insufficient work has been done to suggest a mechanism of herbicidal action. Some acids also modify the transport of IAA (Keitt and Skoog, 1957).

Little work has been carried out on the mode of action of 2,3,6-trichlorophenylacetic acid (fenac) but many of the outward symptoms, e.g. epinasty and bud inhibition, suggest that it has growth-regulating properties.

4. *Halo-aliphatic Acids*

Acids such as dalapon and TCA inhibit plant growth, induce leaf chlorosis and at higher concentrations produce leaf necrosis. Dalapon also increases tillering of barley according to Hilton *et al.* (1959) and interferes with the meristematic activity of root tips. Juniper and Bradley (1958) report reduced wax formation on the leaves of treated plants.

The halo-aliphatic acids affect many plant processes. This may be partly due to the ability of dalapon and TCA to precipitate or effect conformational changes in proteins (Redemann and Hamaker, 1954). One symptom of susceptible plants is a reduction in the amount of amides and β-alanine, and an increase in the amount of free ammonia and protein present (Mashtakov, 1967). This suggests that these herbicides inhibit the enzymes responsible for the conversion of ammonia to amides.

Other effects include interference in the lipid metabolism of leaves and a pronounced reduction in the uptake of phosphate ion ($^{32}PO_4\equiv$) into corn seedling roots (Ingle and Rogers, 1961). Since dalapon has no effect on the oxygen uptake by maize roots (Switzer, 1954) Ingle and Rogers suggest that it does not interfere with respiration or the production of metabolic energy but rather with the utilization of energy. Chlorinated aliphatic acids inhibit the enzymatic synthesis of pantothenate by competing with pantoate for an enzyme site (Hilton *et al.* 1959) but there is reason to believe that this inhibition does not constitute a primary action in higher plants (Åberg and Johansson, 1966).

Dalapon does not affect either respiration or the rate of photosynthesis of leaves and as is so often the case, it is not possible to relate herbicidal activity to any specific biochemical effect of these chemicals.

5. *Amides*

Applied to the foliage, amide damage takes the form of localized or general necrosis, depending on the dose applied. Propanil, dicryl, solan and cypromid also inhibit growth but this is probably a secondary effect. Soil-applied amides such as alachlor, propachlor, diphenamid and naptalam all inhibit root elongation but only naptalam has been found to inhibit seed germination. Seedlings which succeed in emerging are stunted and malformed. Allidochlor (CDAA) and propachlor are the two amides most intensively studied. Both chemicals influence RNA and protein synthesis, amylase and proteinase activity. No consistent pattern of behaviour has emerged. The most consistent effects are inhibition of the rate of photosynthesis and modifications of RNA synthesis and amylase activity.

6. *Nitriles*

Bromoxynil and ioxynil activity first shows within 24 h as necrotic spots on leaves and the damage spreads until the plants ultimately die. Chlorosis develops in untreated leaves in spite of autoradiographic evidence that little movement of ioxynil occurs (in the phloem). Dichlobenil is applied to the soil and its pre-emergence activity is largely damage to the meristems of embryos from germinating seeds. Destruction of phloem, cambium and parenchyma tissue in alligator weed (*Alternanthera philoxeroides*) was observed by Pate (1966).

Ioxynil inhibits the Hill reaction and Smith *et al.* (1966) observe that its effect is comparable to that of diuron. It also uncouples oxidative phosphorylation; Kerr and Wain (1964) found that the effect of ioxynil was much greater than that of dinitro-*o*-cresol. Paton and Smith (1967) conclude that the ability to uncouple oxidative phosphorylation and to block electron flow contribute to the herbicidal activity of ioxynil. Chlorotic symptoms which appear after longer periods from treatment are secondary effects in the phytotoxicity of ioxynil.

7. *Arylcarbamates*

Biochemical responses vary somewhat from member to member of the family but the most striking effects are the inhibition of oxidative phosphorylation, RNA and protein synthesis. The Hill reaction of photosynthesis is also inhibited and the ATP content of tissue sections is reduced.

The modification of protein and RNA synthesis is probably responsible for the strong mitotic effects produced by the arylcarbamates. For example Ennis (1948) using propham found mitotic aberrations in certain root and shoot cells of *Avena* and *Allium*. These consisted of anaphase bridges, blocked metaphases, nuclear fragments, giant vesiculate nuclei and increased numbers of chromosomes. The mitotic disturbances may well account for the herbicidal action of those carbamate herbicides which inhibit meristematic activity. However, Ivens and Blackman (1950) observed that ethylphenylcarbamate inhibited the growth of barley at concentrations that do not affect the spindle. Other growth inhibiting mechanisms must be considered.

Interference of N-phenylcarbamates with photosynthetic activity has been observed (Moreland and Hill, 1959) and Moreland (1958) claims that the relative potency of these chemicals correlates closely with their phytotoxicity on intact plants. It is by no means clear which biochemical actions directly injure plants but those responsible for modifying cell division are most likely to be involved.

8. *Substituted Ureas*

According to Bucha and Todd (1951), the initial effect generally is leaf tip dieback, beginning on the older leaves. This is followed by progressive chlorosis and retardation of growth ending in the death of the plant. Water-logged areas occur 2—3 days after high doses have been applied and this may be followed by leaf wilt, stem collapse, rapid yellowing and ultimately abscission. A general description of symptoms is inadequate to express species variation and the effect of different environments. Ashton and Crafts (1973) illustrate the stages of damage by photographs taken from studies by Minshall (1957).

Substituted urea herbicides strongly reduce the photosynthetic activity of plants. One of the most recent reviews is by Hoffman (1971). Starvation which results is regarded as a likely mode of action of these chemicals as herbicides but Stanger and Appleby (1972) have proposed a different action for diuron under the system of Krinsky (1966). They postulate lethal photo-sensitized oxidations occurring in the cell. These are due to an increased concentration of oxidized chlorophyll resulting from an interrupted electron flow. This mechanism may well account for the chlorosis which is so characteristic of the substituted urea herbicides.

There is evidence however, that substituted ureas damage tissue by processes other than inhibiting photosynthesis. The elongation of detached roots of pea for example is inhibited by monuron as are attached roots of various grasses (Minshall, 1960). Necrotic leaf damage occurring within two days of treatment is moreover, unlikely to be due to a process of starvation.

9. *Thiocarbamates and Dithiocarbamates*

The most obvious symptom following treatment of grasses with thiocarbamates is abnormal emergence of leaves from the coleoptiles. High doses may prevent leaf emergence completely and at low rates leaves may emerge from the base of the coleoptile. After the leaf has elongated for a period, it may form a loop, with the leaf tip fixed within the coleoptile tip. Broad-leaved weeds become necrotic and this is accompanied by a general inhibition of growth. Thio and dithiocarbamates modify a variety of processes in plants such as photosynthesis, respiration, oxidative phosphorylation, protein synthesis and nucleic acid metabolism. Their biochemical action is reviewed by Ashton and Crafts (1973). Banting (1970) using diallate and triallate on wild oats and wheat concluded that although these chemicals are mitotic poisons in young shoots, the major effect is on cell elongation. Shoot growth inhibition occurred in both species at concentrations which have no effect on mitosis.

10. *Bipyridylium Quaternary Ammonium Salts*

Paraquat and diquat damage plant tissue very rapidly. In sunny conditions, leaf discoloration (indicating disruption of plant membranes) may occur within an hour of applying the chemicals. This is followed within 24 h by wilting of the leaves and later still by desiccation. A slower action may then follow in which paraquat moves down the plant presumably in the apoplastic system and chlorosis develops in the unsprayed parts of the plant. The chlorosis, if severe, is followed by death of the affected tissues. Diquat is similar to paraquat though it is translocated less readily in grasses. Morfamquat is selective and controls a wide variety of broad-leaved plants but grasses are highly resistant.

These three bipyridyliums differ in phytotoxic action from conventional quaternary ammonium compounds because they have the ability to form free radicals by reduction in photosystem 1 of the chloroplast (Homer *et al.* 1960). This is an essential stage in the activity of the chemicals since plants are scarcely damaged in the dark (Mees, 1960). It is not however, the only important reaction. Mees showed that no phytotoxicity developed in the absence of oxygen. It has been suggested that hydrogen peroxide produced during re-oxidation of the bipyridylium free radicals is the true toxicant (Black and Meyers, 1966; Calderbank, 1968). However, two further possibilities exist. First peroxide radicals themselves may be toxic. Second, White (1970) proposes that paraquat first reacts with excited chlorophyll to form a free radical/chlorophyll complex. (Molecular spacings in paraquat and of the porphyrin rings in chlorophyll are particularly favourable to this interaction). Oxygen then produces a radical peroxide/chlorophyll complex, whence the porphyrin rings are broken down by the peroxide with regeneration of paraquat. Diquat and paraquat may also be reduced in the electron transport system of respiration (Bozarth *et al.* 1965), but potentials here are less favourable than those of photosynthetic reduction. This however, may be the cause of damage in the dark when high doses of chemical are applied. Morfamquat is similar in action on dicotyledons but its failure to damage cereals and grasses is not understood (Brian, 1972).

11. *Pyridines*

Picloram is a growth-regulating herbicide. It alters leaf morphology at low concentrations and at herbicidal rates, leaf growth is stunted and terminal growth ceases. Tissue proliferation may also occur and many of the effects associated with the phenoxyacetic acids, such as epinasty and stem splitting may be observed. A number of biochemical processes is affected. Picloram inhibits oxygen uptake (Foy and Penner, 1965) and Malhotra and Hanson (1970) found that the nucleic acid content of sensitive plants increased 24 h after picloram treatment. Resistant plants were unaffected. It is reasonable to conclude that picloram-like hormone-weedkillers may modify the synthesis and metabolism of nucleic acid and hence protein synthesis in cells.

12. *Pyridazines*

Maleic hydrazide (MH) and pyrazon are both growth inhibitors, though the latter has a greater tendency to produce abnormal plants. MH is used successfully to inhibit the growth of turf grasses and of hedging plants but in some environmental conditions, it may damage plants. MH has been shown to interfere with cell division (Darlington and McLeish, 1951; Crafts, 1961) and Nooden (1969) found that MH inhibited root elongation through an effect on cell division. A lowering of the levels of dehydrogenases in onion tissue was observed by Isenberg *et al.* (1951) and respiration was affected.

The most obvious biochemical effect of pyrazon is a significant reduction in the rate of photosynthesis (Eshel, 1969; Hilton *et al.* 1969), a reduction which these authors consider is a sufficient basis for the mode of action of pyrazon.

13. *Pyrimidines—Uracils*

The characteristic symptom of bromacil action on oats as found by Ashton *et al.* (1969) is a strong inhibition of root growth mostly at a point just behind the meristem. There is also a number of effects which accompany inhibition such as incomplete, fragmented and branching cell walls. In leaves, chloroplast development is incomplete and the authors conclude from the observed symptoms that bromacil affects the integrity of membranes. Pyrimidines strongly inhibit photosynthesis. Hilton *et al.* (1969) and Hoffman (1971) conclude that this must be a major factor in their herbicidal activity since these chemicals have little effect on non-photosynthetic organisms or on some tissues grown in the dark.

14. *Triazines*

Triazines are inhibitors of growth of most plant organs. Leaves become chlorotic and later, necrotic symptoms develop. At sub-toxic levels however, triazines may increase the chlorophyll content of leaves and the darker green colour of resistant crops can easily be observed. These chemicals inhibit photosynthesis as shown by many workers using isolated chloroplasts (Exer, 1958; Moreland *et al.* 1959) and intact plants. Although triazines also inhibit respiration (Olech, 1966; Nasyrova *et al.* 1968), the growth inhibition they produce in intact plants is attributed to the blockage of photosynthesis within photosystem 2 at the stage where water is photolysed.

A further action but one possibly of minor importance is the mitotic disturbance which results from treatment with certain triazines (Rudenberg *et al.* 1955). More recent work has been carried out by Liang *et al.* (1967) using atrazine on grain sorghum. Pollen cell abnormalities took many forms but there was no effect on yield.

15. *Amitrole*

Amitrole produces albinism in developing leaves and it is a permanent effect if the application rate is high. Severe growth inhibition also occurs followed by death of the plant. Leaves may recover after only low doses of amitrole. Mature leaves may senesce and the herbicide is used to defoliate cotton prior to harvesting. Root inhibition has also been observed (Jackson, 1961).

Albinism may result from a disturbance in the availability of essential metals such as Fe, Mn, Mg and N, P or K. Amitrole may destroy chloroplast pigments but this is unlikely to be the cause since its effects are not usually evident on mature leaves but are restricted to developing leaves. Thus the most likely cause is an inhibition of chloroplast development which, according to Bartels *et al.* (1967), is due to an interference with chloroplast nucleic-acid metabolism. Sund (1961) however, concludes that the inhibition is due to a blocking of riboflavin production which is necessary for chloroplast development.

Peroxidase activity is substantially lowered in nutgrass (*Cyperus rotundus*) tubers and catalase activity reduced to as little as 5% of the untreated controls (Palmer and Porter, 1959). The importance of an altered phosphate economy in plants after amitrole treatment has been emphasized by Wort and Loughman (1961). Evidence of amitrole interference in RNA and DNA synthesis is conflicting (Bartels and Hyde, 1970). The most likely mode of action of amitrole as a herbicide is on chloroplast development but other factors must be considered to account for its ability to defoliate.

16. *Endothal*

Endothal is a soil and contact herbicide. Effects on amylase activity (Penner, 1968) and proteolytic activity (Ashton *et al.* 1968) have been observed. Daniel and Wilson (1956) have also recorded effects on mitosis.

17. *Thiocarbonates*

No leads have arisen on the mode of action of this small family of herbicides.

18. *Glycine Derivative*

Plants treated with MON-0573 ("glyphosate") become severely pigmented about two weeks after treatment. Regrowth is highly chlorotic, growth ceases and the plants eventually die. Glyphosate has been reported by Jaworski (1972) to act by inhibiting the aromatic amino-acid biosynthetic pathway. His studies concerned the aquatic plant *Lemna gibba* and the bacterium *Rhizobium japonicum*; growth inhibition in the presence of the herbicide was alleviated by the addition of phenylalanine to the nutrient medium.

19. *Organo-arsenic Compounds*

Ashton and Crafts (1973) summarize the herbicidal effects of organo-arsenical herbicides as chlorosis, cessation of growth, progressive browning followed by dehydration and death. Buds fail to sprout. Possible modes of action must include effects on membrane permeability. The chemicals also affect the activity of the enzyme fumarase and trivalent arsenic is an uncoupling agent (Dixon and Webb, 1958). No clear picture, however, has emerged of the primary cause of death.

20. *Oils*

There is a body of evidence that oils reduce the rate of photosynthesis of leaves, an effect that has been attributed to a physical interference with gaseous exchange by the layer of oil deposited (Riehl and Wedding, 1959). A destruction of chlorophyll is possible by the breakdown of protoplasm thus exposing the pigment to the destructive action of light. Whilst this may be true, the rate of development and the nature of the damage resulting from the application of oils, suggests rather that the breakdown of protoplasm leading to collapse of the leaf surface (Dallyn and Sweet, 1951) is directly responsible for death. This may well result from the incorporation of oil in the plasma membrane. This has been suggested by van Overbeek and Blondeau (1954) from experiments with maize coleoptiles and pea epicotyls; it would result in a marked increase in membrane permeability and in extreme cases, in complete protoplast disorganization. According to these authors the effect on photosynthesis may result from a similar process since it is known that the lipoid content of chloroplasts exceeds that of the remainder of the cytoplasm. Oils will therefore tend to accumulate there, probably disorganizing the ordered structure of the chloroplast (Granick, 1949).

From experiments on mustard and parsnip leaves, it is claimed that oils interfere with the water relations of the plant. The water supply to the leaves was found to be permanently reduced by phytotoxic oils (Helson and Minshall, 1950). The fact that phytotoxic oils contain fatty acids, also increases the difficulty of elucidating the exact toxic action. The acids themselves have been found to reduce the respiratory rate in plants but it is doubtful if this physiological effect can be a significant factor in the toxic action of oils.

REFERENCES

Åberg, B. (1948). *K. LantbrHögsk. Annlr.* **15**, 37–107.
Åberg, B. and Johansson, I. (1966). *LantbrHögsk. Annlr.* **32**, 245–254.
Antognini, J., Day, H. M. and Tilles, H. (1957). *Proc. N.E. Weed Control Conf.* pp. 2–11.

Arndt, F. and Kötter, C. (1967). *Abstr. 6th int. Congr. Plant Protection, Vienna,* p. 433.
Ashton, F. M. and Crafts, A. S. (1973). "Mode of Action of Herbicides", J. Wiley & Sons, Inc.
Ashton, F. M., Cutter, E. G. and Huffstutter, D. (1969). *Weed Res.* **9,** 198–204.
Ashton, F. M., Penner, D. and Hoffman, S. (1968). *Weed Sci.* **16,** 169–171.
Åslander, A. (1926). *J. Am. Soc. Agron.* **18,** 1101–1102.
Åslander, A. (1927). *J. agric. Res.* **34,** 1065–1091.
Åslander, A. (1928). *J. agric. Res.* **36,** 915–934.
Baird, D. D. and Upchurch, R. P. (1970). *Proc. 23rd Southern Weed Control Conf.* 101–104.
Baird, D. D., Upchurch, R. P., Homesley, W. B. and Franz, J. E. (1971). *Proc. N. cent. Weed Control Conf.,* Kansas City, Missouri, **26,** 64–68.
Banting, J. D. (1970). *Weed Sci.* **18,** 80–84.
Bartels, P. G. and Hyde, A. (1970). *Pl. Physiol. Baltimore,* **46,** 825–830.
Bartels, P. G., Matsuda, K., Siegel, A. and Weier, T. E. (1967). *Pl. Physiol. Lancaster,* **42,** 736–741.
Beatty, R. H. (1958). *Proc. 4th Br. Weed Control Conf.* pp. 86–96.
Behrens, R. (1953). *Proc. 10th ann. Mtg. N. cent. Weed Conf.* p. 61.
Bentley, A. J. (1950). *Nature, Lond.* **165,** 449.
Bissey, R. and Butler, O. (1930). *J. Am. Soc. Agron.* **22,** 124–135.
Black, C. C. (Jr.) and Meyers, L. (1966). *Weeds,* **14,** 331–338.
Bokorny, Th. (1915). *Biol. Zbl.* **35,** 25–30.
Bolley, H. L. (1901). *Rep. N. Dak. agric. Exp. Stn* **11,** 48.
Bolley, H. L. (1908). *Bull. N. Dak. agric. Exp. Stn* **80,** 541–574.
Bozarth, G. A., Funderburk, H. H., Curl, E. A. and Davis, D. E. (1965). *Proc. 18th Southern Weed Conf.* p. 615.
Brebion, G. and Scuflaire, R. (1954). *Méml. Servs chim. État.* **39,** 303–310.
Brenner, W. (1918). *Overs. Finska vet. soc. förhand.* 60A, No. 4.
Brian, R. C. (1972). *Pestic. Sci.* **3,** 409–414.
Brian, R. C., Homer, R. F., Stubbs, J. and Jones, R. L. (1958). *Nature, Lond.* **181,** 446–447.
Bucha, H. C. and Todd, C. W. (1951). *Science, N.Y.* **114,** 493–494.
Bucha, H. C., Cupery, W. E., Harrod, J. E., Loux, H. M. and Ellis, L. M. (1962). *Science, N.Y.* **137,** 537–538.
Calderbank, A. (1968). "Advances in Pest Control Research", Interscience, New York. Vol. **8,** pp. 127–140.
Caluwe, P. de (1900). *Vereen. Oudleerl. Rijks Landbouwschool,* **12,** 103.
Carpenter, K. and Heywood, B. J. (1963). *Nature, Lond.* **200,** 28–29.
Castelfranco, P., Foy, C. L. and Deutsch, D. B. (1961). *Weeds,* **9,** 580–591.
Christiansen, G. S., Kunz, L. J., Bonner, W. D. Jr. and Thimann, K. V. (1949). *Pl. Physiol. Lancaster,* **24,** 178–181.
Cottrell, H. J. and Heywood, B. J. (1965). *Nature, Lond.* **207,** 655–656.
Crafts, A. S. (1935). *Hilgardia,* **9,** 437–457.
Crafts, A. S. (1946). *Pl. Physiol. Lancaster,* **21,** 345–361.
Crafts, A. S. (1961). "The Chemistry and Mode of Action of Herbicides", p. 103. Interscience, New York.
Crafts, A. S. and Raynor, R. N. (1936). *Hilgardia,* **10,** 343–371.
Crafts, A. S. and Reiber, H. G. (1948). *Hilgardia,* **18,** 77–156.
Cupery, W. E. and Tanberg, A. P. (1942). U.S. patent 2,277,744.
Dallyn, S. L. and Sweet, R. D. (1951). *Proc. Am. Soc. hort. Sci.* **57,** 347–354.
Daniel, A. and Wilson, G. B. (1956). *J. Hered.* **47,** 151–155.
Darlington, C. D. and McLeish, J. (1951). *Nature, Lond.* **167,** 407–408.
De Rose, H. R. (1951). *Agron. J.* **43,** 139–142.
Desmoras, J., Jacquet, P. and Métivier, J. (1963). *C.r. Deuxième Conférence du Columa, Paris,* p. 69.
Dixon, M. and Webb, E. C. (1958). "Enzymes", Longman Group Ltd., England.
Dow Chemical Co. (1953). *Dalapon Bull.* No. 2.
Ennis, W. B. (1948). *Am. J. Bot.* **35,** 15–21.
Eshel, Y. (1969). *Weed Res.* **9,** 167–172.

Exer, B. (1958). *Experientia,* **14,** 136–137.

Fischer, E. (1886). *Justus Liebigs Annln. Chem.* **236,** 116–151.

Foy, C. L. and Penner, D. (1965). *Weeds,* **13,** 226–231.

Frankel, S. (1927). "Die Arzneimittel-Synthese", Springer, Verlag, Berlin.

Friesen, G. (1929). *Planta,* **8,** 666–679.

Friesen, G. (1960). *Res. Proc. natn. Weed Committee, Canada.*

Fryer, J. D. and Evans, S. A. (1968). "Weed Control Handbook", 5th edn. Vol. 1. Principles (Brit. Crop Protection Council).

Furtick, W. R. (1970). *Tech. Papers FAO int. Conf. Weed Control* p. 1. Davis, Calif. June 22–July 1, 1970.

Gaur, B. K. and Beevers, H. (1959). *Pl. Physiol. Lancaster,* **34,** 427–432.

Goksøyr, J. (1951). *Physiologia Pl.* **4,** 498–513.

Granick, S. (1949). "Photosynthesis in Plants", Chap. 5. *Monograph Am. Soc. Pl. Physiol., Ames, Iowa.*

Gray, R. A., Curtis, R., Day, H. M. and Klaich, M. (1962). *Proc. N. cent. Weed Control Conf.* **19,** 19–21.

Green, D. H. (1969). *C.r. Troisième Eur. Weed Res. Conf. Symp. sur les Nouveaux Herbicides,* p. 177.

Green, D. H., Schuler, J. and Ebner, L. (1966). *Proc. 8th Br. Weed Control Conf.* pp. 363–371.

Greenfield, S. S. (1941). *Science, N.Y.* **93,** 550–551.

Guthrie, F. B. and Helms, R. (1904). *Agric. Gaz. N.S.W.* **14,** 114 and **15,** 29.

Gysin, H. (1960). *Weeds,* **8,** 541–555.

Gysin, H. (1962). *Chemy Ind.* 1393–1400.

Gysin, H. and Knüsli, E. (1958). *Proc. 4th Br. Weed Control Conf.* pp. 225–233.

Gysin, H. and Knüsli, E. (1960). *Adv. Pest Control Res.* **3,** 289–358.

Hall, W. C., Truchelut, G. B. and Lane, H. C. (1953). *Bull. Tex. agric. Exp. Stn* p. 759.

Halliday, D. J. (1973). *Outlook on Agriculture,* **7**(4), 142–195.

Hamaker, J. W., Johnston, H., Martin, R. T. and Redemann, C. T. (1963). *Science, N.Y.* **141,** 363.

Hamm, P. C. and Speziale, A. J. (1956). *J. agric. Fd. Chem.* **4,** 518–522.

Hamner, C. L. and Tukey, H. B. (1944). *Science, N.Y.* **100,** 154–155.

Hannah, L. M. (1959). *Proc. N. cent. Weed Control Conf.* p. 50.

Hanson, J. B. and Slife, F. W. (1969). *Residue Rev.* **25,** 59–67.

Harvey, E. N. (1911). *J. exp. Zool.* **10,** 507.

Harvey, R. B. (1931*a*). *J. Am. Soc. Agron.* **23,** 481–489.

Harvey, R. B. (1931*b*). *J. Am. Soc. Agron.* **23,** 944–946.

Helson, V. A. and Minshall, W. H. (1950). *Proc. Canada natn. Weed Com. East Sec.* **3,** 19–20.

Hilton, J. L., Ard, J. S., Jansen, L. L. and Gentner, W. A. (1959). *Weeds,* **7,** 381–396.

Hilton, J. L., Scharen, A. L., St. John, J. B., Moreland, D. E. and Norris, K. H. (1969). *Weed Sci.* **17,** 541–547.

Hoffman, C. E. (1971). *2nd. int. Conf. Pestic. Chemy.* pp. 65–85.

Hoffman, O. L. and Smith, A. E. (1949). *Science N.Y.* **109,** 588.

Hofmann, A. W. (1885). *Ber. dt. Chem. Ges.* **18,** 2755.

Holm, L. (1971). *Weed Sci.* **19**(5), 485–490.

Holm, R. E. and Abeles, F. B. (1968). *Planta,* **78,** 293–304.

Homer, R. F., Mees, G. C. and Tomlinson, T. E. (1960). *J. Sci. Fd. Agric.* **6,** 309–315.

Ingle, M. and Rogers, B. J. (1961). *Weeds,* **9,** 264–272.

Isenberg, F. M. R., Odland, M. L., Popp, H. W. and Jensen, C. O. (1951). *Science, N.Y.* **113,** 58–60.

Ivens, G. W. (1952). *Ann. appl. Biol.* **39,** 418–422.

Ivens, G. W. and Blackman, G. E. (1950). *Nature, Lond.* **166,** 954–955.

Jackson, W. T. (1961). *Weeds,* **9,** 437–442.

Jaworski, E. G. (1972). *J. agric. Fd. Chem.* **20,** 1195–1198.

Juniper, B. E. and Bradley, D. E. (1958). *J. Ultrastruct. Res.* **2,** 16–27.

Keitt, G. W., Jr. (1960). *Bot. Gaz.* **122,** 51–62.

Keitt, G. W. and Baker, R. A. (1966). *Pl. Physiol. Lancaster,* **41,** 1561–1569.

Keitt, G. W. and Skoog, F. (1957). (Abs.). *Pl. Physiol. Lancaster*, **32** (sup) xx.
Kerr, M. W. and Wain, R. L. (1964). *Ann. appl. Biol.* **54**, 441–446.
King, L. J. (1966). "Weeds of the World: Biology and Control", Leonard Hill, London. Interscience Publ. Inc. New York, p. 526.
Kögl, F., Erxleben, H. and Haagen-Smit, A. J. (1934). Hoppe-Seyler's *Z. physiol. Chem.* **225**, 215–229.
Koopman, H. and Daams, J. (1960). *Nature, Lond.* **186**, 89–90.
Korsmo, E. (1930). "Umkräuter im Ackerbau der Neuzeit", Springer Verlag, Berlin.
Korsmo, E. (1932). "Undersökelser (1916–23). Over ugressets skadevirkninger og dets bekjempelse. 1. Akerbruket", Johnson and Nielsens Boktrylkeri, Oslo.
Krinsky, N. I. (1966). "Biochemistry of Chloroplasts" (T. W. Goodwin, ed.). Academic Press, London & New York.
Lachman, W. H. (1945). *Mass. State College, Ext. Serv. Spec.* Cir. 120.
Landen, R. H. (1934). *Am. J. Bot.* **21**, 583–591.
Leaf, G. K. (1953). *Proc. Iowa Acad. Sci.* **60**, 176.
Leafe, E. L. (1964). *Proc. 7th Br. Weed Control Conf.* p. 32–37.
Liang, G. H. L., Feltner, K. C., Liang, Y. T. S. and Morrill, J. L. (1967). *Crop Sci.* **7**, 245–248.
Long, H. C. (1934). "Suppression of Weeds by Fertilizers and Chemicals", Caledonian Press, London.
Loomis, W. E., Bissey, R. and Smith, E. V. (1931). *Science, N.Y.* **74**, 485.
Lush, G. B. and Stevenson, H. A. (1959). British Patent, 820,180 Sept. 16th.
Malhotra, S. S. and Hanson, J. B. (1970). *Weed Sci.* **18**, 1–4.
Mallah, G. S. and Dawood, M. M. (1956). *Alex. J. agric. Res.* **4**, 91–105.
Marth, P. C. and Mitchell, J. W. (1944). *Bot. Gaz.* **106**, 224–232.
Martin, H. (1972). "Pesticide Manual", 3rd edn. (Br. Crop Protection Council).
Mashtakov, S. M. and Moshchuk, P. A. (1967). *Agrokhimiya*, **9**, 80–89.
McCall, G. L. (1952). *Agric. Chem.* **7**, 40–42.
McCall, G. L. and Zahnley, J. W. (1949). *Kansas State Coll. Agric. Exp. Stn, Circ.* p. 255.
McDowall, R. K. (1933). "Weed Destruction with Sulphuric Acid", London, Oxford University Press.
Mees, G. C. (1960). *Ann. appl. Biol.* **48**(3), 601–612.
Minarik, C. E., Ready, D., Norman, A. G., Thompson, H. E. and Owings, J. F., Jr. (1951). *Bot. Gaz.* **113**, 135–147.
Minshall, W. H. (1957). *Can. J. Pl. Sci.* **37**, 157–166.
Minshall, W. H. (1960). *Can. J. Bot.* **38**, 201–216.
Monsanto Agricultural Division (1971). *Tech. Bull. MON.* 057.1.71.
Moreland, D. E. (1958). (Abs.) *Pl. Physiol. Lancaster*, **33**, Sup xxx.
Moreland, D. E. (1967). *A. Rev. Pl. Physiol.* **18**, 365–386.
Moreland, D. E. and Hill, K. L. (1959). *J. agric. Fd Chem.* **7**, 832–837.
Moreland, D. E., Gentner, W. A., Hilton, J. L. and Hill, K. L. (1959). *Pl. Physiol. Lancaster*, **34**, 432–435.
Morettini, A. (1915). *Staz. sper. agr. ital.* **48**, 693–716.
Muenscher, W. C. (1932). *Bull. N.Y. St. agric. Exp. Stn.* **542**, 1–8.
Nasyrova, T., Mirkasimova, Kh. and Pazilova, S. (1968). *Uzbek. biol. Zh.* **12**, 23–26.
Neller, J. R. (1931). *J. agric. Res.* **43**, 183–189.
Nobbe, F., Baessler, P. and Will, H. (1884). *Landwn VersStnen.* **30**, 381–395.
Nooden, L. D. (1969). *Physiologia Pl.* **22**, 260–270.
Novikov, V. A. and Barannikova, Z. D. (1954). *Referat. Zhur. Khim.* No. 27290.
Nutman, P. S., Thornton, H. G. and Quastel, J. H. (1945). *Nature, Lond.* **155**, 498–500.
Offord, H. R. (1931). *Tech. Bull. U.S. Dept. Agric.* **240**.
Offord, H. R. and d'Urbal, R. P. (1931). *J. agric. Res.* **43**, 791–810.
Olech, K. (1966). *Annls Univ. Mariae Curie-Skłodowska* (E) **21**, 289–308.
Overbeek, J. van (1964). *In* "The Physiology and Biochemistry of Herbicides" (L. J. Audus, ed.). Academic Press, London and New York.
Overbeek, J. van and Blondeau, R. (1954). *Weeds*, **3**, 55–65.
Palmer, R. D. and Porter, W. K. (Jr.) (1959). *Proc. S. Weed Conf.* **12**, 174.

Pate, D. A. (1966). "Degradation of ^{14}C-dichlobenil in bean, alligator weed, and certain microorganisms". Ph.D. thesis. Auburn University, Auburn, Ala.

Paton, D. and Smith, J. E. (1967). *Can. J. Biochem.* **45,** 1891–1899.

Pearlman, W. M. and Banks, C. K. (1948). *J. Am. chem. Soc.* **70,** 3726–3728.

Peligot, E. (1876). *C.r. hebd. Séanc. Acad. Sci., Paris,* **83,** 686–688.

Penner, D. (1968). *Weed Sci.* **16,** 519–522.

Poignant, P. and Crisinel, P. (1967). *Proc. 4th Columa Conf. Paris,* p. 196.

Rabaté, E. (1911). *J. Agric. prat. Paris,* **21** (new series), **75,** 497–509.

Rademacher, B. (1940). "The Control of Weeds in Germany", Imp. Bur. Pastures and Forage Crops, **27,** 68–112.

Ranjau, S. and Kaur, R. (1954). *J. exp. Bot.* **5,** 414–420.

Raymond, E. L. (1939). French patent, 848,934, Nov. 9th.

Redemann, C. T. and Hamaker, J. (1954). *Weeds,* **3,** 387–388.

Remy, T. and Vasters, J. (1914). *Landw. Jbr.* **46,** No. 4, 627.

Riehl, L. A. and Wedding, R. T. (1959). *J. econ. Ent.* **52,** 88–94.

Robbins, W. W., Crafts, A. S. and Raynor, R. N. (1942). "Weed Control", McGraw-Hill.

Robbins, W. W., Crafts, A. S. and Raynor, R. N. (1952). "Weed Control", 2nd edn., p. 238, McGraw-Hill.

Rudenberg, L., Foley, G. E. and Winter, W. D. (1955). *Science, N.Y.* **121,** 899–900.

Schoene, D. L. and Hoffman, O. L. (1949). *Science, N. Y.* **109,** 588–590.

Schultz, G. (1909). *Arb. dt. Landw. Ges.* p. 158.

Sharp, S. S., Swingle, M. C., McCall, G. L., Weed, M. B. and Cowart, L. E. (1953). *Agric. Chem.* **8,** 56–57.

Simon, E. W. (1953). *Biol. Rev. Cambridge Phil. Soc.* **28,** 453–479.

Singh, B. N. and Das, K. (1939). *J. Am. Soc. Agron.* **31,** 200–208.

Sjollema, B. (1896). *Chemikerzeitung* **20,** 1002–1015.

Slade, R. E., Templeman, W. G. and Sexton, W. A. (1945). *Nature, Lond.* **155,** 497–498.

Smith, J. E., Paton, D. and Robertson, M. M. (1966). *Proc. 8th Br. Weed Control Conf.* **1,** 279–282.

Stanger, C. E. (Jr.) and Appleby, A. P. (1972). *Weed Sci.* **20,** 357–363.

Stewart, J. and Smith, E. S. (1922). *Soil Sci.* **14,** 119–126.

Stoddard, E. M. (1944). *Circ. Conn. agric. Exp. Stn.* p. 160.

Sund, K. A. (1961). *Physiologia Pl.* **14,** 260–265.

Switzer, C. M. (1954). *Pl. Physiol. Lancaster,* **32,** 42–44.

Templeman, W. G. and Sexton, W. A. (1945). *Nature, Lond.* **156,** 630.

Thimann, K. V. and Bonner, W. D., Jr. (1949). *Am. J. Bot.* **36,** 214–221.

Thompson, N. F. and Robbins, W. W. (1926). *Bull. U.S. Dep. Agric.* 1451, 1.

Tischler, N., Bates, J. C. and Quimba, G. P. (1950). *Proc. 4th N.E. Weed Conf.* pp. 51–84.

Truffaut, G. and Pastac, I. (1935). British Patent 424,295, May 29th.

Varner, R. W., Weed, M. B. and Ploeg, H. L. (1964). *Proc. 7th Br. Weed Control Conf.* pp. 38–43.

Wain, R. L. (1957). *Agriculture, Lond.* **63,** 575–579.

Wain, R. L. (1963). *Nature, Lond.* **200,** 28.

Went, F. W. (1928). *Recl. Trav. bot. néerl.* **25,** 1–116.

Went, F. W. and Thimann, K. V. (1937). "Phytohormones", Macmillan, New York.

White, B. G. (1970). *Proc. 10th Br. Weed Control Conf.* pp. 997–1007.

Woodford, E. K., Holly, K. and McCready, C. C. (1958). *A. Rev. Pl. Physiol.* **9,** 311–358.

Wort, D. J. and Loughman, B. C. (1961). *Can. J. Bot.* **39,** 339–351.

Wuu, K. D. and Grant, W. F. (1966). *Can. J. Genet. Cytol.* **8,** 481–501.

Yamasaki, M. (1929). *J. imp. agric. Exp. Stn., Nishigahara,* **1,** 161.

Zimmerman, P. W. and Hitchcock, A. E. (1951). *Contr. Boyce Thompson Inst. Pl. Res.* **16,** 209–213.

Zimmerman, P. W., Hitchcock, A. E. and Kirkpatrick, H. Jr. (1951). *Am. Fertil.* **114**(1), 14–18.

Zimmerman, P. W. and Wilcoxon, F. (1935). *Contr. Boyce Thompson Inst. Pl. Res.* **7,** 209–229.

CHAPTER 2

GENERAL GROWTH RESPONSES OF PLANTS

P. M. CARTWRIGHT

Department of Agricultural Botany
University of Reading, Whiteknights, Reading, England

INTRODUCTION

The growth of a plant can be described in terms of cell division, cell enlargement and cell differentiation. New cells are continuously generated by a regular pattern of cell division in the meristems while the orderly enlargement and differentiation of meristematic cells gives rise to the mature tissues of the plant whose functions integrate to control the activities of the growing regions and supply their necessary substrates. Herbicidal action might result from a direct interference with the division, the enlargement or the differentiation of growing cells, from a loss of control of the balance between those processes or from a disturbance in the functioning of the roots, leaves or vascular tissues on which the growing regions depend.

Only in special situations do cells, once they are differentiated, regain their meristematic state and then they generally do so for only a limited period, e.g. to produce wound callus. Hence a further opportunity for herbicidal action is on the control system which normally restrains such de-differentiation of mature cells.

Thus there are many ways in which a herbicide might operate and it is not surprising that the large number of substances currently in commercial use represent a variety of modes of action. This chapter is primarily concerned with those herbicides which disrupt the growth processes of the plant. Mention is also made of some other inhibitors and growth regulators whose effects resemble those of certain herbicides or which may be employed as adjuvants in herbicide use.

I. Effects of Herbicides on Cell Division

A. MITOTIC POISONS

A number of herbicides can be classed as mitotic poisons since they block mitosis in the primary meristems. They include the carbamates, thiocarbamates, maleic hydrazide, trifluralin and nitralin, and probably also chlorthal dimethyl, bromacil and bensulide. Their action is associated with mitotic aberrations and the precocious vacuolation and enlargement of the cells of the meristem.

The mode of action of the carbamates, especially propham, chlorpropham and barban, has been extensively studied since Templeman and Sexton (1945) reported the selective action of propham on monocotyledonous seedlings. Mitotic cycles are interrupted and cell division ceases in the root and shoot meristems of affected seedlings (Ennis, 1948a). Aberrations in propham-treated oat and barley root tips included the presence of abnormally enlarged meristematic cells with giant nuclei or a number of micronuclei (Ennis, 1948b; Canvin and Friesen, 1959) (Fig. 1), and the appearance of binucleate cells has been reported in barban-treated wheat roots (Burström, 1968). A few dicotyledonous species, are also susceptible to carbamates: responses of cucumber root tips to chlorpropham, similar to those of monocotyledons, have been reported i.e. the inhibition of mitosis and the precocious enlargement and maturation of cells near the apex (Scott and Struckmeyer, 1955). Trifluralin and nitralin superficially resemble the carbamates in their effects on primary meristems (but not in their species selectivity). The action of trifluralin as an inhibitor of cell division in root meristems is widely reported (Talbert, 1965; Bayer et al. 1967; Normand et al. 1968; Schultz et al. 1968; Hacskaylo and Amato, 1968; Mallory, 1969; Shahied, 1970; Lignowski and Scott, 1972; and others). The initial response, within a few hours of treatment, is an arrest of cells in prophase (Talbert, 1965) and this is characteristically followed by other

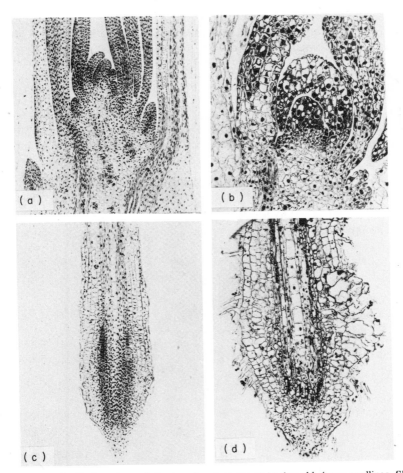

Fig. 1. Longitudinal sections of shoot and root apices of 14-day-old *Avena* seedlings. Shoot apices from seedlings raised in soil. Root apices from seedlings raised in Hoagland mineral nutrient solution. a. Shoot apex from untreated seedling. b. Shoot apex from seedling exposed to propham in the soil (10 mg/½ gal pt) for 144 h. c. Roof apex from untreated seedling. d. Root apex from seedling exposed to propham (9 parts/10⁶) for 168 h (reproduced with permission from Ennis, 1948*b*).

mitotic aberrations, the enlargement of some meristematic cells which have become multinucleate and a progressive decrease in the extent of the meristematic tissue (Bayer *et al.* 1967; Normand *et al.* 1968; Hacskaylo and Amato, 1968). In these respects nitralin closely resembles trifluralin; the blocking of cell division and the appearance of multinucleate cells in the meristem of maize (*Zea*) roots has been reported (Gentner and Burk, 1968; Schieferstein and Hughes, 1968).

Especially in cotton, lateral root primordia are more sensitive to trifluralin than are the primary root meristems. Indeed, a complete inhibition of lateral root development is often effected. Groups of cells in the pericycle region (presumably destined to become lateral roots in the absence of the herbicide) enlarge to produce "primordiomorphs" which are composed of large polyploid or multinucleate cells (Fig. 2) (Bayer *et al.* 1967; Hacskaylo and Amato, 1968).

Some miscellaneous herbicides including chlorthal-dimethyl, bensulide and bromacil probably also act by blocking mitosis. Chlorthal-dimethyl inhibits the rooting of stolons in *Cynodon dactylon*: cell division in the root primordia

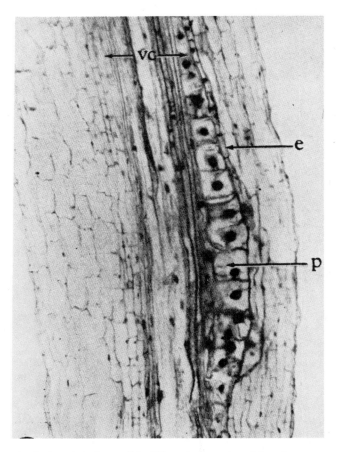

Fig. 2. Longitudinal section of part of the differentiating region of the primary root of a cotton seedling exposed to 10^{-4} M trifluralin for 72 h. e = endodermis; vc = vascular cylinder; p = pericycle, with enlarged cells forming a "primordiomorph". (Reproduced with permission from Bayer *et al.* 1967.)

ceases although cell enlargement may continue for a limited time (Bingham, 1968). Furthermore in maize and onion roots chlorthal-dimethyl blocks mitosis at metaphase and the meristematic tissues become disorganized with the development of many hypertrophied cells (Shaybany, 1970; Shaybany and Anderson, 1972). Bensulide, like chlorthal-dimethyl, inhibits cell division in the tips of root initials on *Cynodon dactylon* stolons (Bingham, 1967). Its effect on oat roots (Cutter *et al.* 1968), like that of bromacil (Ashton *et al.* 1969), is an inhibition of cell division in the meristem linked with the appearance of binucleate cells and a differentiation of tissues nearer to the tip.

Thiocarbamates such as diallate, triallate and EPTC, like the carbamates, are particularly effective in the control of grass weeds and also appear to be cell division inhibitors, although it is not clear whether this is their primary mode of action (Banting, 1970). Unlike the carbamates, however, they seem to be more toxic to the shoots than to the roots. In wild oat seedlings exposed to diallate or triallate cell divisions were fewer, and mitotic abnormalities were observed in the meristematic tissue at the base of the first leaf (Banting, 1970). Selectivity of EPTC towards shoot growth has also been reported in oat, barley, sorghum and corn seedlings: 1 part/10^6 of EPTC fed to the roots inhibited shoot growth while root growth was either unaffected or even enhanced (Gray and Weirich, 1969).

Maleic hydrazide has long been recognized as an inhibitor of plant growth. Its action seems to be on cell division in both the apical and sub-apical meristems of the shoot (Sachs and Lang, 1961). This is consistent with the observations of Greulach and Haesloop (1954) that maleic hydrazide inhibition of stem growth in bean and tomato was confined to the younger internodes in which cell division was contributing to extension growth. Gifford (1956) reported that inhibitory effects on barley seedlings were associated with aberrant nuclei in the intercalary meristems, many of which nuclei showed arrested metaphase plates. Tiller buds were also affected; they showed precocious cell enlargement without any cell division. Cytological studies provide evidence that maleic hydrazide specifically inhibits DNA replication (Scott, 1968).

Further evidence regarding the specific action of carbamates and trifluralin has also been obtained from cytological studies. Mann and co-workers have reported that the carbamates, propham, chlorpropham and barban cause contraction of the chromosomes (Mann, 1967; Storey *et al.* 1968). Since these herbicides, and particularly barban, are also inhibitors of gibberellin-induced α-amylase synthesis in barley endosperm (Yung and Mann, 1967) it is argued that their action at the molecular level is to inhibit gene derepression. Thus they inhibit the onset of metabolic changes but do not affect "on-going" processes. Support for such a mode of action was provided by Keitt (1967) who showed that 10^{-6} M chlorpropham severely inhibited cytokinin-induced cell proliferation in tobacco pith callus cultures but did not affect IAA induced cell elongation in *Avena* first internode segments, i.e. it did not inhibit the elongation of already elongating cells.

Mitosis in the dividing endosperm cells of *Haemanthus katherinae* can be studied *in vivo* by time-lapse cine photomicroscopy (Jackson, 1967). This provides a convenient system for the detailed study of the effects of mitotic poisons. Evidently the effect of propham is to cause a loss of the parallel alignment of the spindle microtubules which, instead, become orientated in radial rays which act as micropoles of the spindle apparatus and thus become focal points for chromosomes which aggregate there into micronuclei. The action of propham thus differs from that of colchicine which destroys the spindle apparatus in the same material (Hepler and Jackson, 1969). A different effect was observed with trifluralin: in treated cells the spindle microtubules were destroyed during prophase and thus the nuclear envelope remained intact at metaphase (Jackson and Stetler, 1973). The authors argue that reports that prophase is not affected by trifluralin (e.g. Lignowski and Scott, 1972) arise from the study of fixed material.

B. AMIDES AND MISCELLANEOUS HERBICIDES

Several amide herbicides including propachlor and allidochlor, as well as siduron, dichlobenil and amitrole, inhibit cell division in primary meristems. They act less specifically than the mitotic poisons such as propham and do not necessarily cause mitotic abnormalities. Also, other growth and differentiation processes are commonly affected.

Propachlor causes an immediate decrease in the proportion of cells in mitosis in onion root tips (Dillon and Anderson, 1972) and allidochlor has a similar effect in barley, but not in pea (Canvin and Friesen, 1959). There is considerable evidence that the inhibitory effects of these herbicides are due to a general blocking of protein synthesis (Mann *et al.* 1965; Duke, 1967; Dillon and Anderson, 1972) (see Chapter 17).

Siduron differs from other substituted urea herbicides in that it does not act as a photosynthetic inhibitor. It prevents seedling establishment in crabgrass and other species by inhibiting the growth of the seedling roots. It decreases the proportion of cells in mitosis in the root tips of germinating barley (*Hordeum*) without causing any obvious structural changes (Splittstoeser and Hopen, 1970).

Both dichlobenil and amitrole interfere with many aspects of growth and differentiation including cell division in the primary meristems. Milborrow (1964) observed that in oat (*Avena*) root tips dichlobenil caused an immediate decrease in the number of cells in prophase followed by a marked decrease of all mitotic figures, indicating that the nuclei were prevented from entering mitosis. In a number of species such blocking of cell division was followed by a blackening and death of the growing points closely resembling the effects of boron deficiency. Amitrole has been reported to inhibit cell division in the primary root meristems of *Pisum* and *Linum*. In *Pisum* mitotic abnormalities and a decrease in the frequency of mitotic figures were noted (Grigsby *et al.* 1955) while in *Linum* an associated increase in the number of lateral primordia initiated was observed (Arntzen *et al.* 1970).

C. AUXIN HERBICIDES

1. *Effects on Callus Growth and the Establishment of Tissue Cultures*

Synthetic auxins such as 2,4-D and NAA are very effective inducers of growth in explants of mature plant tissue. The type of growth induced depends on the nature and physiological state of the tissue and on the level of the auxin and other growth regulating substances. An interaction with cytokinins (or cytokinin-rich fractions or extracts) is commonly observed; for example, Steward and Caplin (1951) found that coconut milk could promote cell proliferation in potato tuber explants only in the presence of low concentrations of a synthetic auxin such as 2,4-D. Also, Jablonski and Skoog (1954) reported that explants composed of mature pith cells from tobacco stem responded to auxins only by a marked enlargement of cells whereas cell proliferation was initiated if extracts from vascular tissue were also added. The synergism between auxins and cytokinins is not however always directed towards cell division: according to the nature of the pretreatment of Jerusalem artichoke tissue and the experimental conditions involved either cell division or cell enlargement may predominate (Adamson, 1962).

It has long been recognized that low concentrations of auxin, with or without added cytokinins, tend to promote DNA synthesis and cell division

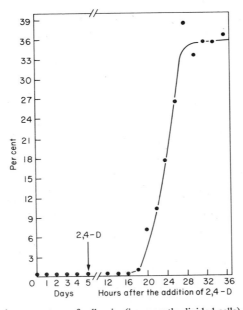

Fig. 3. Change in the percentage of cell pairs (i.e. recently divided cells) in explants of mature Jerusalem artichoke tissue during five days of culture in a sugar and mineral salts medium and after addition of 2,4-D to an overall concentration of 10^{-6} M (reproduced with permission from Yeoman and Mitchell, 1970).

while somewhat higher concentrations favour RNA synthesis and cell expansion (Skoog, 1954). More recent work (Yeoman and Mitchell, 1970) has highlighted the connection between 2,4-D and the start of DNA replication in tissue explants from mature Jerusalem artichokes. In freshly excised tissue cell division takes place only in the presence of 2,4-D (or a similar auxin) and in explants "aged" for several days on the basal medium only a small proportion (5%) of the cells will divide and then by only one division. However if 2,4-D is added, 35% of the cells promptly divide synchronously (Fig. 3) and an extended period of division with some expansion is triggered. While increases in RNA and protein precede the addition of 2,4-D, DNA synthesis is promoted only in association with cell division after 2,4-D has been added. Thus auxins have a specific role in triggering cell division in differentiated cells.

The establishment of tissue cultures from monocotyledons is generally more difficult than from dicotyledons. However, in cases where monocotyledonous callus has been successfully cultured, a synthetic auxin such as 2,4-D has been reported as the agent necessary for the initial induction of callus formation (Atkin and Barton, 1973).

The substituted benzoic acids which have auxin activity (e.g. 2,3,6-TBA) (Bentley, 1950; Zimmerman and Hitchcock, 1951) are also effective in promoting cell division and cell expansion in Jerusalem artichoke tissue explants (Hicks *et al.* 1964).

In isolated stem segments of tobacco and other species callus development takes place only at the basal end. Where phenoxy-auxins such as 2,4-D stimulate such callus growth, as in 2,4-D-susceptible clones of *Convolvulus arvensis* (Whitworth and Muzik, 1967) they do not alter its distribution. However, a number of substituted benzoic acids, including the auxin herbicide 2,3,6-TBA and the "anti-auxin" TIBA, obliterate the polarity of callus development on tobacco stem cuttings, and the effect is related to their action as inhibitors of polar auxin transport (Keitt and Baker, 1966).

2. *Effects on Cell Division in the Intact Plant*

While auxin herbicides such as 2,4-D are commonly observed to cause de-differentiation and the initiation of cell division in mature cells, they generally inhibit cell division in the primary meristems of intact plants. At least in the short term the inhibition may be caused by endogenous ethylene production which is stimulated by auxins (Morgan and Hall, 1964). Thus Apelbaum and Burg (1972a) reported an absence of synergism between the closely similar effects of 2,4-D and exogenous ethylene in the root and shoot meristems of etiolated pea (*Pisum*) seedlings: both substances virtually stopped cell division without causing obvious mitotic abnormalities. However the continued suppression of cell division in the apices of plants treated with lethal doses of 2,4-D is probably caused by changes in the more mature regions which are evidently not directly related to endogenous ethylene (Apelbaum and Burg, 1972a). The marked stimulation of secondary meristematic activity in the cambium and phloem regions of the plant axis may preferentially mobilize substrates for growth or may physically dislocate the vascular strands so that the growing points are starved and/or desiccated. For example, 2,4-D inhibits cell division, cell elongation and RNA and protein synthesis in the apices of soya bean seedlings but a renewal of RNA and protein synthesis in tissues behind the meristem leads to massive tissue proliferation, disorganized growth and ultimately death of the seedlings (Key *et al.* 1966). Similarly in cocklebur (Cardenas *et al.* 1968) abnormal lateral growth of the plant axis, stimulated by 2,4-D treatment, is coupled with the suppression of normal apical growth.

The above arguments do not imply that the cells of the primary meristems cannot be directly affected by auxin herbicides. Apelbaum and Burg (1972*a*) concluded that higher doses of 2,4-D had a "direct herbicidal action" in addition to their ethylene-mediated effects (see Chapter 8). Furthermore, higher doses have been observed to cause cytological abnormalities such as anaphase bridges and the appearance of micronuclei in onion root tip cells (Croker, 1953) and in cells from a number of sources including the staminal hairs of *Tradescantia* (Sawamura, 1964). Chromosomal aberrations due to 2,4-D treatment have also been reported during meiosis in pollen mother cells of *Sorghum* (Liang *et al.* 1969).

Phaseolus vulgaris is typical of many dicotyledonous species which are highly susceptible to auxin herbicides such as 2,4-D. The histological changes which precede death characteristically include massive proliferation of cambium, phloem parenchyma and ray parenchyma cells. These tissues are more readily triggered into cell division than are cortical or pith cells which, if affected, respond by abnormal radial cell enlargement (Swanson, 1946; Beal, 1946; Murray and Whiting, 1947; Kiermayer, 1959). A further characteristic of such 2,4-D-induced cell proliferation is that the derivatives remain meristematic for considerable periods (Swanson, 1946; Eames, 1950).

Similar abnormal tissue proliferation following treatment with phenoxy herbicides has been observed in hypocotyls and stems of many dicotyledonous species. More recent reports include those of Whitworth and Muzik (1967) on *Convolvulus arvensis,* Rubin and Gritsaenka (1968) on *Amaranthus retroflexus* and *Chenopodium album,* Cardenas *et al.* (1968) on *Xanthium pennsylvanicum,* Kiepal (1970) on *Sinapis arvensis,* Srivastava and Sharma (1971) on *Chenopodium album* and Berquist (1971) on *Myriophyllum spicatum.* Also, Meyer (1970) reported typical anatomical abnormalities in *Prosopis juliflora* (honey mesquite) after treatment with 2,4,5-T: sections of the hypocotyl tissues of treated seedlings (Fig. 4) show the presence of a wide zone of "abnormal cortical cells" at the junction of cortex and phloem.

Auxin herbicides grossly modify the pattern of division, expansion and differentiation of cells in leaf primordia and expanding leaves. In cotton, which is highly susceptible, the maximum effect is on leaves which are still in the primordial stage but in which marginal growth has already been initiated. The layer of cells from which the palisade and mesophyll are normally derived (i.e. the middle layer in the marginal initial) undergo precocious divisions and continue dividing without differentiation so that finally the mesophyll region in the much reduced leaves is composed of closely packed undifferentiated cells

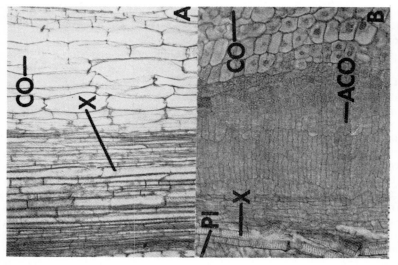

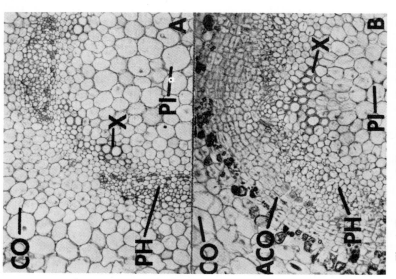

Fig. 4. Transverse (left) and longitudinal (right) sections of the hypocotyls of 5-day-old honey mesquite seedlings. A. Untreated. B. Germinated in the presence of 2,4,5-T (5 parts/10^6 and 10 parts/10^6 for material in T.S. and L.S. respectively). ACO = abnormal cortical cells; CO = cortex; PI = pith; PH = phloem; X = xylem. (Reproduced with permission from Meyer, 1970).

which form a "replacement tissue" (Watson, 1948; Eames, 1951). In already expanded leaves, the effects of auxin herbicide sprays are less marked but some cell division may be stimulated in the mesophyll (Bradley *et al.* 1968).

Petioles also show abnormal cell proliferation in response to auxin herbicides. However, in at least some cases, the differentiation of vascular elements may be promoted rather than delayed as is typical in stems. In cotton, sub-lethal doses of 2,4-D greatly modify the arrangement of vascular tissue and increase the number of differentiated sieve tubes in the petioles of affected leaves (Gifford, 1953). Also, 2,4,5-T has been reported to stimulate secondary phloem and xylem differentiation in apricot petioles (Bradley *et al.* 1968).

Monocotyledonous shoots are generally resistant to auxin herbicides and gross morphological responses such as those of cotton and *Phaseolus* are exceptional. However, some species including nutgrass (*Cyperus rotundus*) and rice are moderately susceptible (Eames, 1949; Kaufman, 1955) and "resistant" species may be susceptible at some developmental stages. When susceptibility does occur, the responses are essentially of the same type as those in dicotyledonous shoots, i.e. a stimulation of cell proliferation in certain weakly differentiated tissues and a modification of leaf anatomy with the associated development of "replacement tissue" in the mesophyll (Eames, 1949). In rice, 2,4-D treatment results in the precocious lengthening of several immature internodes involving the stimulation of both cell division and cell elongation in the rib meristems (Kaufman, 1955).

Probably the unresponsiveness to auxin herbicides of many monocotyledons is accounted for by their peculiar vascular anatomy, i.e. the absence of those relatively undifferentiated tissues, vascular cambium and phloem parenchyma, which are most readily stimulated to proliferate in susceptible dicotyledons (Struckmeyer, 1951).

Roots of both monocotyledons and dicotyledons are susceptible to auxin herbicides, usually at lower concentrations than are required to induce abnormalities in shoots. The inhibition of division and elongation in the tip is accompanied by the stimulation of cell proliferation in the pericycle of the region behind the tip, typically giving rise to abnormally large numbers of grossly modified lateral root primordia (Beal, 1944; Bond, 1947; Wilae, 1951; Callahan and Engel, 1965; Whitworth and Muzik, 1967). Such responses are reported in the roots of *Agrostis tenuis* plants after foliar application of phenoxy herbicides: longitudinal sections of affected root tips (Fig. 5) show typical abnormalities including the proliferation of pericycle cells and the initiation of lateral root primordia within 1 mm of the tip (Callahan and Engel, 1965). In some species, including *Convolvulus arvensis* the abnormal lateral root primordia may fuse to give a thick ring of irregular tissue (Whitworth and Muzik, 1967); in others such as *Phaseolus,* where the initials are largely confined to the radii of the protoxylem strands, their longitudinal fusion gives rise to externally conspicuous wings of tissue (Wilde, 1951) (see Chapter 3).

The substituted benzoic acid herbicides, 2,3,6-TBA, dicamba, amiben and dinoben, closely resemble the phenoxy auxins in their auxin and herbicidal

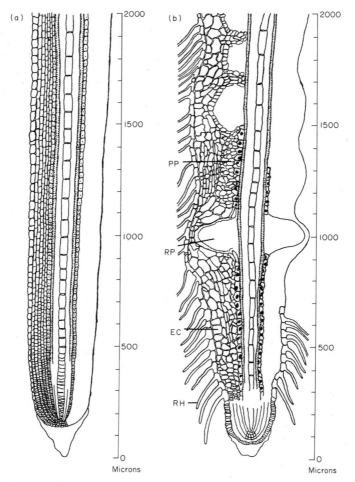

Fig. 5. Camera lucida drawings of median longitudinal sections of root tips of *Agrostis tenuis* plants. a. Untreated. b. Treated with 2-(2,4,5-trichlorophenoxy)-propionic acid (silvex). RP = root primordium; PP = proliferating pericycle cells; EC = enlarged cortical cells; RH = root hairs (reproduced with permission from Callahan and Engel, 1965).

activity (Bentley, 1950; Minarik *et al.* 1951; Zimmerman and Hitchcock, 1951; Keitt and Baker, 1966). Their effects may be dmore persistent than those of 2,4-D, probably because of their slower immobilization or degradation in the plant (Minarik *et al.* 1951).

Picloram also resembles the phenoxy auxins in that it stimulates and prolongs the activity of the vascular cambium in *Phaseolus vulgaris* seedlings (Fisher *et al.* 1968), in tomato and *Cirsium arvense* (Guenther, 1970) and in honey mesquite (*Prosopis juliflora* var. *glandulosa*) (Meyer, 1970). Prolongation of cell division in leaves of western ironweed (*Vernonia baldwini*) followed by destruction of the phloem and mesophyll tissues has also been reported (Scifres and McCarty, 1968). Massive tissue proliferation is not however universally associated with picloram action; cell destruction in the cambium and phloem, and often also in the cortex, are characteristic (Kreps and Alley, 1967) and it is conceivable that this occurs before stimulation of cell proliferation can be effected.

Overall, the "cell-division" activity of the auxin herbicides can be regarded as an ability to stimulate cambial activity, together with the de-differentiation and rapid proliferation of cells in the phloem and pericycle regions and the delay of the differentiation of the derived cells. The action is common to the phenoxy acids, the benzoic acids and picloram; it is essentially similar in all susceptible species.

D. MORPHACTINS AND RELATED SUBSTANCES

Certain derivatives of fluorene-9-carboxylic acid known collectively as the morphactins as well as naptalam are potent plant growth regulators whose effects include the blocking of tropic curvatures (see Section II, C). TIBA and some other substituted benzoic acids have similar but weaker antitropistic properties.

There are several reports that these substances inhibit cell division in primary meristems. Grigsby *et al.* (1955) observed that naptalam effected a rapid reduction in the frequency of mitotic figures in *Pisum* root tips and Ringe and Von Denffer (1967) reported that chlorflurenol caused a temporary inhibition of cell division in *Allium* root tips. However, the marked formative effects which typically result from the treatment of intact plants with morphactins evidently arise largely from a disturbance in the pattern of cell division in the meristems (Schneider, 1970). In onion root tips the normal strictly predetermined polarity of the cell divisions completely disappears in the presence of 0·1–1 mg/l of chlorflurenol (Ringe and Von Denffer, 1967).

Morphactins promote the cell divisions which give rise to lateral primordia in the root pericycle. However, the normal organization of the primordia is greatly disturbed and their further growth is severely inhibited (Ziegler *et al.* 1969).

II. Effects of Herbicides on Cell Elongation and Cell Enlargement

The auxin herbicides have direct regulatory effects on cell elongation and there are important differences between the responses of root and shoot cells. Also, morphactins and other antitropistic agents also modify cell elongation. Finally, herbicides which inhibit primary meristematic activity typically cause precocious enlargement or hypertrophy of some meristematic cells.

A. EFFECTS OF AUXINS ON CELL ELONGATION IN SHOOTS

The marked ability of the auxin herbicides to promote cell elongation in shoot tissues is indicated by their high, positive activity in such auxin bioassays as the Wheat Cylinder Test and the Pea Epicotyl Test and in the semi-quantitative Split Pea Stem Curvature Test. The quantitative nature of the response to 2,4-D in the Wheat Cylinder Test is seen in some typical data represented in Fig. 6. At concentrations up to 10^{-4} M, 2,4-D and IAA have

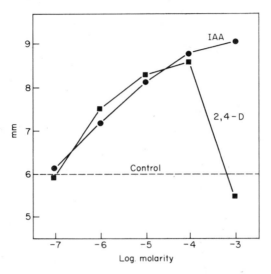

Fig. 6. Mean lengths of batches of 5 mm wheat coleoptile segments after incubation for 24 h with distilled water (control) and with IAA (●) and 2,4-D (■) solutions at concentrations ranging from 10^{-8} to 10^{-3} M.

similar activity but at 10^{-3} M 2,4-D is inhibitory while IAA is more promoting. At a still higher concentration IAA is also usually inhibitory. The minimum concentrations required for promotion and inhibition of elongation vary with the conditions of the test but always the concentration/response curve is of the "optimum" type and usually a lower concentration of 2,4-D than of IAA is needed to cause inhibition. Other phenoxy herbicides, i.e. MCPA and 2,4,5-T, as well as the substituted benzoic acid 2,3,6-TBA (Bentley, 1950) and picloram (Eisinger and Morré, 1971) resemble 2,4-D in such bioassays.

Thus in short-term tests with isolated stem or coleoptile segments composed only of elongating cells a consistent quantitative relationship exists between the elongation and the auxin concentration. However, in intact plants stimulations of elongation are only occasionally observed and then are usually transient and localized. In young *Helianthus* seedlings the growth of the hypocotyls was increased while that of the (younger) first internodes and of the young leaves was depressed by 2,4-D treatment (Blackman and Robertson-Cunninghame,

1954). Also, in rice, Kaufman (1955) reported that 2,4-D stimulated a precocious elongation of certain internodes and that a stimulation of cell expansion contributed to the response. In the longer term (i.e. over periods of several days and longer) or at higher concentrations auxin herbicides inhibit primary growth, both cell division and cell elongation, but at the same time they promote the radial enlargement of cells in the sub-apical region. Recent evidence (Apelbaum and Burg, 1972b) suggests that these effects are mediated through a stimulation of endogenous ethylene production (see Chapter 8).

A striking and prompt response of the shoots of many dicotyledonous species to auxins is epinasty—the pronounced downward bending of leaves and petioles often with random bending of the younger internodes. Bending is visible within a few hours of auxin treatment, suggesting that cell expansion rather than cell division is involved. Also the concentration/response curve for epinasty is of the same "optimum" type as shown for cell elongation (Zimmerman and Hitchcock, 1951). It has been observed that the epinastic bending of the petiole is caused by the hypertrophic swelling of cells near its base and that cell proliferation is not involved (Fröhberger, 1951). All the auxin-type herbicides, including picloram (Eisinger and Morré, 1971), promote epinasty; indeed it is a characteristic symptom of the action of auxin herbicides on susceptible species. Also, certain 2,6-substituted phenols, including 2,6-dichlorophenol, have auxin-like activity in the Split Pea Stem Curvature Test and other tests and are particularly potent promotors of epinasty (Wain and Harper, 1967; Wain, 1968).

B. EFFECTS OF AUXINS ON CELL ELONGATION IN ROOTS

Concentrations of auxin which promote the elongation of coleoptile segments markedly inhibit cell elongation in roots. The degree of inhibition depends on the auxin concentration; the quantitative nature of this relationship is exploited in the Cress Root Bioassay for auxins (Audus, 1949). Some typical concentration/response curves for IAA and 2,4-D are shown in Fig. 7. Over the

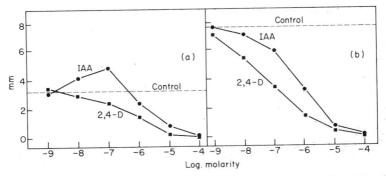

Fig. 7. Mean lengths of primary roots of batches of cress seedlings germinated for (a) two days and (b) four days in distilled water (control) and in IAA (●) and 2,4-D (■) solutions at concentrations ranging from 10^{-9} to 10^{-4} M.

whole effective concentration range, 2,4-D is usually inhibitory in such root growth tests and NAA, MCPA and 2,4,5-T behave similarly. However, lower concentrations of IAA and some less potent phenoxy auxins (Åberg, 1961) frequently stimulate root elongation at least temporarily. Discrepancies in observations on the effects of low auxin concentrations evidently arise from a dual action of auxins on roots: they increase the rate but decrease the duration of cell elongation (Fig. 8). The initial effect is probably accounted for by a change in the mechanical properties of the cell wall while the longer term effect is related to cell wall synthesis (Burström, 1969; Scott, 1972).

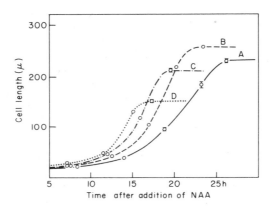

Fig. 8. Elongation of epidermal cells in the growing region of the primary roots of wheat seedlings in mineral nutrient solution after the addition of 1-naphthalene acetic acid (NAA). A. Control (no additive); B. 3×10^{-8} M; C. 10^{-7} M; D. 3×10^{-7} M (overall concentrations) (reproduced with permission from Burström, 1969).

Some growth regulating compounds chemically related to auxins, including several phenoxy acids, e.g. 4-chlorophenoxy-iso-butyric acid (PCIB), and some benzoic acids, e.g. 2,3,5-tri-iodobenzoic acid (TIBA), have effects on cell elongation generally opposite to those of auxins: they stimulate root elongation and inhibit that of coleoptile segments. Some of these, including PCIB, function as auxin antagonists and are thus true anti-auxins: 2-naphthyl-methyl-selenoacetic acid (NMSeA) is such a compound and its effects on the elongation of both roots and coleoptile segments contrast sharply with those of IAA and 2,4-D (Åberg, 1961).

However, TIBA and certain substituted benzoic acids may owe at least part of their anti-auxin-like action to their activity as inhibitors of auxin transport in plants (Niedergang-Kamien and Skoog, 1956; McCready, 1968). They share this property with naptalam (Morgan, 1964) and the morphactins (Schneider, 1970), i.e. with other substances which interfere with geotropic curvatures (see Section II, C).

C. EFFECTS OF MORPHACTINS AND RELATED SUBSTANCES ON TROPIC CURVATURES AND CELL ELONGATION

Morphactins, naptalam and certain substituted benzoic acids have the common property of interfering with geotropic and phototropic curvatures. They also have regulatory effects on cell elongation.

Although the anti-geotropic action of some substituted fluorene-9-carboxylic acids (now known as morphactins) was noted by Jones et al. (1954b), its correlation with other growth regulating properties was not reported until 1964 (Schneider, 1964). The action on germinating seeds of several species was studied by Khan (1967); in treated seedlings the directions of root and shoot growth were independent of the gravitational field and were affected only by the orientation of the seed (Fig. 9). Phototropic curvatures were similarly inhibited.

Morphactins inhibit shoot elongation and cause morphological abnormalities in intact plants (Schneider, 1970). However, they promote the elongation of isolated coleoptile segments and tips (Krelle and Libbert, 1968) and of lentil epicotyl segments (Pilet, 1970). A blocking of endogenous auxin transport is evidently the basis of the inhibition of extension growth and tropic curvatures in intact organs (Schneider, 1970; Pilet, 1970). Stimulatory effects on shoot cell elongation have only been observed in systems with excised segments where auxin transport is not involved in growth regulation.

Naptalam closely resembles the morphactins in its physiological action. Its anti-tropistic effects were reported by Mentzer and Netien (1950), Grigsby et al. (1954), Jones et al. (1954a) and Tsou et al. (1956) and its action as an inhibitor of polar auxin transport by Morgan (1964) and Kiett and Baker (1966). Like the morphactins it promotes the elongation of isolated coleoptile segments (Morgan and Söding, 1958) but inhibits the elongation of intact coleoptiles (Linser and Kiermayer, 1956) and of pea seedlings (Grigsby et al. 1954).

Morphactins and naptalam do not inhibit the elongation of primary roots: indeed they may have a stimulatory effect, as in the primary root of Pisum seedlings (Ziegler et al. 1969). But lateral root growth is strongly inhibited (although non-polar cell divisions in the lateral root initials are stimulated).

The substituted benzoic acids which inhibit tropic curvatures include 2,3,6-TBA which is an auxin herbicide and TIBA which has "anti-auxin" properties (Jones et al. 1954a). In this group, anti-tropistic activity is generally inversely correlated with auxin activity (Van der Beek, 1959; Schrank, 1960) although 2,3,6-TBA is a strong auxin (Bentley, 1950) and a relatively potent inhibitor of tropic curvatures. Like the morphactins and naptalam, the active benzoic acids are inhibitors of polar auxin transport (Keitt and Baker, 1966; Winter, 1968; and others): evidently this is the essential property of all the known anti-tropistic chemicals.

D. EFFECTS OF MISCELLANEOUS HERBICIDES ON CELL ENLARGEMENT

Many herbicides which inhibit cell division in primary meristems (see Section I) also have effects on cell enlargemen⁺

Fig. 9. Seedlings of lettuce germinated for 72 h. Left, in the presence of 6×10^{-5} M n-butyl-9-hydroxyfluorene-(9)-carboxylate; right, control. (Reproduced with permission from Khan, 1967.)

Propachlor is a non-specific inhibitor of cell growth; both cell division and cell elongation are affected. Its primary mode of action is probably an inhibition of protein synthesis (Dillon and Anderson, 1972) (see Chapter 17). These authors observed that the elongation of both roots and coleoptile segments of *Avena* was decreased in proportion to the applied concentration of propachlor. Also, Duke (1967) reported an inhibition by propachlor of auxin-induced cell elongation in cucumber hypocotyl segments as well as an inhibition of root growth correlated with the inhibition of protein synthesis in root tips.

Effects on cell elongation probably contribute to a great extent to the inhibitory effect of thiocarbamate herbicides on the shoot growth of grass seedlings. Thus in wild oat seedlings shoot growth is inhibited by concentrations of diallate and triallate which do not affect root growth or cause mitotic abnormalities (Banting, 1967); also, the mode of action of EPTC on barnyard grass seedlings appears to be an inhibition of the expansion of the mesophyll tissue of the first leaf (Dawson, 1963).

The action of the carbamates, trifluralin, chlorthal-dimethyl and bensulide in blocking cell division in primary meristems has been discussed in Section IIA. Such effects are generally accompanied by the precocious enlargement of cells in the meristem; e.g. in propham-treated *Avena* seedlings (Ennis, 1948b) (Fig. 1) and in the root initials of chlorthal-dimethyl-treated *Cynodon dactylon* stolons (Bingham, 1968). Also, abnormally enlarged multinucleate cells are a characteristic feature of trifluralin-inhibited roots (e.g. Talbert, 1965). In bensulide-treated oat roots meristematic activity is inhibited and the epidermal cells near the tip become markedly elongated radially (Cutter *et al.* 1968).

III. Effects of Herbicides on Tissue Differentiation

Some of these effects have been mentioned earlier in Sections I and II, since they are linked with the herbicidal regulation of cell division and cell enlargement. Thus the precocious differentiation of meristematic cells in trifluralin-treated cotton roots is linked with the blocking of their normal meristematic activity. In addition the action of auxin herbicides in stimulating secondary cell proliferation might alternatively be regarded as a de-differentiation of phloem, pericycle and ray cells followed by a delay in the differentiation of further vascular elements.

Three types of herbicide effects on tissue differentiation have been selected for further comment because of their relevance to herbicide action in the field.

A. EFFECTS ON THE DIFFERENTIATION OF TISSUES IN PRIMARY MERISTEMS

The effects of herbicides which inhibit cell division in the primary meristems has been discussed in Section I. It is a characteristic feature of such inhibited meristems that the cells vacuolate and differentiate precociously in no organized pattern. Such responses in root meristems have been reported with trifluralin (Bayer *et al.* 1967), bensulide (Cutter *et al.* 1968), bromacil (Ashton *et al.* 1969) and pronamide (Peterson and Smith, 1971).

B. EFFECTS OF AUXINS ON THE DIFFERENTIATION OF VASCULAR TISSUE

Perhaps the most important effect of the auxin herbicides on susceptible species is the promotion of secondary meristematic activity in the vascular cylinder (see Section I). The proliferation is not only from the vascular cambium but phloem parenchyma and other cells revert to the meristematic condition. The auxin herbicide also causes a delay in differentiation of the tissue produced by the stimulated cell proliferation. Swanson (1946) observed that "the derivatives of such tissue remain meristematic for considerable periods and differentiation, if any, is not orderly".

An inverse relationship between auxin concentration and vascular differentiation is consistent with observations that in woody plants tension wood develops on the upper side of a horizontally displaced branch, i.e. on the "low auxin" side. Furthermore, when vertical stems of *Acer* seedlings were ringed with TIBA (in lanoline) conspicuous tension wood developed below the ring (Cronshaw and Morey, 1965), where presumably there is a region of auxin deficiency since TIBA inhibits basipetal auxin transport (McCready, 1968).

Although when used as herbicides, or for the initiation of cell division in mature tissue explants, auxins are de-differentiating agents, in other systems they act as xylogenetic agents. For instance, xylem regeneration in wounded vascular strands of *Coleus* internode sections has been reported to occur in response to a very small dose of auxin (less than 4×10^{-3} µg) applied at some distance above the wound (Gee, 1972). Further evidence from this study and from Sachs (1969) suggests that xylem differentiation is associated with the polar transport of auxin through the tissue. Furthermore, morphactins, which are inhibitors of auxin transport, have been observed to counteract the normal auxin effect of causing extensive differentiation of lignified tracheary elements in *Lactuca* pith tissue explants (Roberts and Sankla, 1973).

C. THE EFFECT OF TRIFLURALIN ON XYLEM DIFFERENTIATION

Trifluralin (applied as a pre-emergence herbicide) increases the tendency of soya bean plants to lodging. Kirby *et al.* (1968) reported that in plants on treated plots xylem differentiation in the lowest internode was markedly less during the first 21 days of seedling growth: although by 63 days there was a greater cross-sectional area of xylem tissue, its differentiation lacked uniformity. Furthermore, Kust and Struckmeyer (1971) observed that trifluralin-damaged soya bean plants lacked an organized pattern of xylem differentiation in their stems.

IV. Effects of Herbicides on Seed Germination and Early Seedling Growth

The function of soil-acting pre-emergence herbicides is to prevent or retard the growth of weed seedlings. They, and other herbicides, may be inhibitory in one or more of the phases of seedling growth: (i) early germination, i.e. the emergence of the radicle; (ii) seedling establishment at the expense of endosperm or cotyledonary reserves; (iii) growth of the seedling after the seed reserves are exhausted.

Inhibitory effects on the early germination phase are generally accompanied by a failure of the mobilization of the seed reserves. The carbamates, chlorpropham and barban, as well as dichlobenil and propachlor retard the germination of grass seeds and inhibit the induction by gibberellin of α-amylase synthesis in barley endosperm tissue (Mann *et al.* 1967; Devlin and Cunningham, 1970). Amiben and bromoxynil also inhibit germination and the development of amylase activity in intact seeds (but not endosperm slices) of barley. However the germination of squash and other seeds low in carbohydrate reserves is not inhibited although the development of their normally low level of amylase activity is blocked (Penner, 1968). Presumably the germination of seeds with large oil reserves is independent of carbohydrate mobilization.

Both germination and the development of proteolytic activity in squash seeds are inhibited by ioxynil, chlorpropham and diclobenil but not by atrazine, bromacil or dalapon. Other herbicides including 2,4-D, picloram, dicamba, endothal and bromoxynil are intermediate in their effects on both germination and proteolytic enzymes but the correlation between the responses is not exact and probably the effect on the mobilization of storage protein is not solely responsible for the inhibition of germination (Ashton *et al.* 1968).

Maximum sensitivity to most growth-regulating herbicides (as opposed to photosynthetic inhibitors) is observed in the seedling establishment phase. For example, trifluralin, chlorthal-dimethyl and diphenamid are reported to have no effect on the germination of weed seeds but they retard or kill young seedlings (Grover, 1965). Similarly trifluralin is not inhibitory to the germination of wheat but severely inhibits seedling root growth (Lignowski, 1969). Also species which at later growth stages show resistance to particular herbicides may be sensitive in this phase, e.g. cereals to 2,4-D (Friesen and Olson, 1953). Retarding effects on seedling establishment can be assessed in terms of the resulting decrease in the measurable "emergence thrust" of the seedlings (Burnside, 1971): of a range of herbicides tested on soya bean seedlings 2,4-D and picloram were found to be the most inhibitory, followed by dicamba, 2,3,6-TBA, amitrole, EPTC, propachlor and trifluralin, while chloramben (which, like TIBA, has some "anti-auxin" properties) markedly increased the emergence thrust. The photosynthetic inhibitors, monuron and atrazine had no effect: other work has shown that normal doses of such herbicides inhibit growth only after the seed reserves are exhausted (Olech, 1968).

The marked sensitivity of early seedling growth to herbicides is exploited in many bioassay procedures. Oat, sorghum or cucumber seedlings are commonly employed (Parker, 1964; Horowitz and Hulin, 1971; Kratky and Warren, 1971) and the inhibition of the elongation of radicles or plumules is related to herbicide concentration. Such procedures can detect 10 parts/10^6 of most herbicides and 1 part/10^6 of many including some phenoxy and benzoic acids, amides and carbamates (Kratky and Warren, 1971); e.g. the inhibition of the elongation of oat seedling roots is quantitatively related to the concentration of 2,4-D, chlorpropham or trifluralin in the range 0–4 parts/10^6 (Eshel and Warren, 1967).

The carbamates are particularly inhibitory to seedling establishment in grasses and cereals (Templeman and Sexton, 1945; Allard *et al.* 1946; Ennis, 1948*a*). In addition some dicotyledons, usually members of Cucurbitaceae or Solanaceae, are sensitive to propham (Ennis, 1948*a*), sugar beet is sensitive to chlorpropham (Cullinane, 1966) and tobacco seedlings have stunted hypocotyls and root hair abnormalities in the presence of high doses of carbaryl (a *N*-methyl carbamate) (Fox and Thurston, 1966). Seedlings affected by carbamates are typically severely stunted, swollen and of a darker green colour (Ennis, 1948*a*; Cullinane, 1966).

Auxin herbicides are strongly inhibitory to seedling establishment; dicotyledons are more severely affected but many monocotyledons are also sensitive, e.g. a corn germination test is reported to be very sensitive to 2,4-D and related compounds (Thompson *et al.* 1946). Elongation of the radicle is more sensitive than shoot growth (see Section II, B); as little as 0·01 parts/10^6 MCPA markedly inhibits the growth of cress roots (Yates, 1964). Seedlings inhibited by auxin herbicides are typically severely malformed and may show abnormal swellings and callus growth (e.g. Chang and Foy, 1971).

The inhibitory effect of trifluralin on seedling root growth is discussed in Section I. Shoots are also affected if trifluralin is directly applied or if exposed to trifluralin vapour (i.e. from soil applications) (Swann, 1969; Barrentine and Warren, 1971).

Ureas (except siduron), triazines and pyrazon do not generally affect either germination or seedling establishment. Simazine may even stimulate germination (Jordan and Day, 1968) or promote early seedling growth in some species (Brezny *et al.* 1971). However, the uptake of these photosynthetic inhibitors (see Chapter 16) during germination results in herbicidal action after the seedlings become autotrophic (Olech, 1968; Rodebush and Anderson, 1970).

Still lower doses of herbicides than those required for the inhibition of early seedling growth may have stimulatory effects. In tests with oat seedlings, Wiedman and Appleby (1972) recorded stimulations of 18–28% in the shoot and total dry matter increments by doses of about 1/100 or 1/1000 of the minimum inhibitory dose of barban, dalapon, simazine, diuron and MCPA. Root dry weights were similarly increased by 11 herbicides including barban, diuron, EPTC, MCPA and simazine. The physiological basis of such growth stimulations, by compounds which at higher concentrations have widely differing mechanisms of herbicidal action, remains obscure (see Chapter 11, vol. 2).

V. Effects of Herbicides on the Overall Growth of the Plant

A. EFFECTS ON SUSCEPTIBLE SPECIES

It is obvious that the growth of susceptible species is suppressed by herbicides. The severity of the effect varies with the herbicide dose, the stage of growth and many other factors. Sufficiently large doses of growth regulating herbicides disrupt or modify growth processes to an extent which results in physiological malfunctioning and therefore death of the whole plant. Cardenas *et al.* (1968) reported such a sequence of responses in *Xanthium* plants treated with lethal doses of 2,4-D. Within two days a swelling of the axis had been induced and the primary growth of both the root and the shoot had virtually ceased. Abnormal growth of the axis continued and leaves senesced prematurely until the plants finally collapsed after about 10 days, evidently due to plugging of the phloem.

The doses of herbicides which give effective control of weeds in crops are not necessarily lethal *per se*; they may only retard or modify the growth of the weeds so that they cannot survive in competition with the crop. Thus, wild oat (*Avena fatua*) seedlings stunted by barban can eventually resume growth under favourable conditions ("Weed Control Handbook", 1968). Similarly, the effects of sub-lethal doses of 2,4-D on susceptible plants such as *Phaseolus* are evident only for a limited period: the development of leaf primordia present at the time of treatment is greatly modified but later formed primordia expand normally. By comparison, the benzoic herbicides have a more persistent effect and a single dose affects the expansion not only of the leaf primordia present initially but also of several later formed ones (Minarik *et al.* 1951).

B. EFFECTS ON "RESISTANT" SPECIES

"Resistant" species show varying degrees of susceptibility to growth regulating herbicides depending on the developmental stage of the plant and on the dose, the formulation and perhaps the placement of the herbicide. Thus errors in the application of selective treatments can adversely affect the growth of the crop as well as the target weeds. Wheat, barley and oat crops are particularly susceptible to damage by phenoxy herbicides during two periods of their development; the first from germination up to the completion of the differentiation of the spike initial in the early tillering phase, and the second during stem elongation in the pre-heading phase. MCPA or 2,4-D applied in the earlier of these two periods causes leaf and spike abnormalities and in the later period floret sterility (Friesen and Olson, 1953). Both effects ultimately result in fewer kernels per ear and a lower grain yield (Pinthus and Natowitz, 1967). Similar effects may be caused by benzoic herbicides such as dicamba and by picloram, although these are most severe as a result of applications in the late tillering phase, i.e. during the period of maximum resistance to 2,4-D (Quimby and Nalewaja, 1966; Nalewaja, 1970; Peeper *et al.* 1970) (see Chapter 4).

Herbicides which do not have growth regulating properties may decrease the overall growth and vigour of "resistant" plants without causing obvious injury symptoms. Thus a number of herbicides including diuron and pyrazon are reported to cause decreases in growth parameters such as height and stem diameter of young *Picea* and *Pinus* trees without other visible effects (Grover, 1967).

C. DIFFERENTIAL EFFECTS ON ROOTS AND SHOOTS

The initial responses to growth regulating herbicides may be confined to a particular part of the plant. Thus EPTC acts primarily on the developing foliage leaves of barnyard grass (*Echinochloa crusgalli*) seedlings (Dawson, 1963) and diallate and triallate, which are also thiocarbamates, affect only the shoot growth of wild oat seedlings (Banting, 1967). Root growth of oat seedlings is specificially inhibited by siduron (Splittstoesser and Hopen, 1970) while trifluralin acts on cotton and other species primarily through an inhibition of the growth of lateral roots (Bayer *et al.* 1967). In several species, roots are more sensitive than shoots to diphenamid (Deli and Warren, 1971). However, the carbamates and the auxin-herbicides rapidly disrupt growth in both roots and shoots of susceptible plants (Ennis, 1948*a* and *b*; Cardenas *et al.* 1968).

The death or decreased growth of the primary roots or shoots as a result of herbicide treatment may effect some compensatory growth of lateral or adventitious initials. For example, the inhibition of the growth of the primary roots and/or the older laterals of cotton and soya bean seedlings by soil-incorporated trifluralin often promotes adventitious root development (Anderson *et al.* 1968) or the increased growth of lateral roots in a lower, trifluralin-free, soil zone (Oliver and Frans, 1968). Similarly when the death of the main shoot of wheat seedlings results from an early dose of dicamba, a tiller assumes the role of dominant shoot (Quimby and Nalewaja, 1966). Furthermore, dicamba injury to the shoots of sorghum is associated with the increased growth of suckers (Peeper *et al.* 1970).

D. STIMULATORY EFFECTS ON CROP GROWTH

Generally, the removal of weed competition can account fully for the beneficial effects of herbicides on the long term growth of crops. There are, however, many reports that some herbicides can directly stimulate growth and dry matter accumulation in crops at sublethal concentrations. Such effects are fully discussed in Chapter 11, vol. 2.

VI. REFERENCES

Åberg, B. (1961). *In* "Plant Growth Regulation", IVth int. Conf. Plant Growth Regulation, pp. 219–232. Iowa State Univ. Press.
Adamson, D. (1962). *Can. J. Bot.* **40**, 719–744.
Allard, R. W., DeRose, H.R. and Swanson, C. P. (1946). *Bot. Gaz.* **107**, 575–583.
Anderson, W. P., Richards, A. B. and Whitworth, J. W. (1968). *Weed Sci.* **16**, 165–169.
Apelbaum, A. and Burg, S. P. (1972*a*). *Pl. Physiol., Baltimore,* **50**, 117–124.
Apelbaum, A. and Burg, S. P. (1972*b*). *Pl. Physiol., Baltimore,* **50**, 125–131.
Arntzen, C. J., Linck, A. J. and Cunningham, W. P. (1970). *Bot. Gaz.* **131**, 14–23.
Ashton, F. M., Cutter, E. G. and Huffstutter, D. (1969). *Weed Res.* **9**, 198–204.
Ashton, F. M., Penner, D. and Hoffman, S. (1968). *Weed Sci.* **16**, 169–171.
Atkin, R. K. and Barton, G. E. (1973). *J. exp. Bot.* **24**, 689–699.

Audus, L. J. (1949). *Pl. Soil*, **2**, 31–36.
Banting, J. D. (1967). *Weed Res.* **7**, 302–315.
Banting, J. D. (1970). *Weed Sci.* **18**, 80–84.
Barrentine, J. L. and Warren, G. F. (1971). *Weed Sci.* **19**, 37–41.
Bayer, D. E., Foy, C. L., Mallory, T. E. and Cutter, E. G. (1967). *Am. J. Bot.* **54**, 945–952.
Beal, J. M. (1944). *Bot. Gaz.* **105**, 471–474.
Beal, J. M. (1946). *Bot. Gaz.* **108**, 166–186.
Bentley, J. A. (1950). *Nature, Lond.* **165**, 449.
Berquist, E. T. (1971). *Diss. Abstr. B* **32**, 116.
Bingham, S. W. (1967). *Weeds*, **15**, 363–365.
Bingham, S. W. (1968). *Weed Sci.* **16**, 449–452.
Blackman, G. E. and Robertson-Cunninghame, R. C. (1954). *J. exp. Bot.* **5**, 184–203.
Bond, L. (1947). *Bot. Gaz.* **109**, 435–447.
Bradley, M. V., Crane, J. C. and Marei, N. (1968). *Bot. Gaz.* **129**, 231–238.
Brezny, O., Mehta, I. and Sharma, R. K. (1971). *Ind. J. Weed Sci.* **3**, 1–7 (quoted from *Weed Abstr.* **22**, No. 898).
Burnside, O. C. (1971). *Weed Sci.* **19**, 182–184.
Burström, H. G. (1968). *Physiologia Pl.* **21**, 1137–1155.
Burström, H. G. (1969). *Am. J. Bot.* **56**, 679–684.
Callahan, L. M. and Engel, R. E. (1965). *Weeds*, **13**, 336–338.
Canvin, D. T. and Friesen, G. (1959). *Weeds*, **7**, 153–156.
Cardenas, J., Slife, F. W., Hanson, J. B. and Butler, H. (1968). *Weed Sci.* **16**, 96–100.
Chang, I. K. and Foy, C. L. (1971). *Weed Sci.* **19**, 58–64.
Croker, B. H. (1953). *Bot. Gaz.* **114**, 274–283.
Cronshaw, J. and Morey, P. R. (1965). *Nature, Lond.* **205**, 816–818.
Cullinane, J. P. (1966). *Proc. Irish Crop Protection Conf.*, pp. 88–93 (quoted from *Weed Abstr.* **17**, No. 825).
Cutter, E. G., Ashton, F. M. and Huffstutter, D. (1968). *Weed Res.* **8**, 346–352.
Dawson, J. H. (1963). *Weeds*, **11**, 60–66.
Deli, J. and Warren, G. F. (1971). *Weed Sci.* **19**, 70–72.
Devlin, R. M. and Cunningham, R. P. (1970). *Weed Res.* **10**, 316–320.
Dillon, N. S. and Anderson, H. L. (1972). *Weed Res.* **12**, 182–189.
Duke, W. B. (1967). *Diss. Abstr. B.* **31**, 1315.
Eames, A. J. (1949). *Am. J. Bot.* **36**, 571–584.
Eames, A. J. (1950). *Am. J. Bot.* **37**, 840–847.
Eames, A. J. (1951). *Am. J. Bot.* **38**, 777–780.
Eisinger, W. R. and Morré, D. J. (1971). *Can. J. Bot.* **49**, 889–897.
Ennis, W. B. (1948a). *Bot. Gaz.* **109**, 473–493.
Ennis, W. B. (1948b). *Am. J. Bot.* **35**, 15–21.
Eshel, Y. and Warren, G. F. (1967). *Weeds*, **15**, 115–118.
Fisher, D. A., Bayer, D. E. and Weier, T. E. (1968). *Bot. Gaz.* **129**, 67–70.
Fox, P. M. and Thurston, R. (1966). *Bot. Gaz.* **127**, 70–74.
Friesen, G. and Olson, P. J. (1953). *Can. J. agric. Sci.* **33**, 315–329.
Fröhberger, E. (1951). *Höfchenbr. Wiss.* **4**, 236–287 (quoted from Kiermayer, 1964).
Gee, H. (1972). *Planta*, **108**, 1–9.
Gentner, W. A. and Burk, L. G. (1968). *Weed Sci.* **16**, 259–260.
Gifford, E. M. (1953). *Hilgardia*, **21**, 607–644.
Gifford, E. M. (1956). *Am. J. Bot.* **43**, 72–80.
Gray, R. A. and Weirich, A. J. (1969). *Weed Sci.* **17**, 223–229.
Greulach, V. A. and Haesloop, J. G. (1954). *Am. J. Bot.* **41**, 44–50.
Grigsby, B. H., Tsou, T. M. and Wilson, G. B. (1954). *Proc. 11th N. cent. Weed Control Conf.*, p. 88.
Grigsby, B. H., Tsou, T. M. and Wilson, G. B. (1955). *Proc. 8th sth. Weed Control Conf.* pp. 279–283.
Grover, R. (1965). *Can. J. Pl. Sci.* **45**, 477–486.

Grover, R. (1967). *Weed Res.* **7,** 155–163.
Guenther, H. R. (1970). *Diss. Abstr. B.* **31,** 1015.
Hacskaylo, J. and Amato, V. A. (1968). *Weed Sci.* **16,** 513–515.
Hepler, P. K. and Jackson, W. T. (1969). *J. Cell. Sci.* **5,** 727–743.
Hicks, G. S., Setterfield, G. and Wightman, F. (1964). *Proc. Can. Soc. Pl. Physiol.* **5,** 18.
Horowitz, M. and Hulin, N. (1971). *Weed Res.* **11,** 143–149.
Jablonski, J. R. and Skoog, F. (1954). *Physiologia Pl.* **7,** 16–24.
Jackson, W. T. (1967). *Am. J. Bot.* **54** (5, Pt. 2), 633.
Jackson, W. T. and Stetler, D. A. (1973). *Can. J. Bot.* **51,** 1513–1518.
Jones, R. L., Metcalfe, T. P. and Sexton, W. A. (1954a). *J. Sci. Fd. Agric.* **5,** 32–38.
Jones, R. L., Metcalfe, T. P. and Sexton, W. A. (1954b). *J. Sci. Fd. Agric.* **5,** 38–42.
Jordan, L. S. and Day, B. E. (1968). *Abstr. Meet. Weed Sci. Soc. Am., 1968,* pp. 13–14 (quoted from *Weed Abstr.* **18,** No. 2811).
Kaufman, P. B. (1955). *Am. J. Bot.* **42,** 649–659.
Key, J. L., Lin, C. Y., Gifford, E. M. and Dengler, R. (1966). *Bot. Gaz.* **127,** 87–94.
Khan, A. A. (1967). *Physiologia Pl.* **20,** 306–313.
Kiepal, Z. (1970). *Acta agrobot.* **23,** 73–84 (quoted from *Weed Abstr.* **21,** No. 1426).
Kiermayer, O. (1959). *Physiologia Pl.* **12,** 841–853.
Keitt, G. W. (1967). *Physiologia Pl.* **20,** 1076–1082.
Keitt, G. W. and Baker, R. A. (1966). *Pl. Physiol., Lancaster,* **41,** 1561–1569.
Kirby, C. J., Standifer, L. C. and Normand, W. C. (1968). *Proc. 21st sth. Weed Control Conf.* p. 343.
Kratky, B. A. and Warren, G. F. (1971). *Weed Res.* **11,** 257–262.
Krelle, E. and Libbert, E. (1968). *Experientia,* **24,** 293–294.
Kreps, L. B. and Alley, H. P. (1967). *Weeds,* **15,** 56–59.
Kust, C. A. and Struckmeyer, B. E. (1971). *Weed Sci.* **19,** 147–152.
Liang, G. H. L., Feltner, K. C. and Russ, O. G. (1969). *Weed Sci.* **17,** 8–12.
Lignowski, E. M. (1969). *Diss. Abstr. B* **31,** 992–993.
Lignowski, E. M. and Scott, E. G. (1972). *Weed Sci.* **20,** 267–270.
Linser, H. and Kiermayer, O. (1956). *Mh. Chem.* **87,** 708–719 (quoted from Kiermayer, 1964).
Mallory, T. E. (1969). *Proc. 21st Californian Weed Conf.* p. 127.
Mann, J. D. (1967). *Proc. 19th Californian Weed Conf.* p. 88.
Mann, J. D., Cota-Robles, E., Yung, K. H., Pu, M. and Haid, H. (1967). *Biochim. biophys. Acta,* **138,** 133–139.
Mann, J. D., Jordan, L. S. and Day, B. E. (1965). *Pl. Physiol., Lancaster,* **40,** 840–843.
Mentzer, C. and Netien, G. (1950). *Bull. mens. Soc. linn. Lyon.* **19,** 102 (quoted from Ashton and Crafts, 1973).
Meyer, R. E. (1970). *Weed Sci.* **18,** 525–531.
Milborrow, B. V. (1964). *J. exp. Bot.* **15,** 515–524.
Minarik, C. E., Ready, D., Norman, A. G., Thompson, H. E. and Owings, J. F. (1951). *Bot. Gaz.* **113,** 135–147.
Morgan, D. G. (1964). *Nature, Lond.* **201,** 476–477.
Morgan, D. G. and Söding, H. (1958). *Planta,* **52,** 235–249.
Morgan, P. W. and Hall, W. C. (1964). *Nature, Lond.* **201,** 99.
Murray, M. A. and Whiting, A. G. (1947). *Bot. Gaz.* **109,** 13–39.
McCready, C. C. (1968). *In* "Biochemistry and Physiology of Plant Growth Substances" (F. Wightman and G. Setterfield, eds.), pp. 1005–1023. Runge Press, Ottawa.
Nalewaja, J. D. (1970). *Weed Sci.* **18,** 276–278.
Niedergang-Kamien, E. and Skoog, F. (1956). *Physiologia Pl.* **9,** 60–73.
Normand, W. C., Rizk, T. Y. and Thomas, C. H. (1968). *Proc. 21st sth. Weed Control Conf.* p. 344.
Olech, K. (1968). *Annls. Univ. Mariae Curie-Sklodowska (sect. E),* **23,** 201–211 (quoted from *Weed Abstr.* **20,** No. 1334).
Oliver, L. R. and Frans, R. E. (1968). *Weed Sci.* **16,** 199–203.
Parker, C. (1964). *Proc. 7th Br. Weed Control Conf.* pp. 899–902.
Peeper, T. F., Wiebel, D. E. and Santelman, P. W. (1970). *Agron J.* **62,** 407–411.

Penner, D. (1968). *Weed Sci.* **16,** 519–522.
Peterson, R. L. and Smith, L. W. (1971). *Weed Res.* **11,** 84–87.
Pilet, P. E. (1970). *Experientia,* **26,** 608–609.
Pinthus, M. J. and Natowitz, Y. (1967). *Weed Res.* **7,** 95–101.
Quimby, P. C. and Nalewaja, J. D. (1966). *Weeds,* **14,** 229–232.
Ringe, F. and von Denffer, D. (1967). *Univ. Rostock, Math.-Naturwiss. Reihe* **16,** 693–697 (quoted from Schneider, 1970).
Roberts, L. W. and Sankla, N. (1973). *Pl. Cell Physiol., Tokyo,* **14,** 521–530.
Rodebush, J. E. and Anderson, J. L. (1970). *Weed Sci.* **18,** 443–446.
Rubin, S. S. and Gritsaenka, Z. M. (1968). *Bot. Zh.* **53,** 377–378 (quoted from *Weed Abstr.* **17,** No. 2357).
Sachs, R. M. and Lang, A. (1961). *In* "Plant Growth Regulation", *IVth int. Conf. Plant Growth Regulation,* pp. 567–588. Iowa State Univ. Press.
Sachs, T. (1969). *Ann. Bot.* **33,** 263–275.
Sawamura, S. (1964). *Cytologia,* **29,** 86–102.
Schieferstein, R. H. and Hughes, W. J. (1968). *Proc. 8th Br. Weed Control Conf.* pp. 377–381.
Schneider, G. (1964). *Naturwissenschaften,* **51,** 416–417.
Schneider, G. (1970). *A. Rev. Pl. Physiol.* **21,** 499–536.
Schrank, A. R. (1960). *Pl. Physiol., Lancaster,* **35,** 735–741.
Schultz, D. P., Funderburk, H. H. and Negi, N. S. (1968). *Pl. Physiol., Lancaster,* **43,** 265–273.
Scifres, C. J. and McCarty, M. K. (1968). *Weed Sci.* **16,** 347–349.
Scott, D. (1968). *Mutation Res.* **5,** 65–92.
Scott, M. A. and Struckmeyer, B. E. (1955). *Bot. Gaz.* **117,** 37–44.
Scott, T. K. (1972). *A. Rev. Pl. Physiol.* **23,** 235–258.
Shahied, S. I. (1970). *Diss. Abstr. B.* **31,** 4457.
Shaybany, B. (1970). *Diss. Abstr. B.* **31,** 4302.
Shaybany, B. and Anderson, J. L. (1972). *Weed. Res.* **12,** 164–168.
Skoog, F. (1954). *Brookhaven Symp. Biol.* **6,** 1–21.
Splittstoesser, W. E. and Hopen, H. J. (1970). *Physiologia Pl.* **23,** 964–970.
Srivastava, A. K. and Sharma, V. K. (1971). *Ind. J. Weed Sci.* **3,** 84–87 (quoted from *Weed Abstr.* **22,** No. 895).
Steward, F. C. and Caplin, S. M. (1951). *Science, N.Y.* **113,** 518–520.
Storey, W. B., Jordan, L. S. and Mann, J. D. (1968). *Calif. Agric.* **22,** 12–13.
Struckmeyer, B. E. (1951). *In* "Plant Growth Substances" (F. Skoog, ed.), pp. 167–174. University of Wisconsin Press, Madison.
Swann, C. W. (1969). *Diss. Abstr. B,* **30,** 1877.
Swanson, C. P. (1946). *Bot. Gaz.* **107,** 522–531.
Talbert, R. E. (1965). *Proc. 18th sth. Weed Control Conf.* p. 652.
Templeman, W. G. and Sexton, W. A. (1945). *Nature, Lond.* **156,** 630.
Thompson, H. E., Swanson, C. P. and Norman, A. G. (1946). *Bot. Gaz.* **107,** 476–507.
Tsou, T. M., Hamilton, R. H. and Bandurski, R. S. (1956). *Physiologia Pl.* **9,** 546–548.
Van der Beek, L. C. (1959). *Pl. Physiol., Lancaster,* **34,** 61–65.
Wain, R. L. (1968). *In* "Plant Growth Regulators", *Society for Chemical Industry Monographs No. 31,* pp. 3–21. Soc. chem. Ind., London.
Wain, R. L. and Harper, D. B. (1967). *Nature, Lond.* **213,** 1155.
Watson, D. P. (1948). *Am. J. Bot.* **35,** 543–555.
"Weed Control Handbook", 5th edition, 1968. (J. D. Fryer and S. A. Evans, eds.). Blackwell Scientific Publications, Oxford.
Whitworth, J. W. and Muzik, T. J. (1967). *Weeds,* **15,** 1275–280.
Wiedman, S. J. and Appleby, A. P. (1972). *Weed Res.* **12,** 65–74.
Wilde, M. H. (1951). *Am. J. Bot.* **38,** 79–91.
Winter, A. (1968). *In* "Physiology and Biochemistry of Plant Growth Substances" (F. Wightman and G. Setterfield, eds.), pp. 1063–1076. Runge Press, Ottawa.
Yates, R. J. (1964). *E. Afr. agric. For. J.* **30,** 126–128.
Yeoman, M. M. and Mitchell, J. P. (1970). *Ann. Bot.* **34,** 799–810.
Yung, K. H. and Mann, J. D. (1967). *Pl. Physiol., Lancaster,* **42,** 195–200.

Ziegler, H., Vogt, I., Treichel, S. and Streitz, B. (1969). *Ber. dt. bot. Ges., Vortr. Gesamtgeb. Bot., N.F.* **3,** 43–52 (quoted from Schneider, 1970).
Zimmerman, P. W. and Hitchcock, A. E. (1951). *Contr. Boyce Thompson Inst. Pl. Res.* **16,** 209–213.

CHAPTER 3

EFFECTS ON THE CYTOLOGY AND FINE STRUCTURE OF PLANT CELLS

A. J. LINCK

College of Agriculture, University of Minnesota,
St. Paul, Minnesota, USA

I. INTRODUCTION

The decades of the 1960s and the 1970s brought a veritable explosion of research publication on the biochemistry of growth-regulators including herbicides. The additions to the literature of papers on the biochemistry and physiology of herbicide action, particularly in metabolism of herbicides and the formation of residues attests to the attention given to this area of agricultural and botanical research. While no such rapid increase occurred in research reports on the effects of herbicides on the structure of plants, particularly the ultrastructure of cells, there was nevertheless considerable attention given to this fundamental area of herbicide science. Many research contributions in herbicide physiology also provided clues and information on morphogenetic effects of these growth regulators. This literature and the publications on gross cellular effects, on organ responses and on overall plant growth in response to herbicides are dealt with in Chapters 2 and 4 of this volume. This chapter will report findings of research workers on the effects of growth regulators, naturally-occurring and synthetic, with emphasis on herbicides, as they affect ultrastructure of cells. Included are effects on the nucleus, the nucleolus, organelles, such as the chloroplasts, the endoplasmic reticulum, microtubules, the cell wall and other cell structures.

Some measure of the interest in the field of growth substances and ultrastructure in plants is indicated by the fact that the proceedings of the 4th

International Conference on Plant Growth Regulation held in 1959 had only one paper on structural effects of such substances (Sachs and Lang, 1961). At the 8th International Conference on Plant Growth Substances held in 1973, more attention was given to the relationship of cell organelles and growth substances to cell growth. Several papers (Short *et al.* 1973 and Valdovinos and Jensen, 1973) dealt with studies on the ultrastructure of cells involving hormones or other growth regulators. The first edition of "The Physiology and Biochemistry of Herbicides" published a decade ago (Audus, L. J., ed. 1964) contained two chapters on the broad subject of anatomical and morphogenetic as well as sub-cellular effects of growth regulators. The attention given to this subject, including ultrastructural affects has increased greatly in this 1975 edition. Ashton and Crafts (1973) in their book on the "Mode of Action of Herbicides" devote one chapter to "Morphological Responses to Herbicides". The reader's attention is called to the book by Kihlman (1966), "Actions of Chemicals on Dividing Cells" which deals with the effects of plant and animal inhibitors including a few herbicides.

Research on plant ultrastructure requires expensive and specialized equipment and rather specialized skills and these have probably been factors limiting the growth of this field. In contrast most physiology laboratories in modern research institutions are well-equipped to carry out all necessary analytical procedures in herbicide physiology and metabolism. Apparently there are fewer plant scientists and more limited facilities for ultrastructure research on herbicide-treated plants, but the number of investigations is increasing.

The need to understand the changes in gross or fine structure of plant cells treated with herbicides is to define, if possible, the primary effects of herbicides on plant cells, and to pinpoint the organelles centrally involved, thus giving clues to the biochemical actions. A more complete catalogue of effects on cell structures of the vastly different chemical species, now being tested or in use as herbicides, should lead to more rapid and more successful synthesis of new plant growth regulators. At the present time the "catalogue" of ultrastructure effects of herbicides is still relatively small, with many gaps in our knowledge. Some herbicides have been rather extensively investigated for their effects on fine structure, others of equal importance commercially have not been studied at all. As is true for most plant physiology studies a relatively few species of plants are used for these fine structure investigations. Some herbicides for important crop species have received little attention in regard to ultrastructural effects. As will be seen in this review, the effects of herbicides on weed species is also somewhat limited, so that direct comparisons of the action of a herbicide, under similar environmental conditions, on both the crop and the weed specie or species to be controlled, cannot be made.

II. HISTORICAL COMMENTS

The origin of the interest in the effects of herbicides on plant anatomy and morphology can be traced to the first auxin discoveries at the start of this century. Particularly studies on the formative effects of naturally-occurring indol-3yl-acetic acid (IAA) as well as the synthetic auxins made evident the need to understand how the primordia of organs were formed and how cells and tissues underwent differentiation. One early paper by Friesen (1929) should be mentioned. He studied the effects of phenylurethanes on seed germination and growth and recorded effects on root growth as well as on reduced chlorophyll formation. While light microscopy aided his study, the electron microscope was still almost three decades away from application to herbicide studies.

The mid-1930s brought the classical studies of Zimmerman and Hitchcock including work on root-forming substances (Zimmerman and Hitchcock, 1935 and Pfeiffer, 1937) and of Snow (1935) on cambial growth induced by hetero-auxin. Delisle (1938) studied growth rates, cellular differentiation and auxin relationships in developing leaves of *Aster*. Later studies introduced an ever-widening group of chemicals to these investigations; Mullison (1940) followed the histological responses of bean to tetrahydrofurfural butyrate and Hamner (1941) gave attention to the chemical and anatomical effects of naphth-1yl-acetamide (NAAM) and phenylacetic acid on bean and tomato. Beal (1944) reported on the telemorphic effects of 4-chlorophenoxyacetic acid (4-CPA) on sweet peas.

More detailed anatomical studies followed. Carlton (1943) found that treatment with IAA resulted in a reduction or cessation of root growth in *Allium, Tulipa* and *Narcissus* with histological changes occurring first in the meristematic region, chiefly in the cortex. Watson (1948) followed the morphogenetic changes of bean leaves in intact plants treated with 2,4-dichlorophenoxyacetic acid (2,4-D). Marked leaf distortion was caused by the failure of "normal lateral leaf expansion" and the development of "replacement tissue" which was described as thick-walled turgid, parenchyma-like cells. Swanson (1946) concluded from his systemic study of the major tissues of kidney bean, treated with 2,4-D, that meristematic tissue and tissue capable of division were most readily affected by this compound. Meristematic activity persisted and when differentiation did occur it was always abnormal. Some studies such as that of Struckmeyer, Hildebrandt, and Riker (1949) noted the histological effects of growth regulating substances such as IAA, γ-indol-3yl-butyric acid (IBA), naphth-1yl-acetic acid (NAA) and 4-CPA on bacteria-free, crown-gall tissue from sunflower. Cells were found to be larger from the treatments at high concentrations, but tracheal elements were more frequent at the lower concentration of growth regulator. Wardlaw (1953) compared 2,3,5-tri-iodobenzoic acid (TIBA) with 2,3,6-trichlorobenzoic acid (2,3,6-TBA). TIBA stopped growth of cells most distal to the apex of shoot of *Dryopteris*, and in some plants a ring-fasciated shoot "with a double vascular system" was the result. 2,3,6-TBA suppressed, either totally or partially, apex and leaf formation, but "root formation was greatly increased".

There was considerable interest in this period on the effects of chemicals on mitosis in plants stimulated no doubt by the discovery of polyploidy induction by such compounds as colchicine. Carey and McDonough (1943) and Dermen

(1941) reported on polyploidy in *Allium* from treatments with *p*-dichlorobenzene and in beans from NAA treatments. D'Amato (1948) reported on the cytological effects of 14 compounds on *Allium cepa* root meristems. These studies are some of the forerunners of current cytological and cellular investigations and illustrate the use of a variety of compounds and species and the diversity of the effects reported.

This review will be organized on the basis of the organs of the plant; stem and stem meristems, roots and root meristems, leaves and reproductive structures. In this way all of the literature on the cellular effects of growth regulators on structure of a given organ will be grouped together. First, herbicide effects will be considered followed by pertinent data from other growth regulators particularly those having hormone-like characteristics. A brief section will be devoted to cellular effects of compounds not classified as plant growth regulators.

III. Effects of Herbicides and Other Growth Regulators on Plant Cell Structure

A. STEM AND STEM MERISTEMS

1. *Herbicides*

In the stem of a plant two regions are of particular interest in herbicide physiology. The first is the vascular system because either the xylem or the phloem or both may be essential for translocation of the herbicide to other parts of the plant; disturbance or destruction of the vascular cells is of obvious importance. Secondly, the stem apex or apices controls further aerial growth until lateral buds develop into shoots. Thus growth regulators which alter the cells at the stem apex may greatly affect or even terminate vegetative growth and if the treatment is at a seedling stage may drastically alter formation of reproductive structures.

Dubrovin (1959) found that 4-chlorobut-2ynyl-*N*-(3-chlorophenyl) carbamate (Carbyne) caused a swelling and distortion of the shoot apex of wild oats sprayed about 21 days after emergence. When the shoot apex was sectioned, its cells were found to be swollen and contained numerous groups of chromosomes, having no nuclear membrane or nucleoli. Finally large cells with no cellular material were formed at the apex of the plants. Wheat (Selkirk) reacted in a similar way but to a lesser degree. Whiting and Murray (1946) treated the cut surfaces of red kidney bean plants (*Phaseolus vulgaris*) with phenylacetic acid and found a marked proliferation of inner cortical parenchyma, endodermis and primary phloem parenchyma. The cambium increased in activity and the secondary xylem derivatives differentiated entirely as tracheids. Response to the chemical was small in the pericycle and secondary phloem. In an electron microscope study of foxtail millet (*Setaria italica* (L) Beauv.) treated with DCPA, Chang and Smith (1972) found that while cell division was interrupted, the nucleus and nucleolus did not disintegrate and the chromosomes did not differentiate. Rather, "giant nuclei and giant nucleoli" filled most of the cell volume in meristematic tissue. They found "several nucleolar caps" on the giant nucleolus and later these separated and were encircled by a nuclear membrane, thus forming multiple nuclei. In both mitochondria and chloroplasts the cristae and thylakoid membranes broke down and multiple vacuoles were formed (see Fig. 1a, b, c).

An extensive anatomical study of the effects of TIBA was made by Krause (1971). TIBA causes earlier flowering accompanied by increased activity in the procambium, rapid development of protophloem cells, with thick walls, and the formation of small vessels, in upper nodes. Three weeks after treatment the mid internodes of the plants exhibited less cambial activity than did controls. Two weeks following treatment some lateral shoot apices developed "conical apices" with peripheral cells in a "stacked" arrangement. There was a shrinkage of protoplasmic contents in young pith cells and in some rib meristem cells and thickening of primary walls in young pith cells. Morey and Cronshaw (1968) studied the effect of TIBA on the stem of a woody species, *Acer rubrum*, and found that below the site of application initiation of tracheary elements was reduced and a "complete ring of tension wood" developed. Eames (1950) followed anatomical changes in bean plants treated with 2,4-D, and found that after treatment cell divisions in the hypocotyl began in the endodermis, followed by division of the cells in the inner pericycle, primary phloem and in all "immature cambium derivatives". First divisions in all of these tissues were periclinal, followed by anticlinal and transverse divisions. At the time of treatment the primary phloem was mature, the parenchyma cells proliferated to such an extent that the phloem strands were disrupted and companion cells and sieve tubes were crushed. "After 8—13 days no phloem, as such, is present in the hypocotyl, the region between the photosynthetic organs and the root system," (Eames, 1950). Key et al. (1966) investigated the relationship between aberrant growth caused by 2,4-D and nucleic acid and protein synthesis. While 2,4-D inhibited cell division and cell elongation in the apical regions of the soybean shoot, radial growth was sharply increased below the cotyledons resulting in lateral root formation. They point out in their summary that "the results are consistent with the view that the herbicidal action of 2,4-D is associated with the renewal of RNA and protein synthesis leading to massive tissue proliferation, disorganized growth and finally death of the plant". Baxter and Hanson (1968) have confirmed these findings in their work on soybean. Mitochondria from the lower part of the hypocotyl of 2,4-D-treated plants were larger and incorporated more amino acids into protein, and phosphate into phospholipids and RNA than the non-treated controls.

An interesting effect of a herbicide on stem structure which resulted in a measure of control of a fungus pathogen has been reported by Brener and Beckman (1968). Their study followed an earlier report by Smalley (1962) that 2,3,6-trichlorophenylacetate (Chlorfenac) produced an increased tolerance to Dutch Elm disease in elm trees infected with the pathogen, *Ceratocystis ulmi*. Brener and Beckman (1968) found that chlorfenac retarded springwood

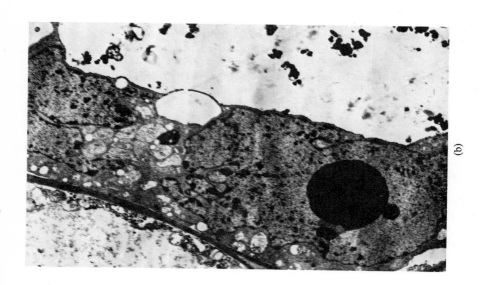

(b)

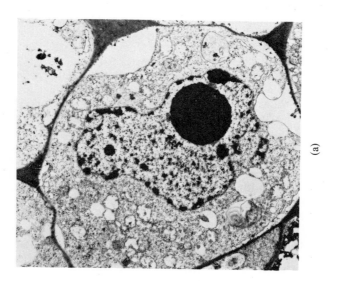

(a)

Fig. 1(a) and (b)

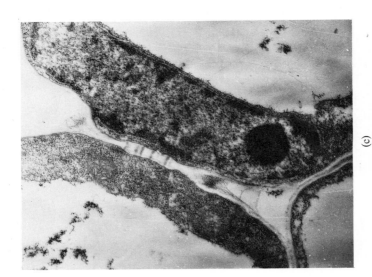

(c)

Fig. 1(a). Shoot apex cell of 7-day old *Setaria italica* plant grown in 20 mg/l DCPA. Nucleolus in cells of treated plants about three times the diameter of that of the controls (×10,000). (b) Nucleolus in cells of DCPA-treated plants is in the process of very abnormal segregation. Two nucleolar caps (lower left) and three nuclei (or two additional lobes) are visible (×10,000). (c) Nucleus and cell wall from untreated Foxtail Millet, 7 days old. Nucleus is spherical with tight entire margins, well defined in contrast to irregular shape and margin of treated cell nuclei (not shown) (×33,500).

TABLE I

EFFECTS OF SEVERAL HERBICIDES ON STEM STRUCTURE

Chemical	Plant species	Effect	Reference
Pronamide	*Agropyron repens* (L.) Beauv., rhizomes	Cell enlargement, necrosis, increase in nuclear volume; abnormal metaxylem elements and phloem necrosis	Peterson and Smith (1971)
Propham	*Avena sativa* L. *Hordeum vulgare* L.	Interruption of mitosis, blocked metaphases, multi-nucleate cells, giant nuclei	Ennis (1948*a*)
Dicamba Dichlobenil	Alligatorweed, (*Alternanthera philoxeroides* (Mart.) Griseb.)	Destruction of phloem, cambium and associated parenchyma above or within nodes treated with either compound	Pate, Funderburk, Lawrence and Davis (1965)
Picloram	*Phaseolus vulgaris*	Vascular cambium remained meristematic, initial divisions periclinal, later anticlinal and transverse, adventitious roots formed	Fisher, Bayer and Weier (1968)
Picloram	Fourwing Saltbush (*Atriplex canescens* (Pursh) Nutt.)	Hypocotyl became "egg-shaped" with vascular tissues distorted; no endodermis visible; small cells with dark nuclei fill "normal cortical zone"	Martin, Shellhorn and Hull (1970)
Picloram	Barley (*Hordeum vulgare* L., var. Wong); Safflower, (*Carthamus tinctorius* L., var. U.S. 10)	Caused mitochondrial swelling and/or changes in mitochondrial membranes	Chang and Foy (1971)
Trifluralin	Soybean (*Glycine max* (L.) Merr.)	At fourth internode xylem elements were less organized, pericycle fibre walls very thickened	Kust and Struckmeyer (1971)
Trifluralin	Corn (*Zea mays* L. var. Dixie 18)	Radial enlargement of cortical cells; multi-nucleate cells in meristematic regions	Schultz, Funderburk and Negi (1968)

Trifluralin	Corn (*Zea mays* L. var. "Michigan 500")	RNA synthesis supported by chromatin greatly reduced	Penner and Early (1972*a*)
Fenac	Mugwort (*Artemisia vulgaris* L.)	Rhizomes enlarged, epidermis disrupted, root primordia formed through activity of inter-fascicular cambium	Rogerson, Bingham, Foy and Sterrett (1972)
Chlorthal dimethyl	Oat (*Avena sativa* L.), Green Foxtail (*Setaria viridis* (l.) Beauv.)	Shoot meristem cells disarranged and hypertrophied differentiation nearer apex in treated plants was irregular	Shaybany and Anderson (1972)
Diallate Triallate	Wild Oats (*Avena fatua* L.) Spring Wheat (*Triticum aestivum* L.)	In Wild Oats, shoot more sensitive than roots with meristem at base of first leaf most affected; reduced mitotic activity in wheat but for either compound cell elongation and enlargement more affected than mitosis	Banting (1970)
Carbyne	Wheat (*Triticum vulgare*, var. Selkirk), and Wild Oats (*Avena fatua* L.)	Shoot apex cells of wild oats swollen, many with numerous groups of chromosomes; these cells had no nuclear membranes or nucleoli; response less but similar in wheat	Dubrovin (1959)
Eptam	Navy Bean, (*Phaseolus vulgaris*)	Marked reduction in wax deposition on hypocotyl surface using scanning EM	Wyse, Meggitt and Penner (1974) (see Fig. 2a, b)
Naptalam	*Fraxinus pennsylvanica*, Marsh	Naptalam or DAA prevented normal periderm formation in hypocotyl below "wounded" tissue	Borger and Kozlowski (1972)
Atrazine	Soybean (*Glycine max* L. Merr. var. "Mark")	Chromatin-enhanced RNA synthesis of as much as 40% at 10^{-6} M found in etiolated seedlings	Penner and Early (1972*b*)

formation and enhanced tylose formation in cells of American elms. Apparently the tylose formation, stimulated by chlorfenac, "provided a cellular barrier to hyphal penetration and spore passage in invaded vessels".

Robnett and Morey (1973) treated stems of honey mesquite (*Prosopis juliflora* var. *glandulosa*) with 2,4-D and 2,4,5-trichlorophenoxyacetic acid (2,4,5-T) at near phytotoxic levels and found "an unusual wood with narrow, thick-walled vessels and axial parenchyma in which cell wall thickening is inhibited". Treatment resulted in a preferential maturation of xylem vessels compared with parenchyma cells. Fusiform initials were transformed into septate parenchyma strands which "resemble the structural changes reported to occur after girdling in the cambial tissue of other arborescent angiosperms". The first effect of 2,4,5-T on *Pinus resinosa* seedlings was in inhibition of cell division and enlargement, followed later by proliferation and expansion of stem and cotyledon parenchyma cells (Chen Wu and Kozlowski, 1972).

An extensive cytological study of the shoot and root meristems of several crop plants (oats, maize and soybeans) to the growth inhibitor, maleic hydrazide (MH) was made by Carlson (1954). Shoot and root growth was retarded and there was an increase in size of both root and plumular organs, due to vacuolation and enlargement of cells of all tissues and an increase in intercellular space. Nuclei in these enlarged cells were abnormally large. Mitosis was inhibited and when it began later, extensive chromosome breakage occurred and chromosomes or their fragments were "lost on the spindle or move into the cytoplasm where they form micronuclei." Chromosomes bridges and chromosome breakage were found. While daughter cells were found to be multinucleate, cytokinesis appeared to be "normal". Effects of other herbicides on stem structure are given in Table I.

2. *Other Growth Regulators*

There has been considerable interest in the effect of phytohormones on cambial activity in stems of plants. Digby and Wareing (1966) reported that IAA applied to the stem of *Robinia pseudacacia* stimulated division of the cambium and the cell derivatives differentiated into xylem tissue. By contrast gibberellic acid (GA$_3$) caused cambial division but the derivative cells "on the xylem side" did not differentiate. They reported that "high IAA/low GA$_3$ concentrations favour xylem formation, whereas low IAA/high GA$_3$ concentrations favour phloem production". New phloem cells formed following hormone treatment were fully differentiated with sieve elements and sieve plates. Wareing (1958) had earlier reported that IAA and GA together greatly stimulate the formation of new xylem "the effect of the gibberellic acid apparently being to stimulate cambial division and that of indoleacetic acid to cause vacuolation and lignification of the resulting cambial derivatives".

When *Coleus blumei* (Benth.) stem explants were pretreated with L-proline and 5 parts/10^6 IAA redifferentiation of very large numbers of wound vessel members took place from pith and cortical parenchyma tissue (Roberts and Baba, 1968). Typically tension wood develops in some woody species placed horizontally, but this phenomenon is inhibited in *Acer rubrum* seedlings by

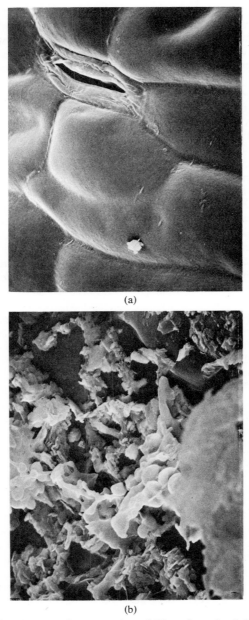

(a)

(b)

Fig. 2(a). Scanning electron microscope view of *Phaseolus vulgaris* hypocotyl below soil surface following treatment with EPTC (10^{-6} M) ($\times 1,400$). (b) Non-treated hypocotyl of bean showing extensive wax deposition in contrast with EPTC effect ($\times 1,400$).

IAA applied to the upper side of the axis (Cronshaw and Morey, 1968). Treatment with either GA_3 or kinetin separately, did not affect the normal development of tension wood. Treatment of tomato (*Lycopersicon esculentum* L.) with GA_3 gave increase in the length and percentage of xylem cells as annular-spiral elements (Davis and Holmes, 1962). In addition treated pitted-scalariform cells were longer and pitted cells more frequent. Roberts and Fosket (1966) reported increased formation of wound vessel cells in transverse slices of tissue from the second internode of *Coleus blumei* (Benth.) stems, supplied with IAA and GA_3. Secondary xylem cells were incompletely differentiated and wound vessel members were found to have thin secondary wall striations. Colchicine was found to block IAA-induced wound vessel member formation in stem explants of *Coleus blemei* (Benth.) but "permitted abnormal xylogenesis associated with the vascular bundles" (Roberts and Baba, 1968). When colchicine was present, the xylem elements which differentiated had abnormal asymmetrical secondary wall sculpturing, suggesting to the authors that the "partial inactivation or reorientation of the microtubules associated with microfibril deposition in prexylem cells (was caused) by colchicine".

Further evidence that IAA may be the hormone which limits xylogenesis comes from the report of Clutter (1960), on the effect of this compound on cell division and vascular differentiation in tobacco (*Nicotiana tabacum*) pith callus cultures. The study of Fosket and Roberts (1964) confirmed earlier work of Jacobs (1952) on the induction of wound-vessel differentiation in *Coleus* stem segments (*in vitro*). At low concentrations of either IAA or 2,4-D (0.5 and 0.01 parts/10^6 respectively) they found a 100% increase in the numbers of wound-vessel members differentiated. Wound vessel differentiation was inhibited by relatively high concentrations of 2,4-D, TIBA or kinetin. Both ethylene and 2,4-D were found to inhibit xylogenesis and "completely prevent fibre lignification" in etiolated pea seedlings of *Pisum sativum,* var. Alaska (Apelbaum, Fisher and Burg, 1972). The effect of ethylene is reversible with normal differentiation resuming after 72 h following removal. They concluded that the "inhibition of xylogenesis and fibre lignification caused by 2,4-D is partly or wholly due to auxin-induced ethylene production". Sorokin, Mathur and Thimann (1962) also studying pea (*Pisum sativum* L., var. Alaska) found that either IAA or 2,4-D activated the vascicular cambium and caused the initiation of some interfascicular cambium in the epicotyl and the "partial or total occlusion of proto- and metaxylem". By contrast, kinetin alone or with IAA led to the initiation of a very active cambium, forming several layers of secondary xylem. They observed that "the effect of kinetin is to make the xylem more normal and to alter the epicotyl structure from herbaceous to more-or-less woody".

In an extensive study of the effect of GA_3 on the anatomy of soybean (*Glycine max,* var. Hawkeye) Bostrack and Struckmeyer (1964) found that this hormone increased cell elongation resulting in an early elongation of the internodes of the treated plants. GA_3 caused more lignified xylem parenchyma to be produced and resulted in a partial collapse of the relatively thin-walled vessel elements and tracheids.

1. *Herbicides*

Since many herbicides are applied to the soil and hence are absorbed by roots or subterranean parts of plants, these below-ground organs are of considerable importance in effective weed control. Partly because of the difficulty with which intact root systems can be observed and studied, our knowledge of the physiology of root growth and the response of roots to growth regulators is not as complete as is desirable.

Ashton *et al.* (1969) have reported on their studies on bromacil, a non-selective herbicide effective in the control of annual weeds on non-cropped land. They found that root elongation was inhibited by this herbicide and that this inhibition was restricted largely to the terminal 5 mm of the root which is the region of maximum elongation. Bromacil caused the roots to become necrotic and to collapse, resulting in dead cells in the meristem, procambium and epidermis. Multiple nuclei were found in cells, where cell wall formation was inhibited and "precocious vacuolation" resulted.

Studies have been conducted on trifluralin, used for weed control in safflower and cotton (Bayer, Foy, Mallory and Cutter, 1967). These authors report that the most striking effect of trifluralin on root development in cotton, safflower, watergrass and onion was on the radial expansion near the root tip. As treatment with the herbicide continued, the extent of the meristematic tissue zone decreased because of vacuolation and differentiation of the tissue. Trifluralin disrupted mitosis although mitosis was not altered in every cell. Cells of the pericycle were greatly enlarged in the areas opposite the protoxylem and lateral root formation was noted. The authors referred to these enlarged pericyclic cells as primordiomorphs. Hacskaylo and Amato (1968) have also reported on studies on trifluralin on roots of corn and cotton. Cells at the extreme root tip, after treatment with trifluralin, were small and dense and multinucleate. Adjacent to this apical region the cells were abnormally large, thin-walled and aberrant. They report that "cell plate and cell wall formation were apparently rare". Lignowski and Scott (1972) found that trifluralin inhibited mitosis in onion and wheat root tips, in a manner very similar to colchicine. They reported "arrested metaphases, c-pairs, micronuclei, amoeboid nuclei, and polyploidy in the cells studied". The metaphase appeared to be blocked with the prophase not affected. From centrifugation studies they concluded that the spindle apparatus had been disrupted since chromosomes were displaced in arrested metaphase cells. Inhibition of lateral roots of field-grown pecan (*Carya illinoensis* (Want) K. Koch) by soil treatment with trifluralin (Norton, Walter and Storey, 1970) and inhibition of primary and lateral roots in sugarbeet seedlings (*Beta vulgaris* L.) in greenhouse tests with this herbicide (Schweizer, 1970) have also been reported.

Arntzen *et al.* (1970) have reported on the effect of amitrole on flax root apices. In addition to the inhibition of root growth they found an increase in vacuolation in cortical and root cap cells, the precocious differentiation of vascular elements and root primordia initiation. Elongating cells of the root exhibited severe ultrastructural damage including deterioration of the cell membranes. Irregular dense deposits in the plastids appeared to be starch granules. Where the inner and outer plastid membranes had separated, membranous vesicles were present (see Fig. 3a, b).

The herbicide DCPA used for weed control in turf, has been studied by Bingham (1967). These studies show that although cell division was arrested by DCPA, cell enlargement continued with the cells becoming excessively large and irregularly shaped. In one of the species studied, *Zea mays* var. Pioneer 310, while cell size in the radicle was essentially unchanged the "treated tissue contained six times as many binucleate cells as untreated radicle tissue".

Hess and Bayer (1974) treated roots of cotton (*Gossypium hirsutum* L. "Acala 4-42") with an aqueous saturated solution of trifluralin. For treatments longer than 2 h the authors found only prophase and blocked metaphase mitotic figures in the meristem area. Here microtubules were completely absent and in the "cells began division in a normal manner". The chromosomes in these cells were not aligned along the metaphase plate and subsequent nuclear envelope formation resulted in cells with polyploid, polymorphic nuclei (see Fig. 4a, b).

Additional studies on the effects of herbicides on root structure are summarized in Table II.

2. Other Growth Regulators

Relatively few studies have been made on the effect of growth regulators other than herbicides on the structure of plant roots. Torrey and Fosket (1970) examined the effect of kinetin on the growth of excised root tips of *Pisum sativum* on artificial media. If no kinetin was added to the medium, proliferation of the pericycle occurred in contrast to no proliferation of the cortex which "sloughed off" forming a callus of diploid cells. With kinetin in the medium ($0 \cdot 1 - 1 \cdot 0$ parts/10^6) the cortical cells divided, were found to be polyploid and at $5-7$ days from the beginning of the treatment were found to have undergone cytodifferentiation with the formation of tracheary elements.

In an earlier study Torrey (1957), using isolated pea roots on sterile medium, found that IAA changed the vascular-tissue pattern from triarch to hexarch and that the latter pattern was present as long as the hormone was in the medium. When the roots were transferred to intermediate auxin concentrations pentarch or tetrarch roots were formed after regeneration. Svensson (1971) has reported on extensive studies on the effects of coumarin

---

Fig. 3(a). Ultrastructural organization of a root cap cell of flax (*Linum usitatissimum* L.). Active cellular metabolism is suggested by numerous organelles and an extensive cytomembrane system (endoplasmic reticulum, dictyosomes, etc.) ($\times 31,000$). (b) A flax root cap cell observed after five days of amitrole treatment. A general reduction in metabolic activity concomitant with abnormal cellular aging is suggested by the reduced level of active dictyosomes associated secretory vesicles and the presence of a large, well-defined vacuole ($\times 28,500$). N, nucleus; M, mitochondria; P, plastid; D, dictyosome; S, starch; ER, endoplasmic reticulum; CW, cell wall; V, vacuole.

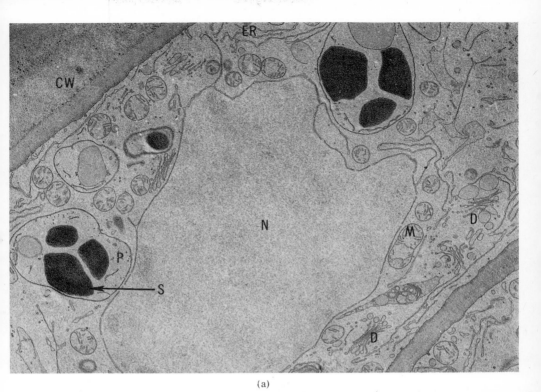

(a)

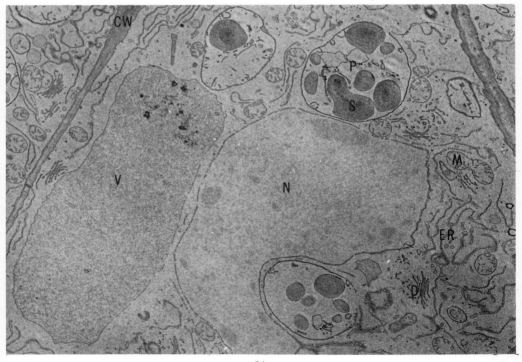

(b)

TABLE II

HERBICIDE EFFECTS ON CELLS OF PLANT ROOTS

Chemical	Plant species	Effect	Reference
Pronamide	*Agropyron repens* (l.) Beauv.	Precocious differentiation and maturation of rhizome tissues, vacuolation of meristem, nuclei enlarged and had several nucleoli	Peterson and Smith (1971)
Propham	*Avena sativa* L., *Allium cepa* L.	Anaphase bridges, blocked metaphases, nuclear fragments, binucleate cells and giant, vesiculate nuclei found in some parts of primary roots	Ennis (1948b)
Propham	Onion (*Allium cepa*), Rye (*Secale cereale*)	Abnormal centromere action, spindle suppression, paired chromosomes and polyploid nuclei for both species; multi-polar spindles in onion and chromatid fragmentation in rye	Doxey (1949)
Propham	*Vicia faba* L. and *Cycas circinalis* L.	Contraction of chromosomes in prophase, metaphase and anaphase	Storey and Mann (1967)
Avadex	Barley (*Hordeum vulgare* L.); Wheat (*Triticum vulgare* L.)	Caused abnormal, swollen cells irregularly-shaped nuclei or micronuclei; polyploid cells found	Morrison (1962)
Ethyl phenyl-carbamate	Barley (*Hordeum vulgare*)	Reduction in anaphase and telophase figures with an increase in metaphase figures at concentrations affecting the spindle	Ivens and Blackman (1950)

Herbicide	Plant	Effect	Reference
Nitralin	Corn (*Zea mays* L., var. U.S. 13)	Swelling in zone of active cell division at root tip, cell wall formation arrested, extensive replication of nuclei	Gentner and Burk (1968)
Silvex 2,4-DB	Colonial Bentgrass (*Agrostis tenuis* Sibth.)	Proliferation of pericycle, hypertrophy of cortical cells increased lateral root and root hair development	Callahan and Engel (1965)
Propachlor	Onion (*Allium cepa* L.), Oat (*Avena sativa* L.)	Cell division inhibited in *Allium* and stopped at 16 parts/10^6 cell elongation induced by auxin inhibited in proportion to herbicide concentration	Dhillon and Anderson (1972)
Trifluralin	Corn (*Zea mays*) Wild Oats (*Avena fatua*)	Root tip swelling and inhibition of root elongation growth appeared to be non-polar	Lignowski and Scott (1971)
2,4-D	Soybean (*Glycine max*)	Dense, meristematic tissue produced by pericycle	Sun (1956)
Chlorthal-dimethyl 2,4-D	Oat (*Avena sativa* L.), *Setaria viridis* (l.) Beauv. Bean (*Phaseolus vulgaris*)	Root meristem cells disarranged and hypertrophied Four "shoulders" of tissue formed on primary root tips opposite the protoxylem points from proliferated pericycle	Shaybany and Anderson (1972) Wilde (1951)
2,4-D	Onion (*Allium cepa* L.)	Long treatments of several weeks destroyed main meristem; mitosis with diplochromosomes in cortex and central cylinder	D'Amato and Avanzi (1948)

Table II—cont.

Chemical	Plant species	Effect	Reference
2,4-D 2,4,5-T MCPA	Onion (*Allium cepa* L.)	Spindle disturbance and formation of c-pairs, chromosome "stickiness" and chromatin bridges in anaphase and telephase found frequently; polyploid nuclei and micronuclei found	Nygren (1949)
2,4-D 2,4,5-T	Onion (*Allium cepa* L.)	Condensation and "stickiness" of chromosomes; retardation of spindle formation; chromatid breaks observed	Croker (1953)
2,4-D	*Narcissus* sp. and *Allium* sp.	Disintegration of mitochondria, cell division inhibited with c-mitoses found but no polyploidy	Ryland (1948)
MCPA	Onion (*Allium cepa* L.)	Swelling and "stickiness" of chromosomes, chromatin bridges at anaphase with micronuclei and enlarged nucleoli	Doxey and Rhodes (1949)
2,4-D	Soybean (*Glycine max.*)	Treatment almost doubled number of cells, cells were 58% smaller with mitosis inhibited 28%	Rojas-Garciduenas and Kommedahl (1958)
Picloram	Canada Thistle (*Cirsium arvense* (L.) Scop.)	Disintegration of cambial and phloem cells, destruction and destortion of cortical cells	Kreps and Alley (1967)

Compound	Plant species	Effect	Reference
Chlorpropham	Squash (Cucurbita maxima) Cucumber (Cucumis sativus L.)	In squash 15 parts/10⁶ resulted in radial elongation of pericycle and endodermis cells in mature region of root; similar result in cucumber with cells in meristematic region hypertrophied, some cells enlarged with several nucleoli; abnormal wall thickenings	Scott and Struckmeyer (1955)
CDDA Propham	Peas (Pisum sativum, var. Chancellor) Barley (Hordeum vulgare var. Parkland)	CDAA inhibited cell division and root elongation; IPC at higher concentrations resulted in endopolyploidy and binucleate cells	Canvin and Friesen (1959)
Bensulide	Oat (Avena sativa L. var. Kanota)	Epidermal cells elongated radially, tracheary elements differentiated near root tip were short pitted; mitosis not completely inhibited	Cutter, Ashton, Huffstutter (1968)
Maleic hydrazide	Corn (Zea mays) Oat (Avena sativa L.) Soybean (Glycine max.)	Acquire an abnormally mature appearance with root hairs and mature vascular tissues up to 0·1 mm from tip; nucleolar volume increases, no visible change in mitochondria	Carlson (1954)
Substituted phenols	Pisum sativum	All compounds tested gave quantitative and qualitative reactions similar to colchicine; inhibition of cell division; some compounds produced an overcontracted late prophase	Muhling, Van't Hof, Wilson and Grigsby (1960)

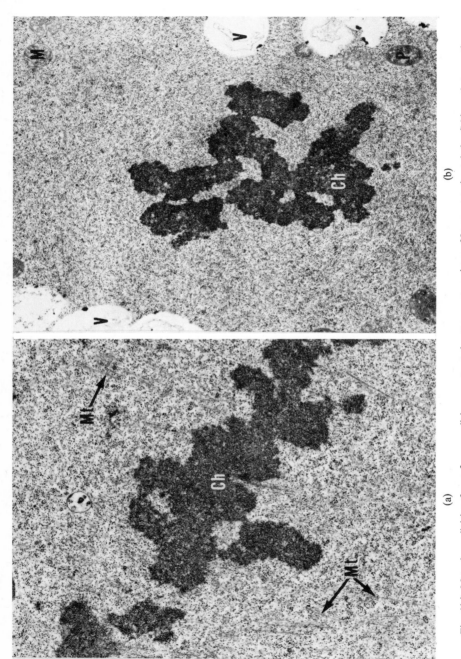

(a)

(b)

Fig. 4(a). Metaphase division figure from a cell in an untreated cotton root meristem. Numerous microtubules (Mt) are between the chromosomes (Ch) and the polar region (×17,000). (b) Arrested metaphase division figure. Trifluralin treatment duration of 48 h. Microtubules are absent and vacuolation (V) has increased. The chromosomes (Ch), mitochondria (M) and plastids (P) are cytologically normal (×11,000).

on root growth and root anatomy. He found that coumarin inhibited transverse divisions in all cell layers with the perivascular layers exhibiting the most sensitivity to this compound. Mitotic activity associated with the initiation of lateral roots was inhibited while longitudinal divisions within the stele were enhanced. Svensson states that "coumarin causes an inhibition of the longitudinally directed processes and a stimulation of the radially directed ones". He interpreted this as an indication that the "formative system is disengaged or reorientated, i.e. the polarity of the cells is changed". He developed the hypothesis that the microtubuli are a main component in the formative system and that they are common to both cell division and cell elongation. In a second study Svensson (1972) reported on the interactions of coumarin with several growth regulators. Kinetin reduced the inhibitory effect of coumarin on the growth in length of the zone of elongation of the roots of maize and wheat. GA$_3$ reduced the "swelling" caused by coumarin in the zone of elongation. Ivanov (1966) found that for maize roots (var. Sterling) 6-benzylaminopurine and NAA caused "thickenings" because of the growth of meristematic cells, chiefly the cork tissue. These cells grew both in length as well as in width becoming more nearly isodiametric.

An interesting effect of a naturally-produced growth regulator called malformin from *Aspergillus niger* van Tiegh has been reported by Curtis (1961) and Izhar, Bevington and Curtis (1969). Malformin caused root curvatures in *Zea mays,* stimulated root hair and lateral root formation, enhanced radial expansion, but inhibited cell division and cell wall synthesis. Griseofulvin, a fermentation product of *Penicillium griseofulvin* and several other species of *Penicillium* was studied by Paget and Walpole (1958) for its effect on root tips of *Vicia faba*. Multipolar mitosis was induced and in some cells the total complement of chromosomes segregated into "irregular groups, up to five in number, which are then reconstructed as nuclei of varying size within the original cell".

C. LEAVES

1. *Herbicides*

Research on the effects of herbicides on leaf structure, both gross and fine, is important for two fundamental reasons. Control of many weed species is through aerial application of the herbicide and entry of the chemical is essentially through foliar absorption. Particularly for those compounds which do not kill leaf cells on contact, but disrupt cell processes, and/or are translocated to other parts of the plant, the study of the changes in leaf cell structure is of obvious importance. Secondly, the leaves of autotrophic plants provide the food for maintenance of life and for further growth by means of photosynthesis. It is not surprising then that many workers have investigated the effects of herbicides on chloroplast structure, particularly in leaf cells.

Studies on the effects of 2,4-D on leaf structure span a period of over a quarter of a century. Felber (1948) noted protuberances on both surfaces of leaves of Red Kidney bean, following treatment with 2,4-D while Eames

(1949*a*) compared structural differences of leaves of a monocot (nut grass) with a dicot (bean) treated with 2,4-D. In nut grass leaves, 2,4-D sharply reduced normal differentiation of tissues and a marked increase in cell vacuolation with little cytoplasm was noted. Vascular bundles were also reduced in size and showed some distortion. In bean, Eames (1949*b*) also found no differentiation of leaf tissue and the "replacement tissue" had cells which were highly vacuolate, without chloroplasts and with "scanty" cytoplasm. He pointed out that even though external differences between a monocot and a dicot arising from 2,4-D treatment did not appear very striking, substantial internal structural changes may have occurred. Friesen and Olson (1953) found that in barley 2,4-D caused deformity of leaf initials in seedling stages and that in oats the same deformities resulted but these extended over a longer period of growth. Bradley *et al.* (1968) studied apricot treated with a compound related to 2,4-D, namely 2,4,5-T. They found an increase in petiole diameter caused by the enhanced growth of the secondary xylem and phloem. In the leaf blades of these plants the increase in thickness was a result of increased depth of the palisade tissue and the bundle sheaths. At concentrations of 2,4-D of 100 and 200 parts/10^6 many leaf cells and their nuclei were larger than the controls. They state that "distributions of frequencies of cell widths and of nuclear diameters in palisade-cell populations indicated that the general degree of endopolyploidy increased following 2,4,5-T treatment and that some of the comparable cells in controls were also polyploid".

Hallam (1970) studied the effect of 2,4-D and related compounds (2-CPA, 4-CPA, and 2,6-D) on the ultrastructure of primary leaves of *Phaseolus vulgaris* L. var. Canadian Wonder. While some changes in leaf cells were reported for 2-CPA, 4-CPA and for 2,6-D, the major ultrastructural alterations were found to be caused by 2,4-D. Only 2,4-D caused changes in the morphology and internal structure of the chloroplasts and these occurred primarily in the light-treated leaves. Just 4 h after application, 2,4-D caused a breakdown in the membranes of cells of the epidermis, palisade and mesophyll. Hallam states that "after 8 hours, the chloroplasts are distorted, the granularity of the stroma is more marked, vesicles have appeared in the stroma, outer chloroplast membrane is broken in many places and invaginations of the inner membrane into the stroma appear. At this time the plasmalemma has moved away from the cell wall and the cytoplasm." The effects of the 2,4-D are even more pronounced 24 h after treatment. At this time the cytoplasm was even more condensed giving the appearance of a plasmolysed cell and the cytoplasm was "densely packed with ribosomes". By this time the extreme distortion of the chloroplasts made the identification of internal structure very difficult and "membranes appear to bound the vesicles enlarging within the stroma and the osmiophilic granules have aggregated". White and Hemphill (1972) studied the effect of 2,4-D on tobacco leaves (*Nicotiana tabacum* L "Samsun N.N.") and found that mature fully expanded leaves were very sensitive and that the herbicide caused a rapid breakdown of the mesophyll. The tonoplast, plasmalemma and membranes of the chloroplasts and mitochondria were found to be ruptured and disintegrating within 30 min after treatment with 500 mg/l 2,4-D. At 1 h after treatment no tonoplast was found in many cells and the "plasmalemma was highly convoluted". There was also a significant swelling of the granal compartments of the chloroplasts and a general dissolution of the cytoplasmic matrix, including the ribosomes." At the 2 h sample time the chloroplasts become more spherical and the lipid storage bodies, the plastoglobuli had increased in size. White and Hemphill caution that under the conditions of their experiments the amount of 2,4-D arriving at individual cells was probably higher than under field conditions (see Fig. 5a, b, c).

Amitrole treatment of many plants results in a bleaching or whitening of leaf tissue thus causing a number of researchers to study chlorophyll changes and the effects of this herbicide on chloroplasts. Shiue and Hansen (1958) studied the effect of this herbicide on red pine (*Pinus resinosa*) and on white spruce (*Picea glauca*) and, using the light microscope, found absence of chlorophyll and increased vacuolation of leaf cells. The analytical study of Wolf (1960) revealed that the net rate of chlorophyll synthesis was markedly reduced and at least suggested that in the dark the direct action of amitrole was not on chlorophyll destruction. Jacobson and Rogers (1961) reported large differences in the ultrastructure of chloroplasts of maize following treatment with amitrole. Earlier (Rogers, 1957) had found that no plastids in corn tissue were made chlorotic by amitrole.

Bartels (1965) found that the plastids of wheat seedlings (*Triticum vulgare* L. var. Vigo) grown in the dark had clearly defined concentric lamellae in contrast with light-grown plants which developed disorganized lamellae after treatment with amitrole. Bartels *et al.* (1967) studying the same species treated with amitrole confirmed the earlier work that this herbicide prevents the normal formation of the chloroplasts and further brings about the loss of 70 S chloroplastic ribosomes. Amitrole did not seem to have any effect on the proplastids of seedlings which had developed in the dark. In a continuation of this study, Bartels and Weier (1969) using wheat seedlings (*Triticum vulgare* L. "Federation") grown in the light and treated with amitrole reported leaf chlorosis. The plastids were found to lack normal grana-fret membrane systems and chloroplast ribosomes. Some disorganization of membranes was found in these chloroplasts. Dark-grown amitrole-treated seedlings were found to contain proplastids with non-crystalline prolamellar bodies and ribosomes. Subsequent treatment of these seedlings with light did not cause the development of normal grana-containing chloroplasts from the proplastids. Guillot-Salomon (1966) has also reported an irreversible alteration of the ultrastructure of plastids of *Zea* following treatment with amitrole.

General growth inhibition and foliar chlorosis in some species treated with triazines have been assumed to be causally related to blockage of photosynthesis. Several authors have reported the direct effects of the triazines on the chloroplasts. Ashton *et al.* (1963a) studied the effect of atrazine on mature fully expanded primary leaves of bean (*Phaseolus vulgaris* L.) and found that in the light the chloroplasts were markedly changed in contrast to

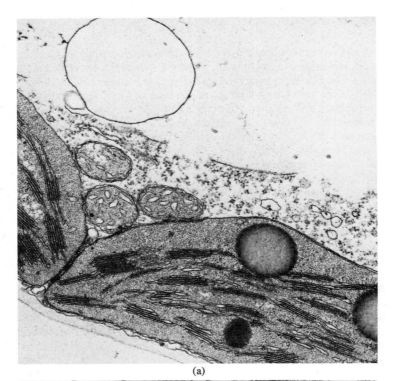

(a)

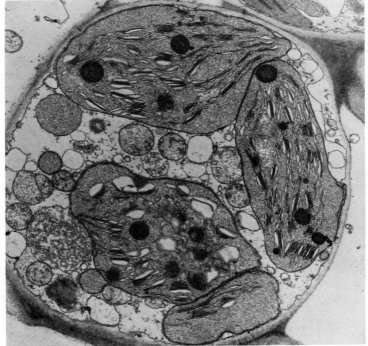

(b)

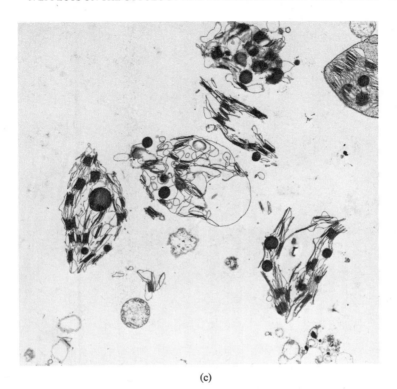

(c)

Fig. 5(a). Part of a mesophyll cell from the lower fully-expanded leaf of tobacco (*Nicotiana tabacum*) 30 min following treatment with 2,4-D. Tonoplast shown as ruptured in three places (diagonally across centre of photo) and swelling of mitochondrial membranes (centre left) (×26,000). (b) Changes in mesophyll cell from lower, fully-expanded leaf after 60 min showing swollen granum (×9,500). (c) Chloroplasts from lower, fully-expanded leaf, following 6 h of treatment with 2,4-D. Note various stages of the breakdown of the chloroplasts (×7,000).

no change in plants which were kept in the dark. They report a progression of changes in the chloroplasts starting with a change in shape from discoid to spherical, accompanied by the disappearance of starch from the lamellar system. Further they found that "the frets or parts of frets are destroyed, leading to the disorganization of the granal arrangement; the compartments of the grana swell; and, the envelope and the swollen compartments ultimately disintegrate". In a continuation of this work Ashton *et al.* (1963*b*) found that in developing bean leaves vacuolation of these cells was hastened by atrazine treatment compared with control plants in the light. The interaction of light and atrazine on leaf structure was further noted in the reduced "air space system" of mature primary leaves treated with atrazine but kept in the light in contrast to plants in the dark with or without the herbicide. Ectoplast and tonoplast integrity was changed under light and atrazine treatment but no

TABLE III

EFFECTS OF SEVERAL HERBICIDES ON CELL STRUCTURE OF LEAVES

Chemical	Plant species	Effect	Reference
Bromacil	Oat (Avena sativa L., var. Kanota)	Chloroplast grana and fret system inhibited; loculi of grana and fret vesicles swelled; changes in chloroplast envelope	Ashton, Cutter and Huffstutter (1969)
Pyriclor	Tobacco (Nicotiana tabacum L., var. NC-402)	Progressive disruption of chloroplast fine structure; first change to spherical form, swelling of fret system and loss of starch; later swelling and disruption of granal disc membranes and breakage of chloroplast envelope; mitochondria unchanged but increased in number	Geronimo and Herr (1970)
Sirmate	Wheat (Triticum vulgare L. var. Federation)	Grana-fret membranes and ribosomes absent in chloroplasts of treated leaves grown in light	Bartels and Pegelow (1968)
Paraquat	Honey Mesquite (Prosopis juliflora (Swartz) DC. var. glandulosa (Torr.) Cockerell)	Disintegration of plasmalemma and rupture of chloroplast membranes; effect similar in light and dark	Baur, Bovey, Baur and El-Seify (1969)
Pyrazon	Bean (Phaseolus vulgaris L., var. Tendercrop)	Chloroplasts became spherical and swollen; clumped not located at periphery; devoid of starch; grana formation stopped; thylakoids swollen, perforated and finally disintegrated; some rupture of outer chloroplast membrane	Anderson and Schaelling (1970)

Pyrazon	Bean (*Phaseolus vulgaris* L., var. Tendercrop)	Chlorosis at leaf margin first, progressing inward; chloroplasts became round, swollen and had affinity for safranin	Rodebush and Anderson (1970)
Picloram	Western Ironweed (*Vernonia baldwini* Torr.)	Rapid pro-cambium activity in leaves; destruction of phloem parenchyma, sieve elements, and companion cells	Scifres and McCarty (1968)
Picloram plus 2,4-D	*Pinus radiata*	Protoplasts of cells from needle segments shrunk severely, with loss of plasmalemma integrity	Bachelard and Ayling (1971)
Picloram (K-salt) and 2,4,5-T	Honey Mesquite (*Prosopis juliflora* (Swartz) D.C. var. *glandulosa* (Torr.) Cockerell)	Both herbicides caused radial enlargement of phellem cells; inner cortex and phloem parenchyma proliferated and xylem vessels lignified	Meyer (1970)
N-alkanes (hexane to dodecane)	Spinach (*Spinacia oleracea*)	Isolated chloroplasts treated for several seconds resulted in chloroplast thylakoid sacs swollen but intact; swelling inversely proportional to chain length	Mukohata, Mitsudo, Nakae and Myojo (1971)
Maleic hydrazide	Tennessee Bush Beans (*Phaseolus vulgaris* L.)	Few mitotic figures in MH-treated buds compared to controls, apical cells enlarged and vacuolated	Greulach and Haesloop (1954)
Maleic hydrazide	Cotton, TPSA, Stoneville 2B (*Gossypium* sp.)	Leaves increased in thickness due to cell enlargement	McIlrath (1950)
Maleic hydrazide	Red Kidney Bean (*Phaseolus vulgaris* L. var.)	Cell enlargement inhibited, epidermal cells smaller with palisade parenchyma not elongated but isodiametric; same number of chloroplasts in spongy parenchyma but smaller compared with non-treated	Watson (1952)

changes were found in plants placed in the dark either with or without this herbicide. Other atrazine effects reported were the stopping of cambial activity and reduced thickness of cell walls of sieve and tracheae elements in the plant stems. The same relationship to light and darkness was found in these effects (Ashton et al. 1963b). It is interesting to note that Ashton et al. (1966) found little if any effect of atrazine on the structure of the chloroplast of Chlorella ellipsoidea. In an electron-micrographic study of the cotyledons of the bush bean, Singh et al. (1972) reported that the cells of this organ, after treatment with several s-triazines, exhibited a two-fold increase in the number of cisternae of rough endoplasmic reticulum. Further there was a greater number of vesicles and an increase in the amount of cytoplasmic ribosomes. Hill et al. (1968) studied barnyard grass (Echinochloa crus-galli (l.) Beauv.) at the two- and three-leaf stage following treatment with atrazine. Chloroplast degradation in this species began with a swelling of the fret system followed by "swelling and disruption of the granal discs". Later breakdown stages revealed ruptured grana membranes and the chloroplast envelope. The mitochondria were not effected either by the duration of the treatment or by the concentration of atrazine studied (2, 5, 10 and 20 parts/10^6). The results of additional studies on the efforts of herbicides on leaf structure are summarized in Table III.

2. Other Growth Regulators

The control of abscission of leaves by the naturally occurring hormone system of plants as well as the synthetic externally applied growth regulators has been the subject of considerable research. Considerable progress has been made in describing the cells and tissues involved in abscission at the ultrastructural level, particularly as a result of the work of Jensen and Valdovinos and their co-workers. In the first study in their series of papers (Jensen and Valdovinos, 1967) they have studied the cells in an area of indentation of epidermal tissue which delineates the abscission zone of Lycopersicon esculentum and Nicotiana tabacum petioles. Cells in this zone are characterized by frequent invagination of the plasmalemma and in these invaginations a fibrillar material of the density of the cell wall was observed. Cells in the abscission zone often contained microbodies with crystalloid cores having a cubical shape and composed of "parallel sheets of osmiophilic material". These bodies were found to be about three times larger in tobacco than in tomato. They noted a curious granular component in the chloroplasts enclosed by a membrane. Microtubules were found to be adjacent and parallel to the plasmalemma particularly in the corners of the cells. A second study (Valdovinos and Jensen, 1968) focused on changes in the cell walls during the abscission process. Initiation of separation appeared to begin in the middle lamella followed by disintegration of the primary wall which appeared to become swollen and highly flexible. Some cell walls invaginated during this stage and the cells finally collapsed. At the time of cell separation the number of microbodies with crystalloid cores found in the abscission zone cells decreased markedly (Jensen and Valdovinos, 1968). The authors hypothesize that changes in the fine structure of the crystalloid core "may represent the release of latent enzymes into an active form which then induces cell separation in the abscission tissue". Also before and during cell wall disintegration much rough endoplasmic reticulum was observed. Other changes including a spatial segregation of the nucleoli were observed just before the cell-separation phase, in some cells. The effect of the exposure of the pedicels of flowers of tobacco, *Nicotiana tabacum* L., cv. Little Turkish to ethylene was studied by Valdovinos *et al.* (1971 and 1972) (see Fig. 6a, b, c, d). Two hours after exposure to ethylene, rough endoplasmic reticulum accumulated in the abscission cells and increased in abundance in the period 3–5 h after treatment. A loss in the integrity of the membranes of microbodies also occurred at the 5 h period and as cell wall degradation follows, fibrous material, vesicular structures and electron-dense bodies some with striations appear in the walls. At this same period after treatment (5 h) there was little change in the nuclei, mitochondria, chloroplasts and in the crystalloid cores of microbodies. An ultrastructural study of the localization of peroxidase in tobacco abscission zone tissue has also been made by Henry and Jensen (1973). Webster (1973) has also followed the ultrastructural changes in abscission zone cells of leaves of *Phaseolus vulgaris* L. cv. Red Kidney bean, with and without ethylene treatment. This author also reported the invagination of the plasma membrane of cells prior to cell wall break. Chloroplasts and mitochondria were found to be structurally altered but recognizable and many unidentifiable inclusions of different size and shape were found in the cell wall region. Plasmodesmata were often observed in abscission cells.

The effects of kinetin on leaf structure with changes which are observable with the light microscope and ultrastructural changes recorded with the electron microscope have been reported. Powell and Griffith (1960) found that while both kinetin and red light increased the expansion of etiolated bean leaf discs, only kinetin caused a sharp increase in cell size. Srivastava and Arglebe (1968) found that leaves of barley floated on water rapidly lost polyribosomes and ribosomes but that this phenomenon was greatly retarded by the addition of kinetin. They suggest that "kinetin by stimulating RNA synthesis and by suppressing the activities of ribonuclease and peptidase may preserve the ribosomes in excised leaves". Kinetin has also been reported to increase the amount of endoplasmic reticulum and ribosomes in detached senescing wheat leaves whereas excised leaves floating on water without this hormone had lost endoplasmic reticulum and cytoplasmic ribosomes in mesophyll cells by 4–5 days (Shaw and Manocha, 1965). Several reports in the literature indicate that cytokinins increase the dimensions of the nucleus and nucleoli of detached leaves, cause increases in the number of mitochondrial cristae, the synthesis of grana and lamellae in the chloroplasts and the numbers of ribosomes (Kulaeva, 1967; Kursanov *et al.* 1964; Sveshnikova *et al.* 1966; and Shaw and Manocha, 1965). Yoshida (1970) studied the effect of benzimidazole on wheat leaf chloroplasts which were senescing. Benzimidazole counteracted the alterations in colour, shape and fine structure of the chloroplasts which occurred in the controls. This compound caused "extreme curvature of the

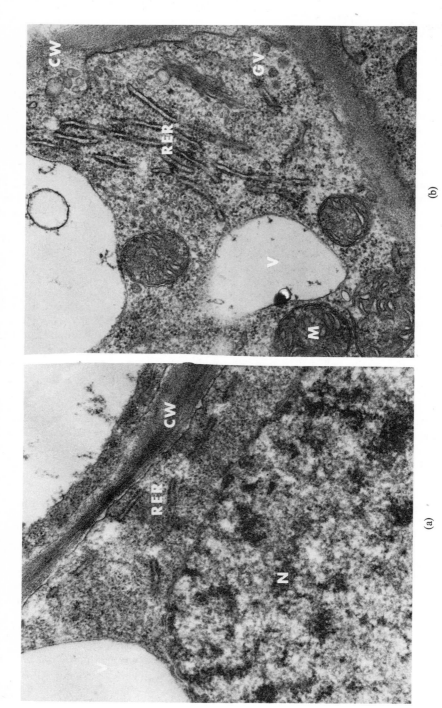

Fig. 6(a), (b), (c) and (d). See page 115 for description.

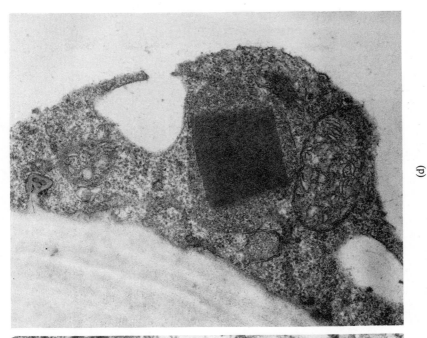

(d)

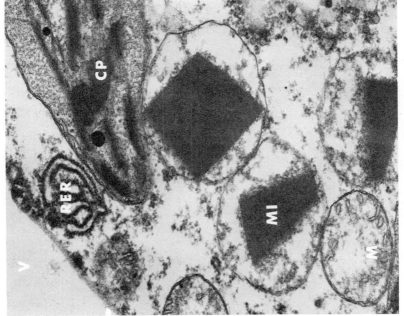

(c)

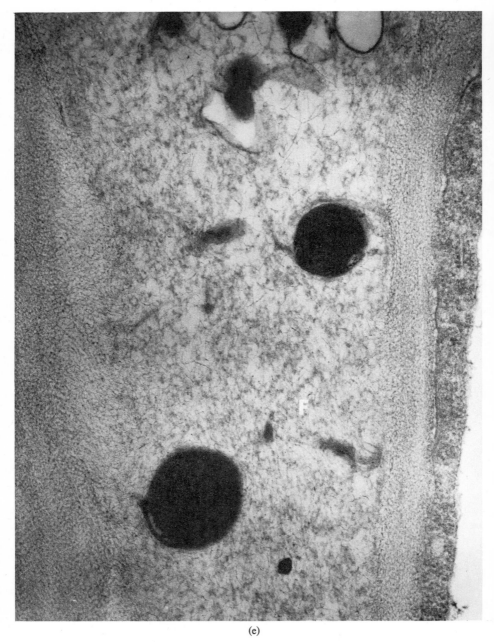

(e)

Fig. 6(e). See facing page for description.

grana-fretwork skeleton of the chloroplasts forming the keel of a boat," hence the chloroplasts were referred to as "boat shaped".

D. ORGANS OF REPRODUCTION

1. *Herbicides*

The study of the action of herbicides on the organs of reproduction of plants has received relatively little attention in terms of physiological, biochemical or of gross morphological, cellular or subcellular responses. This is not surprising for several reasons. First the control of weed species is of necessity the control of seedling vegetative plants in many instances. Elimination or control of the undesirable plants should occur long before flower and seed formation. However the flower primoridia are initiated early in the phylogeny of most species and thus the effects of herbicides can and should be ascertained at these embryonic stages. Further, for those weed species which complete a reproductive cycle, even after treatment with herbicides, the effect of these chemicals on the seed and its germination are of considerable importance. In so far as the research literature provides information on these aspects of the effects of herbicides and other growth substances, particularly on the fine structure of plants, this subject will be dealt with here. The effect of herbicides on seed germination has been discussed elsewhere in this volume (Chapter 5).

An early paper reporting the deleterious effect of 2,4-D on a reproductive structure is that of Friesen and Olson (1953). They found two critical periods of injury following application of 2,4-D to barley. The first was at a seedling stage and the second was at the advanced boot stage just before spike emergence. Sterility was induced at the time when the anthers and stigma were differentiating. Rehm (1952) also found that 2,4-D produced sterile pollen in tomato and found abnormal pollen grains with vacuolization, plasma coagulation and a few empty grains. Pollen mother cells in grain sorghum (*Sorghum vulgare* Pers.) treated with either atrazine or the propylene glycol butyl ether ester of 2,4-D had chromosomal aberrations, mostly aneuploidy and polyploidy. Liang *et al.* (1967) also studied the effect of atrazine on grain sorghum and found that many microsporocytes were affected by the herbicide. The microspore mother cells were multinucleate cells, and showed bridges and increased chromosome numbers. The dyads and quartets were found to contain micronuclei. However they reported no apparent relationship between

---

Fig. 6(a). Cortical cell of the abscission layer of tobacco flower pedicel, treated for 2 h with ethylene. Rough endoplasmic rectulum (RER), cell wall (CW) and nucleus (N) ($\times 35,000$). (b) Cortical cell 5 h after ethylene treatment. RER accumulation has occurred in cells where walls have separated. Golgi vesicles (GV), mitochondria (M) and vacuoles (V) can be seen ($\times 48,500$). (c) Electron micrograph of part of a cortical cell following 5-hour treatment with ethylene. Where walls are in an advanced stage of breakdown the cytoplasm and microbody (MI) matrices are less electron dense. Mitochondria (M) and chloroplast (CP) ($\times 48,500$). (d) Microbody in a cortical cell with no ethylene treatment and no abscission taking place ($\times 48,500$). (e) Wall of cortical cell after 5 h of ethylene treatment showing electron dense bodies and fibrous material (F) of the disintegrating wall ($\times 50,000$).

the frequency of chromosomal aberrations and agronomic responses to atrazine treatment.

In a time-of-treatment study using 2,4-D on curly dock (*Rumex crispus* L.) Maun and Cavers (1969) found that the treatment at 12 days before anthesis prevented the development of viable seeds. If treatment occurred at anthesis only 2% of the seeds had minute embryos and none were viable. At seven days after anthesis 91% of the seeds had embryos but only 5–15% germinated. By 34 days after anthesis spraying with 2,4-D had no effect on number of seeds or viability. Dicamba produced a response in wheat and barley similar to 2,4-D and "grossly disturbed" the arrangement of the florets and cell development in treated plants (Friesen *et al.* 1964).

A detailed study of ovule development following 2,4-D treatment of tradescantia (*Tradescantia paludosa* Anders and Woodson) was made by Swanson *et al.* (1949). These authors found a marked contrast in the effect of 2,4-D on vegetative parts of the plant where cell proliferation resulted. The effect on the reproductive organs was inhibition and disintegration of parts of the ovule. Endosperm development was inhibited severely and in mature ovules the endosperm consisted of only a few nuclei. The chalazal region of the ovule rapidly disintegrated and this, plus the lack of development of the endosperm, resulted in the "collapse" of the ovule.

Unrau and Larter (1952) have reported meiotic irregularities in spikes of barley, durum wheat and common wheat treated with 2,4-D. Abnormal anaphases and metaphases of, for example, up to 35% in barley, and chromosomal aberrations such as bridges, fragmentation, asynapsis, aneuploidy, polyploidy, and chromosome "stickiness" and chain and ring formation were found in all three cereals studied. Unrau (1953) also reported that 2,4-D caused considerable abnormal chromosome behaviour in the pollen mother cells of barley, similar to effects observed following X-ray treatment. In studies with five herbicides, 2,4-D, 2,5-D, 2,4,5-T, 2,4-DES and MCPA, Sawamura (1964) found that all of these compounds caused abnormal mitoses. Such effects as chromosome bridges due to chromosome "stickiness", slowing down of chromosome movement in anaphase, occurrence of binucleate and multinucleate cells, incomplete cell wall formation and inhibition of differentiation in meristematic tissues were reported. This study included staminal hair cells and pollen grain cells of *Tradescantia*, stipular cells of *Vicia faba*, petal cells of *Allium fistulosum* and *Allium cepa* and root tip cells of *Triticum vulgare*.

The work of Jackson (1972) raises questions about the effects of 2,4,5-T on cytological changes in plant cells. Jackson found that mitotic inhibition and cytological abnormalities in the dividing cells of African blood lily (*Haemanthus katherinae*, Baker) were due to 2,3,7,8-tetrachlorodibenzo-*p*-dioxin, a contaminant in 2,4,5-T. In this study, highly purified 2,4,5-T treatment for 2 h caused little change compared with controls, while dioxin alone or in combination with 2,4,5-T brought about the formation of dicentric bridges, chromatin fusion, "with formation of multinuclei or a single large nucleus".

Hepler and Jackson (1969) had earlier used the same test materials, endosperm cells from African blood lily, to study the effect of propham on the fine structure of these plant cells. They report that the microtubules lost their parallel alignment and were found to be arranged in radial arrays. "These radial arrays are interpreted as micropoles of the spindle apparatus and are thought to be the focal points for the chromosomes as they aggregate into micronuclei," (Hepler and Jackson, 1969). Propham did not affect the structure of chromosomes, mitochondria, plastids, dictyosomes, parts of the endoplasmic reticulum or the ribosomes.

Jackson (1969) also reported that propham treatment of the endosperm cells inhibited chromosome movement and resulted in a loss of birefringence of the mitotic spindle. Melatonin partly reversed the effects of propham in these tests. While all of these studies involved the reproductive organs of entact plants, Boulware and Camper (1972) separated the protoplasts from immature tomato fruits for their tests with several herbicides. While several herbicides produced no detectable effects on the chloroplasts, paraquat caused a "segregation" of the protoplasm into discrete areas on the inner membrane surface of the protoplast. With longer treatment this membrane ruptured and the protoplast collapsed. Sprays of 2,4,5-T on apricot, at a stage when the fruit pits were hardening, brought about an increase in fruit volume which Bradley and Crane (1955) found was due to increased cell volume not to increased cell division. An average cell volume 57 times greater than the controls was reported and the difference (increase) in nuclear size was even greater. Treated fruits had many more cells of high polyploid classes compared with controls including some 64-ploid dividing cells.

Wuu and Grant (1967) have reported on their study of some 15 pesticides including the herbicides, alanap-3 (Naptalam), atrazine, Banvel-D*, Cytrol*, Embutox E*, Hyvar X*, Lorox*, monuron and simazine for their cytological effects on barley microspore mother cells. The reader is referred to this paper for specific details but all of the compounds studied induced chromosomal aberrations.

2. Other Growth Regulators

Relatively little research has been carried out on the ultrastructural effects of growth regulators on reproductive organs. An exception has been the interest in the effect of the gibberellins on the aluerone layer of grain. The effect of gibberellin on the amylolytic activity of aleurone layers in barley has been well established by a number of workers and it is logical that fine structure studies would follow the physiological investigations. Jones (1969a) reported that in barley (Hordeum vulgare L., cv. Himalaya) just 2 h after treatment with GA_3 the aleurone grains lost their spherical appearance and increased in volume

---

* Banvel-D (57% 2-methoxy-3,6-dichlorobenzoic acid and dimethylamine salts of related acids); Cytrol (38·2 oz/Imperial Gallon 3-amino-1,2,4-triazole); Embutox E (64 oz/Imperial Gallon γ-(2,4-dichlorophenoxy)-butyric acid); Hyvar X (80% 5-bromo-3-sec-butyl-6-methyluracil); Lorox (50% 3-(3,4-dichlorophenyl)-(methoxy-1-methylurea).

during the first 10 h of treatment. The rough endoplasmic reticulum also increased during this period, while the mitochondria, dictyosomes, microbodies and leucoplasts did not appear to be affected by GA₃. In a continuation of this study using longer treatment times (up to 22 h) Jones (1969*b*) found that there was additional proliferation of the endoplasmic reticulum (ER), a distention of the ER cisternae and a proliferation of vesicles from the ER and the dictyosomes. He noted a reduction in aleurone grain size and a lower number of spherosomes. Paleg and Hyde (1964) studied changes in the aleurone of barley grains incubated in GA₃. The variety of *Hordeum vulgare* L., was Naked Blanco Mariout. They found that after 42 h of treatment the walls of many cells showed marked signs of erosion; there was fusion of the aleurone grains resulting in the formation of larger vacuolar areas, and spherosomes almost completely disappeared. Eb and Nieuwdorp (1967) also studied the fine structure of barley aleurone but hormone treatment was not involved.

IV. EFFECTS OF COMPOUNDS NOT CLASSIFIED AS PLANT GROWTH REGULATORS

Brief mention will be made of a number of compounds which are not herbicides or plant growth regulators as this latter term is normally used by plant physiologists. This part of the review, while not complete, will call the reader's attention to reports of ultrastructural effects of a variety of compounds many of present or potential importance in agriculture such as fungicides, insecticides and bacteriocides. A number of antibiotics have been studied for their cytological effects on plant species. Fuller (1947) examined the effect of sulphanilamide and other sulfa compounds on onion roots and found that the mitotic process was markedly lengthened. Wilson and Bowen (1951) reported a number of chromosomal aberrations in onion root tips following treatment with aureomycin, terramycin, streptomycin and chloromycetin.

Tanaka and Sato (1952) found that streptomycin caused "every kind of cytological disruption, comparable to those inducible by X-ray irradiation or the action of mutagenic chemicals" when used on mitotic cells of *Tradescantia paludosa*. Siegesmund, Rosen and Gawlik (1962) have followed the effect of streptomycin on the fine structure of *Euglena gracilis*. Proplastids were unaffected by this chemical as evidenced by the "greening" of the cells after transfer from darkness into light. However, cells "bleached" by streptomycin developed many bodies which were made up of concentric lamellae which differed from chloroplasts in several structural features.

Increasing attention is being given to fine structural effects of compounds which alter cell metabolism such as chloramphenicol (Ben-Shaul and Markus, 1969; Klein *et al.* 1971). The latter authors also studied the fine structural effects of actinomycin D and cycloheximide on lima bean. Dobel (1963) has also reported on studies on the effects of streptomycin, chloramphenicol and 2-thiouracil on chloroplast structure in *Lycopersicon esculentum*, Miller (var. Condine Red). Merz (1961) found chromatid aberrations in *Vicia faba* root tips following treatment with mitomycin-C., known to effect DNA synthesis.

Risueno *et al.* (1972) found marked cytological changes in *Allium cepa* L. root meristems caused by ethidium bromide, a compound reported to inhibit nucleic acid synthesis. Nass and Ben-Shaul (1973) have also studied ethidium bromide as it causes changes in fine structure in green and bleached mutants of *Euglena gracilis*. The pyrimidine, 2-thiouracil, caused chlorosis in hemp (*Cannabis sativa*) and affected the fine structure of chloroplasts (Heslop-Harrison, 1962). Damage to chloroplasts from the air pollutant, peroxyacetyl nitrate, were reported by Thomson *et al.* (1965), on pinto bean, *Phaseolus vulgaris* L. Black Valentime. Signol (1961*a* and 1961*b*) has carried out extensive work on the effects of 3(α-imino-ethyl)-5-methyltetronic acid on corn chloroplast fine structure.

Ehrenberg and co-workers have carried out extensive studies on mutagenic effects of ethyleneimine, on several crop plants (Ehrenberg *et al.* 1959, 1961; and Blixt *et al.* 1960). Pickett-Heaps (1967) studied the effect of colchicine on plant cells with particular reference to effects on cytoplasmic microtubules.

A study by Vaarama (1947) was one of the first to provide data on the effect of DDT on cells of plants (*Allium cepa* and *Trigonella foenum-graecum*). DDT brought about a number of changes in the chromosomes which differed somewhat between the two species studied. Scholes (1955*a*, 1955*b*) has also used *Allium cepa* root tips to study the effects of the insecticides, chlorodane, toxaphene, aldrin, dieldrin, isodrin, endrin and DDT on mitosis. Amer (1965) investigated the effect of the insecticide sevin (*N*-methyl-1-naphthylcarbamate) on roots of *Allium cepa* (var. Giza 6). Treatment for 24 h almost completely arrested mitosis, but the roots "recovered" after 48 h in water and the mitotic rate became normal. In an earlier study Scholes (1953) reported on the effects of hexachlorocyclohexane on mitosis in onion and strawberry roots. Gimenez-Martin and Lopez-Saez (1960, 1961) have reported on the effects of lindane (γ-hexachlorocyclohexane) on onion root tips. A number of chromosomal aberrations were noted. In studies with rather high concentrations of an alkylating agent, ethyl methane sulphonate, Rieger and Michaelis (1960) found that this compound induced chromosal aberrations of the chromatid type. They compared this chemical with Myleran and found the latter "50 times more active in producing chromosome aberrations". Their test material was *Vicia faba* primary root tips.

V. CONCLUSIONS

While it is difficult to draw any very specific conclusions about the effects of herbicides and other growth regulators on plant cell structure due to the diversity of the chemicals and species involved and to the wide concentration ranges and environmental conditions under which usage occurred, some general observations can be made.

Most herbicides have an inhibiting or growth-arresting effect on plants, often expressed first in slowing down of the growth of one of the apical meristems; root, stem or both. Somewhat paradoxically while these organs may be decreasing in overall growth, some cells may be temporarily enlarging. Enlargement of cells frequently is accompanied by greater vacuolization, enlargement and multiplication of the nucleus and drastic effects on the genetic apparatus of the cell.

After the effects on the cell nuclei, the most striking changes reported in the literature are those involving the chloroplasts. While meristem disruption and malformation of the nuclei typically bring irreversible and often lethal consequences to the plant, eventually disruption of the photosynthetic apparatus drastically alters the "food supply" to the remaining functioning parts of the plant. The combination of these various structural disturbances, translated into physiological and metabolic events leads to a slowing down of plant growth and vigour and a loss of competitive advantage with neighbouring plants. The events just described are thus part of the basis for selective control of weeds in crop plants. Other biologic factors such as disease organisms and insects, coupled with adverse environmental factors are often sufficient to bring on the demise of the undesirable weed following herbicide treatment.

The goal of all research on herbicides is to increase the understanding of how such compounds act and to improve the efficiency, selectivity and safety of their use in nature. The study of such data should result in more accurately designed herbicide molecules for specific purposes for the control of weeds, with minimal or no deleterious impact on the environment. Ultrastructure research in the next decade will have to be increased many-fold to include more compounds and more plant species if these goals are to be reached. For valid reasons the test plant for many studies on cells as reported in this review have been on the monocot onion. Other representative test species for both monocots and dicots will need to be used to permit a greater generalization from the results.

Several general areas and certain aspects of this research should receive more attention. First the purity of the compounds used in the studies is of paramount importance. Even a trace contaminant may invalidate the results or at least confound the interpretations. Second, more attention should be paid to the determination, or at least the estimation, of the actual concentration of the compound being studied at the cell level. Dosages applied to the surface of a plant may have little relationship to the actual concentration bringing about a response at tissues and cells remote from the site of entry. Use of concentrations both far higher or far lower than the actual levels experienced in practical usage can occur in laboratory tests. Conclusions can thus be drawn from massive changes due to excessively high dosages to which the subcellular organelles are exposed as well as from dosages so low that the herbicide might be acting more like a naturally-occurring hormone than a synthetic toxicant. The author is well aware, however, of the difficulty in accurately assaying levels of herbicides in individual or small groups of cells and of the analysis for contaminants where these occur.

A further complication in this research area is likely, in applications where more than one agricultural chemical is used simultaneously in plant treatments.

One compound could in fact reverse the ultrastructural effect of the other, could be synergistic, or result in other effects. Perhaps ultrastructural researchers investigating plants treated with growth-regulating compounds, such as herbicides, will have to develop new preparative techniques for the study of plant cells, physically and biochemically altered in these treatments. Mohr and Cocking (1968) have reported on a method for the preparation of highly vacuolated tissue which was senescent or damaged prior to electron microscopic study. Their results should be given study by others working in this area of abnormal cell development following chemical treatment. An example of another relatively new approach to the study of cell physiology is the paper of Anderson, Roels, Dreher and Schulman (1967). These authors reported on the effect of two compounds, retinol and alpha-tocopherol on the stability and structure of "artificial" lipid bilayers. The extension of such *in vitro* model systems to the study of herbicide effects on cell structure—particularly the all-important cell membranes—is worth further investigation.

The coming decade should bring a sieable increase in research in this area of herbicide science. The integration of these findings with the physiological and biological changes observed should greatly enhance our understanding of herbicide activity and extend their application.

Research of the author has been supported in part by the Minnesota Agricultural Experiment Station. The support of the Public Health Service under Research Service Grant No. UI 00110–07, National Center for Urban and Industrial Health is acknowledged. Misc. Paper Series No. 1513. Review of the manuscript by Dr. Mark Brenner, University of Minnesota, is gratefully acknowledged.

Reproduction of figures by permission as follows:
Figure 1a, b, c: *Weed Science,* **20,** 220–225 (1972), C. T. Chang and D. Smith.
Figure 2a, b: D. L. Wyse, W. F. Meggitt and D. Penner (unpublished photographs).
Figure 3a, b: *Botanical Gazette,* **131,** 14–23 (1970), C. J. Arntzen, A. J. Linck and
  W. P. Cunningham (copyright 1970 by the University of Chicago Press).
Figure 4a, b: *Journal Cell Science* (1974, **15,** 429–441), F. D. Hess and D. E. Bayer.
Figure 5a, b, c: *Weed Science,* **20,** 478–481 (1972), J. A. White and D. D. Hemphill.
Figure 6a, b, c, d, e: *Planta (Berl.)* **102,** 324–333 (1972), J. G. Valdovinos, T. E.
  Jensen and L. M. Sicko.

REFERENCES

Amer, S. (1965). *Cytologia,* **30,** 175–181.
Anderson, O. R., Roels, O. A., Dreher, K. D. and Schulman, J. H. (1967). *J. Ultrastruct. Res.* **19,** 600–610.
Anderson, J. L. and Schaelling, J. P. (1970). *Weed Sci.* **18,** 455–459.
Apelbaum, A., Fisher, J. B. and Burg, S. P. (1972). *Am. J. Bot.* **59,** 697–705.
Arntzen, C. J., Linck, A. J. and Cunningham, W. P. (1970). *Bot. Gaz.* **131,** 14–23.
Ashton, F. M., Bisalputra, T. and Risley, E. B. (1966). *Am. J. Bot.* **53,** 217–219.
Ashton, F. M. and Crafts, A. S. (1973). "Mode of Action of Herbicides", John Wiley & Sons, New York.
Ashton, F. M., Cutter, E. G. and Huffstutter, D. (1969). *Weed Res.* **9,** 198–204.
Ashton, F. M., Gifford, Jr., E. M. and Bisalputra, T. (1963a). *Bot. Gaz.* **124,** 329–335.

Ashton, F. M., Gifford, Jr., E. M. and Bisalputra, T. (1963b). *Bot. Gaz.* **124,** 336–343.
Audus, L. J. (1964). "The Physiology and Biochemistry of Herbicides" (L. J. Audus, ed.). Academic Press, London and New York.
Bachelard, E. P. and Ayling, R. D. (1971). *Weed Res.* **11,** 31–36.
Banting, J. D. (1970). *Weed Sci.* **18,** 80–84.
Bartels, P. G. (1965). *Pl. Cell Physiol.* **6,** 361–364.
Bartels, P. G., Matsuda, K., Siegel, A. and Weier, T. E. (1967). *Pl. Physiol., Lancaster,* **42,** 736–741.
Bartels, P. G. and Pegelow, Jr., E. J. (1968). *J. Cell Biol.* **37,** C1–C6.
Bartels, P. G. and Weier, T. E. (1969). *Am. J. Bot.* **56,** 1–7.
Baur, J. R., Bovey, R. W., Baur, P. S. and El-Seify, Z. (1969). *Weed Res.* **9,** 81–85.
Baxter, R. and Hanson, J. B. (1968). *Planta,* **82,** 246–260.
Bayer, D. E., Foy, C. L., Mallory, T. E. and Cutter, E. G. (1967). *Am. J. Bot.* **54,** 945–952.
Beal, J. M. (1944). *Bot. Gaz.* **105,** 471–474.
Ben-Shaul, Y. and Markus, Y. (1969). *J. Cell Sci.* **4,** 627–644.
Bingham, S. W. (1967). *Weed Sci.* **16,** 449–454.
Blixt, S., Ehrenberg, L. and Gelin, O. (1960). *Agric. hort. Genet.* **18,** 109–123.
Borger, G. A. and Kozlowski, T. T. (1972). *Weed Res.* **12,** 190–194.
Boulware, M. A. and Camper, N. D. (1972). *Physiologia. Pl.* **26,** 313–317.
Bostrack, J. M. and Struckmeyer, B. E. (1964). *Am. J. Bot.* **51,** 611–617.
Bradley, M. V. and Crane, J. C. (1955). *Am. J. Bot.* **42,** 273–281.
Bradley, M. V., Crane, J. C. and Marei, N. (1968). *Bot. Gaz.* **129,** 231–238.
Brener, W. D. and Beckman, C. H. (1968). *Phytopathology,* **58,** 555–561.
Callahan, L. M. and Engel, R. E. (1965). *Weeds,* **13,** 336–338.
Canvin, D. T. and Friesen, G. (1959). *Weeds,* **7,** 153–156.
Carey, M. A. and McDonough, E. S. (1943). *J. Hered.* **34,** 238–240.
Carlson, J. (1954). *Proc. Iowa Acad. Sci.* **29,** 105–128.
Carlton, Wm. (1943). *Bot. Gaz.* **105,** 268–281.
Chang, I. K. and Foy, C. L. (1971). *Weed Sci.* **19,** 54–58.
Chang, C. T. and Smith, D. (1972). *Weed Sci.* **20,** 220–225.
Chen Wu, C. and Kozlowski, T. T. (1972). *Weed Res.* **12,** 229–233.
Clutter, M. E. (1960). *Science, N.Y.* **132,** 548–549.
Croker, B. H. (1953). *Bot. Gaz.* **114,** 274–283.
Cronshaw, J. and Morey, P. R. (1968). *Protoplasma,* **65,** 379–391.
Curtis, R. (1961). *Pl. Physiol., Lancaster,* **36,** 37–43.
Cutter, E. G., Ashton, F. M. and Huffstutter, D. (1968). *Weed Res.* **8,** 346–352.
D'Amato, F. (1948). *Caryologia,* **1,** 49–78.
D'Amato, F. and Avanzi, M. G. (1948). *Nuovo G. bot. ital.* N.S., **55,** 161–213.
Davis, E. L. and Holmes, J. J. (1962). *Phyton, B. Aires,* **19,** 31–34.
Delisle, A. C. (1938). *Am. J. Bot.* **25,** 420–430.
Dermen, J. (1941). *J. Hered.* **32,** 133–138.
Dhillon, N. S. and Anderson, J. L. (1972). *Weed Res.* **12,** 182–189.
Digby, J. and Wareing, P. F. (1966). *Ann. Bot.* **30,** 539–548.
Dobel, P. (1963). *Biol. Zbl.* **82,** 275–295.
Doxey, D. (1949). *Ann. Bot.* **13,** 329–335.
Doxey, D. and Rhodes, A. (1949). *Ann. Bot.* **13,** 105–111.
Dubrovin, K. P. (1959). *Proc. N. cent. Weed Control Conf.* **16,** 15.
Eames, A. J. (1949a). *Am. J. Bot.* **36,** 571–584.
Eames, A. J. (1949b). *Science, N.Y.* **110,** 235–236.
Eames, A. J. (1950). *Am. J. Bot.* **37,** 840–847.
Eb, A. A. van der, and Nieuwdorp, P. J. (1967). *Acta bot. neerl.* **15,** 690–699.
Ehrenberg, L., Gustafson, A. and Lundquist, U. (1959). *Hereditas,* **45,** 351–368.
Ehrenberg, L., Gustafson, A. and Lundquist, U. (1961). *Hereditas,* **47,** 243–282.
Ennis, W. B. (1948a). *Bot. Gaz.* **109,** 473–493.
Ennis, W. B. (1948b). *Am. J. Bot.* **35,** 15–21.
Felber, I. M. (1948). *Am. J. Bot.* **35,** 555–558.

Fisher, D. A., Bayer, D. E. and Weier, T. E. (1968). *Bot. Gaz.* **129,** 67–70.
Fosket, D. E. and Roberts, L. W. (1964). *Am. J. Bot.* **51,** 19–25.
Friesen, G. (1929). *Planta,* **8,** 666–679.
Friesen, G. and Olsen, P. J. (1953). *Can. J. agric. Sci.* **33,** 315–329.
Friesen, H. A., Baenziger, H. and Keys, C. H. (1964). *Can. J. Pl. Sci.* **44,** 288–294.
Fuller, T. C. (1947). *Bot. Gaz.* **109,** 177–183.
Gentner, W. A. and Burk, L. G. (1968). *Weed Sci.* **16,** 259–260.
Geronimo, J. and Herr, J. W. (1970). *Weed Sci.* **18,** 48–53.
Gimenez-Martin, G. and Lopez-Saez, J. F. (1960). *Phyton, B. Aires,* **14,** 61–78.
Gimenez-Martin, G. and Lopez-Saez, J. F. (1961). *Phyton, B. Aires,* **16,** 45–55.
Greulach, V. A. and Haesloop, J. G. (1954). *Am. J. Bot.* **41,** 44–50.
Guillot-Salomon, T. (1966). *C.r. hebd. Séanc. Acad. Sci. Paris,* Ser. D. **262,** 2510–2513.
Hacskaylo, J. and Amato, V. A. (1968). *Weed Sci.* **16,** 513–515.
Hallam, N. D. (1970). *J. exp. Bot.* **21,** 1031–1038.
Hamner, C. L. (1941). *Bot. Gaz.* **103,** 374–385.
Henry, E. W. and Jensen, T. E. (1973). *J. Cell Sci.* **13,** 591–601.
Hepler, P. K. and Jackson, W. T. (1969). *J. Cell Sci.* **5,** 727–743.
Heslop-Harrison, J. (1962). *Planta,* **58,** 237–256.
Hess, F. D. and Bayer, D. E. (1974). *J. Cell Sci.* **15,** 429–441.
Hill, E. R., Putala, E. C. and Vengris, J. (1968). *Weed Sci.* **16,** 377–380.
Ivanov, V. B. (1966). *Dokl. Acad. Nauk. SSR.* **167,** 1184–1186.
Ivens, G. W. and Blackman, G. E. (1950). *Nature, Lond.* **166,** 954–955.
Izhar, S., Bevington, J. M. and Curtis, R. W. (1969). *Pl. Cell Physiol., Tokyo,* **10,** 687–698.
Jackson, W. T. (1969). *J. Cell Sci.* **5,** 745–755.
Jackson, W. T. (1972). *J. Cell Sci.* **10,** 15–25.
Jacobs, W. P. (1952). *Am. J. Bot.* **39,** 301–309.
Jacobson, A. B. and Rogers, B. J. (1961). *Pl. Physiol., Lancaster,* **36** (suppl), xi.
Jensen, T. E. and Valdovinos, J. G. (1967). *Planta,* **77,** 298–318.
Jensen, T. E. and Valdovinos, J. G. (1968). *Planta,* **83,** 303–313.
Jones, R. L. (1969a). *Planta,* **87,** 119–133.
Jones, R. L. (1969b). *Planta,* **88,** 73–86.
Key, J. L., Lin, C. Y., Gifford, Jr., E. M. and Dengler, R. (1966). *Bot. Gaz.* **127,** 87–94.
Kihlman, B. A. (1966). "Actions of Chemicals on Dividing Cells", Prentice-Hall, Inc., New Jersey.
Klein, S., Barenholz, H. and Budnik, A. (1971). *Pl. Cell Physiol., Tokyo,* **12,** 41–60.
Krause, B. F. (1971). *Am. J. Bot.* **58,** 148–159.
Kreps, L. B. and Alley, H. P. (1967). *Weeds,* **15,** 56–59.
Kulaeva, O. N. (1967). *Usp. sovreen Biol.* **63,** 28–53.
Kursanov, O., Kulaeva, O. N., Sveshnikova, I. N., Popova, E. A., Bolyakina, Yu P., Klyachko, N. L. and Vorobeva, I. P. (1964). *Soviet Pl. Physiol.* **11,** 710–719.
Kust, C. A. and Struckmeyer, B. E. (1971). *Weed Sci.* **19,** 147–162.
Liang, G. H. L., Feltner, K. C., Liang, Y. T. S. and Morrill, J. L. (1967). *Crop Sci.* **7,** 245–248.
Lignowski, E. M. and Scott, E. G. (1971). *Pl. Cell Physiol.* **12,** 701–708.
Lignowski, E. M. and Scott, E. G. (1972). *Weed Sci.* **20,** 267–270.
Martin, S. C., Shellhorn, S. J. and Hull, H. M. (1970). *Weed Sci.* **18,** 389–392.
Maun, M. A. and Cavers, P. B. (1969). *Weed Sci.* **17,** 533–536.
McIlrath, W. J. (1950). *Am. J. Bot.* **37,** 816–819.
Merz, T. (1961). *Science, N.Y.* **133,** 329–330.
Meyer, R. E. (1970). *Weed Sci.* **18,** 525–531.
Mohr, W. P. and Cocking, E. C. (1968). *J. Ultrastruct. Res.* **21,** 171–181.
Morey, P. R. and Cronshaw, J. (1968). *Protoplasma,* **65,** 287–313.
Morrison, J. E. (1962). *Can. J. Pl. Sci.* **42,** 78–81.
Muhling, G. N., Van't Hof, J., Wilson, G. B. and Grigsby, B. H. (1960). *Weeds,* **8,** 173–181.
Mukohata, Y., Mitsudo, M., Nakae, T. and Myojo, K. (1971). *Pl. Cell Physiol.* **12,** 859–868.
Mullison, W. R. (1940). *Bot. Gaz.* **102,** 373–381.
Nass, M. M. K. and Ben-Shaul, Y. (1973). *J. Cell Sci.* **13,** 567–590.

Norton, J. A., Walter, J. P. Jr. and Storey, J. B. (1970). *Weed Sci.* **18**, 520–522.
Nygren, A. (1949). *K. LantbrHögsk. Annlr.* **16**, 723–728.
Paget, G. E. and Walpole, A. L. (1958). *Nature, Lond.* **182**, 1320–1321.
Paleg, L. and Hyde, B. (1964). *Pl. Physiol., Lancaster,* **39**, 673–680.
Pate, D. A., Funderburk, Jr., H. H., Lawrence, J. M. and Davis, D. E. (1965). *Weeds,* **13**, 208–210.
Penner, D. and Early, R. W. (1972*a*). *Weed Sci.* **20**, 364–366.
Penner, D. and Early, R. W. (1972*b*). *Weed Sci.* **20**, 367–370.
Peterson, R. L. and Smith, L. W. (1971). *Weed Res.* **11**, 84–87.
Pfeiffer, N. E. (1937). *Contrib. Boyce Thompson Inst. Pl. Res.* **8**, 493–506.
Pickett-Heaps, J. D. (1967). *Devl. Biol.* **15**, 206–236.
Powell, R. D. and Griffith, M. M. (1960). *Pl. Physiol., Lancaster,* **35**, 273–275.
Rehm, S. (1952). *Nature, Lond.* **170**, 38–39.
Rieger, R. and Michaelis, A. (1960). *Kulturpflanze,* **8**, 230–243.
Risueno, M. C., Fernandez-Gomez, M. E., De La Torre, C. and Gimenez-Martin, G. (1972). *J. Ultrastruct. Res.* **39**, 163–172.
Roberts, L. W. and Baba, S. (1968). *Pl. Cell Physiol.* **9**, 353–360.
Roberts, L. W. and Fosket, D. E. (1966). *New Phytol.* **65**, 5–8.
Robnett, W. E. and Morey, P. R. (1973). *Am. J. Bot.* **60**, 745–754.
Rodebush, J. E. and Anderson, J. L. (1970). *Weed Sci.* **18**, 443–446.
Rogers, B. J. (1957). *Pl. Physiol., Lancaster,* **32** (Suppl.), vi–vii.
Rogerson, A. B., Bingham, S. W., Foy, C. L. and Sterrett, J. P. (1972). *Weed Sci.* **20**, 445–449.
Rojas-Garciduenas, M. and Kommedahl, T. (1958). *Weeds,* **6**, 49–51.
Ryland, A. G. (1948). *J. Elisha Mitchell scient. Soc.* **64**, 117–125.
Sachs, R. M. and Lang, A. (1961). In "Plant Growth Regulation", Iowa State University Press, Ames, Iowa.
Sawamura, S. (1964). *Cytologia,* **29**, 86–102.
Scholes, M. E. (1953). *J. hort. Sci.* **28**, 49–68.
Scholes, M. E. (1955*a*). *J. hort. Sci.* **30**, 12–24.
Scholes, M. E. (1955*b*). *J. hort. Sci.* **30**, 181–187.
Schultz, D. P., Funderburk, Jr., H. H. and Negi, N. S. (1968). *Pl. Physiol., Lancaster,* **43**, 265–273.
Schweizer, E. E. (1970). *Weed Sci.* **18**, 131–134.
Scifres, C. J. and McCarty, M. K. (1968). *Weed Sci.* **16**, 347–349.
Scott, M. A. and Struckmeyer, B. E. (1955). *Bot. Gaz.* **117**, 37–45.
Shaw, M. and Manocha, M. S. (1965). *Can. J. Bot.* **43**, 747–756.
Shaybany, B. and Anderson, J. L. (1972). *Weed Res.* **12**, 164–168.
Shiue, C. J. and Hansen, H. C. (1958). *Hormolog.* **2**, 9–10.
Short, K. C., Fosket, D. E. and Davey, M. R. (1973). *Abstr. 8th int. Conf. Pl. Growth Substances, Japan,* p. 151.
Siegesmund, A., Rosen, G. and Gawlik, S. R. (1962). *Am. J. Bot.* **49**, 137–145.
Signol, M. (1961*a*). *C.r. hebd. Séanc. Acad. Sci. Paris,* **252**, 1645–1646.
Signol, M. (1961*b*). *C.r. hebd. Séanc. Acad. Sci. Paris,* **252**, 1993–1995.
Singh, B., Campbell, W. F. and Salunkhe, D. K. (1972). *Am. J. Bot.* **59**, 569–572.
Smalley, E. B. (1962). *Phytopathology,* **52**, 1090–1091.
Snow, R. (1935). *New Phytol.* **34**, 347–360.
Sorokin, H. P., Mathur, S. N. and Thimann, K. V. (1962). *Am. J. Bot.* **49**, 444–454.
Srivastava, B. I. S. and Arglebe, C. (1968). *Physiologia Pl.* **21**, 851–857.
Storey, W. B. and Mann, J. D. (1967). *Stain Technol.* **42**, 15–18.
Struckmeyer, B. E., Hildebrandt, A. C. and Riker, A. (1949). *Am. J. Bot.* **36**, 491–495.
Sun, C. N. (1956). *Science, N.Y.* **123**, 1129–1130.
Svensson, S. (1971). *Physiologia, Pl.* **24**, 446–470.
Svensson, S. (1972). *Physiologia, Pl.* **26**, 115–135.
Sveshnikova, I. N., Kulaeva, O. N. and Bolyakina, Y. K. (1966). *Soviet Pl. Physiol.* **13**, 681–686.
Swanson, C. P. (1946). *Bot. Gaz.* **107**, 522–531.

Swanson, C. P., LaVelle, G. A. and Goodgal, S. H. (1949). *Am. J. Bot.* **36,** 170–175.
Tanaka, N. and Sato, S. (1952). *Cytologia,* **17,** 124–133.
Thomson, W. W., Dugger, Jr., W. M. and Palmer, R. L. (1965). *Bot. Gaz.* **126,** 66–72.
Torrey, J. G. (1957). *Am. J. Bot.* **44,** 859–870.
Torrey, J. G. and Fosket, D. E. (1970). *Am. J. Bot.* **57,** 1072–1080.
Unrau, J. (1953). *Rep. Can. Seed Grow. Ass.* 1952–1953, pp. 25–28.
Unrau, J. and Larter, E. N. (1952). *Can. J. Bot.* **30,** 22–27.
Vaarama, A. (1947). *Hereditas,* **33,** 191–219.
Valdovinos, J. G. and Jensen, T. E. (1968). *Planta,* **83,** 295–302.
Valdovinos, J. G. and Jensen, T. E. (1973). *Abstr. 8th int. Conf. Pl. Growth Substances, Japan,* p. 169.
Valdovinos, J. G., Jensen, T. E. and Sicko, L. M. (1971). *Pl. Physiol., Lancaster,* **47,** 162–163.
Valdovinos, J. G., Jensen, T. E. and Sicko, L. M. (1972). *Planta,* **102,** 324–333.
Wardlaw, C. W. (1953). *New Phytol.* **52,** 210–217.
Wareing, P. F. (1958). *Nature, Lond.* **181,** 1744–1745.
Watson, D. P. (1948). *Am. J. Bot.* **35,** 543–555.
Watson, D. P. (1952). *Bull. Torrey bot. Club* **79,** 235–241.
Webster, B. D. (1973). *Am. J. Bot.* **60,** 436–447.
White, J. A. and Hemphill, D. D. (1972). *Weed Sci.* **20,** 478–481.
Whiting, A. G. and Murray, M. A. (1946). *Bot. Gaz.* **107,** 312–331.
Wilde, M. H. (1951). *Am. J. Bot.* **38,** 79–91.
Wilson, G. B. and Bowen, C. C. (1951). *J. Hered.* **42,** 251–255.
Wolf, F. T. (1960). *Nature, Lond.* **188,** 164–165.
Wuu, K. D. and Grant, W. F. (1967). *Cytologia* **32,** 31–41.
Wyse, D. L., Meggitt, W. F. and Penner, D. (1974). Abstr. *Weed Soc. Am.* p. 77.
Yoshida, Y. (1970). *Pl. Cell Physiol., Tokyo,* **11,** 435–454.
Zimmerman, P. W. and Hitchcock, A. E. (1935). *Contrib, Boyce Thompson Inst. Pl. Res.* **7,** 439–445.

CHAPTER 4

MORPHOGENETIC RESPONSES OF PLANTS

O. M. VAN ANDEL

Botanical Laboratory, Utrecht, The Netherlands

W. VAN DER ZWEEP

*Institute of Biological and Chemical Research on Field Crops
and Herbage, Wageningen, The Netherlands*

CHR. J. GORTER

Englaan 22, Wageningen, The Netherlands

I. INTRODUCTION

The appearance of a plant depends on the shape, size and physical characteristics of its organs, and on the way they are orientated in relation to each other. Growth and development of a plant follows a certain pattern, which is genetically determined and thus characteristic for a species, but which may be modified to a certain extent by external factors.

The growth pattern is determined by the activities of the meristematic tissues, as will be described in Section II, A. Many herbicides may affect these tissues, thus causing abnormalities in the development and subsequently in the appearance of the plant. Within one plant, different organs, different tissues in the same organ or even different cells in the same tissue do not necessarily react in the same way to the herbicide treatment. This may be due to "innate" differences in susceptibility or to differences in the stage of development of the various organs, tissues or cells. For instance, a root may be killed by a concentration of an auxin herbicide to which the shoot hardly responds at all; application of a herbicide to an expanding leaf will not affect its shape, but application at an early stage of development may even change a compound leaf to a simple one as is shown in Fig. 11 for leaves of *Erodium cicutarium* sprayed with 2,4-D. Moreover local differences in dosage of the compound may involve differences in response of the various parts of a plant. Practical applications involving foliar or root uptake never ensure an even distribution over the whole surface of the plant. With systemic herbicide application not only the extent of the transport, but also the pathway it follows, determine the response of those parts of the plant, like the roots, which are not hit at spraying.

The type of response depends on the species of plant concerned, on the stage of development and the organ or tissue involved, on the kind of herbicide used and on the environmental conditions. These factors also determine when the abnormalities become apparent after exposure to the herbicides. Such abnormalities may constitute a phase of the herbicidal action of the compounds and be ultimately followed by the death of the plant, or they may arise from sublethal dosages and be considered side effects. Such side effects will, under certain conditions also occur in crop plants, and they are in general considered undesirable although minor distortions may be tolerated. On the other hand, the ability of some herbicides, in particular those of the auxin type, to evoke abnormal but desirable developmental responses, has found use in agricultural and horticultural practice. Although some of these will be mentioned the reader should refer to Audus (1959) and Weaver (1972) for a detailed discussion.

Not only meristematic but also mature tissues may respond to herbicide treatment resulting in the production of typical symptoms. The response of the tissue itself may be more or less typical, consisting of chlorosis or loss of colour, as in the leaves of amitrole-treated plants, necrosis, etc. The characteristics of the symptoms may also be due largely to a pattern resulting from differences in susceptibility of the various organs or tissues, or from a particular distribution of the herbicide, such as the preferential retention in the veins or the mesophyll of leaves. Such problems will not be discussed here.

In some cases mature cells will be induced to become meristematic, a phenomenon called dedifferentiation. This may, but does not have to, result from partial death or heavily impaired physiological activity of the plant; in this case it is called regeneration. If, for instance, the growing tip of a plant is removed, one of the upper lateral buds grows out. In practical crop growing the art of pruning is founded on this phenomenon.

Even when all parts of a plant are reached and affected by the herbicide, the difference in susceptibility of the various organs may cause in some a more serious disturbance than in others. Consequently the usual correlative relationships between organs are partially or completely upset and the shape of the plant may be partially or completely changed.

Finally, not only the size and shape but also the orientation of various organs may be affected by herbicides; petioles may curve downwards and roots may grow upwards instead of downwards, etc.

In this chapter only a few examples are given to illustrate various possible visible responses to herbicides. A more extensive survey of symptoms may be found in various handbooks (Ashton and Crafts, 1973; Arlt and Feyerabend, 1973).

II. Morphogenetic Responses

A. MORPHOGENESIS IN NORMAL PLANTS

It may be useful at this point to indicate the difference between morphology and morphogenesis. The first deals with the shapes of plants and plant organs, whereas the latter is concerned with how these shapes originate. The physiological and biochemical processes involved will not be considered here but a few characteristics of the growth of plants will be given. For a more detailed discussion we refer to Chapter 2.

Growth results from cell division and cell extension. In a very young embryo, cell division may occur throughout the whole volume of the organism but in the later stages of development only certain parts of the plant (the meristems) are capable of cell division. Newly divided cells may increase in volume; this extension, which is responsible for the greater part of growth seen as an increase in size and volume of an organ, takes place usually in a zone bordering the meristem. Meristematic cells are often more or less isodiametric and thin-walled, and may possess a dense protoplasm and conspicuous nuclei. In the various meristems the divisions proceed according to a pattern of very striking regularity in space as well as in time. It is this regularity which causes each plant organ to develop its characteristic shape. Some of the cells formed at division differentiate and develop into various elements, which constitute the mature tissue; others stop dividing at a certain time but remain meristematic, i.e. capable of further division.

Esau (1965) classifies meristems according to their position in the plant organ as apical and lateral meristems. A meristem is considered an apical meristem when it is located at the tip of an organ no matter whether the main axis or a lateral organ is concerned. In other words lateral roots or axillary shoots have an apical meristem as well as the main root and shoot. The distribution of apical meristems in dicotyledonous and monocotyledonous plants is shown in two diagrams (Figs 1 and 2).

A lateral meristem occurs between differentiated or mature tissues; it is arranged parallel to the side of the organ (Esau, 1965) and causes growth in a radial direction. Thus the vascular cambium, for instance, may be considered a lateral meristem and causes thickening of the stem.

A meristematic zone situated at some distance from the apex and also inserted between two more or less differentiated tissue regions, causing longitudinal growth of the organ, is called an intercalary meristem. This is

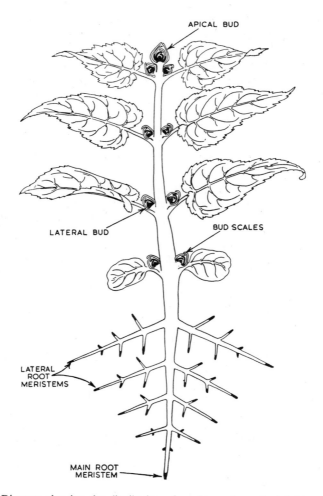

Fig. 1. Diagram showing the distribution of meristems in a dicotyledonous plant. The meristems are shown in black (Audus, 1959).

found, for instance, at the base of the internodes and of leaves of monocotyledons (Fig. 2).

An embryo has two apical meristems at opposite ends of the axis, which give rise to the root and the shoot respectively.

The *root* of the embryo of dicotyledonous plants grows out into a primary or tap root. At some distance from the apex and in internal tissue, usually the pericycle, lateral or secondary roots are initiated by the formation of protrusions, the lateral root primordia. These penetrate the more external tissue layers of the primary root by apical growth. These lateral roots may branch again and so on. In most monocotyledons the primary root of the embryo dies early and is replaced by adventitious roots, initiated on the nodes of the stem (Fig. 2). Sometimes adventitious roots are already developed in the embryo, for instance in the grasses where they are called seminal roots. These may also die when later adventitious roots develop. When the internodes elongate so that the adventitious roots on the various nodes appear as separate whorls they are called crown roots. Adventitious roots branch in a way similar to primary roots.

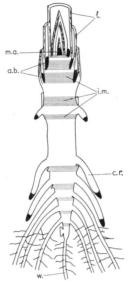

Fig. 2. Meristems in a monocotyledonous plant (schematically after Sachs). l, leaves; ma, main apical growing point; ab, axillary buds; im, intercalary meristems; cr, crown roots; w, main root. Apical meristems are black. Intercalary meristems are shaded.

Near the tip of either primary or adventitious roots, behind the zone of cell division and of cell extension, root hairs may be formed as long, hair-like pro-tuberances of epidermal cells. They are usually short-lived, so that only a zone a few centimetres long near the root tip is provided with root hairs.

The *shoot* in the embryo consists of the cotyledon(s) and the first bud or plumule which consists of a stem with very short internodes and one or more leaf primordia. The bases of the leaf primordia develop into more or less well defined nodes. Although initiated at the apical meristem, further stem growth depends mainly on the activity of a subapical meristematic region (Sachs, 1965). In the flower stalks of monocotyledons and some dicotyledons this is situated at the base of the internode and has the nature of an intercalary meristem.

Internode length may vary considerably, depending on the stage of development of the plant and the species, and as a result the successive leaves may be clearly separated, on a well developed stem, or so close together and to the soil that a rosette is formed.

In dicotyledons lateral protuberances are formed at the shoot apex at regular intervals: they are the leaf initials or primordia. At first, leaf growth occurs mainly at the apex, later throughout the whole leaf. The fundamental structure of most leaves is often determined when the leaf is still minute, so that the small leaves in the buds already resemble fully grown leaves into which they develop later mainly by rapid cell extension. At that time cell division contributes little to leaf growth although in some tissues differentiation may still be continuing. In the course of the phase of cell division the leaf becomes dorsiventral, viz. the lower and upper sides develop different structures (Eames and MacDaniels, 1947; Esau, 1965).

Leaves consist of three parts: the blade, the petiole or stalk, and the stipules. In many plants the latter are absent, in others they abscise at a very early stage, and in others again, e.g. in pea, they persist for the whole life of the plant. The petiole may vary in length or be practically lacking. It is the shape of the leaf which is particularly characteristic of the species. In general all leaves of any one plant have the same shape; this, however, may change with the position of the leaf. Thus older leaves of *Solanum nigrum* are simpler than younger ones (Fig. 3). The opposite may also occur. The arrangement of the veins contributes much to the appearance of the leaves. In dicotyledons they show a reticulate pattern, the larger veins branching into and being interconnected by a larger number of smaller veinlets. The ground tissue between the epidermis at the leaf surfaces, and the veins, consists of parenchymatous mesophyll cells. Two mesophyll tissue types can often be distinguished, i.e. palisade parenchyma and spongy parenchyma. Palisade parenchyma cells are cylindrical in shape, with their long axis perpendicular to the leaf surface. Spongy parenchyma cells may have various shapes; typical of this tissue is the air-filled intercellular-space system.

Fig. 3. Polymorphy in leaves of *Solanum nigrum*. Leaves are arranged in order of decreasing age from left to right (Kiermayer, 1961).

The regularity of the production of leaf initials on the shoot apex results in a characteristic insertion of leaves along the stem, a phenomenon known as phyllotaxis. The phyllotaxis is important for the appearance of a plant, and typical for the species.

Leaf development of monocotyledons differs in some aspects from that of dicotyledonous plants. The primordia are again produced by the apical meristems. Leaf growth, however, results from the activity of a meristem at the base instead of at the tip of the organ, i.e. an "intercalary" meristem. Petioles and stipules are lacking, but the base of the leaf often consists of a sheath, which surrounds the younger leaves or the stem and the apex. The leaves are strap-shaped, and parallel-veined. As in dicotyledons, buds develop in the leaf axils.

At a certain time, and often under certain conditions, the apical meristems begin to produce flower primordia instead of leaf primordia. This is accompanied by a change in the shape of the apex. The production of flowers consumes the whole apex and the stem ends with a flower or inflorescence. In addition to or instead of the terminal apex the apices of lateral buds may produce flowers. In the flowering apex the parts of the flower, viz. sepals, petals, stamens and carpels arise in this sequence but in quick succession. The processes in the flowering apex are much more complex than in the vegetative apex, where the organs are produced in a more steady sequence and at a lower rate.

The final stage in plant development is seed and fruit formation following fertilization. The normal sequence is pollination, germination of the pollen grain, growth of the pollen tube down into the style, release and fusion of the pollen tube nuclei with the nuclei in the embryo sac, and finally growth of the embryo. Fruit growth starts only after embryo growth has begun, the embryo providing the necessary hormones in some cases.

B. CORRELATION

The various parts of a plant influence each other in their development. The most spectacular example is the influence exerted by the apical bud on subjacent buds. The latter is called *apical dominance*: as long as the apical bud is present within a certain distance and in an active stage of growth the subjacent axillary buds remain dormant. As soon as the apical bud is removed or becomes physiologically less active, the subjacent buds begin to grow. Similar control mechanisms exist mutually between other parts, e.g. between all buds, and between shoot and root. Such phenomena of mutual growth regulation are called correlations.

Correlative growth phenomena are ascribed both to competition for nutrients between the organs and to effects of plant hormones. In this context the latter aspect is of much importance. The concept "plant hormone" is to a certain extent similar to that of animal hormones: a hormone is defined as an organic chemical substance, produced at a certain place in the organism and exerting its physiological and morphogenetic effects elsewhere. Hormones regulating growth are specified as growth hormones, but synthetic chemicals, which, after external application, induce effects similar to those hormones, should be called growth substances or growth regulators. Those herbicides, whose action is similar to that of the plant growth hormone auxin, are often referred to as auxin herbicides.

At present four native compounds or groups of compounds are considered to be of particular importance in plant growth regulation: the "auxin", indol-3yl-acetic acid (IAA) and related compounds, the gibberellins, the cytokinins, and abscisic acid. Research on tissue cultures of mostly undifferentiated callus, has contributed much to an understanding of the role of such compounds. Besides affecting growth *sensu strictu*—increase in cell size and/or number—the hormones determine to a considerable extent the organization and development of the plants. Here it is often not the presence of a single hormone, but the balance between the various types of hormones and their mutual synergistic or antagonistic effects which appear to determine the pattern of development (Steward and Krikorian, 1971; Phillips, 1969). Thus, a callus may develop roots when the ratio of the concentrations of IAA/cytokinin is high; shoots may develop when it is low (Steward, 1968; Steward and Krikorian, 1971). Abscisic acid may induce root formation in cuttings by antagonizing cytokinin action. IAA, gibberellins and cytokinins may all interact in apical dominance (Phillips, 1969). Abscisic acid is assumed to be involved in dormancy of buds and seeds, its action being antagonized by cytokinins in some processes (Addicott and Lyon, 1969) (see also Chapters 5 and 6).

In general, morphogenetic effects are induced in particular by herbicides with growth substance activity or those affecting cell division. Phenoxyalkyl-carboxylic acid derivatives and benzoic acid derivatives (Audus, 1972; Wain and Fawcett, 1969) belong to the former group, as well as picloram, which, although having a different chemical structure, has been shown to behave like indol-3yl-acetic acid in various bioassays (Eisinger and Morré, 1971; Goodin and Fletcher, 1967; Kefford and Caso, 1966). Trifluralin, carbamates (Steward and Krikorian, 1971) and maleic hydrazide (Noodén, 1969) are compounds inhibiting cell division. From the above it will be clear that morphogenetic responses, in which regeneration phenomena are involved, may be induced in various ways. First, as already mentioned, death of or heavy injury to a more susceptible part of the plant may interfere with apical dominance. Second, the balance of the native growth hormones may be upset, either because a herbicide interferes with the action or metabolism of one or more of the native hormones, as has been suggested for some triazines (Ebert and Van Assche, 1969) or because it behaves as a hormone itself, as in the case of the auxin herbicides (see Chapter 14).

C. ABNORMALITIES OF ROOTS

1. *Interference with Root Growth*

a. *Primary roots.* Many herbicides inhibit the elongation of primary roots. Growth inhibition alone cannot be considered a morphogenetic response. It is, however, often accompanied by other effects.

Roots are much more sensitive to auxin herbicides than shoots. The degree of inhibition of primary root elongation is so closely related to the herbicide concentration in the environment, that this is used for a bioassay of these compounds (Linser and Kiermayer, 1957; Mitchell *et al.* 1958).

Growth inhibition by auxin herbicides resulting from inhibition of cell elongation may be accompanied by radial expansion of the cells, and the production of short thick roots, as shown for pea seedling in Fig. 4. Inhibition of root elongation with simultaneous swelling, sometimes only at the tip, have been reported for picloram (Malhotra and Hanson, 1966), trifluralin (Schultz *et al.* 1968), bromacil (Ashton *et al.* 1969) and bensulide (Cutter *et al.* 1968).

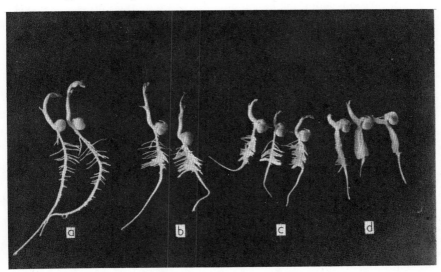

Fig. 4. Seedlings of garden pea (*Pisum sativum*) showing the effect on root growth of short exposure to auxins. (a) Ten-day old plants grown continuously in water; (b) 10-day old plants that had their main roots immersed for 24 h on the third day of germination in 1 p.p.m. 2,4-D and were subsequently grown in water; (c) as (b) but treated with 3 p.p.m. 2,4-D solution; (d) as (b) and (c), but treated with 10 p.p.m. 2,4-D solution (Audus, 1959).

Inhibition of elongation and stunting of primary roots are usually associated with the production of large numbers of lateral primordia, which may develop into lateral roots, as is often the case with auxin herbicides and has been found for amitrole (Arntzen *et al.* 1970), or they may not develop beyond the primordial stage as with trifluralin (Bayer *et al.* 1967; Mallory, 1969). Sometimes lateral root production is very prolific, and the roots burst through the external tissues of the root, disrupting it completely. These laterals may be situated so close together that they form one single fused sheet (see Fig. 4d). Several such sheets may run along the length of the roots. In other cases tumour-like structures occur, resulting from an increase in the number of lateral initials and a simultaneous inhibition of their elongation (Watson, 1950).

Such inhibition of root elongation, coupled with increased ramification results in the production of very bunched structures with impaired functions (Fig. 4). Increased root ramification may be due to a decrease in apical dominance as the result of growth inhibition of the root tip. In this case it is a correlative phenomenon. It may also be a direct effect of the herbicide, inducing cell division. Herbicides like the carbamates propham and chlorpropham, CDAA, amitrole, dichlobenil and MH also inhibit root growth but without inducing the formation of laterals.

Stunting and thickening can also occur when cell division is affected. Irregular divisions along the longitudinal axis of the organ produce irregular thickening of radish, which develop from primary roots and the adjacent hypocotyledonary tissue, as shown in Fig. 5, or other root crops, such as sugar beet, carrot, and parsnip (Way, 1963a, b, c, 1964a, b). Figures 6 and 7 demonstrate such effects in parsnip and carrot treated with MCPA and chlorpropham respectively.

Fig. 5. Hour-glass shaped radishes produced by soaking seed for 20 h in 200 p.p.m. NAA before sowing (Audus, 1959).

Contrary to the compounds mentioned above trifluralin inhibits the development of laterals without affecting growth of the primary roots of various plants (Bayer *et al.* 1967).

Stimulation of root elongation by herbicides is rare, but was observed for the roots of seedlings of maize and wheat, treated respectively with simazine or propazine (Gräser, 1967) and DMPA (Holmsen, 1969).

b. *Adventitious Roots.* Premature death of primary roots caused by herbicides, may promote the development of adventitious roots. Auxin herbicides sometimes increase the degree of branching of the adventitious roots of grasses, which usually have few lateral roots. On the other hand early initiation and further development may be strongly inhibited by those herbicides, which inhibit cell division. This happens for instance in *Echinochloa* and some rice varieties treated with chlorpropham (Baker, 1960) and *Avena fatua* and other grass weeds (Pfeiffer and Holmes, 1961) treated with barban.

Fig. 6. Effects of MCPA on roots of parsnip (left) and carrot (right) (Way, 1961).

c. *Root hairs.* Root hair elongation is stopped by high concentrations of auxin herbicides, antiauxins (e.g. TIBA) and amitrole. Instead of elongating the hairs swell at the tip, but sometimes they accumulate thick masses of cell wall material at the tip and their function is impaired (Gorter, 1949). 2,4-D, 2,4-DB and fenoprop induced formation of root hairs (as well as of lateral roots) in *Agrostis tenuis* (Callahan and Engel, 1965); bensulide does the same in oats (Cutter *et al.* 1968).

Fig. 7. Effects of chlorpropham on root development in carrots. Left, untreated plant (Orth, 1957).

2. *Nodulation*

The phenomena described above result mainly from effects of herbicides on growing cells. An exception is the induction of root primordia, but, as already mentioned, this may, at least in some cases, be an indirect effect. However the formation of root nodules as induced in legume roots by infection with *Rhizobium* species, is a matter of dedifferentiation of mature cells. Auxin is involved in this process, so an effect of auxin herbicide application is not surprising. The number of nodules on the roots of *Vicia faba* was greatly increased after foliar application of 2,4-D, but was decreased by root applications (Kumar and Dube, 1962). The different results arising from different application methods may be ascribed to differences in concentration of the herbicides in the tissues. According to Geranmayah (1964), however, the effect of 2,4-D on nodulation would be associated with a drastic change in plant growth and be considered an indirect effect. Trifluralin reduced nodulation in soybean but this effect was not considered responsible for the inhibition of plant growth (Kust and Struckmeyer, 1971).

D. ABNORMALITIES OF STEMS

Abnormalities of stems mainly result from effects on the apical, subapical, intercalary or lateral meristems, responsible for stem development. Correlative effects may produce abnormal branches; regeneration is limited to adventitious root formation.

1. *Fasciation*

The apex or growing point is normally a dome-shaped hemisphere, in which most cells are in a state of active division; it gives rise to leaf primordia in regular succession. If compounds interfering with cell division enter the apex, its shape may change and morphological irregularities, which are called fasciations, may result. A fasciated organ usually shows very regular growth, which indicates that at least in the beginning orderly development prevails (Gorter, 1965).

The most common type of fasciation is a band-shaped distortion of a normally cylindrical organ like the stem. Plants may for instance have flat, often fluted stems, arising from a ridge-shaped growing point.

Fasciated plants occasionally occur in nature. In general the cause of the phenomenon is unknown. Sometimes it is caused by infection with pathogens (Gorter, 1965), for instance *Corynebacterium fascians*. In the latter case a cytokinin may be responsible for the disturbed growth (Thimann and Sachs, 1966). Treatment with compounds with auxin activity may cause consistent changes in the apex, which are exactly the same as in naturally occurring fasciations. Instead of becoming ridge-like, as described above, the apex may become stellate with three or four arms, giving rise to a radiate or stellate fasciation. Finally it may become annular and give rise to a ring fasciation. The last two types of fasciation are extremely rare in nature, but by application of compounds like TIBA that stunt growth, ring fasciation has been induced in tomatoes (Gorter, 1951), lupin (Van Steveninck, 1956) and *Solanum nigrum* (Kiermayer, 1961). The ring-shaped apex of such plants produces organs, leaves and lateral shoots on the inside as well as on the outside of the hollow stem. In fields treated with auxin herbicides various types of fasciations occur quite frequently. Fasciation of *Avena* coleoptiles was found after treatment with trifluralin (Hendrix and Muench, 1969).

In addition to their abnormal shape the fasciated organs are characterized by a greater volume and weight. However, the additional tissues produced are otherwise normal.

Fasciation is not limited to stems. Shoots, whether above or below ground, leaves, flowers, inflorescences, fruits and roots have all been recorded as becoming fasciated. One or more organs of a plant may fasciate (White, 1948; Gorter, 1965).

Stem fasciation may adopt various forms. In rosette plants the entire stem may be fasciated over its whole length. In annuals such as pea and tobacco the stems usually retain their normal circular cross-section at the base, but they broaden gradually as the tip is approached.

The "true" fasciation arising from one changed growing point is hard to distinguish from the "connation" arising from the fusion of several growing points whose growth in a longitudinal direction has been inhibited by some herbicidal compound.

It seems that any vascular plant can become fasciated under suitable conditions and the full range of fasciations together with connations may occur in one and the same plant.

2. *Abnormalities in Stem Elongation*

Stem elongation of dicotyledons may be increased, in particular by auxin herbicides at concentrations of 10^{-4} to 10^{-8} M in case of the more active compounds like IAA and 2,4-D. Stimulated stem growth has also been reported for rice (Kaufman, 1955). Higher concentrations may cause a reduction in growth as is also found for roots. The lack of stimulation in the latter case is probably due to greater sensitivity of the roots. Various other herbicides, for instance propanil (Hofstra and Switzer, 1968), dalapon (Funderburk and Davis, 1960), atrazine and simazine (Goren and Monselise, 1965) inhibit growth by merely shortening the internodes. Similar results have been obtained with morphactins (Humphries and Pethiyagoda, 1969), which are not herbicides *sensu stricto*.

Inhibition of stem elongation may be due to inhibition of cell elongation or cell division. Often inhibition of cell elongation is accompanied by increased lateral expansion, resulting in thickening of the organ. This is very common in young seedlings, for instance in *Cupressus macrocarpa* treated with 2,4,5-T (Fig. 8). Dichlobenil when applied to *Echinochloa* sp. (Shimizu, 1967) and

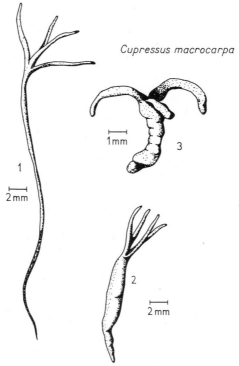

Fig. 8. Lateral swellings in stems of *Cupressus macrocarpa* caused by 2,4,5-T and IAA.(1) normal; (2) treated with 10 mg/l of 2,4,5-T; (3) treated with 500 mg/l IAA (Fromantin, 1957).

locally to bean stems (Koopman and Daams, 1960) and chlorpropham applied to maize (Jacques-Felix and Chezeau, 1962) have similar effects.

Many non-auxin herbicides inhibit intercalary meristematic growth in grasses. The stunting effect of some auxin herbicides, e.g. fenoprop, is probably also due to influences on cell division in the intercalary regions (Sen, 1960). MH affected growth mainly of the apex of grasses (Foote and Himmelman, 1971); in tobacco the terminal bud appeared more sensitive than axillary buds, so that spraying in low concentrations stimulated the growth of laterals (Peterson and Naylor, 1953).

Stimulation of the activity of the cambium may also cause swelling of the stem, for instance in bean, where after treatment with picloram the cambium remained active whereas the meristems causing stem, leaf and bud growth were inhibited (Fisher *et al.* 1968).

Auxin herbicides and DCPA caused a cylindrical swelling if the compound was uniformly distributed in the stem, otherwise a unilateral swelling or localized strip of proliferating tissue was formed (Nishimoto and Warren, 1971). Symptoms arising from enhanced cambial activity can also be observed in underground organs, as, for instance, the rhizomes of *Pteridium aquilinum* (Conway and Forrest, 1961).

3. *Gall and Tumour Formation*

As mentioned before some compounds may induce mature cells to divide (dedifferentiation). Treatment with such compounds causes callus or tumour formation in all stem tissue, resulting in disorganization of the tissue. Figure 9 shows a stem of celeriac densely covered with little warts induced by the action of auxin herbicides. Tumours occurring in the outer layers of the stem rupture the cortex and thus create an entrance for pathogens. In potatoes auxin herbicides may cause hypertrophy of epidermal or cortical tissues, especially under the lenticels. Tumours are formed and consist of large cells devoid of starch. These cells separate, and cracks and pits are formed. Wound periderm which would normally seal off the wounds does not develop at all or only incompletely, and thus the tuber can easily be infected, in particular by *Streptomyces scabies,* the main cause of potato scab. Curiously, this organism, which normally enters the tuber through the lenticels, itself causes effects similar to those of the herbicides (Bradbury and Ennis, 1953). Scab-like phenomena have also been reported as induced by dinoseb acetate, linuron and monuron (Arlt and Feyerabend, 1973). Lesions induced by 2,4-D in the bark of *Macadamia* trees resemble the tumours caused by the pathogen *Phytophthora cinnamomi* (Fukunaga, 1965).

4. *Adventitious Root Formation*

If a plant loses its whole root system the base of the stem is capable of regenerating new roots. This regeneration process is strongly promoted by auxins, and in the practice of rooting cuttings, auxins are widely used. As with tumour formation this effect is due to induction of cell division in mature cells.

Fig. 9. Effects of auxin herbicide on stem of celeriac (Zonderwijk, 1959).

Cell division starts in the endodermis, in particular in regions opposite the vascular bundles. Subsequent divisions appear in the pericycle and phloem parenchyma. In extreme cases every living cell becomes meristematic. Especially at high auxin concentrations a mass of more or less undifferentiated cells develops a tumour from the enlarged pith cells, and thus remains undifferentiated for a long time, continuing to grow and forming the well-known tumours that accompany the formation of root initials in the cutting. Adventitious root formation as a consequence of auxin herbicide sprays on stems of hempnettle with an intact root system is shown in Fig. 10. The formation of many, still invisible root initials may cause a certain brittleness of the stem (Rodgers, 1952).

Adventitious root formation in *Phaseolus vulgaris* induced by picloram caused ruptures of cortical cells and stretching of the epidermis with the development of fissures (Fisher *et al.* 1968). In mung bean adventitious root formation was promoted by atrazine and ametryne (Copping *et al.* 1972).

Fig. 10. Adventitious root formation in *Galeopsis tetrahit* (hempnettle) after treatment with 2,4,5-T and MCPA mixtures (Holz, 1957).

In monocotyledonous plants the pattern is different, because their stems and roots lack an endodermis and pericycle. Here cell divisions may start in the parenchymatous tissue outside and between the outer vascular bundles, as in species of *Lilium*. In others, e.g. *Asparagus officinalis,* first a tumour develops and in this tumour any cell may divide and be the starting point of a root initial.

Rhizomes of both dicotyledons and monocotyledons are also induced to form adventitious roots, as has, for instance, been described for *Agropyron repens* (Struckmeyer, 1951).

5. *Induction of Tuber Formation*

In some plants the action of herbicides leads to the formation of tubers. In potato tubers generally occur at the end of underground stems (stolons) only. Under the influence of auxin herbicides, MH and others, the apical dominance of the tuber is broken and more tubers develop on the same stolon. Tubers may even be found in leaf axils (Bodlaender, personal communication).

E. ABNORMALITIES OF LEAVES

1. *Effects on Primordia Development*

It has been stated in the introduction that some of the morphogenetic effects of herbicides are due to differences either in sensitivity of the various tissues, or in the stage of development when the herbicide acts, or in the local concentration in various parts of the plant. This may be illustrated by some observations of herbicide effects on leaf development.

The most severe effect of an auxin herbicide on the growing point is to stop leaf primordia formation completely. This does not mean however, that the growing point is killed, as the formation of flower primordia may continue.

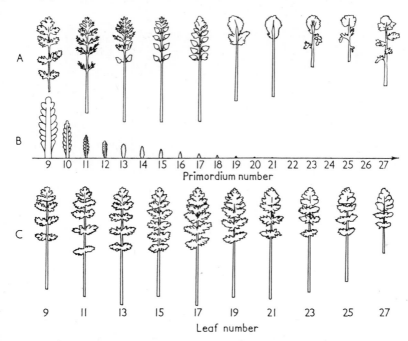

Fig. 11. Leaf malformations in *Erodium cicutarium* after spraying with 2,4-D. (A) Malformation of leaves when sprayed at the primordium stage indicated in (B); (C) the normal shape of leaves numbered 9–27 (Linser and Kirchner, 1957).

Thus application of halogen-substituted benzoic acids (TIBA) on tomatoes may result in a stem without leaves bearing an inflorescence at the tip (Gorter, 1951). Depending on the stage at which the growing point is reached by the compound and how many leaf primordia were already present, reduction in number of leaves is observed, for instance in citrus trees treated with simazine and atrazine (Goren and Monselise, 1965). Leaf shape is sometimes drastically changed: leaves of lettuce treated with MCPA at an early stage are tongue-shaped or more or less slipper-shaped (Arlt and Feyerabend, 1973).

The effect of herbicide treatment may be apparent only in leaves whose primordia were already existent at the time of treatment, as in *Phaseolus vulgaris* (Eames, 1949; Watson, 1948). As shown for cotton by Gifford (1953) and McIlrath and Ergle (1953), leaf primordia laid down after the time of treatment can also be affected. The most typical phenomena are:

a. *Fusion of Leaflets of Compound Leaves.* Every leaflet of a compound leaf may show the same abnormalities as a simple leaf. Moreover some or all of the leaflets may fuse, as is illustrated by Fig. 11, representing leaves of *Erodium cicutarium* sprayed with 2,4-D at different stages of development. When the spray was applied at a very early stage of development of the leaf apex (leaves 19–21) all leaflets fused to one simple blade; applications at a later stage produced various degrees of fusion. The degree of fusion may also depend on other factors: the effect of 2,4-D on fusion in potato leaves, for instance, was increased by the presence of Zn, but partially reversed by Cu, Fe or B (Wasnik, 1969).

b. *Funnel-shaped or Tubular Leaves.* The development of a leaf into a tubiform or cup-like organ is frequently observed. Three types of funnel- or cup-like leaves are shown schematically in Fig. 12. In Type I the margins of

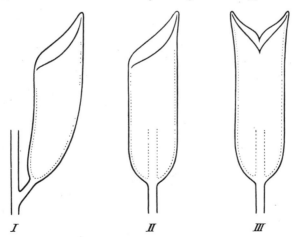

I   *II*   *III*

Fig. 12. Schematic presentation of tubular leaf formation. The dotted central portion in II and III represents the stem. (I) Ascophylly (a single leaf primordium); (II) coleophylly (a single leaf primordium); (III) gamophylly (fusion of more than one leaf primordium—two in this case). See text (Haccius and Schneider, 1958).

one leaf have fused after the leaf and its petiole had developed to a certain degree. Type II, where the stem is entirely encircled by the leaf is provoked by herbicide at an earlier stage, as is Type III, which shows fusion of more than one leaf. Cup formation was, for instance, induced in lettuce leaves by CDEC (Zink and Agamalian, 1965) and MCPA (Arlt and Feyerabend, 1973) CDEC and EPTC caused the fusion of the cotyledons of *Pinus resinosa* (Sasaki *et al.* 1968).

If the primordia of two or more organs fuse at an early stage of development, the two organs grow together and form a "connation". This phenomenon has already been referred to in the discussion on fasciation, and is also represented by Type III of the cup-like leaf mentioned in the previous section (Fig. 12). Connations are common after application of auxin herbicides and carbamates, which is understandable because of their stimulation of growth in the transverse direction together with stunting of longitudinal growth. Connation of cotyledons of *Eranthis hiemalis* seedlings induced by 2,4-D is shown in Fig. 13: the resulting structure may show all transitions from a flat to a cup-shaped organ. All such structures may be asymmetrical. The same malformations may happen in normal leaves, when the active compound reaches the growing point in the stage of active leaf initiation. Two or more primordia may fuse giving rise to a variety of leaf connations, the exact type being related to the stage of development of the growing point at the time of treatment. In general the effect is most severe when the growing point is very young. The more advanced the development of a leaf the smaller is the influence of the compound.

Another type of connation has been found in MCPA- or 2,4-D-treated carrots. In this case the petioles had fused to form a tubular stalk, which carried tufts of leaves on top (Way, 1961).

c. *Fasciation.* As mentioned before fasciation arises from a change in the shape of the growing point of an organ and is thus different from a connation, although the result may look very similar.

A change of the stem apex may affect the influence of the apical meristem on the developing leaf primordia (White, 1948; Van Steveninck, 1956). The result is often a change in the number and position of the leaves, and thus abnormal phyllotaxis (see 1d). Other phenomena are the appearance of two leaves or semi-double leaves on the same stem node and arising from one primordium as found in *Hymenocallis senegambia,* or an increase in the number of pinnae in the leaves of *Trifolium pratense* (White, 1948). Seedlings of yellow lupin, sprayed with 2,4-D showed funnel-shaped leaflets of a type similar to that described for a mutant or for plants infected with pea mosaic virus. In some cases the whole compound leaf formed one funnel (Van Steveninck, 1956). 2,4-D caused fusion of the leaflets of those leaves of *Vicia faba* plants, which were initiated after the treatment, leaves already in the primordial stage being merely smaller and narrower than normal (Geranmayah, 1964).

d. *Effect on Phyllotaxis.* Chemicals may interfere with the regularity of the production of leaf primordia and thus affect the characteristic arrangement of

Fig. 13. Abnormalities of the cotyledons of *Eranthis hiemalis* seedlings after treatment with 2,4-D. (I) Untreated; (II–XII) various connations (Haccius and Trompeter, 1960).

the leaves or phyllotaxis (Schwabe, 1971). An apparent effect on phyllotaxis can be caused by inhibition of stem elongation in a very early stage so that leaf insertions are close together and whorls of leaves result.

2. *Effects on Leaf Expansion*

a. *General Changes in Size.* It has been pointed out in Section A that the shape of the leaf (blade), at least in most of the dicotyledons, is largely determined at an early stage of development, and that growth in later phases consists mainly of cell enlargement and differentiation. Compounds inhibiting cell expansion may thus reduce the dimensions of the leaves, a rather atypical symptom, which may also be caused by insufficient supplies of water, nutrients etc. An increase in leaf thickness was found in tobacco and *Salvia* sp. treated with dichlobenil (Koopman and Daams, 1960) or cotton treated with prometryne (Foy and Bisalputra, 1964). In the former case the diameter of the mesophyll cells was increased; in the latter the number of tiers of cells increased as well. Interference with the activity of the intercalary meristem of cereals by carbamates, however, may produce corkscrew-like distortions of the leaves (Arlt and Feyerabend, 1973). A special case is "hollow nose" of hyacinths, which after treatment with propham, showed broadened and shortened leaves with thickened leaf bases, from which discoloured bulb scales were formed. The tips of these scales became necrotic and desiccated, producing a cavity in the apex of the bulb during storage (Muller and Van der Boon, 1972). In *Salvinia natans* the size of the leaves and the number of cells per leaf was reduced by DCPA. The latter caused a reduction in density of epidermal leaf hairs which affected the ability of the leaves to float (Prasad, 1965).

b. *Effects on Unrolling.* Incomplete unrolling may cause changes in leaf shape comparable with those caused by fusion (see D 1). In peanuts treated with vernolate the adaxial surfaces of the two halves of a leaf stuck together because of a lack of a wax layer (leaf sealing) (Upchurch *et al.* 1968). Certain thiocarbamates caused such tightly rolled leaves in *Echinochloa crus-galli,* that emergence of subsequent leaves was prevented (Koren, Foy and Ashton, 1968). Similar effects were caused in wheat or *Beta* by TCA and dalapon (Arlt and Feyerabend, 1973). "Onion leaf" was found in *Agropyron desertorum* treated with siduron, propham, chlorpropham and 2,3,6-T (Klomp and Hull, 1968).

Picloram induced tubular leaves in various grasses by causing the edges to roll downwards (Arnold and Santelmann, 1966). As in the case of the inrolling of the margins of young leaves of cucumber caused by 2,3,6-TBA and, to a lesser degree, by MCPA (Kingham and Fletcher, 1963), this may be considered an epinastic phenomenon (see Section III, B).

c. *Differential effects on mesophyll and vein growth.* Growth of the mesophyll is, at least to a certain extent, regulated by mechanisms different from those controlling vein growth (Engelke *et al.* 1973). It is therefore not surprising that mesophyll and vein development may be affected by herbicides to different extents, leading often to a more drastic change in shape than described in 2a.

Reduction of the amount of mesophyll by auxin herbicides such as 2,4-D leads to the formation of pointed or even strap-like leaves, as shown for cotton in Fig. 14. In extreme cases the mesophyll can be completely lacking. Picloram reduced the lamina of tomato leaflets. The mid-vein was broadened at the same time, with the principal lateral veins standing out from it at right angles (Fletcher, 1968).

When, on the other hand, vein growth is more inhibited than mesophyll growth, the mesophyll bulges out between the veins, and "crisped" leaves are formed. Various kinds of crisped leaves have been described (Watson, 1948; Gifford, 1953). Crisping or crinkling of leaves by herbicides very much resembles the effects to be observed in virus-diseased plants. It is caused by auxin herbicides as well as by TCA, dalapon, chlorpropham and diallate. The latter compound caused young plants of *Beta* to look like heads of lettuce (Arlt and Feyerabend, 1973).

Fig. 14. Effects of 2,4-D on cotton leaves. Left, untreated leaf (McIlrath, 1955).

Besides a general inhibition of vein growth an abnormal vein pattern may occur as a typical symptom of auxin herbicide damage. The bandlike spreading of the tissues of the central vein and sometimes of veins of the second or third order is also commonly combined with inhibited development of the mesophyll. For different types of abnormal vein pattern see Hanf (1957).

F. ABNORMALITIES OF FLOWERS AND INFLORESCENCES

A flower or inflorescence is to be considered as an axis with appendages, with the apical meristems producing the various parts of the flower instead of leaf primordia. As with stems and leaves the strongest responses must be expected with herbicide application at an early stage of development and effects generally similar in type will result, viz. changes in size, shape or arrangement of the various parts of the flower. The following phenomena have been reported:

a. *Reduction or Multiplication of Flower Parts.* In the most extreme case development and growth of all flower parts is inhibited. This happened for instance in Canada thistle sprayed with MCPA (Weaver and Nyland, 1965). If only one or two whorls are missing, flowers without sepals, petals, stamens or carpels may occur. The absence of one or both of the last two whorls clearly leads to sterility. Dicamba affected development of the female flower parts of *Sorghum* but had no effect on production of pollen (Peeper *et al.* 1970).

Furthermore parts of one whorl may not come to full development and this will result in a subnormal number of the part concerned. Reduction in the number of petals of cucumber is shown in Fig. 15. On the other hand a whorl may also be replicated either regularly, or irregularly, i.e. instead of, for example, a normal number of 5 petals, 10, 15, 6 or 7 may develop. Obviously, such effects are due to disturbance of cell-division patterns in the apex at the time of formation of the whorl primordia.

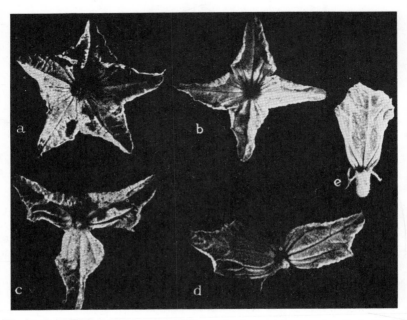

Fig. 15. Normal (a) and abnormal (b–e) flower development in cucumber. Reduction of the number of petals (Haccius and Massfeller, 1959).

b. *Fusion of Flower Parts.* Reduction of the normal number of parts of a whorl may be caused by the fusion of the primordia on the growing point. This occurs when excessive transverse growth of primordia cells is induced by high auxin concentrations. This kind of fusion of the sepals of tomato flowers results in formation of a single lobe (Audus, 1959).

c. *Changes in the Shape of Petals and Sepals.* With treatment at an advanced stage of development, broadening or narrowing of the organs may occur as well as a general reduction or increase in size. Such a variety of different shapes may be exhibited that the range cannot be described here.

d. *Changes in the Morphology of Stamens.* Stamens may be reduced in number or show changes in shape. A very common and conspicuous phenomenon is the development of a filament into a flattened petal-like structure. An intermediate structure may also be formed, i.e. half petal, half filament. On this kind of organ anthers can be present or absent. Sometimes half of the anthers fail to develop.

The development of the anther may be stopped at any stage and hence pollen may be absent or may never mature, with sterility as the consequence. 2,3-dichlorobutyric acid, inhibiting mitosis in the filaments and thus preventing seed-set, has been successfully applied to prevent spread of *Tussilago farfara* (Orth and Leuchs, 1961). Preventing pollen formation may also be useful in plants producing pollen which induces allergics in man. Spraying ragweed, *Ambrosia* species, with 2,4-D prevented the release of pollen without killing the plants, which would incur the risk of soil erosion (Grigsby, 1945). Gametocidal effects are ascribed to mendok (Eaton, 1957) and cacodylic acid (Taylorson, 1966). In plant breeding chemical induction of male sterility is sometimes also desirable (Eaton, 1957).

e. *Changes in Development of Carpels.* A considerable number of abnormalities can be observed in carpels. The many separate carpels of some species may fuse completely or partially. One or more carpels may be absent. In many-loculed fruits any number of locules may fuse. In some species the single carpel may multiply under the influence of 2,4-D, so that instead of one fruit, bunches of fruit are formed at the tip of the stem (Kiermayer, 1956). In these abnormal carpels ovules may fail to develop and a reduction in seed number is the result.

Not only was the number of seeds reduced by spraying *Rumex crispus* with 2,4-D during anthesis, but also they were smaller and the development of the embryo was inhibited (Maun and Cavers, 1969).

f. *Change of Sex of Flowers.* Most flowering plants have flowers with both female (carpels) and male (stamens) organs. The flowers of some plants, however, contain either female or male organs only, for instance cucumber and maize. In some species both kinds of flowers are found on the same plant (monecious plants), in others on different plants (dioecious plants). In plants with both female and male flowers the former are usually produced first. The transition from the vegetative to the generative stage is probably correlated with a change in plant hormone balance, but also, depending, at least partially, on the ratio between various hormones, either a male or female flower will develop. Application of auxins to the leaf axils of *Cucumis sativus* induced an increase in the number of female and a decrease in the number of male flowers (Laibach and Kribben, 1950). Similar data were obtained with pumpkin (Laibach and Kribben, 1951). In genetically male plants of hemp, female flowers developed after application of naphth-1yl-acetic acid (Heslop-Harrison, 1956). According to Stryckers (1960) the concentration of applied 2,4-D determined whether hemp plants would develop male or female flowers. After a seed treatment of maize with propham some plants showed terminal as well as lateral cobs, whereas the number of the latter per main shoot was increased (Jacques-Félix and Chezeau, 1962).

Suppression of male flowers can be of importance for breeding purposes just as the induction of male sterility mentioned above.

g. *Effects on Inflorescences.* Any flower of an inflorescence may show the

Fig. 16. Branched ears in barley after treatments with 2,4-D (Audus, 1959).

abnormalities described under a–g. Moreover, the habit of the inflorescence as a whole may be changed. One of the more common phenomena is a change in the number of flowers per inflorescence. An increase in the number of flowers is often found in umbelliferous plants like *Daucus carota* or Compositae like *Matricaria* or *Taraxacum*. Since the primordia of the individual flowers of an inflorescence originate more or less simultaneously at the growing point, it is understandable that compounds affecting cell division produce an increase or reduction in the number of flowers.

Inflorescence damage by herbicide treatment has been very extensively studied in cereals. The formative effect of MCPA on the development in barley was, for instance, analysed by studying the development of the inflorescence after treatment at various stages (Luxová and Lux, 1964). The main effects were found to be a reduction in length of the rachis internodes to various degrees, failure of initiation or development of lateral buds in part of one or both orthostiches, sometimes stimulation of elongation of lateral stems, and a disturbance of the normal arrangement of the flowers.

Fig. 17. Incomplete "heading" (1) and "bunched" ears (2) in wheat after treatment with 2,4-D (Audus, 1959).

The effect of auxin herbicides on inflorescence development of pasture grasses has been found to be basically the same as in cereals (Jeater, 1958; Sen, 1960).

Such disturbances in development may result in the following types of abnormalities.

(a) Branched ears. Instead of producing one rachis the primordium apparently splits into several primordia, as is shown for barley treated with 2,4-D (Fig. 16).

(b) Ears with opposite instead of alternating spikelets along the rachis. This abnormality is caused by a shifting of the growth initials during the development of the ear.

(c) Ears with multiple spikelets ("bunched ears") caused by multiplication of primordia in early stages of development (Fig. 17(2)).

(d) "Tweaked ears", in which part of the rachis is devoid of spikelets,

Fig. 18. "Tweaked" ears in barley after treatment with 2,4-D (Large and Dillon Weston, 1951).

caused by a temporary inhibition of the initiation of flower primordia on the apex (Fig. 18).

(e) Incomplete heading. A very conspicuous and rather frequent abnormality. The ear remains partly or completely enclosed in the leaf which may be tubular. In barley treated with 2,4-D in the 4 to 5 leaf stage, a "constricted sheath" (Derscheid, 1952) or tubular leaf (Andersen, 1952; Andersen and Hermansen, 1950) is common. The collar of the sheath is so constricted that only the awns may emerge. Elongation of the upper internode may make the ear emerge from the leaf, but if it fails to do so, incomplete heading results (Fig. 17(1)). Afterwards such temporarily enclosed ears usually remain bent ("bowed ears").

The number of spikelets per ear may be reduced or increased, giving it a peculiar appearance; moreover they may lack any number of any flower part. Complete or partial sterility of an ear often occurs (Friesen and Olsen, 1953).

Gramineae, which form panicles or tassels instead of ears also show a variety of deviations from the normal pattern under the influence of herbicide action at an early stage of inflorescence development. The number of lateral branches of a tassel, or of flowers per spikelet may be reduced (Fig. 18), or no tassel may be produced. Cob development in maize can be reduced as well and cobs of any size and with any number of kernels may be found (Staniforth, 1952; Rodgers, 1952). An increased number of cobs per main shoot was observed in maize plants, whose grain had been soaked in chlorpropham before planting. Dicamba caused delayed ripening, ear deformation and reduced seed production in wheat and barley (Friesen et al. 1964).

h. *Fasciations*. Just as in stems and leaves fasciations may occur in inflorescences and flowers. Fasciation of an inflorescence, due again to a change in shape of the apex of the main axis and its branches, may cause either an increase or decrease in the number of flower-bearing branches. In the former case a witches broom arises, as is for instance known in *Erigeron, Nicotiana, Sonchus spp* (White, 1948). A well-known example is the cockscomb, *Celosia cristata,* whose inflorescence is constituted of a broadening main axis with many very short pedicels resulting in a fan-shaped structure (Gorter, 1965). Part of the flowers of a fasciated inflorescence may be sterile.

When stem fasciation occurs flowers are often also fasciated, but sometimes they develop normally, for instance in *Pisum umbellatum* (Gorter, 1965). Fasciation of flowers consists mainly of a change in the number and arrangement of the parts of one or more of the whorls of sepals, petals, stamens and carpels. Besides deviating in number the parts may show deformities, or parts of different whorls may be united, for example a stamen filament and the corolla. Various examples of the many possibilities are given by White (1948). Many of the phenomena described in F (a)–(e) may arise from fasciation, but this cannot be ascertained from a mere description of the symptoms.

G. EFFECTS ON FLOWERING

In many plants transition from the vegetative to the reproductive phase requires very special external conditions: the plants may have to be exposed to low temperatures for a certain time, or to day/night cycles of specific duration (day length) (flower induction). Such conditions presumably induce a shift in hormone metabolism and perhaps production of special hormones. Treatments of seeds or whole plants with various herbicides, especially those of the auxin type, have been shown to affect flowering, that is induce flowering in plants that would have remained vegetative under the conditions applied, or delay or accelerate the onset of the reproductive stage. The example known best is flower induction in pineapple by 2,4-D. Delay of flowering was caused in *Phaseolus vulgaris* by picloram (Fisher et al. 1968).

Reduction of the number of flowers or inflorescences by herbicide treatment is not uncommon, as mentioned in a former paragraph. MH had such an effect on barley and soybean (Klein and Leopold, 1953), picloram on *Dactylis glomerata* and *Festuca* (Lee, 1970). Experiments of Klein and Leopold (1953) with barley and chrysanthemum indicate that the effect of MH was probably due to its inhibition of primordia growth, as has also been found for shoot growth, and that the compound does not typically interfere with induction of flowering. This subject, as well as the effect of herbicides on flower (and leaf) abscission, is outside the scope of this paper, and for a detailed discussion we refer to the Chapters 6, 13 and 14 of this book, and to Steward and Krikorian (1971), Wain and Fawcett (1969) and Weaver (1972).

H. ABNORMALITIES OF FRUITS

A short description of the normal processes of flower fertilization and fruit set has been given in the introduction. In some cases fruits may develop in the absence of seed formation i.e. parthenocarpic fruits. Most of the cell divisions in fruit formation take place in early stages of development, whereas the increase in the size of a fruit is mainly the result of cell enlargement. As with the other organs discussed, the more typical and drastic changes in appearance must be expected from early applications of herbicides. For their normal development fruits are dependent on a regular supply of hormones produced in the seeds. When seeds develop at one side of the fruit only, flesh growth predominates on the side containing the seeds and a lop-sided fruit is formed (Fig. 19). Uneven distribution of auxin herbicides over growing fruits might similarly produce distorted fruits (Fig. 20).

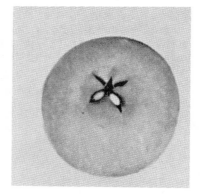

Fig. 19. Lopsided apple, with seeds developed only at one side. On the right the same apple cut in half (Luckwill, 1954).

Fasciation of fruits as a consequence of herbicide treatment may also be expected to occur, since fasciation is quite frequent in fruits (Gorter, 1965; White, 1948). Some commercial cultivars of pineapple, tomato, strawberry and others have fasciated fruits: these are very large with a great number of locules.

Fig. 20. Contortion of young banana fruits induced by 2,4-D (Hendrix, 1952).

III. EFFECTS ON BEHAVIOUR

A. TROPISMS

The orientation of a plant organ is in most cases determined by external stimuli. If the direction of the orientation depends on the direction of the stimulus, it is called a tropic response (Audus, 1969). Stimuli may be gravity (geotropism), light (phototropism), or contact (thigmotropism).

The main axis of many plants are orientated parallel to the pull of gravity; they are orthogeotropic. Roots which grow in the direction of the gravity vector are positively geotropic, stems which grow in the opposite direction, negatively geotropic. A lateral growing at an angle with the gravity vector is called plagiogeotropic (branches); or diageotropic (rhizomes), if it is a right angle. The mechanism of tropic responses is linked with hormone action, in particular with that of auxins. It is thus not surprising that herbicides, especially those with growth regulatory activity interfere with these responses.

The geotropic response of roots is reported to be affected by amitrole and naptalam (Counts, 1961), causing the root tips to grow upwards. The geotropic response of the axis of inflorescences of *Antirrhinum majus,* brought to a horizontal position, was also affected by naptalam (Teas and Shehan, 1957). TIBA and morphactins affect both phototropism and geotropism (Schneider, 1970).

Tendrils of *Vicia* sp. plants, injected with 2,4-D, lacked thigmotropic response (Schnell and Dauphin, 1962). Leaf movement of *Prosopis juliflora* and *Acacia farnesiana,* induced mechanically or by light, appeared to be prevented by picloram (Morgan and Baur, 1970).

B. NASTIC RESPONSES

The orientation of an organ is also affected by nastic movements or phenomena. These occur as a response to either internal or external stimuli, but the direction of the movement is independent of that of an external stimulus; in fact the latter may come from various sides at the same time, as for example in the case of a change of air temperature.

Well known nastic responses are epinasty and hyponasty. In the former case a stem or petiole curves downwards as a result of faster growth on the upper or adaxial side than on the lower or abaxial side of the organ.

Fig. 21. Nastic curvature and twisting of fronds of bracken, *Pteridium aquilinum* L. Kuhn, after treatment with 2,4-D and 2,4,5-T esters (by courtesy of J. Stryckers).

Hyponasty, an upward curvature results if the opposite occurs. Here, too, auxins are considered to play an important part in the underlying mechanism, whereas ethylene is also involved, possibly by affecting auxin distribution in the organ (Pratt and Goeschl, 1969).

Induction of nastic curvatures resulting in characteristic bending or twisting of stems, is one of the most spectacular effects of auxin herbicides (Fig. 21). They may be produced by various herbicides like picloram (Eisinger and Morré, 1971), dicamba (Friesen *et al.* 1964), siduron, carbamates or certain triazines (Klomp and Hull, 1968). The curvatures may be temporary, but frequently they become more or less fixed by changes in anatomical structure. In flax, for example, fibres may become less thick-walled and retain larger lumens on the convex side. Curled, twisted structures occur most strikingly in actively growing organs.

Leaves may also show epinasty. This has been reported for many plants as tomato, pear (Arlt and Feyerabend, 1973), various grasses, as *Echinochloa crus-galli* treated with EPTC and other thiocarbamates (Koren *et al.* 1968), and *Agropyron desertorum* treated with siduron or propham (Klomp and Hull, 1968). Dicamba caused twisting of leaves of perennial grasses (Lee, 1965).

On the other hand, dichlobenil was found to prevent induction of epinasty in tomato, induced by 2,4-D (Milborrow, 1964).

IV. Consequences for the Whole Plant or Crop

In the previous sections we described effects of herbicides which are more or less localized in, and which thus impair the function of a particular organ, although obviously different organs may be affected simultaneously. Even a localized effect, however, may have far-reaching direct or indirect consequences for the whole organism.

In the first place correlative phenomena may be expected, as mentioned in Section II, B. One of the most conspicuous of these is the result of interference with apical dominance. Several examples have already been given, such as lateral root formation, induction of stolon formation, etc. Increased growth of axillary buds of dicotyledons produced a bushy growth as in dinoseb-treated *Vicia faba* plants (Roberts and Wilson, 1961) or "brooming" of *Melilotus* spp. sprayed with MH (Foote and Himmelman, 1971). Increased tillering in grasses or cereals is also very common, for instance after treatment with MH or mendok (Mohan Ram and Rustagi, 1969) or of *Setaria glauca* and *Eleusine indica* with cacodylic acid (Taylorson, 1966) and may cause a drastic change in the appearance of the plant as well as affect its yield in seeds or fruit.

Other kinds of correlative effects may also be involved. We have emphasized the differences in susceptibility of different organs, which would in extreme cases, cause death of certain organs whereas others would remain unaffected. In the long run, however, the development of the whole plant must be affected. Reduction of size of organs—leaves, flowers, fruits—may result from direct inhibition of their enlargement by herbicides, but also, indirectly, from damage of the root system, causing a decrease in supply of water, nutrients, or cytokinins, or from leaf injury and the resulting decreased supply of carbohydrates. Premature leaf abscission or delayed senescence, as for instance that caused in maize by bromacil, atrazine and fluometuron (Hiranpradit and Foy, 1973) may also be expected to influence development of other parts of the plants. A complicated case is presented by DNOC-treated winter rye (Bruinsma, 1962). Vegetative growth was retarded at first but became more vigorous later; senescence and ripening were delayed. The enlargement of the photosynthetic apparatus and the longer period of generative development were considered responsible for the increase in yield consisting of a larger number both of ears and kernels per ear. Similarly, the increased number of kernels in ears of wheat treated with mendok after the emergence of the flag leaf might be due to the larger number of flowers producing seeds, associated with the reduction in stem growth and resulting availability of more nutrients for seed development (Mohan Ram and Rustagi, 1969).

Some examples of indirect effects of a non-correlative kind have already been given, for instance the increased chance of infection by pathogens of potato tubers in which herbicide-induced tumour formation had caused lesions. Such effects may be of a very different nature, and importance. Inhibition of the elongation of flax stems at a very early stage caused fusion of leaves (see Section II, E), with the apical ends of many stem fibres coinciding with the level of the fused leaves. This resulted in breakable stems and post-harvest damage during processing (Jacquemart, 1961). The formation of many, still invisible, root initials after spraying may cause a certain brittleness of the stem of maize (Rodgers, 1952). In rice stems brittleness was caused by dichlobenil (Shimizu, 1967). Kinking of the elongated first internode of *Echinochloa crus-galli* seedlings treated with EPTC (Dawson, 1963) was explained by the decreased ability of the injured coleoptile to penetrate the soil. Changes in the tendency to lodge have also been reported; the decrease of yield of soybeans after treatment with 2,4-D or 2,4,5-T was not considered an effect on flower or seed development, but was ascribed to increased lodging (Smith, 1965). In wheat, on the other hand, mendok appeared to decrease lodging (Mohan Ram and Rustagi, 1969).

When flower development is affected and seed production reduced, seed viability may be affected as well. Decreased germination was found for soybeans produced by plants treated with 2,4,5-T in the early bloom stage (Smith, 1965); picloram induced a similar effect in various grasses (Lee, 1970).

The extent to which plant development is affected by direct or indirect effects of herbicide treatment depends among other factors on their duration. This will be determined by several factors, the most important ones being probably the specific susceptibility of the plant species concerned and the persistence of the herbicide in the tissue. The latter, too, may be different for different species. For a more extensive discussion we refer to Chapters 11, 12 and 13.

In some plants the effects of herbicides may show up only after considerable time. In grapes, for instance, where buds remain dormant for considerable periods, auxin effects may become apparent after several years (Hanf, 1957). Potato plants showed no symptoms after being treated with mecoprop, but their tubers gave rise to stunted plants and deformed leaves in the next year (Arlt and Feyerabend, 1973).

REFERENCES

Addicott, F. T. and Lyon, J. L. (1969). *A. Rev. Pl. Physiol.* **20,** 139–164.
Andersen, S. (1952). *Physiologia Pl.* **5,** 321–333.
Andersen, S. and Hermansen, J. (1950). *K. Vet.-og Landbohøisk. Aarsskr.* **26.**
Arlt, K. and Feyerabend, G. (1973). "Herbizide und Kulturpflanzen", Akademie Verlag, Berlin.
Arnold, W. R. and Santelman, P. W. (1966). *Weeds,* **14,** 74–76.
Arntzen, C. J., Linck, A. J. and Cunningham, W. P. (1970). *Bot. Gaz.* **131,** 14–23.
Ashton, F. M. and Crafts, A. S. (1973). "Mode of Action of Herbicides", John Wiley and Sons, New York.
Ashton, F. M., Cutter, E. G. and Huffstutter, D. (1969). *Weed Res.* **9,** 198–204.
Audus, L. J. (1959). "Plant Growth Substances", Leonard Hill, London.
Audus, L. J. (1969). *In* "Physiology of Plant Growth and Development" (M. B. Wilkins, ed.), pp. 205–242. McGraw Hill Publ. Co. Ltd., New York.
Audus, L. J. (1972). "Plant Growth Substances", 2nd Ed. Vol. 1. Leonard Hill, London.
Baker, J. B. (1960). *Weeds,* **8,** 39–47.
Bayer, D. E., Foy, C. L., Mallory, T. E. and Cutter, E. C. (1967). *Am. J. Bot.* **54,** 945–952.
Bodlaender, K. B. A. Personal communication.
Bradbury, D. and Ennis, W. B. (1953). *Am. J. Bot.* **40,** 827–834.
Bruinsma, J. (1962). *Weed Res.* **2,** 73–89.
Callahan, L. M. and Engel, R. E. (1965). *Weeds,* **13,** 336–338.
Conway, E. and Forrest, J. D. (1961). *Weed Res.* **1,** 114–130.
Copping, L. G., Davis, D. E. and Pillai, C. G. P. (1972). *Weed Sci.* **20,** 274–277.
Counts, B. (1961). *Weeds,* **9,** 329–330.
Cutter, E. G., Ashton, F. M. and Huffstutter, D. (1968). *Weed Res.* **8,** 346–352.
Dawson, J. H. (1963). *Weeds,* **11,** 60–67.
Derscheid, L. A. (1952). *Pl. Physiol., Lancaster,* **27,** 121–134.
Eames, A. J. (1949). *Science, N.Y.* **110,** 235–236.
Eames, A. J. and MacDaniels, L. H. (1947). "An Introduction to Plant Anatomy", 2nd edn. McGraw Hill Publ. Co. Ltd., New York.
Eaton, F. M. (1957). *Science, N.Y.* **126,** 1174–1175.
Ebert, E. and van Assche, Ch. J. (1969). *Experientia,* **25,** 758.
Eisinger, W. R. and Morré, D. J. (1971). *Can. J. Bot.* **49,** 889–897.
Engelke, A. L., Hamzi, A. Q. and Skoog, F. (1973). *Am. J. Bot.* **60,** 491–495.
Esau, K. (1965). "Plant Anatomy", Wiley and Sons, New York; Chapman and Hall, London.
Fletcher, J. T. (1968). *Weed Res.* **8,** 153–155.
Fisher, D. A., Bayer, D. E. and Weier, T. E. (1968). *Bot. Gaz.* **129,** 67–70.
Foote, L. E. and Himmelman, B. F. (1971). *Weed Sci.* **19,** 86–90.
Foy, C. L. and Bisalputra, T. (1964). *Pl. Physiol., Lancaster,* **39,** suppl. lxviii.
Friesen, H. A., Baenziger, H. and Keys, C. H. (1964). *Can. J. Pl. Sci.* **44,** 288–294.
Friesen, G. and Olsen, P. J. (1953). *Can. J. agric. Sci.* **33,** 315–329.
Fromantin, J. (1957). *Rev. gén. Bot.* **64,** 279–298.
Fukunaga, E. T. (1965). *Weed Abstr.* **14,** 23.
Funderburk, H. H. and Davis, D. E. (1960). *Weeds,* **8,** 6–11.
Geranmayah, R. (1964). *Planta,* **62,** 66–87.
Gifford, E. M. (1953). *Hilgardia,* **21,** 607–644.
Goodin, J. R. and Fletcher, F. L. A. (1967). *Pl. Physiol., Lancaster,* **42,** suppl. 23.
Goren, G. and Monselise, S. P. (1965). *Pl. Physiol., Lancaster,* **40,** Suppl. xv.
Gorter, Chr. J. (1949). *Nature, Lond.* **164,** 800–801.

Gorter, Chr. J. (1951). *Proc. K. Ned. Akad. Wet.* **54,** 181–190.
Gorter, Chr. J. (1965). *In Encycl. Pl. Physiol.* XV/2 (W. Ruhland, ed.), pp. 330–351. Springer Verlag, Berlin.
Gräser, H. (1967). *In Wachstumsregulatoren bei Pflanzen. Wiss. Zeit. Univ. Rostock, Math. Naturwiss. Reihe* **16,** 565–568.
Grigsby, B. H. (1945). *Science, N.Y.* **52,** 99–100.
Haccius, B. and Massfeller, D. (1959). *Naturwissenschaften,* **46,** 585–586.
Haccius, B. and Schneider, W. (1958). *Planta,* **52,** 206–229.
Haccius, B. and Trompeter, G. (1960). *Planta,* **54,** 466–481.
Hanf, M. (1957). *Beitr. Biol. Pfl.* **33,** 177–218.
Hendrix, J. W. (1952). *Bull. Hawaii agric. Exp. Sta.* **106,** 1–20.
Hendrix, D. L. and Muench, S. R. (1969). *Pl. Physiol., Lancaster,* **44,** Suppl. xxvi.
Heslop-Harrison, J. (1956). *Physiologia Pl.* **9,** 588–597.
Hiranpradit, H. and Foy, C. L. (1973). *Bot. Gaz.* **134,** 26–31.
Hofstra, G. and Switzer, C. M. (1968). *Weed Sci.* **16,** 23–28.
Holmsen, T. W. (1969). *Weed Sci.* **17,** 187–188.
Holz, W. (1957). *Mitt. biol. Zent. Anst. Berl.* **87,** 99–103.
Humphries, E. and Pethiyagoda, U. (1969). *Ber. dt. bot. Ges. Vortr. Gesamtgeb. Bot.* N.F. **3,** 139–147.
Jacquemart, J. (1961). *C.r.4me Congr. Techn. int., Conf. int. du Lin et du Chanvre, Vienne,* 1960. *Wageningen,* 1961, pp. 104.
Jacques-Félix, H. and Chezeau, R. (1962). *J. Agric. trop. Bot. appl.* **9,** 291–296.
Jeater, R. S. L. (1958). *J. Br. Grassld. Soc.* **13,** 7–12.
Kaufman, P. B. (1955). *Am. J. Bot.* **42,** 649–659.
Kefford, N. P. and Caso, O. H. (1966). *Bot. Gaz.* **127,** 159–163.
Kiermayer, O. (1956). *Phyton,* **7,** 183–185.
Kiermayer, O. (1961). *Öst. bot. Z.* **108,** 101–156.
Kingham, H. G. and Fletcher, J. T. (1963). *Weed Res.* **3,** 242–245.
Klein, W. H. and Leopold, A. C. (1953). *Pl. Physiol., Lancaster,* **28,** 293–298.
Klomp, G. J. and Hull, A. C. (1968). *Weed Sci.* **16,** 315–317.
Koopman, H. and Daams, J. (1960). *Nature, Lond.* **186,** 89–90.
Koren, E., Foy, C. L. and Ashton, F. M. (1968). *Weed Sci.* **16,** 172–175.
Kumar, K. and Dube, S. D. (1962). *J. scient. Res. Benaras, Hindu Univ.* **13,** 201–211.
Kust, C. A. and Struckmeyer, B. E. (1971). *Weed Sci.* **19,** 147–151.
Laibach, F. and Kribben, F. J. (1950). *Ber. dt. bot. Ges.* **62,** 53–55.
Laibach, F. and Kribben, F. J. (1951). *Beitr. Biol. Pfl.* **28,** 131–144.
Large, E. C. and Dillon Weston, W. A. R. (1951). *J. agric Sci.* **41,** 338–349.
Lee, W. J. (1965). *Weeds,* **13,** 205–208.
Lee, W. J. (1970). *Weed Sci.* **18,** 171–173.
Linser, H. and Kiermayer, O. (1957). "Methoden zur Bestimmung pflanzlicher Wuchsstoffe", Springer Verlag, Wien.
Linser, H. and Kirchner, R. (1957). *Planta,* **50,** 211–237.
Luckwill, L. C. (1954). *In* "Plant Regulators in Agriculture" (H. B. Tukey, ed.), pp. 81–98. John Wiley and Sons, New York.
Luxová, M. and Lux, A. (1964). *Biologia Pl.* **6,** 258–264.
Mallory, T. E. (1969). *Proc. 21st Calif. Weed Conf.,* 127.
Malhotra, S. S. and Hanson, J. B. (1966). *Pl. Physiol., Lancaster,* **41,** suppl. vi.
Maun, M. A. and Cavers, P. B. (1969). *Weed Sci.* **17,** 533–536.
Milborrow, B. V. (1964). *J. exp. Bot.* **15,** 515–523.
Mitchell, J. W., Livingston, G. A. and Marth, P. C. (1958). "U.S.D.A. Agriculture Handbook", **126.**
McIlrath, W. J. (1955). *Cott. Ginner,* November.
McIlrath, W. J. and Ergle, D. R. (1953). *Pl. Physiol., Lancaster,* **28,** 693–702.
Moham Ram, H. Y. and Rustagi, P. H. (1969). *Agron. J.* **61,** 198–201.
Morgan, P. W. and Baur, H. R. (1970). *Pl. Physiol., Baltimore,* **46,** 655–659.
Muller, P. J. and Boon, J. van der (1972). *Neth. J. Pl. Pathol.* **78,** 198–203.

Nishimoto, R. K. and Warren, G. F. (1971). *Weed Sci.* **19,** 343–345.
Noodén, L. D. (1969). *Physiologia Pl.* **22,** 260–270.
Orth, H. (1957). *Mitt. biol. Zent. Aust.* Berlin, **87,** 73–76.
Orth, H. and Leuchs, F. (1961). *1e Conf. du Columa, Paris,* 314–318.
Peeper, T. F., Weibel, D. E. and Santelman, P. W. (1970). *Agron. J.* **62,** 407–411.
Peterson, E. L. and Naylor, A. W. (1953). *Physiologia Pl.* **6,** 816–828.
Pfeiffer, R. K. and Holmes, H. M. (1961). *C.r. Symposium sur les Herbicides,* Paris, EWRC, 144–154.
Phillips, I. D. J. (1969). *In* "Physiology of Plant Growth and Development" (M. B. Wilkins, ed.), pp. 165–207. McGraw Hill, New York.
Prasad, R. (1965). *J. exp. Bot.* **16,** 86–106.
Pratt, H. K. and Goeschl, J. D. (1969). *A. Rev. Pl. Physiol.* **20,** 541–584.
Roberts, H. A. and Wilson, H. J. (1961). *Weed Res.* **1,** 289–300.
Rodgers, E. G. (1952). *Pl. Physiol., Lancaster,* **27,** 153–171.
Sachs, R. M. (1965). *A. Rev. Pl. Physiol.* **16,** 73–96.
Sasaki, S., Kozlowski, T. T. and Torrie, J. H. (1968). *Can. J. Bot.* **46,** 255–262.
Schneider, G. (1970). *A. Rev. Pl. Physiol.* **21,** 499–536.
Schnell, R. and Dauphin, R. (1962). *J. Agric. trop. Bot. appl.* **9,** 423–430.
Schultz, D. P., Funderburk, H. H. and Negi, N. S. (1968). *Pl. Physiol., Lancaster,* **43,** 265–273.
Schwabe, W. W. (1971). *Symp. Soc. exp. Biol.* **25,** 301–322.
Sen, K. M. (1960). *Studies on the influence of some auxin herbicides on grass-seed crops.* Ph.D. Thesis. LandbHogeschool, Wageningen, 70 pp.
Shimizu, M. (1967). *Weed Abstr.* **15,** 296.
Smith, R. J. (1965). *Weeds,* **13,** 168–169.
Staniforth, D. W. (1952). *Pl. Physiol., Lancaster,* **27,** 803–811.
Steveninck, R. F. M. van (1956). *Ann. Bot.* **28,** 385–392.
Steward, F. C. (1968). "Growth and Organization in Plants", Addison-Wesley, Reading, Mass.
Steward, F. C. and Krikorian, A. D. (1971). "Plants, Chemicals and Growth", Academic Press, London and New York.
Struckmeyer, B. Esther (1951). *In* "Plant Growth Substances" (F. Skoog, ed.), pp. 166–174. Univ. Wisc. Press, Madison.
Stryckers, J. (1960). *Med. LandbHogesch. Opz. Stations Gent* **15,** 735–760.
Taylorson, R. B. (1966). *Weeds,* **14,** 207–210.
Teas, H. J. and Sheehan, T. J. (1957). *Pl. Physiol., Lancaster,* **32,** suppl. xliii.
Thimann, K. V. and Sachs, T. (1966). *Am. J. Bot.* **53,** 731–739.
Upchurch, R. P., Selman, F. L. and Webster, H. L. (1968). *Weed Sci.* **16,** 317–323.
Wain, R. L. and Fawcett, C. H. (1969). *In* "Plant Physiology" (F. C. Steward, ed.), Vol. Va, pp. 231–296. Academic Press, London and New York.
Wasnik, K. G. (1969). *Weed Abstr.* **18,** 342.
Watson, D. P. (1948). *Am. J. Bot.* **35,** 543–555.
Watson, D. P. (1950). *Am. J. Bot.* **37,** 424–431.
Way, J. M. (1961). *World Crops,* **13,** 455–458.
Way, J. M. (1963a). *Weed Res.* **3,** 11–25.
Way, J. M. (1963b). *Weed Res.* **3,** 98–108.
Way, J. M. (1963c). *Weed Res.* **3,** 312–321.
Way, J. M. (1964a). *Weed Res.* **4,** 12–23.
Way, J. M. (1964b). *Weed Res.* **4,** 319–337.
Weaver, M. L. and Nylund, R. E. (1965). *Weeds,* **13,** 110–113.
Weaver, R. J. (1972). "Plant Growth Substances in Agriculture", Freeman and Co., San Francisco.
White, E. D. (1948). *Bot. Rev.* **14,** 319–358.
Zink, F. W. and Agamalian, H. (1965). *Weeds,* **13,** 19–22.
Zonderwijk, P. (1959). *Versl. Meded. plziektenk. Dienst Wageningen* **111,** 218.

CHAPTER 5

EFFECTS ON THE DORMANCY OF PLANT ORGANS

C. PARKER

A.R.C. Weed Research Organization, Begbroke Hill, Yarnton, Oxford, England

INTRODUCTION

Although this volume is concerned primarily with herbicides, the following discussion will include considerable reference to non-herbicidal growth regulators and to a lesser extent mineral salts. There is no clear cut boundary between herbicides and growth regulators and it could even be argued that any dormancy-breaking compound could be used indirectly to achieve control of weeds and could therefore very loosely be termed a herbicide.

"Dormancy" in its broadest sense may refer to any state of "suspended development" in plant organs (Shorter Oxford English Dictionary) and could be used to describe the condition of mature seeds or buds whose development is suspended simply by lack of water or low temperature. A narrower meaning is usually implied, however, and for the purposes of this chapter, dormancy is defined as a lack of growth "under conditions normally considered to be adequate . . . viz. when provided with a suitable temperature, adequate water and oxygen" (Roberts, 1972).

Two main classes of dormancy will be distinguished namely, correlative (usually due to apical dominance and of relevance only to bud dormancy) and inherent. Inherent dormancy will include categories such as "innate", "induced" and "enforced" as used by Harper (1957). These three terms describe different ways in which seeds or buds (and also such organs as corms, bulbs and tubers) acquire their dormancy, but are not considered to represent any consistently different physiological mechanism. The factors responsible for such dormancy may include (a) seed coats or scales impermeable to water and/or gases, (b) seed coats or scales providing mechanical resistance to expansion of the embryo or bud, (c) excess growth inhibitor or (d) lack of essential nutrient or other stimulating substance, or almost any combination of these factors.

Mention of a dormancy-breaking effect in this chapter should not be taken as an indication that dormancy is completely overcome. In many instances only a small proportion of seeds may respond, such that germination is increased significantly but as many as 50% may still remain dormant. Such partial effects may not be of great practical value but are of academic interest. So far as possible some indication will be given of the completeness of response (responsiveness) and the relative level of concentration required to cause the response (sensitivity).

Bud or seed dormancy is a characteristic of almost all successful weed species. Although some serious weeds have little or no seed dormancy e.g. *Galinsoga parviflora,* such species are generally characteristic of intensive horticultural situations where little time is allowed for cultivation and destruction of weeds between crops. Their persistence as weeds depends on their ability to mature and produce a new crop of seed in a very short time. *Agrostemma githago* also has little dormancy but persisted as a weed of cereals in Britain for centuries simply because its seed was not successfully separated from the cereal seed by primitive seed cleaning procedures. As soon as more efficient seed cleaning was introduced it declined rapidly as a weed of cereals (Salisbury, 1961; Chancellor, 1968b; Thompson, 1973).

The correlation between lack of dormancy and lack of weediness is most apparent in crop species. After the harvest of a cereal crop in Britain, an autumn cultivation will often result in dense growth from the shed seeds of the crop. This growth may then be destroyed by further cultivation and there is little or no seed persisting in the soil to provide a "weed" problem in the following year. An interesting exception to this convenient behaviour of crop plants is in the potato whose "ground-keepers" occur as a weed in subsequent crops (Kadijk, 1972). The problem has become sufficiently serious in Britain to warrant a special research project at the ARC Weed Research Organization (Lutman and Elliott, 1973). This problem is ascribable very largely to the dormancy of the tubers which do not sprout for several months and hence cannot be destroyed before the planting of the following crop. In this way it is closely analogous to the sub-tropical weed *Cyperus esculentus* which has a similar tuber system and similarly will sprout only after a period of chilling.

Among annual weeds *Avena fatua* well exemplifies the typical behaviour of a weed, contrasting with that of the cereal crops. After harvest very few seeds will germinate in the stubble and a great many will persist to create a weed problem in the crop in the next, or subsequent, two or three years. Even a whole season of cultivated fallow will not result in complete elimination of this species. Indeed, Roberts (1963) has shown that the burden of viable dormant seeds of a mixture of species in the soil is reduced by less than 50% per year, even with repeated cultivations to stimulate growth and with no seed production allowed. Hence the action of any herbicide or other compound in reducing dormancy is of potential value towards reducing long term weed problems.

Dormancy is a phenomenon common to most wild plants, having evolved as a means of ensuring the suspension of development during periods of climatic stress, mainly cold and drought, which could be lethal to the growing plant. Unfortunately, however, the characteristics of dormant organs which make them resistant to extremes of moisture and temperature also make them very largely immune to control by cultivation and by herbicides.

The failure of most herbicides to kill dormant seeds or buds has not been discussed or analysed in any depth but is generally assumed to be due to the lack of metabolic activity in such organs. This results in reduced sensitivity to metabolic interference which is assisted by a lack of flow of water or other materials into the organ. In a growing plant there is ample evidence that the movement of herbicides is dependent on either the transpiration stream or the movement of photosynthates in the phloem. So long as either of these transport systems is operating, herbicides may be translocated over long distances and be accumulated in the appropriate sinks. For instance, in a rhizomatous species such as *Agropyron repens,* there is no difficulty in killing the apices of rhizomes up to 2 m from the point of application but most, if not all of the dormant buds along the rhizome will be undamaged and regrowth may occur from any of these, or even from buds at the base of the treated shoot itself. Death of dormant buds may apparently occur but this often follows release from dormancy resulting indirectly from injury to the apex by the herbicide.

Even when dormant seeds become imbibed in the presence of herbicide, the amount of herbicide entering the seed may not be adequate to cause damage. Mitchell and Brown (1947) were able to show that after holding seed of *Trifolium subterraneum* at 30°C subsequent exposure to soil treated with 2,4-D caused no harm. At this temperature the seed remains dormant. After 19 days, by which time the 2,4-D had been degraded, the seed could be moved to 20°C (at which it is not dormant) and would germinate normally. When set up at 20°C from the start the seed germinated immediately and was damaged. The reasons for the lack of effect on dormant seeds could be attributed to lack of penetration through the seed coat, lack of necessary accumulation, or low physiological sensitivity. The only compounds which appear capable of killing dormant seeds (rather than persisting in the seed coat and killing the seed after dormancy is lost) are the highly volatile fumigants such as methyl bromide and methyl isothiocyanate. Even these are more effective on non-dormant than on dormant seeds (Bush and Staniforth, 1966). Whether this class of compound represents an exception to the general rule or whether compounds of this type are able to kill dormant seeds or buds by first breaking their dormancy, will be discussed below.

II. Effects on Seed Dormancy

A. BREAKING SEED DORMANCY

Although it has been noted above that dormant seeds are generally resistant to non-volatile herbicides, there are of course many examples of non-volatile substances successfully entering the seed in sufficient quantity to break dormancy. Thiourea (Mayer and Poljakoff-Mayber, 1963; Holm and Miller, 1972) and potassium nitrate are but two examples. Thiourea substitutes for light in lettuce and for chilling in a number of tree species but has not found any practical use in this sphere. Nitrate on the other hand increases the germination of a large number of species, especially, but not exclusively those with a light requirement, and has been proposed as a practical means of increasing the emergence of *Avena fatua* prior to seed bed cultivations (Sexsmith and Pittman, 1963). Another important weed species also responding to nitrate is *Chenopodium album* (Williams and Harper, 1965).

Among the more complex organic materials, the gibberellins (GA) and cytokinins are capable of breaking the dormancy of a wide range of seeds. Of the weed species responding to GA, *Thlaspi arvense* (Corns, 1960a; Chancellor and Parker, 1972), *Argemone mexicana* and *Asclepias fruticosa* (Tager and Clarke, 1961), *Avena fatua* (Simpson 1965; Corns, 1960b) and *Digitaria (?) sanguinalis* (Gray, 1958) show the greatest response (from less than 10% to over 50%). *A. fruticosa* has responded to concentrations as low as 1 and 10 mg/l, *T. arvense* to 100 mg/l but *Avena fatua* usually requires about 500 mg/l. These differences presumably reflect different ease of entry into the dormant seeds. GA has also been shown to have a moderate effect on *Galium aparine* (Ueki and Shimizu, 1967), *Brassica kaber, Datura stramonium, Abutilon theophrasti* and *Amaranthus retroflexus* (Holm and Miller, 1972), and *Orobanche* spp. (Nash and Wilhelm, 1960; Abu-Shakra *et al.* 1970). Gray (1958) also showed an effect of GA in soil, increasing emergence of *Digitaria* sp. by a factor of 3·6 with a 1,000 mg/l drench (volume unstated).

Cytokinins were shown first to be active in lettuce seed germination. Among weeds, *Xanthium pennsylvanicum* is stimulated by kinetin at 0·1 mg/l, but only with the assistance of light (Khan, 1966). Small effects have been shown with kinetin on a few species by Holm and Miller (1972) and on *Striga asiatica* (Worsham *et al.* 1959). No herbicide has been observed to have direct effects on seed dormancy comparable to the cytokinins or GA. The substituted ureas include a range of compounds which have shown cytokinin-like activity (Bruce and Zwar, 1965; Kefford *et al.* 1966) but the activity of the herbicidal ureas such as monuron and diuron has been shown to be very much lower than that of *N,N'*-diphenylurea and some other non-herbicidal analogues. The action of thiourea is also apparently unrelated to that of the cytokinin-like substituted ureas. Kefford *et al.* (1965) showed that although thiourea and other urea derivatives would break the dormancy of lettuce seeds, this activity was not necessarily associated with cytokinin-like effects on cell division in callus tissue. Bokarev and Satarova (1957) considered that the action of thiourea results from conversion to thiocyanic acid.

Auxins have rarely been shown to have any stimulatory action in seed germination, and 2,4-D and related herbicides have likewise generally been found to be inhibitory rather than stimulatory but there are some exceptions. Rojas-Garciduenas and Kommedahl (1960) reported a small increase in germination of *Amaranthus retroflexus* (from 66% to 82%) when seeds had been soaked in 2,4-D amine (20 mg/l) for 20 h. In a later paper Rojas-Garciduenas *et al.* (1962) again increased germination of *A. retroflexus* (from 43% to 85%) this time by exposing the seeds continuously to a solution of 2,4-D butyl ester at 1 mg/l. MCPA butyl ester had no significant stimulating effect at low concentrations but its toxicity was equal to that of 2,4-D at higher levels of 1,000 to 10,000 mg/l. Milyi (1972) also succeeded in accelerating germination of *Amaranthus retroflexus* and *Setaria lutescens* by drenching soil with a 800 mg/l solution of 2,4-D. He found a spray of 2,4-D at 0·5−1 kg/ha to be 2−3 times more effective than disc harrowing as a means of stimulating the germination of "summer" weeds.

Arai and colleagues in Japan (Arai *et al.* 1967; Chisaka *et al.* 1967a, b) have also reported some influence of MCPA allyl ester, 2,4-D acetonitrile and α-(2,4-D) propionitrile in increasing germination of *Echinochloa crus-galli* after soaking for two days at 100−1,000 mg/l. The sodium salt of MCPA was not effective.

Arai and his colleagues also found dinoseb and three members of the substituted diphenyl ether group—nitrofen, chlornitrofen (2,4,6-trichloro-phenyl-4-nitrophenyl ether) and KK-60 (2,4dinitro-4-chlorodiphenyl ether) to be among the most active of the 27 compounds in their tests after two-day soaking at 100−1,000 mg/l. Chancellor (personal communication) found fluorodifen to have a similar effect in increasing germination of *Chenopodium album* from 5% to 52% at 500 mg/l.

Ueki and Shimizu (1968, 1969) and Shimizu and Ueki (1972a, b) have found *Echinochloa crus-galli* to respond to a range of inhibitors of respiration. Compounds active in breaking dormancy include sodium and potassium cyanide, 8-hydroxyquinoline, mercuric chloride, sodium sulphide, 8-oxyquinoline, and sodium diethyldithiocarbamate, whilst inactive compounds included sodium azide and thiourea.

In yet another series of papers from Japan, Inoue *et al.* (1970a, b, 1971) have confirmed the effects of the cyanides and also calcium cyanamide and hydrogen sulphide on *E. crus-galli*. Complex cyanides, covalent cyanide and ionic cyanates were not effective. Results were even better under anaerobic conditions and they again conclude that it is their action in inhibiting oxygen uptake into normal respiratory processes which is responsible for their dormancy breaking effect.

There are some differences between the range of compounds breaking dormancy in *Echinochloa* and those doing the same in rice (*Oryza sativa*) but the conclusions reached by Roberts (1969) concerning the effects of respiratory inhibitors on rice seed are probably of relevance to *Echinochloa*. Briefly, "It is necessary for some oxidation reaction to occur before germination can take place. The conditions within the seed are relatively anaerobic. . . . Conventional respiration is a strong competitor for the available oxygen since the oxygen affinity of cytochrome oxidase is extremely high. Consequently any treatment which decreases this competition for oxygen also speeds up the breaking of dormancy—for example, the application of inhibitors of cytochrome oxidase or the provision of alternative hydrogen acceptors which can be used in respiratory processes, such as nitrate, nitrite or methylene blue." Roberts further proposes that the "oxidation reaction" which is benefited by the inhibition of the main respiratory system is the pentose phosphate pathway which apparently has an essential function in the early stages of germination though the precise nature of this function is not yet clearly understood.

The mode of action of the substituted diphenyl ethers as described by Moreland *et al.* (1970) is not a simple inhibition of respiration but the end result is apparently somewhat related to that of the respiratory inhibitors. Unfortunately the concentrations required to stimulate germination of *Echinochloa* are relatively high and Chisaka *et al.* (1967*b*) quote quantities of 40 kg/ha of nitrofen or dinoseb to produce an effect under field conditions. Dinoseb is an uncoupler rather than an inhibitor of respiration and although Ballard and Lipp (1967) have shown a stimulatory effect of 2,4-dinitrophenol (at 0·2 mM) and some other uncouplers on the seed of *Trifolium subterraneum*, this effect may be restricted to seeds which also have the unusual characteristic of germinating in response to high levels of CO_2. No other instances of dormancy breaking by dinoseb have been reported.

A small but significant effect of atrazine and ametryne in increasing germination of lettuce and tobacco, has been reported by Copping *et al.* (1972) at very low concentrations.

For sodium azide there have been some other indications of useful dormancy-breaking effects. Colby and Feeny (1967) studied combinations of calcium cyanamide and potassium azide for weed control and found a synergism between them on several species. An increase of weed emergence was incidentally observed with azide alone at levels of 56–168 kg/ha but not with calcium cyanamide at levels of 560–2,240 kg/ha. The weeds included *Panicum* spp. and *Allium vineale*.

It is interesting to note that several of the more highly volatile compounds used as fumigants have been demonstrated to have some dormancy-breaking effects. Arai and colleagues found a number of thiocyanato-bromopropanes and -propenes to be effective at 50 mg/l over a four-day period. Dibromoethane (ethylene dibromide or EDB) and 2-chloroethanol (ethylene chlorhydrin) are other compounds which may be effective fumigants at high doses but which break dormancy at lower concentrations. Gianfagna and Pridham (1952) reported an increase in germination of *Digitaria sanguinalis* from 0·5 to 96% as a result of treatment with 2-chloroethanol 500 mg/l for

48 h. Miller *et al.* (1965) reported that increased emergence of *Digitaria* spp. has been observed by farmers who had treated potato fields with EDB at 50 l/ha for eelworm control. They conducted experiments which confirmed that EDB could increase germination of *D. sanguinalis* from 1% to 20% and *Chenopodium album* from 3% to 15%. 1,1-Dibromo-3-chloropropane (DBCP) had somewhat greater effects on these two species at a lower dose of 17 l/ha and also increased germination of *Echinochloa crus-galli* from 4% to 18% and *Amaranthus retroflexus* from 14% to 25%.

There are no definite reports for methyl bromide having similar effect on any weed seeds but there are reports of increases in speed and completeness of germination of groundnut (Somade, 1955) and of onion (Lubatti and Blackith, 1956) when low doses have been tested for control of diseases. There are also indications of a stimulatory action of low doses of methyl bromide on the resting stages of nematodes (Hague, 1958). Emergence of potato root eelworm was stimulated by exposure to concentrations of 18 to 84 mg/l for 1 h, approximately one-tenth the levels required for 90% kill.

The fumigant metham-sodium has not been reported to cause any stimulation of germination. This compound is sodium N-methyldithiocarbamate but the active ingredient which is released by decomposition in soil is methyl isothiocyanate. The nematocide dazomet also releases the same compound; again, there are no reports of dormancy breaking, though dormant seeds are killed.

The fact that these compounds are highly volatile may be of significance in permitting more rapid, continuous diffusion into the seed from the surrounding soil. It is also of considerable interest to speculate whether the success of the fumigants such as EDB and methyl bromide in killing dormant seeds may depend on an initial dormancy-breaking action. There have not been reports of "failures" in fumigation treatments resulting in increased weed emergence but reduced doses would be worth testing for the possibility of observing such effects. It could be more economical to break dormancy and destroy mechanically or with conventional herbicide than to rely on the fumigant to do the whole job.

The one class of volatile compound which is being seriously considered, if not already used as a practical control measure, is ethylene. The effects of ethylene in breaking dormancy have been demonstrated in a number of species such as *Trifolium subterraneum* (Esashi and Leopold, 1969) and groundnut (Ketring and Morgan, 1970). Egley and Dale (1970) initially found (2-chloroethyl) phosphonic acid (CEPA) to be active in stimulating seeds of *Striga asiatica* to germinate and subsequently determined that the activity was due to release of ethylene in the medium around the seeds. At low pH which prevented spontaneous release of ethylene from the compound, there was no activity. Subsequently they investigated the activity of ethylene gas and found it highly stimulating at gaseous concentrations of $0 \cdot 1$ μl/l. Eplee (1972) has reported development of field techniques for the injection of ethylene gas into *Striga*-infested soils and has demonstrated the excellent lateral and vertical distribution which occurs in soils of suitable texture and moisture content. It has been demonstrated that over 90% of *S. asiatica* seed germinate where the seed has been suitably "pre-treated" and introduced artificially into the soil. Unfortunately, the seed has an essential requirement for a 7—10 day imbibition period before it will respond to any dormancy-breaking stimulant and with more prolonged imbibition, especially at high temperatures, a secondary dormancy may be induced (at least in *S. hermonthica* as shown by Vallance, 1950). Hence there is no certainty that *Striga* seed occurring naturally in the soil will be in a state to respond to the ethylene and no estimates have been published of the proportion of the seed stimulated in soil. Apart from the period of imbibition which may not be optimal, there is also the possibility of unfavourable oxygen/carbon dioxide balance which could interfere with response to ethylene. Now that one of the naturally occurring *Striga*-germinating substances, strigol, from the roots of cotton, has been identified (Cook *et al.* 1972), there is interest in the use of this substance also, for freeing soil of viable dormant seed, in the absence of a crop. The work with ethylene will be a valuable pointer to the degree of success that may be achieved by such an approach. Correct pre-conditioning of the seed in the soil may be critically important and greatly restrict the practical usefulness of the principle. The work on *Striga* in the USA is being pursued in spite of such reservations, because of the declared policy of eradication of *Striga* from the areas in which it occurs in South and North Carolina. As an aid to normal control measures, even a 50% reduction in viable seed in the soil would not be of great value, but as part of an eradication campaign it acquires much greater significance.

Effects of CEPA on dormant seeds have also been reported by Chancellor *et al.* (1971). Effects on *S. hermonthica* were comparable to those obtained by Egley and Dale on *S. asiatica*. However, another parasitic species, *Orobanche aegyptiaca* with somewhat related germination requirements to the *Striga* spp., was not affected by CEPA (Kasasian, 1973).

Chancellor *et al.* (1971) found some indication of beneficial interaction between CEPA and light on *Matricaria recutita* and between CEPA and chilling on *Chenopodium album*. Olatoye and Hall (1972) also found comparable interaction between CEPA and light on *Spergula arvensis* and have discussed the way in which such interactions may come about. Attempts by Chancellor *et al.* to use CEPA as a practical soil treatment at 4 kg/ha sprayed on the soil surface, resulted in some increase in emergence of naturally occurring *Matricaria recutita* but the possibilities for commercial use do not appear especially striking so far.

Species whose germination has been stimulated to a significant extent by 2-chloroethylphosphonic acid include *Matricaria recutita* and *Datura stramonium* (Chancellor *et al.* 1971), *Amaranthus retroflexus* and *Brassica kaber* (Holm and Miller, 1972), *Spergula arvensis*, *Hypochaeris radicata*, *Rumex crispus* and *Trifolium repens* (Olatoye and Hall, 1972) and *Chenopodium album* (all these authors). Species not stimulated by 2-chloroethylphosphonic acid in these three series of experiments included *Avena fatua*, *Sinapis arvensis*, *Thlaspi arvense*, *Echinochloa crus-galli*, *Abutilon theophrasti*, *Ipomoea purpurea*, *Agropyron repens*, *Sida spinosa*, *Polygonum pennsylvanicum, Sonchus oleraceus, Silene dioica, Senecio jacobea, Taraxacum officinale.* All were exposed to 10 and 100 parts/10^6 and some to a much wider range of doses. Holm and Miller also failed to confirm any activity on *D. stramonium.*

B. PREVENTING INNATE DORMANCY

It has been reported for a number of species, that immature seeds while still on the plant may be non-dormant. In *Avena* spp. dormancy is only acquired with full maturity and desiccation (Thurston, 1963; Morgan and Berrie, 1970). Compounds causing premature abscission and shedding of seed might therefore be expected to influence dormancy, but no reports are known of effects attributable to this particular cause. On the other hand a few compounds have been shown to have an influence on the dormancy of seeds as a result of application to the parent plant.

Åberg *et al.* (1948) reported that spraying *Galium aparine* with MCPA or 2,4-D resulted in the production of seeds which would germinate on the soil soon after harvest. Åberg (1956) further reported that confirmation of this result had been obtained with 2,4-D and 2,4,5-T on *G. aparine* and the effect could be demonstrated also on *Spergula arvensis* and *Polygonum lapathifolium.*

Aamisepp (1959) showed that the seed from plants of *A. fatua* which had been sprayed with MCPA, 2,4-D or 2,4,5-T gave increased germination compared with seed from untreated plants. Immediately after harvest in one year, seed from plants treated with 2,4-D (4·4 kg/ha) at the 4–5 leaf stage gave 24% germination compared with 2–3% in controls. Equivalent doses of MCPA and 2,4,5-T and lower doses of 2,4-D were much less effective.

In two other years seed was stored for 4–5 months before being tested for germination. By this time there was 35–50% germination from untreated plants, while as little as 0·8–0·9 kg/ha of MCPA and 2,4-D applied at the 4–5 leaf stage increased germination to 70–80%. 2,4,5-T was less effective but caused considerable increase in germination when sprayed at 2·5 and 5 kg/ha.

In a later publication the same author (Aamisepp, 1961) showed how 2,4-D could be detected in the seeds from plants of peas sprayed with 2,4-D, so it could be that the reduction in dormancy of *A. fatua* is due to the presence of herbicide in the seeds. On the other hand, there has been no direct stimulation of *A. fatua* germination reported from application of 2,4-D or related compounds to seed after harvest.

Khan *et al.* (1970) showed that GA_3 sprayed on to barley at 500 and 1,000 mg/l 7–28 days after awn emergence, prevented the normal post-harvest dormancy.

Parker (unpublished) tested a range of compounds for their effect on dormancy of *A. fatua* treated at several stages of flowering in a glasshouse experiment. Germination tests conducted within a few weeks of harvesting indicated a small increase in germination from 2,4-D at 1·1 and 11 kg/ha, fenoprop at 11 but not at 1·1 kg/ha, dicamba at 0·27 and 2·7 kg/ha, picloram at $0 \cdot 1$ but not at $1 \cdot 1$ kg/ha, chlorflurecol at $0 \cdot 27$ but not at $2 \cdot 7$ kg/ha and gibberellic acid (GA$_3$) at $1 \cdot 1$ but not at $0 \cdot 1$ kg/ha. Unsprayed plants gave $3 \cdot 6\%$ germination. The highest germination from any treatment was $13 \cdot 8\%$ (2,4-D 11 kg/ha) but this was achieved only with seeds from the main inflorescence which had already flowered at the time of application. Inflorescences flowering after application did not respond. Five months later, only GA$_3$ continued to show a distinct effect, cumulative germination then being 53% versus $31 \cdot 6\%$ for untreated inflorescences.

Black and Naylor (1959) obtained somewhat greater effects from GA$_3$ on *A. fatua* by cutting off immature inflorescences and dipping their stems into solutions of 1, 10 and 100 mg/l. Germination was increased from 0% up to 50%. Morgan and Berrie (1970) obtained similar results with *A. ludoviciana* by a similar technique, increasing germination from 14 to 100% with GA$_3$ at 285 mg/l. They also obtained substantial reduction in dormancy by similar treatments with the metabolic inhibitors coumarin, p-chloromercuribenzoate and iodoacetate.

Rojas-Garciduenas and Kommedahl (1960) sprayed *Amaranthus retroflexus* with 2,4-D $1 \cdot 1$ kg/ha at an early flowering stage and showed a 5-fold increase in germination of seed from these plants both at 15 days $(2\% \rightarrow 11\%)$ and at 5 months $(10\% \rightarrow 50\%)$ after harvest. The direct effects of 2,4-D on seeds of *Amaranthus* noted under the previous section, suggest that the increased germination following spray application to the parent plant may have been the result of 2,4-D entering the seeds before maturity, rather than any more subtle effect via the parent plant.

The size of the increases in germination reported here are clearly not of immediate practical significance, but the principle could conceivably be exploited if the mechanism of the effect could be more clearly understood and the degree of effect correspondingly increased.

A technique which has provided some more information on the effects of herbicides and growth regulators on dormancy of *Avena fatua* is the "roguing glove" developed by Holroyd (Holroyd, 1972, May, 1972). This glove is used to wet inflorescences of *A. fatua* with herbicide by hand where a sparse population does not warrant overall spraying. It has advantages over conventional hand roguing in that the plants do not need to be uprooted and carried off the field, so saving about 50% of labour costs. Most herbicides tested have had (as intended) an inhibitory effect on subsequent germination but there has been an indication of a slight stimulatory effect of fluorodifen. CEPA on the other hand applied to inflorescences at 1 and 5% had no stimulatory effect (Holroyd, personal communication).

C. INDUCING DORMANCY

There are many reports of reduced germination of crop and weed seeds resulting from herbicide treatment to the parent plant. There is rarely, however, any indication whether the failure to germinate is due to a loss of viability or to an induced "dormancy". Most often it is probably the former and this may be obvious from the rotting of the seed. With certain herbicides, however, particularly dalapon and maleic hydrazide, there is a failure to germinate which is not accompanied by rotting, and this effect has sometimes been described as "induced dormancy". It is an effect observed when dalapon is applied by "roguing glove" to *A. fatua* (Holroyd, 1972). Holroyd, however, considers that the seeds which do not germinate have definitely lost their viability. The small proportion that do grow from a marginal dose are seriously deformed and one may probably assume that those not germinating are more severely affected and would never be able to germinate normally.

For maleic hydrazide (MH) the evidence of Carder (1959) is particularly convincing. He reported on the use of MH sprayed at the "milk stage" on to crops of wheat, barley and flax infested with *A. fatua*. After such treatment germination of crop and weed seeds was reduced or prevented altogether. Even after two years there was no recovery from this "induced dormancy" and it may be assumed that the effect involved a destruction of viability rather than a reversible induction of dormancy. (Untreated seed was still viable after this period.) It has been shown in physiological studies that the effects of MH can be prevented or "reversed" by IAA or other auxin (e.g. Leopold and Klein, 1952) but such interactions have not been shown where the application of the auxin is delayed long after the application of MH. This question is discussed further in the following section on bud dormancy.

III. Effects on Bud Dormancy

A. BREAKING BUD DORMANCY

1. *Correlative Dormancy*

The mechanism of apical dominance is still the subject of considerable controversy. It is not completely resolved whether the lateral buds which are suppressed by a dominant apex fail to grow because of an excess of an inhibitor, such as IAA or abscisic acid (ABA) or because of a shortage of essential materials in the form of growth substances (e.g. cytokinin or gibberellin), mineral nutrient (e.g. nitrogen) or elaborated nutrient (e.g. sugars). As with other similar controversies the truth may turn out to depend on a balance of inhibitory and promotive substances.

Interference with apical dominance is in fact very frequently demonstrated by herbicides but usually in an indirect way as a result of damage to the dominant apex itself. Once the apex is destroyed, e.g. in the rhizome of *Agropyron repens,* there is a tendency for lateral buds to grow out, usually those closest to the apex. Unless these are themselves damaged by the herbicide, one or more of them will become dominant and maintain the dormancy of the more basal buds. Chancellor (1968a) has made a detailed study of this resumption of apical dominance in 7- and 15-node fragments of *A. repens* rhizome. The general observations are somewhat similar where the rhizome apex is removed from an otherwise intact plant system (McIntyre, 1970). The early re-imposition of a new correlative inhibition system means that any dormancy-breaking effect resulting from damage to an apex is liable to be temporary or incomplete, at least in long rhizome systems. In shoot systems the loss of dormancy of lateral buds may be somewhat more complete. Maleic hydrazide has been used to some extent to overcome apical dominance and promote lateral growth (Schoene and Hoffman, 1949). The apical bud is apparently more susceptible to the herbicide than the laterals, presumably because it is making active growth at the time of application (see, Molero and Blackhurst, 1956). A number of other compounds used to control the growth form of ornamentals have a similar mode of action, i.e. damaging the active apical bud but having no residual effect on laterals. Cathey *et al.* (1966) described a range of alkyl esters of fatty acids which could be used as such "chemical pruning" agents.

Many herbicides may show such indirect effects of interference with apical dominance but more direct interference without damage to the apex is much more rarely observed.

Meyer and Buchholtz (1963) showed that it could be achieved by local application of TIBA in a ring around the rhizome of *A. repens*. Such treatment resulted in sprouting of a few buds basal to the site of application, and the effects could be interpreted in terms of the known physiological effects of TIBA, i.e. interference with polar movement of auxin, if one assumes that apical dominance in this case is due to an effect of excess auxin or other inhibitor originating in the rhizome apex and moving in a basipetal direction.

Comparable effects were later observed by ringing *A. repens* rhizomes with chlorflurecol-methyl (Parker, unpublished), and Chancellor (1970) obtained very striking effects from the same compound on 7-node rhizome fragments.

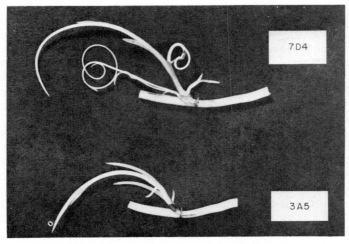

Fig. 1. An effect of chlorflurecol-methyl at 20 mg/l on the sprouting of *Agropyron repens* rhizome buds. This compound has interfered with apical dominance within the lateral shoot in such a way that secondary laterals have also developed (WRO photo).

He not only delayed or prevented the re-imposition of an apical dominance system but also caused outgrowth of axillary buds on the lateral buds of the rhizome themselves. Chlorflurecol is also known to interfere with polar movement of auxin, so again its effects could be interpreted in accordance with the "auxin" theory of apical dominance.

Chlorflurecol-methyl has also shown comparable dormancy-breaking effects on the tubers of *Cyperus rotundus* (Rehm, 1969; Parker and Dean, 1972). Isolated tubers of *C. rotundus* behave in a way analogous to that of multi-node fragments of *A. repens* rhizome. Several buds may sprout initially but one or two soon acquire dominance and the remainder return to a dormant stage. Parker and Dean found that after soaking tubers in chlorflurecol-methyl (1 mg/l) for 24 h, the re-imposition of a dominance system is delayed and a rather larger number of buds continue to grow. TIBA was also active but only at a much higher concentration of 100 mg/l. It was also possible to show an effect on the intact plant by exposing the root and rhizome system to a solution of chlorflurecol-methyl (10 mg/l) or by spraying the plants with a dose of 2 kg/ha. In this case the effect was represented by an increase in the number of tubers resuming growth from their apical buds. "Dormant" tubers on the intact *C. rotundus* plant cannot be said to demonstrate "apical dominance", as it is the apical buds of these tubers which remain dormant while lateral buds on the lower side of the tubers grow out to form further links in what may become chains of dormant tubers. Although not due to apical dominance in the normal sense, it is probable that the dormancy of the tuber apices is a form of correlative inhibition; however, the nature, origin and direction of movement of the inhibitory influence is still under investigation.

The activity of chlorflurecol in interfering with apical dominance has also been observed in a number of other species (see Schneider, 1970).

A further herbicide with action related to TIBA and chlorflurecol-methyl is naptalam. Morgan (1964) and McCready (1968) confirmed that this compound also inhibits polar movement of auxin and Rehm (1969) emphasized the close similarity between naptalam and flurenol in their effects on *C. rotundus,* including effects of increased sprouting of buds on "basal bulbs". Parker and Dean (1972) also found naptalam to cause some increase in sprouting of *C. rotundus* tubers.

The most striking interference with apical dominance in *A. repens* has been achieved with CEPA. Chancellor (1970) and Caseley (1970) have described how, in the intact plant system, a large proportion of the rhizome lateral buds are stimulated to sprout, usually to form branch rhizomes rather than to emerge as shoots. The final effect can be to increase rhizome weights substantially. In isolated rhizome fragments, Chancellor found no corresponding interference with the new dominance system—rather the contrary. A similar combination of effects is reported by Beasley (1969) and Hull (1970) for the comparable rhizomatous grass *Sorghum halepense.*

The mode of action of CEPA in these rhizomatous species has not been explained but is presumed due to the release of ethylene. The effect of ethylene in breaking apical dominance has sometimes been attributed to an inhibition of

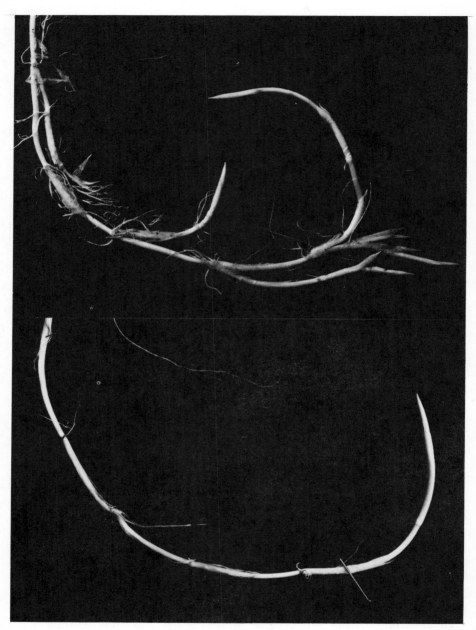

Fig. 2. The effect of 2-chloroethylphosphonic acid (CEPA) on *Agropyron repens* rhizome. An application of CEPA to the soil around an established plant at 4·5 kg/ha (above) has resulted in the development of branch rhizomes at most nodes. On an untreated plant (below) the rhizome buds have remained dormant. (WRO photo reproduced with acknowledgement to *Pesticide Science*.)

the apex but such inhibition is not always apparent in *Agropyron* and some other mechanism is assumed.

No herbicides have shown any effects comparable to CEPA with the possible exception of glyphosate (see below).

A further group of growth regulators with well-established reputation for breaking dormancy are the cytokinins. Direct application to suppressed lateral buds can result in immediate outgrowth (Miller, 1961). Their action has been particularly strikingly demonstrated on *Cyperus rotundus* (Teo *et al.* 1971, 1973; Parker and Dean, 1972). After soaking for 24 h in solutions of benzylaminopurine (BAP) or kinetin at 100 mg/l or SD 8339 [(6-benzylamino-9-(tetrahydro-2-pyranyl)-9H-purine] at 10 mg/l, five or six sprouts per tuber continue to grow concurrently. Effects can also be demonstrated on the intact plant by dipping the root and rhizome system of a potted *C. rotundus* plant into a 50 mg/l solution of BAP. A range of substituted phenyl urea herbicides have failed to show effects comparable to those of BAP on tubers of *Cyperus rotundus* (Dean, unpublished).

The cytokinins have not shown such great effects in rhizomes of *A. repens* (Chancellor and Leakey, 1972).

Gibberellins have not generally been shown to interfere with correlative inhibitions but rather to reinforce them. This has been observed to be the case in *C. rotundus* (Teo *et al.* 1971; Parker and Dean, 1972).

Apart from mild effects from some new experimental herbicides on isolated tubers of *C. rotundus* at 100 mg/l, Parker and Dean also found some apparent stimulation of sprouting from treatment with three of the diphenyl ether group of herbicides which have been observed to influence the dormancy of seeds of

Fig. 3. The effect of the synthetic cytokinin, 6-benzylamino-9-(tetrahydro-2-pyranyl)-9H-purine (SD 8339) at 10 mg/l on the sprouting of a *Cyperus rotundus* tuber. Normally correlative effects would prevent all but one or two sprouts developing to this extent (WRO photo).

Echinochloa. These were nitrofen, fluorodifen and chlornitrofen but the activity was apparent only at 500 mg/l and was barely significant. There were also some mild effects from the benzothiadiazole compounds DU 13594 (5,7-dichloro-4-ethoxycarbonyl methoxy-2,1,3-benzothiadiazole) and DU 16333 (6-chloro-5-ethoxy-carbonylmethoxy-2,1,3-benzothiadiazole). This class of compound has been reported by van Daalen and Daams (1970) to have some influence on apical dominance.

A new herbicide with interesting effects on apical dominance, is glyphosate. This herbicide has proved to be lethal to a wide range of perennial weeds at doses of 1–4 kg/ha. At sub-lethal doses, mainly below 1 kg/ha, however, there is often a very pronounced degree of extra tillering at the base of the treated shoots of for example *A. repens, C. rotundus, Convolvulus arvensis* and on the stolons of *Agrostis stolonifera.* This sprouting may to some extent be secondary as a result of damage to the shoot apices, but it has not been observed to the same extent with any other herbicide, and it seems possible that, as suggested for fumigants above, the extreme success of the compound may be at least partly attributable to its ability to break the dormancy of buds and hence make them more susceptible. Some similar symptoms have been observed with aminotriazole but not to the same degree.

All the perennial species mentioned so far have underground stem structures in the form of rhizomes and tubers which have preformed buds in limited numbers. Once these buds are killed no new buds can be formed, except by new development from terminal meristems.

A further distinct group of perennials, however, has root systems on which almost unlimited numbers of buds can develop. Important examples are *Convolvulus arvensis* and *Cirsium arvense.* Where the buds have not already

Fig. 4. An example of the effect of glyphosate in causing proliferation of shoots from nodes on the stolons of *Agrostis stolonifera* (WRO photo).

developed it is questionable whether they can be referred to as dormant, but there are recognizable phases in the life cycle when the roots may have a greater or lesser tendency to produce new adventitious shoots. On isolated fragments of such roots there is a tendency for new roots to develop from the distal end of the fragment and buds on the proximal end. Bonnett and Torrey (1965) have shown that this is associated with a polar re-distribution of auxin downwards within the fragments. In the intact plant, there is corresponding evidence that high auxin levels from the shoot system tend to reduce adventitious sprouting in the roots as in *Populus tremula* (Eliasson, 1971). Applications of 2,4-D tend also to suppress such sprouting, as in *Chondrilla juncea* (Caso and Kefford, 1973). Caso and Kefford on the other hand, found aminotriazole to cause some stimulation. It might be expected that the morphactins would inhibit downward movement of auxin and hence increase sprouting from roots but this has not been generally observed.

In woody species there is a potential for buds to develop around the base of the stem. These buds may normally be suppressed by some form of apical dominance and be released on cutting. Vogt and Cox (1970) showed that in oak (*Quercus alba* and *Q. palustris*) stump-sprouting after decapitation could be prevented by application of IAA but that MH would counteract this effect and allow sprouting. Morgan *et al.* (1969) sprayed *Prosopis juliflora* and *Acacia farnesiana* with CEPA at 8 kg/ha and obtained increased sprouting of buds around the stem base.

Shafer and Monson (1958) reported dormancy-breaking effects of GA_3, applied as sprays of 25–500 mg/l to the foliage, on the crown buds of *Euphorbia esula* and *Vernonia baldwini*. Although some of this effect was attributable to a breaking of apical dominance, there is also a tendency to inherent rather than correlative dormancy in these species and the result may be attributable to an effect on such innate dormancy rather than to a correlative one. Monson and Davis (1964) showed seasonal variations in degree of dormancy in both these species.

2. *Attempts at Exploitation of Dormancy-breaking Effects*

Parker and Dean (1972) have attempted to exploit the dormancy-breaking effect of the cytokinins and chlorflurecol-methyl to increase the effectiveness of herbicides on *C. rotundus*. Following on the assumption stated above that bud dormancy is one of the main sources of difficulty in killing perennial weeds, it was further assumed that by causing extra sprouting of buds at or around the time of application of a herbicide, a greater degree of control should be achieved. Unfortunately they were unable to confirm this hypothesis with the various combinations of compounds that they tested. Plants growing from tubers pre-treated with BAP proved equally if not more resistant to a range of pre-emergence or post-emergence herbicides. Likewise simultaneous applications of chlorflurecol-methyl with terbacil or glyphosate showed no advantage. Horowitz (1972) did not look directly at bud dormancy but observed additional sprouting of several perennial species following treatments with CEPA, chlorflurecol-methyl and flurecol-methyl. Combination sprays of CEPA with dalapon showed useful synergism on *Cynodon dactylon* but other combinations with aminotriazole did not prove a particular advantage on this species, nor on *Sorghum halepense* or *Cyperus rotundus.*

Akhavein (1971) also claimed some benefit from combination of CEPA with aminotriazole on *Agropyron repens* but the results were not altogether consistent. Caseley (1972) failed to find any advantage from combinations of CEPA with paraquat or pronamide on *A. repens,* but chlorflurecol-methyl showed synergism with paraquat and dalapon under certain circumstances. With paraquat a simultaneous application of chlorflurecol-methyl was synergistic at 20°C but not at 10°C. With dalapon, chlorflurecol-methyl was synergistic when applied after the dalapon but not when applied before or simultaneously.

Parker (1970) reported the results of spray applications in which CEPA and chlorflurecol-methyl were added separately and together to dalapon, paraquat, asulam and aminotriazole and applied to the tropical rhizomatous grass, *Imperata cylindrica.* No synergism was demonstrated by any two-way combinations but the three-way combinations of CEPA and chlorflurecol-methyl with either dalapon or paraquat showed some advantage, which might be due to the great stimulation of extra sprouts resulting from the combination of CEPA (4 kg/ha) and chlorflurecol-methyl (2 kg/ha). Soerjani (1970) claims to have found some advantage for combinations of herbicide with CEPA on *I. cylindrica* but has not yet published full details.

3. *Breaking Inherent Dormancy*

In *A. repens* the dormancy of rhizome buds is usually lost on fragmentation and is therefore assumed to be correlative. Under certain conditions, however, the buds acquire an innate dormancy, referred to in USA as "late-spring" dormancy (Johnson and Buchholtz, 1962). This type of dormancy has been shown to be associated with very low internal nitrogen levels, and to be overcome by suitable nitrogen treatments (Buchholtz, 1962; Leakey, 1974). No herbicides have been tested for their effects on this type of dormancy.

Bulbs and tubers are commonly found to have an innate dormancy which is overcome under normal conditions by cold and/or drought. The tuber of *Cyperus esculentus* is analogous in most ways to that of a potato (*Solanum tuberosum*). The whole tuber becomes dormant on maturity and no lateral buds grow out as is the case in *C. rotundus.* The dormancy of these tubers has been observed to be broken by 2-chloroethanol, thiourea, potassium thiocyanate and ethyl ether (Durfee, 1960, reported in Bell *et al.* 1962). 2-Chloroethanol caused only the apical bud to grow. Potassium thiocyanate and thiourea broke apical dominance and caused up to four sprouts per tuber, but these were retarded in growth. Jackson *et al.* (1971) reported loss of dormancy in *C. esculentus* tubers with both ethylene (3–10 μl/l) and CEPA (10–100 mg/l). It may be of significance that Tames and Vieitez (1970) were able to break dormancy in *C. esculentus* var. *aureus* by scarification, especially near the apical bud. Such scarification might be expected to result in a wound response which would include production of ethylene.

Innate dormancy has been most intensively studied in woody species. Although the dormancy of tree buds is probably induced by a correlative effect from the foliage, once imposed it persists until broken under natural conditions by either chilling or increasing day length (see Wareing, 1969 for a general review). GA_3 and kinetin have been shown to induce bud-break in a number of woody species (Vegis, 1964) but no herbicides have been reported to have any equivalent effects.

Innate dormancy is a common occurrence in the resting buds or turions of a number of aquatic species. Those of *Hydrocharis morsus-ranae* respond to kinetin (Kunz and Kummerov, 1957).

The winter buds of *Potomogeton nodosus* apparently have a somewhat different mechanism. Dormancy may be broken by as little as three days chilling at 0°C or by removing the bud scales. They do not respond to kinetin, thiourea, 2-chloroethanol or to the herbicide dichlobenil but do respond to auxins. Soaking in IAA at 1,000 mg/l for 18 h completely broke dormancy, while NAA and the herbicide chlorfenac were almost fully effective in the range 100–1,000 mg/l. GA_3 was also partially effective (Frank, 1966). In *P. crispus* Sahai and Sinha (1969) again found IAA to be effective, and also thiourea, potassium nitrate, ammonium sulphate and phosphoric acid.

Hydrilla verticillata has two types of resting organ—axillary turions and subterranean turions which differ slightly in their response to chemical dormancy-breaking agents. The subterranean type is most readily stimulated to grow by GA_3 (1–1,000 mg/l), IAA (1–100 mg/l), NAA (100 mg/l) and CEPA (100–1,000 mg/l). For axillary turions higher doses of GA_3 (100–1,000 mg/l) and IAA (1,000 mg/l) were required; CEPA was hardly active at 1,000 but 2,4-D was effective at 100–1,000 mg/l. Herbicides endothal and paraquat were ineffective but there was a slight stimulation of axillary turions only, by diquat at 0·5 and 1 mg/l (Steward, 1969).

B. PREVENTING INNATE DORMANCY

Allium vineale produces strongly dormant underground "minor offset bulbs" Håkansson (1963) observed that treatment of the parent plants with MCPA shortened the dormant period of these offset bulbs. Parker (1966b) confirmed that a number of growth-regulator herbicides including MCPA, 2,4-D, 2,3,6-TBA, chlorfenac and dicamba and especially picloram and the phenoxy propionic acids, sprayed on to the parent plants at a time when offset bulbs were being formed (April/May) resulted in these bulbs having little or no dormancy. Sprouting occurred during the following few months whereas offset bulbs of unsprayed plants remained dormant for at least six months.

Håkansson (1963) and Davis *et al.* (1965) have described morphological differences in minor offsets after treatment with 2,4-D and 2,3,6-TBA which might explain the effect and Håkansson also reports that a shortened dormancy period can result from spring cultivations. The fact that it can also be caused by paraquat (Parker, 1966b) suggests that it may be the result of a relatively non-specific effect such that the outer scales fail to attain their full degree of impermeability upon which dormancy is believed largely to depend.

GA_3 may prevent or delay or break dormancy in woody species as in blackcurrant (Modlibowska, 1960) and in potatoes (Lippert et al. 1958) or it may occasionally have the opposite effect of prolonging or inducing it, as in Prunus avium (Brian et al. 1959), vines (Weaver, 1959) and in aerial tubers of Begonia evansiana (Okagami and Esashi, 1972).

C. INDUCING DORMANCY

Dalapon is sometimes regarded as having a dormancy-inducing effect on perennial weeds. Application to Agropyron repens in autumn may result in a delay in emergence of the weed in the Spring. This regrowth may be from buds that have been delayed in growth (but otherwise undamaged) by dalapon, or it may be that lateral or axillary buds have eventually sprouted after the major rhizome buds have made abortive growth and then died. In most instances it seems likely that the latter mechanism is the cause of the observed delay in emergence.

Similarly in potato, and in Agropyron repens, there is no evidence that the "dormancy" induced by maleic hydrazide is reversible. Attempts have been made to overcome the maleic hydrazide "dormancy" in potatoes by application of GA_3 but this has not proved possible (Rappaport et al. 1957; Rao, 1954). MH is believed to persist for some months in plant tissue (Smith et al. 1959) but even after prolonged storage, up to 12 months, there is still no growth from MH-treated potatoes (Wittwer, 1953). As in seeds there may be some antagonism between MH and auxin at the time of application. Rao (1954) found 2,4-D to nullify the effect of MH applied to potatoes before harvest and the effects of MH may be temporary and reversible to some extent—Greulach and Atchison (1950) found that mitosis and cell division inhibited by MH at low concentrations, was resumed after removal from MH—but it appears very doubtful indeed whether one can regard the effect of MH as an induced dormancy. Any prolonged inhibition of growth appears to be associated with a loss of viability of the buds. Smith et al. (1959) conclude "biological data do not permit differentiation between the actual presence of MH and growth inhibition caused by irreversible changes in the plant which might persist after the chemical itself has disappeared".

Dormancy induced artificially by the natural inhibitor abscisic acid (ABA) on the other hand has been shown to be reversible by a washing treatment or by GA_3 (Eagles and Wareing, 1963; Wareing and Saunders, 1971).

With propham and chlorpropham which are also used to prevent sprouting of potatoes there is evidence (van Vliet and Sparenberg, 1970) that the higher the dose of carbamate, the more prolonged the effect of suppressed sprouting. It is possible, here, however, that the carbamates are not so much inducing and maintaining dormancy as suppressing the growth of the sprouts as they emerge naturally from their dormancy. The higher the initial dose, then the longer the persistence of the herbicide in the atmosphere of the store or perhaps in the tissues of the buds.

IV. TECHNIQUES FOR TESTING THE EFFECTS OF HERBICIDES ON DORMANCY

To test herbicides for their effects in breaking dormancy of seeds or buds is relatively straightforward. If it is a particular species that is of concern, then the seeds or buds may be treated by temporary soaking or continuous exposure and subsequent germination or sprouting observed (e.g. Chancellor and Parker, 1972; Parker and Dean, 1972). Preferably some standard dormancy-breaking technique should be available as a standard of comparison and also as an assurance that the seeds or buds are indeed dormant rather than non-viable. There is usually some suitable physical or chemical treatment available, whether it be chilling, scarification, light, GA_3 etc.

If there is an interest in the possibility of *killing* dormant seed, then it is vital to have a means of overcoming the dormancy and testing for viability after treatment. This requirement dictates either the use of seed or buds for which there is a simple dormancy-breaking technique known (e.g. alternating temperature or H_2SO_4 as used by Parochetti and Warren (1970) with *Polygonum persicaria* and *Gleditsia triacanthus* respectively) or the use of material which is not strictly dormant in the sense used in this chapter but in which dormancy can be temporarily enforced by conditions of, for example, low temperature, high temperature, high nitrogen or carbon dioxide atmosphere, etc. Temperature has been the technique most commonly used. Pieczarka and Warren (1959) used low temperatures (9–10°C) to maintain tomato, watermelon, musk melon, *Amaranthus retroflexus* and *Portulaca oleracea* dormant while they were exposed to different fumigants. The seeds were then ventilated for some hours before the temperature was raised to assess viability. Mitchell and Brown (1947) on the other hand used a high temperature of 30°C to maintain dormancy in seed of *Trifolium subterraneum* during exposure. The temperature was then reduced to 20°C for germination.

"Tetrazolium salt" (2,3,5-triphenyl tetrazolium chloride) could perhaps also be used in conjunction with truly dormant seed but this test requires careful interpretation which would involve comparisons of colour reaction in seeds of different (known) viability. Hence there would still be the need for some means of artificial breaking of dormancy. Some of the possibilities and pitfalls of this technique are given by Moore (1973) and in the International Rules for Seed Testing (1966).

More serious difficulties arise if the objective of a test is not to study the effects on particular species but to determine whether a compound has any useful dormancy-breaking effects at all. There is then the need to decide how many different mechanisms of dormancy there may be and hence how many different organisms should be included in the tests. Unfortunately the state of knowledge is still far too incomplete for such a decision to be made with any certainty. There are overlaps between many different dormancy-breaking agents; for instance GA_3 replaces a chilling requirement in some seeds or a light requirement in others but the overlap is never complete. There are always certain exceptions, such that a seed will respond to light but not to GA_3 and so on (Mayer and Poljakoff-Mayber, 1963). At present one can only suggest that a range of seeds be used which will respond to the more important dormancy-breaking agents such as light (e.g. *Matricaria* sp.) chilling (e.g. *Polygonum aviculare*), increased oxygen (or respiratory inhibitor) (e.g. *Avena fatua*) and perhaps sulphuric acid (to cover those with seed coats preventing imbibition such as *Convolvulus arvensis*). *Striga* sp. would also be of interest as an example of a seed responding to ethylene and not to any of the above agents. Innate bud dormancy could be represented by potato or *Cyperus esculentus* (both responding to chilling and ethylene) and perhaps by turions of one of the aquatic weeds such as *Hydrilla verticillata* (responding to auxin). The bulbs of *Oxalis latifolia* could also provide a good test for chemical dormancy breakers. Dormancy can be broken by heat and by peeling (Parker, 1966a) and by chilling (Chawdhry, 1971) but no chemical treatment has been found effective, other than with the cytokinins and then to a limited extent (Parker and Dean unpublished).

Buds (inhibited correlatively) are more difficult to provide, for a whole system must be included. Shoots of pea (Sachs and Thimann, 1967), flax (Gregory and Veale, 1957) or *Xanthium strumarium* (Tucker and Mansfield, 1972, 1973) have been much used in classical work and may be suitable, but the freshly isolated tubers of *Cyperus rotundus* could prove convenient as a multi-bud system in which apical dominance is normally re-established quite rapidly, as well as the 7-node rhizome fragments of *Agropyron repens* as used by Chancellor.

The testing of highly volatile fumigants poses extra problems of containment, especially where numerous experimental treatments are to be held in a confined space. However, the self-sealing silicone-rubber caps now available for glassware, in conjunction with syringes, offer a very simple and safe way of evading this problem, and hopefully will provide encouragement for more intensive study of such compounds on dormant seeds and buds.

V. CONCLUSION

This chapter will have revealed that there are few herbicides that are known to have striking effects on dormancy. For several major groups of herbicides, there is little or no published evidence and it is quite possible that useful effects have yet to be discovered among the established compounds. It does, however, appear unlikely that any one compound will have effects on a wide spectrum of weed species. Even the most active of the growth-regulatory chemicals such as GA_3 or CEPA have a limited range of activity. For individual weed species, however, there is the prospect of finding compounds with dormancy-breaking effects which could be utilized for improved control measures.

REFERENCES

Aamisepp, A. (1959). *Växtodling,* **10,** 58–67.
Aamisepp, A. (1961). *K. LantbrHögsk. Annlr.* **27,** 445–451.
Åberg, E. (1956). *Proc. 3rd Br. Weed Control Conf.* 141–164.
Åberg, E., Hagsand, E. and Väärtnöu, H. (1948). *Växtodling,* **3,** 8–64.
Abu-Shakra, S., Miah, A. A. and Saghir, A. R. (1970). *Hort. Res.* **10,** 119–124.
Akhavein, A. A. (1971). *Effects of 2-chloroethylphosphonic acid (Ethrel) and selected environmental factors on growth of quackgrass (Agropyron repens L. Beauv.) and field bindweed (Convolvulus arvensis L.).* Ph.D Thesis, Oregon State University, pp. 129.
Arai, M., Chisaka, H. and Kataoka, T. (1967). *Proc. Crop Sci. Soc. Japan,* **36,** 321–325.
Ballard, L. A. T. and Lipp, A. E. G. (1967). *Science, N.Y.* **156,** 398–399.
Beasley, C. A. (1969). *Abstr. 9th Meet. Weed Sci. Soc. Am.* 190.
Bell, R. S., Lachman, W. H., Rahn, E. M. and Sweet, R. D. (1962). *Bull. Rhode Isl. agric. Exp. Stn.* **364,** pp. 33.
Black, M. and Naylor, J. M. (1959). *Nature, Lond.* **184,** 468–469.
Bokarev, K. S. and Satarova, N. A. (1957). *Fiziol. Rast. (Pl. Physiol., USSR),* **4,** 361.
Bonnett, H. T. and Torrey, J. G. (1965). *Pl. Physiol., Lancaster,* **40,** 1228–1236.
Brian, P. W., Petty, J. H. P. and Richmond, P. T. (1959). *Nature, Lond.* **184,** 69.
Bruce, M. I. and Zwar, J. A. (1965). *Proc. R. Soc. B.* **165,** 245–265.
Buchholtz, K. P. (1962). *Proc. 16th N. East Weed Control Conf.* 16–22.
Bush, L. P. and Staniforth, D. W. (1966). *Abstr. Meet. Weed Sci. Soc. Am.* 59.
Carder, A. C. (1959). *Weeds,* **7,** 141–152.
Caseley, J. C. (1970). *Pestic, Sci.* **1,** 114–115.
Caseley, J. C. (1972). *Proc. 11th Br. Weed Control Conf.* 736–743.
Caso, O. H. and Kefford, N. P. (1973). *Weed Res.* **13,** (2), 148–157.
Cathey, H. M., Steffens, G. L., Stuart, N. W. and Zimmerman, R. H. (1966). *Science, N.Y.* **153,** 1382–1383.
Chancellor, R. J. (1968a). *Proc. 9th Br. Weed Control Conf.* 125–130.
Chancellor, R. J. (1968b). *Proc. 9th Br. Weed Control Conf.* 1129–1135.
Chancellor, R. J. (1970). *Proc. 10th Br. Weed Control Conf.* 254–260.
Chancellor, R. J. and Leakey, R. R. B. (1972). *Proc. 11th Br. Weed Control Conf.* 778–783.
Chancellor, R. J. and Parker, C. (1972). *Proc. 11th Br. Weed Control Conf.* 772–777.
Chancellor, R. J., Parker, C. and Teferedegn, T. (1971). *Pestic. Sci.* **2,** 35–37.
Chawdhry, M. A. (1971). *Studies of some species of Oxalis.* Ph.D Thesis, University College of North Wales, Bangor, pp. 124.
Chisaka, H., Kataoka, T. and Arai, M. (1967a). *Proc. Crop Sci. Soc. Japan,* **36,** 326–331.
Chisaka, H., Kataoka, T. and Arai, M. (1967b). *Proc. Crop Sci. Soc. Japan,* **36,** 332–337.
Colby, S. R. and Feeny, R. W. (1967). *Weeds,* **15,** 163–167.
Cook, C. E., Whichard, L. P., Wall, M. E., Egley, G. H., Coggon, P., Luhan, P. A. and McPhail, A. T. (1972). *J. Am. chem. Soc.* **94,** 6198–6199.
Copping, L. G., Davis, D. E. and Pillai, C. G. P. (1972). *Weed Sci.* **20,** 274–277.
Corns, W. G. (1960a). *Can. J. Bot.* **38,** 871–875.
Corns, W. G. (1960b). *Can. J. Pl. Sci.* **40,** 47–51.
Davis, F. S., Peters, E. J. and Fletchall, O. H. (1965). *Weeds,* **13,** 210–214.
Durfee, J. W. (1960). *Life history and the control of northern nutgrass, Cyperus esculentus L.* M.Sc. University of Massachusetts.
Eagles, C. F. and Wareing, P. F. (1963). *Nature, Lond.* **199,** 874–875.
Egley, G. H. and Dale, J. E. (1970). *Weed Sci.* **18,** (5), 586–589.
Eliasson, L. (1971). *Physiologia Pl.* **25,** 118–121.
Eplee, R. E. (1972). *Abstr. Meet. Weed Sci. Soc. Am.* 22.
Esashi, Y. and Leopold, A. C. (1969). *Pl. Physiol., Lancaster,* **44,** 1470–1472.
Frank, P. A. (1966). *J. exp. Bot.* **17,** 546–555.
Gianfagna, A. J. and Pridham, A. M. S. (1952). *Proc. 6th N. East Weed Control Conf.* 321–326.
Gray, R. A. (1958). *Pl. Physiol., Lancaster,* **33,** suppl., xl–xli.

Gregory, F. G. and Veale, J. A. (1957). *Symp. Soc. exp. Biol.* **11**, 1–20.
Greulach, V. A. and Atchison, E. (1950). *Bull. Torrey bot. Club,* **77**, 262–267.
Hague, N. G. (1958). *Emergence of encysted potato root eelworms responding to controlled and variable stimuli.* Thesis, Diploma of the Imperial College of Science and Technology, London, pp. 111.
Håkansson, S. (1963). *Växtodling,* **19**, pp. 208.
Harper, J. L. (1957). *Proc. 4th int. Congr. Crop Prot.* **1**, 415–420.
Holm, R. E. and Miller, M. R. (1972). *Weed Sci* **20**, 150–153.
Holroyd, J. (1972). *Proc. N. cent. Weed Control Conf.* 74–76.
Horowitz, M. (1972). *Weed Res.* **12**, 11–20.
Hull, R. J. (1970). *Weed Sci.* **18**, 118–121.
Inoue, K., Higashi, T. and Yamasaki, K. (1970a). *J. Sci. Soil Manure Tokyo,* **41**, 377–382.
Inoue, K., Higashi, T. and Yamasaki, K. (1970b). *Soil Sci. Pl. Nutrition,* **1**, 20–26.
Inoue, K., Higashi, T. and Yamasaki, K. (1971). *J. Sci. Soil Manure, Tokyo,* **42**, 157–162.
International Seed Testing Association (1966). *Proc. int. Seed Test. Ass.* **31**, pp. 152.
Jackson, E. K., Jangaard, N. O. and James, A. L. (1971). *Pl. Physiol., Baltimore,* **47**, suppl., Abstr. 87.
Johnson, B. G. and Buchholtz, K. P. (1962). *Weeds,* **10**, 53–57.
Kadijk, E. J. (1972). *Proc. 2nd int. Meet. Selective Weed Control in Beet Crops, Rotterdam 1970,* 208–209.
Kasasian, L. (1973). *Proc. Eur. Weed Res. Coun. Symp. Parasitic Weeds,* 68–75.
Kefford, N. P., Bruce, M. I. and Zwar, J. A. (1966). *Planta,* **68**, 292–296.
Kefford, N. P., Zwar, J. A. and Bruce, M. I. (1965). *Planta,* **67**, 103–106.
Ketring, D. L. and Morgan, P. W. (1970). *Pl. Physiol., Baltimore,* **45**, 267–273.
Khan, A. A. (1966). *Am. J. Bot.* **53**, 607.
Khan, R. A., Hashmi, S. H. and Ahmad, S. (1970). *Pakist. J. scient. ind. Res.* **13**, 294–298.
Kunz, L. and Kummerow, J. (1957). *Naturwissenschaften,* **44**, 121.
Leakey, R. R. B. (1974). *Factors affecting the growth of shoots from fragmented rhizomes of Agropyron repens (L.) Beauv.* Ph.D Thesis, University of Reading, pp. 197.
Leopold, A. C. and Klein, W. H. (1952). *Physiologia Pl.* **5**, 91–99.
Lippert, L. F., Rappaport, L. and Timm, H. (1958). *Pl. Physiol., Lancaster,* **33**, 132–133.
Lubatti, O. F. and Blackith, R. E. (1956). *J. Sci. Fd Agric.* **7**, 149–159.
Lutman, P. J. W. and Elliott, J. G. (1973). *Arable Farmer,* **7**, 19, 21.
McCready, C. C. (1968). *In* "Biochemistry and Physiology of Plant Growth Substances" (F. Wightman and G. Setterfield, eds), pp. 1005–1023. Runge Press, Ottowa.
McIntyre, G. I. (1970). *Can. J. Bot.* **48**, 1903–1909.
May, M. J. (1972). *Proc. 11th Br. Weed Control Conf.* 294–300.
Mayer, A. M. and Poljakoff-Mayber, A. (1963). "The Germination of Seeds", Pergamon Press, N.Y.
Meyer, R. E. and Buchholtz, K. P. (1963). *Weeds,* **11**, 4–7.
Miller, C. O. (1961). *A. Rev. Pl. Physiol.* **12**, 395–408.
Miller, P. M., Ahrens, J. F. and Stoddard, E. M. (1965). *Weeds,* **13**, 13–14.
Milyi, V. V. (1972). *Trudy Khar'kov. sel'.-khoz. Inst.* **172**, 163–166.
Mitchell, J. W. and Brown, J. W. (1947). *Science, N.Y.* **106**, 266–267.
Modlibowska, I. (1960). *Ann. appl. Biol.* **48**, (4), 811–816.
Molero, F. J. and Blackhurst, H. T. (1956). *Proc. Am. Soc. hort. Sci.* **67**, 416–420.
Monson, W. G. and Davis, F. S. (1964). *Weeds,* **12**, (3), 238–239.
Moore, R. P. (1973). *In* "Seed Ecology" (W. Heydecker, ed.) *Proc. 19th Easter Sch. agric. Sci., University of Nottingham,* 347–366.
Moreland, D. E., Blackmon, W. J., Todd, H. G. and Farmer, F. S. (1970). *Weed Sci.* **18**, 636–642.
Morgan, D. G. (1964). *Nature, Lond.* **201**, 476–477.
Morgan, P. W., Meyer, R. E. and Merkle, M. G. (1969). *Weed Sci.* **17**, 353–355.
Morgan, S. F. and Berrie, A. M. M. (1970). *Nature, Lond.* **228**, 1225.
Nash, S. M. and Wilhelm, S. (1960). *Phytopathology,* **50**, 772–774.
Okagami, N. and Esashi, Y. (1972). *Planta,* **104**, 195–200.

Olatoye, S. T. and Hall, M. A. (1972). *In* "Seed Ecology" (W. Heydecker, ed.) *Proc. 19th Easter Sch. agric. Sci., University of Nottingham*, 233–249.
Parker, C. (1966a). *Proc. 8th Br. Weed Control Conf.* 126–134.
Parker, C. (1966b). *Proc. 8th Br. Weed Control Conf.* 553–562.
Parker, C. (1970). *Proc. 4th E. Afr. Herbicide Conf.* 265–278.
Parker, C. and Dean, M. L. (1972). *Proc. 11th Br. Weed Control Conf.* 744–751.
Parochetti, J. V. and Warren, G. F. (1970). *Weed Sci.* **18,** 555–560.
Pieczarka, S. J. and Warren, G. F. (1959). *Weeds,* **7,** 133–140.
Rao, S. N. (1954). *Certain physiological and morphological responses in potato and onions induced by maleic hydrazide.* Ph.D Thesis, Michigan State College, pp. 85.
Rappaport, L., Lippert, L. F. and Timm, H. (1957). *Am. Potato J.* **34,** 254–260.
Rehm, S. (1969). *Ber. dt. chem. Ges.*: Vorträge aus dem Gesamtgebiet der Botanik N.F.3, Symposium Morphaktine, 131–137.
Roberts, E. H. (1969). *Symp. Soc. exp. Biol.* **23,** 161–192.
Roberts, E. H. (1972). *In* "Viability of Seeds" (E. H. Roberts, ed.) 321–359, Chapman and Hall Ltd.
Roberts, H. A. (1963). *In* "Crop Production in a Weed-free Environment" (Symposium of British Weed Control Council No. 2, E. K. Woodford, ed.), 73–82. Blackwell, Oxford.
Rojas-Garciduenas, M. and Kommedahl, T. (1960). *Weeds,* **8,** 1–5.
Rojas-Garciduenas, M., Ruiz, M. A. and Carrillo, J. (1962). *Weeds,* **10,** 69–71.
Sachs, T. and Thimann, K. V. (1967). *Am. J. Bot.* **54,** 136–144.
Sahai, R. and Sinha, A. B. (1969). *Experientia,* **25,** 653.
Salisbury, E. J. (1961). "Weeds and Aliens", Collins, London.
Schneider, G. (1970). *A. Rev. Pl. Physiol.* **21,** 499–536.
Schoene, D. L. and Hoffman, O. L. (1949). *Science, N.Y.* **109,** 588–590.
Sexsmith, J. J. and Pittman, U. J. (1963). *Weeds,* **11,** 99–101.
Shafer, N. E. and Monson, W. G. (1958). *Weeds,* **6,** 172–178.
Shimizu, N. and Ueki, K. (1972a). *Proc. Crop Sci. Soc. Japan,* **41,** (4), 480–487.
Shimizu, N. and Ueki, K. (1972b). *Proc. Crop Sci. Soc. Japan,* **41,** (4), 488–495.
Simpson, G. M. (1965). *Can. J. Bot.* **43,** 793–816.
Smith, A. E., Zukel, J. W., Stone, G. M. and Riddell, J. A. (1959). *J. agric. Fd Chem.* **7,** 341–344.
Soerjani, M. (1970). *Biotrop Bulletin,* **1,** pp. 88.
Somade, H. M. (1955). *J. Sci. Fd Agric.* **6,** 799–804.
Steward, S. K. (1969). *Weed Sci.* **17,** 299–301.
Tager, J. M. and Clarke, B. (1961). *Nature, Lond.* **192,** 83–84.
Tames, R. S. and Vieitez, E. (1970). *An. Edafol. Agrobiol.* **29,** 775–781.
Teo, C. K. H., Bendixen, L. E. and Nishimoto, R. K. (1971). *Proc. N. cent. Weed Control Conf.* **26,** 85–86.
Teo, C. K. H., Bendixen, L. E. and Nishimoto, R. K. (1973). *Weed Sci.* **21,** 19–23.
Thompson, P. A. (1973). *Ann. Bot.* **37,** 133–154.
Thurston, J. M. (1963). *Rep. Rothamsted exp. Stn 1962*, 236–253.
Tucker, D. J. and Mansfield, T. A. (1972). *Planta,* **102,** 140–151.
Tucker, D. J. and Mansfield, T. A. (1973). *J. exp. Bot.* **24,** (81), 731–740.
Ueki, K. and Shimizu, N. (1967). *Zasso Kenkyu, Tokyo,* **6,** 30–33.
Ueki, K. and Shimizu, N. (1968). *Zasso Kenkyu, Tokyo,* **7,** 110–115.
Ueki, K. and Shimizu, N. (1969). *Proc. Crop Sci. Soc. Japan,* **38,** (2), 261–272.
Vallance, K. B. (1950). *Ann. Bot.* **14,** 347–363.
van Daalen, J. J. and Daams, J. (1970). *Naturwissenschaften,* **8,** 395.
van Vliet, W. F. and Sparenberg, H. (1970). *Potato Res.* **13,** 223–227.
Vegis, A. (1964). *A. Rev. Pl. Physiol.* **15,** 185–224.
Vogt, A. R. and Cox, G. S. (1970). *Forest Sci.* **16,** 165–171.
Wareing, P. F. (1969). *In* "Dormancy and Survival" (H. W. Woolhouse, ed.), 241–262. *Symp. Soc. Exp. Biol. Norwich 1968.*
Wareing, P. F. and Saunders, P. F. (1971). *A. Rev. Pl. Physiol.* **22,** 261–288.
Weaver, R. J. (1959). *Nature, Lond.* **183,** 1198–1199.

Williams, J. T. and Harper, J. L. (1965). *Weed Res.* **5,** 141–150.
Wittwer, S. H. (1953). *Am. Veg. Grow.* pp. 13–14, May 1953.
Worsham, A. D., Moreland, D. E. and Klingman, G. C. (1959). *Science, N.Y.* **130,** 1654–1656.

CHAPTER 6

ACTIONS ON ABSCISSION, DEFOLIATION AND RELATED RESPONSES

F. T. ADDICOTT

Department of Botany, University of California, Davis, California, USA 95616

I. INTRODUCTION

A. NATURAL HISTORY OF ABSCISSION

Abscission is one of the important physiological processes utilized by plants in the control and coordination of their growth and development. Patterns of abscission are critical determinants of both form and function in many plant species, especially those patterns involving the abscission of buds, leaves, flowers, and fruits. As one of the more delicately poised processes in plants, abscission often is quite susceptible to the influence of regulator chemicals. The object of the use of such regulators is seldom that of a drastic effect on the plant, but rather a relatively mild acceleration or retardation of normal abscission.

Competition among organs for light, water, nutrients, and similar factors usually results in the abscission of the weaker organs. Differing levels of hormones within the organs determine in large measure which organs become vigorous sinks and develop to maturity, and which organs weaken and are abscised. Thus the abscission of an organ is preceded by important changes in its hormone metabolism (see Addicott, 1970); such changes occur whether abscission is the result of senescence or of abortion. Some of the most effective practices for the regulation of abscission are those in which the applied regulator appears to augment, reduce, or otherwise modify the activity of endogenous hormones.

The process of abscission is sensitive to ecological conditions, and changes in such factors modify the plant's ability to respond to abscission regulators. Abscission is favoured by low light intensity, shortening photo-period, high temperature, moisture stress, mineral (especially nitrogen) deficiency, atmospheric pollution, and infection by certain insects and micro-organisms. Abscission tends to be delayed by high light intensity, long photo-period, moderate temperature, ample moisture, and high nitrogen (see Addicott and Lyon, 1973). Sometimes, although not always, these tendencies facilitate abscission responses to applied chemicals.

B. MORPHOLOGY

A wide variety of plant organs and tissues can be shed through the process of abscission including leaves, flowers, fruits, seeds, buds, branches, stipules, prickles and bark, among others. Typically, the process of abscission is restricted to an *abscission zone* at the base of an organ (Fig. 1). The cells of the zone remain relatively undifferentiated; the cells are smaller, richer in cytoplasm, and have thinner walls than cells in adjacent regions. Within the abscission zone one or more rows of cells develop into a *separation layer* (Fig. 2). As abscission proceeds there are many signs of increased metabolic activity, particularly within the separation layer, as observed by microscopy and the use of histochemical reagents (Carns, 1966; Jensen and Valdovinos, 1968; Webster, 1968; Henry and Jensen, 1973; Henry *et al.* 1974; Sutcliffe *et*

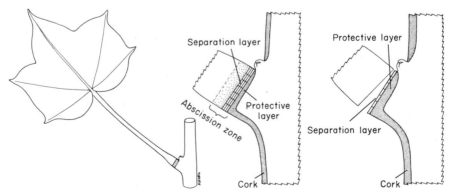

Fig. 1. Diagram of a leaf abscission zone showing location of separation layer and protective layer.

al. 1969). The metabolic activity culminates in the dissolution of cell walls and sometimes protoplasm as well. The early plant anatomists demonstrated that cellulose and pectins are rapidly hydrolyzed during abscission (see Eames and MacDaniels, 1947). It is now apparent that hydrolysis can affect all of the cell wall components including lignin, and protoplasmic constituents as well. The anatomical pattern of hydrolysis varies with species and determines the nature of the scar that remains after abscission (see Addicott, 1965). In most cases one or more *protective layers* develop in the abscission zone, usually involving some cell division. The protective layer is often continuous with the cork (periderm) that commonly develops simultaneously in adjacent portions of the stem (Fig. 2; and see Eames and MacDaniels, 1947).

The electron microscope has enabled observations of the fine structure of cells of the separation layer during abscission. So far there has been confirmation of the swelling and solubilization of cellulosic and pectic components of the cell wall (Valdovinos and Jensen, 1968). Rough endoplasmic reticulum increased, microbodies decreased, and there were substantial alterations in the nucleolus during flower pedicel abscission (Jensen and Valdovinos, 1968). Such changes are indicative of the greatly modified metabolism in active abscission cells.

C. PHYSIOLOGY

The onset and rate of abscission can be influenced by such a wide variety of ecological and physiological factors (see Addicott and Lyon, 1973) that it is not surprising that the mechanism of control remained a mystery for many

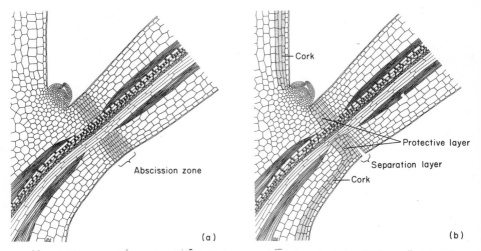

Fig. 2. Diagram of tissues in a leaf abscission zone. (a) Before abscission. Note small compact cells and absence of fibres in the abscission zone. (b) During abscission. Note continuity of protective layer and cork.

years. However, in the past two decades (i) the essentiality of respiration was demonstrated, and (ii) it was recognized that the final steps in abscission include the synthesis and/or release of hydrolytic enzymes. With establishment of the involvement of these two processes, the roles that other factors must play have become more apparent, although the complete picture is still far from clear.

In 1951 Carns et al., showed that oxygen was essential for abscission and that the process could not proceed anaerobically. Carns further showed that a variety of inhibitors of respiratory reactions were all inhibitory to abscission (Carns, 1951). Additional indication of the involvement of respiration came from the observation that abscission proceeded slowly at best in explants (excised abscission zones) depleted of photosynthate. In such explants application of sucrose insured a rapid rate of abscission (Biggs and Leopold, 1957). Also it can be noted that other physiological factors affect abscission in the same manner that they affect respiration. For example, carbon dioxide is generally inhibitory to abscission, as it is to respiration; and abscission response curves to temperature parallel respiration response curves to temperature (see Addicott and Lyon, 1973). Plants that have been exposed to extreme temperatures (that irreversibly damage the respiratory mechanism) are no longer able to abscise leaves or other organs.

While a moderate supply of photosynthate appears essential to abscission, higher amounts tend to retard or prevent abscission. Sucrose applied to explants that already contain adequate levels of photosynthate retarded abscission (Brown and Addicott, 1950; Biggs and Leopold, 1957). High levels of carbohydrate reserves increase fruit set, i.e. reduce abscission of young fruit, in many horticultural crops (Chandler, 1951). The interpretation of these observations is that an abundant supply of carbohydrate makes for thicker cell walls which are resistant to hydrolysis, and the additional soluble carbohydrate such as sucrose would tend to maintain carbohydrate equilibria favourable to the integrity of the existing walls.

The mineral plant nutrients are also important to abscission and influence the responsiveness of the plant to abscission regulators. The tendency of nitrogen-deficient plants to abscise leaves and fruits is well known. Conversely plants supplied with an abundance of nitrogen retain leaves longer and are resistant to chemical defoliation (Table I). The levels of auxin in the plant are also correlated with nitrogen supply (Avery et al. 1937; and see Table I). In view of the strong ability of auxin to influence abscission, the correlation between nitrogen supply and abscission behaviour can be ascribed largely (but not solely) to the intermediate effect that nitrogen has upon the levels of auxin within the plant. A similar situation appears to exist in the phenomenon of apical dominance where differences in the supply of nitrogen and other nutrients are correlated with differences in responses to regulators (McIntyre, 1973). In addition to auxin the levels of two other hormones are presumably influenced by nitrogen: the cytokinin molecule contains six nitrogen atoms, and ethylene is derived from the amino acid, methionine. Each of these hormones has its effects upon abscission as discussed in other sections of this chapter (and see Chapter 8 for a full discussion of ethylene metabolism).

TABLE I

RELATION OF SOIL NITROGEN, PETIOLE AUXIN AND RESPONSE TO
A DEFOLIANT IN COTTON[1]

Soil nitrogen[2]	Petiole auxin[3]	Percent defoliation[4]
0	0·1	95
75	0·4	80
105	0·8	70
150	1·4	50

[1] Note that as soil nitrogen was increased, petiole auxin increased and defoliation response fell sharply (data of H. R. Carns, unpublished).
[2] Pounds N per acre, applied as NO_3^-.
[3] IAA equivalents in μg per kg fr. wt. measured 60 h after defoliant application.
[4] Percent leaf abscission following application of 20 pounds per acre (22·5 kg/ha) of calcium cyanamide.

Deficiencies of elements other than nitrogen also induce abscission. These elements include P, K, S, Ca, Mg, Zn, B, and Fe. Their probable involvement with mechanisms of abscission has been discussed elsewhere (Addicott and Lyon, 1973). For the purpose of the present discussion it is sufficient to note that such deficiencies tend either to weaken the cell wall chemically, or to reduce the physiological restraints on abscission.

Abscission is also sensitive to the water relations of the plant. Prolonged flooding can induce abscission, presumably by limiting the ability of the plant to absorb nitrogen. Moisture stress is perhaps the most common ecological factor inducing abscission. Among the many physiological changes brought on by moisture stress are several hormonal changes each of which has a promotive effect on abscission. These include a rapid rise in abscisic acid (ABA) to many times its normal level (Fig. 3; Wright and Hiron, 1972). There is also a decrease in diffusible auxin, decrease in cytokinin activity, and increase in release of ethylene (see Addicott and Lyon, 1973).

Abscission of leaves and other discrete organs is clearly a correlation phenomenon. A healthy leaf inhibits its own abscission. Normally leaf abscission is initiated with the senescence of the leaf blade. Senescence of any particular leaf will be accelerated if it loses out in competition with other leaves for light, water, mineral nutrients, etc. Physical injury can induce abscission but an appreciable portion of the blade must be removed before abscission is accelerated. For example, in the bean (*Phaseolus vulgaris*) removal of 90% of the blade was required before abscission was accelerated (Swets and Addicott, 1955).

In addition to the nutritional factors described above one of the most important ways in which an organ controls its abscission is by the secretion of hormones. Each of the five major plant hormones has one or more significant influences on abscission. Auxin, approaching the abscission zone from the

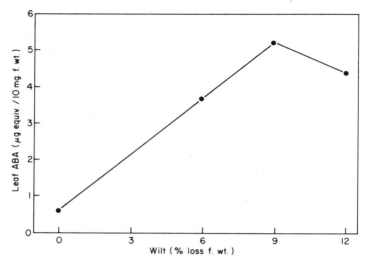

Fig. 3. Increase in abscisic acid (ABA) content of wilting wheat leaves (after Wright, 1969).

distal, subtending organ, appears to be the most important factor inhibiting abscission. In contrast, auxin approaching the abscission zone from a proximal direction tends to accelerate abscission. Hormone levels *in* the abscission zone are also important factors. For example, when cytokinin was applied directly to the abscission zone of bean leaflet explants abscission was retarded, while applications either distal or proximal to the abscission zone accelerated abscission (Osborne and Moss, 1963). Ethylene is a potent accelerator of abscission, and increased release of it has been found to precede or accompany abscission in many (but not all) situations. Further, applications of regulators and hormones that accelerate abscission also accelerate the release of ethylene by the tissues. In the case of hormonal accelerations each of the hormones has one or more other effects on abscission beyond increasing the release of ethylene (e.g. Cracker and Abeles, 1969; Jackson and Osborne, 1972; Cooper and Henry, 1973). Levels of ABA have been found to increase before abscission, particularly before the abscission of young fruit. Applied ABA strongly promotes abscission in most explant systems, but larger dosages are required to accelerate abscission under field conditions. On young fruit, local applications of auxins, GA (gibberellin), or cytokinin almost invariably stimulate development of the treated fruit and greatly reduce young fruit abscission. For example, in cotton 100% of the young fruit individually treated with GA are retained (Walhood, 1957). Apparently the increased levels of these hormones make the fruit better sinks, better able to mobilize nutrients to themselves, and the resulting more vigorously growing fruit are better able to inhibit their own abscission. Some of the effects of auxin applied proximal to the abscission zone, and of cytokinin applied away from the abscission zone, would appear due to the mobilization of nutrients away from the abscission

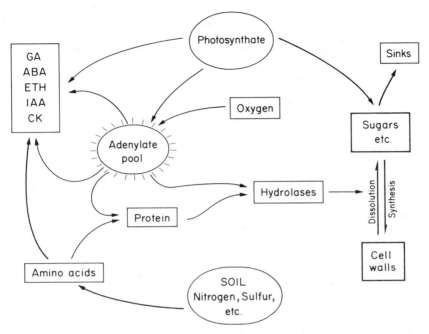

Fig. 4. Relationships of some of the principal physiological factors in abscission. Abbreviations: ABA, abscisic acid; CK, cytokinins; ETH, ethylene; GA, gibberellins; IAA, indol-3-yl-acetic acid and related auxins. See text for explanation.

zone to the site of application, thus facilitating the hydrolysis of cell walls. Relationships of some of the major factors in the physiology of abscission are shown schematically in Fig. 4. (For references and further discussion of the role of hormones in the control of abscission see Addicott, 1970.)

In conclusion, the control of abscission is clearly a complex function of plant hormones and nutrients acting and interacting (i) at the abscission zone, (ii) in the subtending organ, and (iii) at even more distant sites. The influence of an applied regulator chemical on abscission is to modify one or more of these actions or interactions. The following sections outline current practices of abscission regulation and present current views of the physiology and biochemistry of this regulation. For broader treatment of historical and applied aspects of the subject see Audus (1959), Avery and Johnson (1947), Tukey (1954), and Weaver (1972).

II. Leaf Abscission

A. DEFOLIATION

The application of regulator chemicals to defoliate or at least to accelerate leaf abscission is an important agricultural practice. Defoliation is especially valuable in the culture of crops such as cotton where the removal of leaves greatly facilitates harvest by machine (Fig. 5). The leaves not only obstruct operation of the picking machines, they also contribute stain and trash which are difficult to clean from the fibre. When cotton plants are especially dense and leafy the lower portions only may be defoliated (bottom defoliation) before the top portions, to reduce losses from boll rots (Elliott, 1969; Hoover, 1972; Mullendore *et al.* 1972).

Fig. 5. Experimental plot of chemically defoliated cotton ready for machine harvest. Note leafy, undefoliated control plots on each side (photograph courtesy of V. T. Walhood).

Defoliation of young nursery trees prior to digging, storage, and shipping, helps prevent both moisture loss and development of diseases that start on foliage during storage. Further, defoliation enables earlier digging which is advantageous especially in cool, rainy climates (see Addicott and Lynch, 1957; Larsen, 1967).

The use of defoliation to control insects or disease has been explored in a number of situations. For example, the causative organism of bacterial canker of stone fruit trees enters through freshly exposed leaf scars and infection may occur repeatedly during autumnal leaf abscission. Early defoliation by chemicals followed by a protective spray to the unhealed leaf scars has been investigated as a means of controlling the disease (W. H. English, unpublished). In the case of diseases of foliage such as the South American leaf blight of *Hevea,* the spread of the disease can be prevented by defoliating nearby healthy trees (Osborne, 1968). While the treated trees themselves may be badly injured, chemical defoliation offers a practicable method of containing and eradicating such serious diseases. Control of phyllophagous insects such as the pine needle borer may be feasible if defoliation can be achieved at times when the insects are not highly mobile, and if the trees can survive the loss of foliage. The Tsetse fly of Africa thrives in shaded, leafy habitats and the removal of vegetation or at least the defoliation of areas near human habitations is an important public health objective. Aerial applications of 2,4,5-T were effective in killing or defoliating a number of plant species in Tsetse fly habitats, but the applications did not defoliate enough species to render the habitats untenable to the flies (Osborne, 1968). Thus, defoliation is a potentially valuable means for the control of certain insects and diseases, but in many instances difficulties remain to be overcome.

Defoliation is also practised for the purpose of increasing visibility in forested areas. Such practices have obvious military applications and are credited with saving many lives from ambush during recent military operations in the tropics. Military "defoliation" is more closely related to herbicidal action than to defoliation proper, since the chemicals employed usually kill vegetation rather than merely promote leaf abscission from otherwise unaffected trees. As a military weapon the chemicals used in defoliation are unusual in that they rarely affect humans; however the ecology of the treated areas can be seriously disturbed (see Chapter 12, vol. 2; Kozlowski, 1973).

A wide variety of chemicals can function as defoliants. However, because of official regulations and limitations in the chemicals themselves, only a few are at present in commercial use in the United States. These include: sodium chlorate (with fire suppressant, usually a borate); tributyl phosphorotrithioate; tributyl phosphorotrithioite; ethephon; and ammonium nitrate. Other chemicals whose use for defoliation is not at present officially approved include: calcium cyanamide, ammonium thiocyanate, sodium monochloracetate, sodium ethyl xanthate, aminotriazole, butynediol, endothal, zinc chloride, sodium hypochlorite, thiourea, octynediol, and phenyl mercuric compounds. Dozens of other chemicals have been demonstrated to have defoliant action in certain circumstances (Addicott and Lynch, 1957; Elliott, 1969; Hoover, 1972; Mullendore et al. 1972).

The related agricultural practice of *desiccation* intergrades with defoliation. The two practices have the same general objective, to prepare a crop for harvest. Desiccants are essentially contact herbicides whose application results in rapid loss of moisture from the treated foliage. Usually the entire plant is killed including abscission zones, so that leaf abscission is prevented. However, when the abscission zones are not killed, the application can lead to considerable leaf abscission. For example, when 2,4,5-T is applied to a forest containing many species, commonly some species respond with defoliation (and soon resprout) others are killed outright. Light dosages of desiccants to crop plants can result in defoliation, while heavy dosages of defoliants can result in desiccation. Desiccants currently approved for commercial use in the United States include arsenic acid, thionoformate, ammonia (requires a special machine), diquat dibromide, paraquat, DNBP, endothal, hexachloroacetone,

PCP, and petroleum fractions (Elliott, 1969; Hoover, 1972; Mullendore *et al.* 1972).

It should be noted in passing that while the use of regulator chemicals is the common method of achieving defoliation, it is not always the most practicable method. For example, in some situations sheep are the most effective defoliators available. In other circumstances a useful degree of defoliation can be induced merely by withholding irrigation water. Also, use of a chemical "wiltant" such as neodecanoic acid can be a valuable means of preparing cotton for harvest. A few hours after application of neodecanoic acid cotton can be machine harvested; by that time the wilted leaves offer reduced mechanical resistance to the picking machine and contribute little stain to the fibre (Miller *et al.* 1971).

Knowledge of the physiological action of defoliants is yet far from complete. Responses to defoliants duplicate to some extent (and in an accelerated fashion) the normal preabscission changes that take place in senescent leaves. These changes include degradation of chlorophyll and other leaf pigments. Polysaccharides and polypeptides are hydrolyzed and there is some export of sugars and amino acids from the senescent leaf blade. The chemical changes are facilitated by increased activity of a number of enzymes, especially RNAase, protease, peroxidase, and several hydrolytic enzymes. There is a gradual loss of moisture. Some of the soluble mineral elements such as phosphorus, potassium, iron, and magnesium are exported. Levels of auxin and cytokinin decline in the senescent leaf while ABA and ethylene have been found to increase (see Addicott, 1965, 1970; Osborne, 1973; Sacher, 1973). The extent to which any particular preabscission change stimulates abscission is not yet clear. Experiments with debladed petioles and with explants indicate that the isolated abscission zone is capable of all the metabolic activity required to complete the process of abscission. Present evidence indicates that the most important preabscission change in the leaf blade is the lowering of the auxin level releasing the abscission zone from the auxin-imposed inhibition of abscission. Probably the next most important changes are increases in ABA and in ethylene, both of which are promotive of abscission. While sugars and amino acids influence abscission in various ways, there is only limited evidence that these substances from the senescent leaf blades might influence the relatively rapid abscission responses to defoliants.

The pattern of injury following application of a defoliant varies with the chemical employed (Johnson, 1955). The initial symptoms may be mild, almost imperceptible changes (such as the darkening and reddening that follow application of calcium cyanamide), or the symptoms may develop rapidly (as does wilting after application of sodium chlorate). In cotton defoliation, hydrolysis of carbohydrate and protein, and reduction in available sulfhydryl groups occur (Hall and Lane, 1952; Leinweber and Hall, 1959; Katterman and Hall, 1961).

Rapid changes in hormone metabolism are among the important responses to defoliant applications. Auxin levels in bean leaf blades fell 30% in the first hour after application of ammonium thiocyanate (Table II; Swets and

Addicott, 1955). Similar changes have been observed in cotton (Table I). Abscission-promoting applications of ethylene reduced auxin transport in cotton petioles by 70% (Beyer and Morgan, 1971; Beyer, 1973). Further, auxin-transport inhibitors such as TIBA, naptalam, and DPX-1840 were quite promotive of abscission (Fig. 6; Morgan and Durham, 1972, 1973; and see Chapter 8).

TABLE II

AUXIN IN BEAN LEAF BLADES AFTER DEFOLIANT APPLICATION[1]

	Auxin	
Time[2]	*IAA equivalents in* *μg per kg fr. wt.*	*Percent of control*
0[3]	2·6	100
1	1·8	69
24	1·2	46
72	1·2	46

[1] Swabbed with 3% NH_4SCN (from Swets and Addicott, 1955).
[2] Hours after application.
[3] Control: no application.

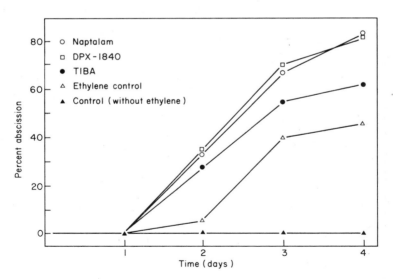

Fig. 6. Augmentation of ethylene-induced leaf abscission in cotton by auxin transport inhibitors. Cotton seedlings, 26 days old were sprayed with 1 mM of a transport inhibitor and kept in chambers with 5 μl/l ethylene. There was no abscission response to the transport inhibitors in the absence of ethylene (closed symbols) (after Morgan and Durham, 1972).

Ethylene is readily released by plant tissues treated with defoliants (Abeles, 1967), but the extent to which such wound ethylene influences abscission in the field remains to be established. Certainly much of the ethylene released by blades and petioles will escape to the atmosphere. However, even transitory amounts of ethylene would be expected to promote in some degree the oxidative destruction of IAA and to reduce its transport (Chapter 12). Further, if the defoliant application resulted in an appreciable increase of ethylene in the abscission zone proper, the ethylene would stimulate directly the metabolic changes that are essential to abscission (see Addicott, 1970). The picture is somewhat confused by the fact that senescent leaves, those most susceptible to defoliants, normally release much less ethylene than do vigorous mature leaves (Hall *et al.* 1957). Clearly, much more work is required to elucidate the role of defoliant-induced ethylene in defoliation responses.

As yet there have been no investigations of the effects of defoliants on ABA. However the wilting that typically follows a defoliant application should lead to a rapid increase in ABA (see Fig. 3). In wilting leaves of wheat, ABA increased ten-fold in 4 h (Wright, 1969). Such changes coupled with auxin and ethylene changes would result in a hormonal combination strongly promotive of abscission.

Amino acids, released in the accelerated senescence that follows defoliant application could conceivably contribute to the promotion of abscission. Such amino acids could provide some of the precursors for the synthesis of the hydrolytic enzymes of abscission. Further the amino acid, methionine, is both the source of ethylene (Abeles, 1973) and a donor of methyl groups that can replace calcium bridges, thus weakening the pectic substances of the cell walls (Valdovinos and Muir, 1965).

B. PREVENTION OF LEAF ABSCISSION

The need to prevent leaf abscission arises in various circumstances, particularly in the florist and nursery trades. In some instances the auxin regulators have been notably successful. For example, NAA applied as a dip is quite effective in preventing the abscission of leaves and berries of holly during shipment (Roberts and Ticknor, 1970). Leaf abscission of *Euonymus* has been similarly delayed (Worley and Grogan, 1941). However, in many cases regulator treatment is not necessary; premature leaf abscission can be prevented by careful attention to mineral nutrition, to water relations, and to atmosphere during storage and shipment.

Needle abscission from conifers used as Christmas trees can be a serious problem. The auxin-regulators (e.g. NAA, 2,4-D) have not been effective in preventing this abscission; indeed such regulators more often accelerate needle abscission of conifers (Worley and Grogan, 1941; F. T. Addicott, unpublished). With Christmas trees, the maintenance of relatively high levels of moisture within the tree, by use of antitranspirants or by other means, can prevent appreciable premature needle abscission, and also reduce the flammability of the trees (R. W. Dingle, personal communication). Possibly the relatively rapid abscission of conifer needles following the imposition of moisture stress indicates the presence of an efficient mechanism for release or synthesis of ABA.

Although there has been no investigation of the action of auxin-regulators in preventing leaf abscission, it is reasonable to suppose that they act by helping to maintain high auxin levels in the abscission zone (as well as in other tissues).

III. FLOWER AND FRUIT ABSCISSION

A. FLOWER ABSCISSION

The abscission of flowers and flower parts is often under rather precise physiological control. For example, petals of many species are abscised promptly after fertilization is effected, but if fertilization is delayed or prevented the petals will be retained appreciably longer. Presumably hormones have a central role in the control of petal abscission. However, application of hormone-regulators has had only limited success in delaying petal abscission or prolonging flower life. Of 15 species of flowering trees and shrubs sprayed with NAA, NOXA, and 4-CPA only the Japanese flowering cherries and white flowering dogwood showed delayed petal abscission (Wester and Marth, 1950). Flower abscission of lupin and *Begonia* was delayed for several days by application of NAA (Warne, 1947; Wasscher, 1947). *Bougainvillea* can be grown as an attractive pot plant; however the flowers abscise soon after the plants are removed from the greenhouse. This abscission can be delayed for two or three weeks by a spray of NAA (Fig. 7; Hackett *et al.* 1972). Similarly,

Fig. 7. Retardation of flower abscission by naphth-1yl-acetic acid (NAA) in *Bougainvillea*. The plants were sprayed at full bloom, three weeks prior to taking this photograph. All flowers have abscised from the untreated check plants (from Hackett *et al.* 1972).

abscission of the showy bracts (as well as leaves) of potted *Poinsettia* can be delayed by application of 2,4,5-trichlorophenoxyacetamide (Carpenter, 1956).

There are numerous situations in which flower parts or fruit falling from ornamental trees are a nuisance. When the flowers of such trees are not attractive or desirable, sprays that lead to the early abortion and abscission of flowers can be applied. If the flowers are desired but the fruit are not, sprays of NAA applied toward the end of the flowering period can prevent fruit development and the aborted fruit will be abscised (Batjer, 1954).

B. YOUNG FRUIT ABSCISSION

1. *Prevention*

The prevention of young fruit abscission is a matter of some importance (i) in the production of certain crops such as tomatoes and green beans, and (ii) after hand pollination in hybridization experiments.

When days are short and light intensity is low, greenhouse-grown tomatoes commonly abscise most of their young fruit. Similarly, field-grown tomatoes usually abscise the young fruit of the first flower clusters that develop when weather is still cool (see Wittwer, 1954). Under such circumstances auxin-regulators are used commercially to increase the number of fruit set. The present practice in southern California is to spray 25 mg/l 4-CPA to the first flower clusters, four or five times at 10-day intervals (B. J. Hall, personal communication).

Green beans (and related kinds of *Phaseolus vulgaris*) abscise most of the

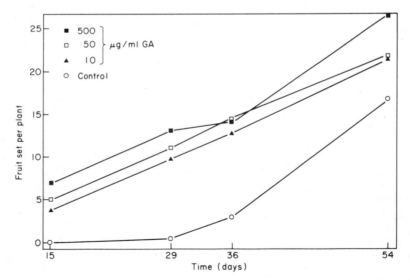

Fig. 8. Stimulation of fruit set in tomato from sprays of gibberellin (GA) applied twice weekly to flower clusters (after Rappaport, 1957).

flowers and young fruit during periods of hot, dry weather. Under these conditions also, the auxin-regulators can greatly reduce losses from young fruit abscission. One or two grams per acre of 4-CPA have given very good results; NAA, NOXA, and α-(2-chlorophenoxy)propionic acid have also been effective (Wittwer, 1954).

Two other classes of hormone-regulators, GA and cytokinins, have also increased fruit set in various experiments. One or both of these are effective on apples, Calimyrna figs, *Citrus,* grapes, tomatoes (Fig. 8), muskmelons, and *Pinus* (e.g. Rappaport, 1957; Hield *et al.* 1958; Crane, 1969; Krezdorn, 1969; Sweet, 1973; Wertheim, 1973).

In addition to the crops mentioned above many others have responded to regulator chemicals with increased fruit set, but various limitations have so far precluded use of those regulators on a commercial scale. Principal limitations encountered are: (i) the delay of abscission may be only transitory, the effects not persisting long enough to improve yields; (ii) in some situations hormone-regulators accelerate rather than retard abscission; (iii) the regulator may injure fruit or vegetative tissues; and (iv) the abscission retardation may persist until maturity, delay normal abscission, and make harvest more difficult (see Audus, 1959; Wittwer, 1954).

Regulator chemicals have proven of great value where pollination and fertilization are essential for the success of a research investigation. It is common knowledge that flowers manipulated by emasculation and pollination are much more likely to be abscised than untouched flowers. Also in the case of some kinds of pollen incompatibilities, the ovary may be abscised before the slowly growing pollen tube can reach the embryo sac. Such unwanted abscission can be prevented or at least greatly reduced by application of regulator chemicals at the time of pollination (Emsweller, 1954). For example, NAA (0·0025% in lanolin) applied to tomato ovaries at the time of hand pollination reduced young fruit abscission and enabled double the number of seeds to be harvested from a cross than would otherwise be possible (C. M. Rick, personal communication). With cotton, a spray of GA (50 to 100 mg/l) is applied to the flower at the time of hand pollination; the treatment is highly effective in assuring the development of a young fruit until it is past the period of abscission (V. T. Walhood, personal communication).

The physiological action of the hormone-regulators in preventing abscission appears to be to make the fruit a sink which is better able to attract nutrients from the rest of the plant (see Crafts and Crisp, 1971). As a result the treated fruit are more vigorous and better able to inhibit their own abscission. Presumably inhibition is accomplished by greater production of growth hormones, particularly auxin (see Addicott, 1970). An important consequence of this physiological action is that when all the sinks of the plant are treated equally there may be no overall response. For example, with both cotton and peaches when the entire plant was sprayed with GA there was some increase in number of fruit set, but the total yield of the plant was not different from the controls (e.g. Stembridge and Gambrell, 1972).

Prevention of young fruit abscission has sometimes followed application of the hormone ABA, and of synthetic regulators such as CCC, Phosphon-D, SADH (succinic acid-2,2-dimethylhydrazide), TIBA, naptalam, and 4-phthalimido-2,6-dimethylpyrimidine (see Crane, 1969; Anderson et al. 1965). For the most part these abscission responses were correlated with inhibition of vegetative growth and promotion of early flowering and fruitfulness. (For other attributes of inhibitor-regulators see Weaver, 1972.)

2. *Fruit Thinning*

The value of thinning young fruit from certain varieties of trees has been apparent for centuries; the most obvious benefits are increased size of remaining fruit and maintenance of regular annual bearing (see Edgerton, 1973; Martin, 1973). Commencing 40 years ago chemical sprays gradually replaced the less economical methods of hand thinning. As often happens the first indications that chemicals might be useful were fortuitous. Insecticides and fungicides such as calcium polysulphide, copper sulphate, and oil emulsions, applied at flowering, induced abscission of young fruit. Because those chemicals had a number of injurious side effects, their use never became established. However, it was soon learned that dinitrophenols, especially dinitro-o-cyclohexylphenol and sodium dinitro-o-cresylate were less hazardous. These chemicals found wide commercial use.

Again fortuitously, in the early 1940s it was discovered that the auxin-regulators, NAA and naphth-1yl-acetamide were also effective fruit thinning agents and even less injurious than the phenolics (Burkholder and McCown, 1941; Thompson, 1957; Hield et al. 1966). Since then a growing variety of chemicals have been investigated. Some have already found use in commercial practice, e.g. 1-naphthyl-N-methylcarbamate (Batjer and Thomson, 1961). Others show good promise and include the auxin 3-chloro-phenoxypropionamide, naptalam (Beutel et al. 1969; but see Martin, 1973), and the ethylene-releasing chemical, ethephon (Buchanan and Biggs, 1969; de Wilde, 1971; Edgerton, 1971; Sweet, 1973; Thompson and Rogers, 1972). Still other regulators that have effectively thinned young fruit, but which because of high cost or other reasons may not become accepted commercially, include the hormone, ABA (Edgerton, 1971), chlorpropham, 1,1,5,5-tetramethyl-3-dimethylaminodithiobiuret (Martin, 1973), morphactin (Weaver and Pool, 1969), and TIBA (Stahly and Williams, 1972; and see Sweet, 1973).

None of the regulator chemicals mentioned above is a panacea. At best, each is effective only on certain cultivars and under certain conditions. Each is a herbicide or potential herbicide and is capable of producing injury at high concentrations or to susceptible tissues.

Progress has been made toward understanding the physiological action of some of the fruit thinning regulators. For example, the dinitro compounds are essentially pollenicides, destroying pollen either before or after pollination (Hildebrand, 1944). Fruit thinning by dinitro regulators is thus the consequence of a reduced number of pollen tubes reaching the ovules. Responses to NAA and related regulators have been studied in more detail.

There is usually an initial cessation of abscission that lasts about two weeks (Luckwill, 1953); this is a "direct" effect of NAA on abscission processes. After that period rates of abscission for the treated fruit increase well above the rates for the untreated fruit. The action of NAA in promoting fruit thinning involves several possible intermediate steps. Thus this influence of NAA on abscission is considered "indirect". When applied at full bloom, NAA affected the styles and stigmas of apples so that most of the germinated pollen tubes seemed incompatible, i.e. they grew slowly and had characteristically swollen tips (Luckwill, 1953). Fruit growth ceased soon after application of NAA (Crowe, 1965; Teubner and Murneek, 1955) and levels of diffusible auxin declined sharply during the first 24 h (Crowe, 1965). Considerable embryo abortion resulted from NAA applications (Luckwill, 1953; Teubner and Murneek, 1955), and translocation of nutrients to the affected young fruit appeared blocked soon after spraying (see Crowe, 1965; Martin, 1973). Almost certainly NAA sprays lead to an increased release of ethylene, but whether this ethylene persists long enough to affect abortion and abscission of the fruit has not been ascertained. Investigation of the role of NAA in fruit thinning has not been made easier by recognition that NAA readily undergoes photodecomposition (Shindy et al. 1973) and that within the plant NAA is easily conjugated and rapidly altered chemically (Luckwill and Lloyd-Jones, 1962; Veen, 1966). Of the photodecomposition products tested only 1-naphthoic acid induced abscission responses in cotton explants that resembled those of NAA, i.e. 1-naphthoic acid accelerated abscission at low dosages and retarded abscission at high dosages. Other photodecomposition products (1-methylnaphthalene, 1-naphthaldehyde, phthalic acid, and naphthalene) had no effect on cotton explant abscission (Shindy et al. 1973). However, the foregoing results do not preclude the possibility that some photodecomposition products could have a role in fruit thinning responses to NAA.

From the foregoing it is evident that the action of NAA on young fruit involves, first, a temporary inhibition of abscission that is presumably the result of the spray directly augmenting the levels of endogenous auxin at the fruit abscission zone. Secondly, NAA causes injuries of several kinds that result in impaired nutrient translocation and in embryo abortion. A major symptom of these injuries is greatly reduced export of endogenous auxin (and presumably other growth hormones) from the young fruit to their abscission zones. The reduced auxin alone is sufficient to account for the increased abscission, but its effect is almost certainly aided by increased ethylene and ABA from the injured, wilting fruit.

The physiological action of other regulator chemicals used in fruit thinning has received little attention. An investigation by Williams and Batjer (1964) suggested that the thinning action of 1-naphthyl-N-methylcarbamate resulted from blockage of nutrient transport to developing fruits. The ethylene released from ethephon undoubtedly has some direct effect upon abscission (see Abeles, 1973). But of equal or greater importance may be the influence of the ethylene on the physiology of endogenous auxin, reducing auxin transport and increasing activity of IAA-oxidase (Morgan et al. 1972; Hall and Morgan,

1964). In a possibly similar way the action of TIBA in fruit thinning appears related to its ability to interfere with the transport of auxin (see Goldsmith, 1968, 1969).

C. PREVENTION OF SEED SHATTERING

In nature plants that shed their seeds in a dry condition are usually adapted to abscise them gradually over an extended period of time. In contrast, it is almost imperative that cultivated plants shed their seeds over a brief period so that harvest can be conducted with efficiency. Established cultivars of seed crops have been selected in part because of an inherently low tendency to seed shattering. From time to time, however, a new cultivar is produced in which extended seed shattering is combined with otherwise quite desirable characteristics. Attempts to prevent such shattering by the use of regulators have met with limited success, e.g. a spray of NAA increased seed retention of a new cultivar of *Phalaris tuberosa* from 50% to 76% (Mullett, 1966). However, subsequent large scale trials gave such variable results that no commercial practice could be established. In that case, in the case of veldtgrass (*Ehrharta calycina*), and most similar cases non-shattering forms are eventually found and established within the cultivar (Love, 1963; R. N. Oram, personal communication).

Many legumes are strongly inclined to shatter seed with the explosive dehiscence of the pods. In some instances, e.g. alfalfa (*Medicago sativa*) grown for seed, the careful coordination of desiccant chemical application and harvest operations can greatly reduce losses from shattering. Harvest can be accomplished successfully by combine after the foliage is dry but before the pods have opened. Desiccants used for this purpose include: diquat, dinoseb, endothal, hexachloroacetone, and petroleum solvents (Hoover, 1972; Pederson *et al.* 1972).

D. MATURE FRUIT ABSCISSION

1. *Prevention*

Preharvest drop, abscission of full-size fruit shortly before they are ready for harvest, afflicts a number of fruit cultivars, particularly of apples and pears. Following the discovery that auxin could retard abscission of debladed petioles (Laibach, 1933; LaRue, 1936), Gardner *et al.* (1939) showed that the auxin-regulator, NAA, and its derivatives could delay the abscission of apples until they were ready for harvest (Batjer, 1954). Since that discovery other auxin-regulators also have been shown to be effective on various cultivars of apples, pears, prunes, oranges, grapefruit and lemons. These regulators include 2,4-D, MCPA, and 2,4,5-TP (Edgerton and Hoffman, 1953; Hield *et al.* 1964; and see Batjer, 1954 and Weaver, 1972).

In commercial practice auxin-regulator sprays are applied to the entire tree, from one to a few weeks before onset of preharvest drop. The role of the applied regulators has been investigated only to a limited extent (e.g. Batjer and Thompson, 1948), but it appears clear that the net effect of the spray must be to maintain a high level of auxin in the pedicel abscission zone. With high dosages or with susceptible cultivars the auxin-regulators accelerate colour development and fruit maturation. Presumably such responses are related to an increased release of ethylene induced by the regulator.

Recently it has been found that the growth retardant, SADH, applied early in the season can reduce substantially the preharvest drop of apples (Batjer and Williams, 1966) and of pears (Martin and Griggs, 1970). On apples such applications have increased anthocyanin colouration and firmness of the mature fruit, and reduced incidence of the defects "water core" and storage scald (Edgerton, 1971; Batjer and Williams, 1966). On pears effective applications can be made any time from shortly after flowering until a month before harvest (Martin and Griggs, 1970). However, application late in the season can lead to the undesirable side effect of increased fruit set the next season (Edgerton, 1971). The responses to SADH can be attributed at least in part to retardation of vegetative growth and consequent higher levels of nutrients and hormones.

2. *Promotion*

In recent years the rising costs of fruit harvest have stimulated intensive research directed to increasing harvest efficiency. Mechanical devices are now available that are capable of shaking fruit from the trees, if the abscission processes are fairly well advanced. However, with many cultivars the fruit are not yet abscising at the optimum time for harvest, and cannot be removed by the limited degree of shaking that can be safely tolerated.

The requirements of a successful promotor of fruit abscission are so rigorous that only a few chemicals have shown any indication of becoming commercially acceptable. Such a regulator must meet the obvious requirements of reasonable cost, of practicability of application and of not affecting the quality of the fruit nor leaving a toxic residue. But perhaps the most stringent requirement is that the regulator must be capable of inducing the abscission of a substantial number of fruit while inducing the abscission of few, if any, of the leaves. None of the regulators examined to date fully meet the requirements, except possibly in restricted circumstances. What is known of their action is discussed in the following paragraphs.

Maleic hydrazide (MH) was the first among the growth retardants to show ability to promote abscission of mature fruit. In olives, fruit abscission was increased from 32% (for the control) to 97% by application of 1% MH (Hartmann, 1955). However, the spray also induced abscission of 20% of the leaves and retarded shoot growth the following season. Subsequent flowering and fruit set were not affected. However, MH was effective during periods of high humidity only, a condition that is not common at the time of olive harvest in California. Although several inhibitory effects of MH on cytological and biochemical processes are known, information on its mode of action as a plant regulator is still quite fragmentary (see Ashton and Crafts, 1973).

The growth retardant, SADH, can hasten maturity and promote abscission of cherry, peach and *Citrus* (see Cooper and Henry, 1973). Possibly SADH functions via a mechanism similar to that described by Reid and Carr (1967) in growth retardation by CCC. Their results suggested that CCC blocked normal GA production and diverted precursors to the synthesis of abnormal GAs. Such an effect by SADH in cherry, peach, and *Citrus* could account for the observed acceleration of maturity and abscission. In *Citrus,* pretreatment with SADH enhanced the ability of the fruit to release ethylene after application of cycloheximide (Cooper and Henry, 1973).

A number of organic acids have shown some ability to promote mature fruit abscission. Some of these participate in normal respiration and metabolism (e.g. ascorbic acid, citric acid) and others are known to affect metabolic processes (e.g. salicylic acid, iodoacetic acid). With such chemicals relatively high dosages are usually required to induce moderate promotion of abscission, up to one or two pounds per tree. In these cases it appears probable that the action of the chemical is simply to injure the fruit, leading to increased release of ethylene and to other changes favouring early abscission (Ben-Yehosua and Eaks, 1969; and see Wilson and Coppock, 1969). In olives a favourable response to these acids required high humidity, and iodoacetic acid induced a heavy leaf abscission (Hartmann *et al.* 1967, 1968). Iodoacetate acted at lower dosages than the other acids, possibly via its well known ability to inhibit sulfhydryl enzymes. Possibly also there are abscission effects from iodide ions released from the iodoacetate. The iodide readily separates in plant tissues (Facteau *et al.* 1968); its promotion of abscission has been known for some time (see Addicott and Lyon, 1973).

Abscisic acid has been investigated in several situations and has induced a variety of responses. In apples ABA strongly promoted abscission of mature fruit with no noticeable fruit injury but with a small amount of leaf abscission (Edgerton, 1971). In cherries ABA has also promoted fruit abscission, including the development of the separation layer (Zucconi *et al.* 1969). In olives, however, ABA had no effect on fruit abscission, but did induce complete defoliation (leaf abscission) (Hartmann *et al.* 1968). In *Citrus* ABA sprays had little effect on fruit abscission, although it sometimes promoted colour development and maturation. However, when followed by sprays of cycloheximide, ABA pretreatment enhanced the release of ethylene by the cycloheximide. In experiments with explants of (i) *Citrus* fruit abscission zones, and (ii) cotton cotyledonary nodes, ABA promoted abscission under conditions apparently free of ethylene, i.e. volatile substances released by the explants were not permitted to accumulate (Cooper and Henry, 1973; M. C. Marynick, unpublished). Abscisic acid can have a number of biochemical effects in the tissues into which it is absorbed, e.g. in the abscission zone it promotes the development of hydrolytic enzymes. Such responses have already been discussed elsewhere (Addicott and Lyon, 1969; Addicott, 1970).

Cycloheximide is an antibiotic inhibitor of protein synthesis. Among its many physiological properties is the ability to accelerate abscission of mature fruit. It has a number of side effects that make its use on some fruits impractical, and it requires favourable water relations to be effective. For example, after cycloheximide sprays on olive trees, leaf abscission was often above the limit the tree could tolerate without damage. Favourable responses in olives required the additional use of agents such as anti-transpirants to maintain high moisture levels in the trees (Hartmann *et al.* 1972). In apples dosages sufficient to induce significant fruit abscission were usually associated with undesirable leaf abscission and injuries to both fruit and leaves (Edgerton, 1971). In *Citrus* cycloheximide can induce a valuable degree of mature fruit abscission. It is useful particularly in the harvest of cultivars such as Valencia orange that is produced primarily for juice. With such varieties the rind injury induced by cycloheximide is of little commercial importance. Effective promotion of *Citrus* fruit abscission is correlated with the release of substantial amounts of ethylene by the injured rind tissues. Presumably ethylene is the intermediate agent through which cycloheximide promotes abscission (Cooper and Henry, 1973). For the present the practical use of cycloheximide appears limited to regions of high humidity, such as Florida; in regions of low humidity cycloheximide has not yet proved reliable.

Ethephon is a regulator chemical that has the remarkable ability to break down and release ethylene in physiologically significant amounts. Ethephon applied in the field is able to induce many responses previously observed only with ethylene under laboratory conditions. It has been extensively tested for the ability to promote abscission of mature fruit in *Citrus* (Hield *et al.* 1968, 1969; Cooper and Henry, 1973), in olives (Hartmann *et al.* 1968), in cherries and plums (Bukovac *et al.* 1969), in grapes (Weaver and Pool, 1969), and in apples (Edgerton, 1971). While the promotion of fruit abscission was sometimes satisfactory in these tests, serious side effects also resulted including acceleration of leaf senescence, excessive leaf abscission, and injury to young shoots. Such side effects appeared correlated in part with the ease with which ethephon is absorbed by leaves and translocated from them (Edgerton and Hatch, 1972). Simultaneous application of NAA, 2,4,5-T, or SADH with ethephon has been beneficial in some instances. For example, NAA with ethephon reduced the olive-leaf abscission action of ethephon alone (Hartmann *et al.* 1968) and in apples the combination of a pretreatment with SADH followed by application of NAA or 2,4,5-TP with ethephon resulted in highly satisfactory ripening and abscission responses (Edgerton and Blanpied, 1970). The physiology and biochemistry of responses to ethephon are discussed in detail in Chapter 8.

As indicated above promotion of mature fruit abscission in olives is particularly difficult because of the low humidity that generally prevails at harvest time and because of the tendency of leaves to be even more susceptible than the fruit to abscission promotors. A new regulator chemical, 2-chloroethyl-*tris*-(2-methoxyethoxyl)-silane, shows high promise . of surmounting the difficulties; in tests extending over three years this chemical was active when humidities were low, and it was free of undesirable side effects (H. T. Hartmann, personal communication).

IV. MISCELLANEOUS

A. STEM AND BRANCH ABSCISSION

Shoot tip abortion and abscission is a common event in the sympodial woody species that do not form true terminal buds (Romberger, 1963; Kozlowski,

1973; Millington and Chaney, 1973). Abscission of lower, shaded branches by means of a separation layer within an abscission zone is a regular occurrence in *Agathis* (Licitis-Lindbergs, 1956) and in several tropical genera (Kozlowski, 1971). Responses of such stem and branch abscission to regulator chemicals have not been investigated.

The species that regularly abscise branches appear to intergrade with species that abscise them only sporadically or with difficulty. Examples of this condition are found in species of *Quercus* and *Populus* that have a swollen abscission zone at the base of the branch twigs. In these species the separation layer often fails to develop fully so that the twigs may not fall of their own weight, but may require the assistance of a force such as wind to complete abscission. Applied to the bases of twigs of *Quercus alba* NAA reduced twig abscission on three of four application dates and ethephon increased twig abscission (Table III) (Chaney and Leopold, 1972). Similar treatments to *Juglans nigra* were without effect on twig abscission (W. R. Chaney, unpublished).

TABLE III

PERCENT TWIG ABSCISSION FROM WHITE OAK TREATED WITH NAA OR ETHEPHON[1]

Treatment[3]	Percent abscission[2]			
	30 Aug	*11 Sept*	*28 Sept*	*11 Oct*
Control	28	36	36	18
1% NAA	36	20	20	12
1% Ethephon	44	56	48	4
2% Ethephon	52	80	48	8

[1] Note that NAA applied on the first date increased abscission, but retarded abscission on the last three dates. While ethephon retarded abscission on the last date and increased it on the first three dates (from Chaney and Leopold, 1972).

[2] Cumulative abscission from date of treatment to 31 December.

[3] Chemicals in lanolin were smeared around the base of the twig.

In *Taxodium, Sequoia,* and related trees the leafy branchlets are abscised as a unit (cladoptosis). Weekly applications of GA during the late summer and early autumn accelerated branchlet abscission in *Taxodium* (Brian *et al.* 1959).

The natural pruning of the lower branches of forest species is not abscission in the strict sense. It is the result of death of the shaded and otherwise disadvantaged lower branches, followed by their decay and eventual falling away from the tree (Kozlowski, 1973). The fewer the number of persistent lower branches the more valuable the tree is for lumber. Pruning of lower branches can be effected by hand tools, by careful use of fire, and by herbicidal chemicals (Millington and Chaney, 1973). Lower branches of oak and other hardwood trees can be readily killed by sprays of 2,4-D and 2,4,5-T. After such sprays branches up to 5 cm in diameter decayed and fell away in four to seven years (McConnell and Kenerson, 1964).

B. BARK REMOVAL

Abscission of bark is a widespread phenomenon, common to numerous species of trees with notable examples in *Pinus, Eucalyptus, Arbutus,* and *Acacia.* In bark abscission, outer dead layers are abscised by the rupture of thin-walled cells. The underlying healthy layers remain to continue the protective function of bark (Eames and MacDaniels, 1947).

Removal of the entire bark is often advantageous in the logging of trees for timber. Such removal eliminates tissues that foster decay and insect attack, and enables the wood to dry, lowering the weight of a log by as much as one-third (Kozlowski, 1973).

In chemical bark removal a herbicide is applied to the sapwood in a band around the base of the tree, after cutting away a ring of bark. The herbicide moves upward in the trunk, initially at least, in the sapwood. Ultimately the cambium and adjacent thin-walled cells collapse and permit the bark to separate from the more mature sapwood (Kozlowski, 1960). The intensity of the response varies with many factors such as season of application and tree vigour, and not all tree species are capable of a commercially acceptable response. The most satisfactory herbicides for bark removal have been arsenic compounds, especially sodium arsenite (Wilcox *et al.* 1956). Other herbicides such as 2,4-D and 2,4,5-T may be satisfactory in some circumstances (Raphael *et al.* 1954).

C. ANOMALOUS ABSCISSION

Abscission zones and separation layers sometimes develop and function in positions where normally none would appear. This is considered *anomalous abscission.* [Pierik (1973) has used the term secondary abscission to indicate one kind of anomalous abscission.] Some of the more common instances of anomalous abscission appear in herbaceous species that can abscise the stump of an internode after injury or excision of the subtending portion. Such abscission can be retarded by auxin-regulators (Beal and Whiting, 1945; Webster and Leopold, 1972).

Application of TIBA to the terminal bud of a bean seedling (*Phaseolus vulgaris*) resulted in anomalous abscission in the subtending internode a few mm above the lower node. In this case the separation layer developed in about the same position as it did following the excision of the terminal bud (Weintraub *et al.* 1952; and see Whiting and Murray, 1948). In related experiments with excised stems of bean, excision induced the development of a separation layer above the node in a similar position to that in the Weintraub experiments. However, upon exposure to ethylene, development of the layer above the node was inhibited, and instead a separation layer developed at the very base of the internode, immediately adjacent to the node (Webster and

Leopold, 1972). Application of IAA with the TIBA, or of NAA with the ethylene, prevented the development of the anomalous separation layers. Such results give further support to the view that both TIBA and ethylene have among their major physiological effects the disruption of normal auxin physiology.

In addition to the foregoing experiments, GA induced anomalous abscission of the stem stump of excised cotyledonary nodes of cotton (Carns *et al.* 1961; Bornman *et al.* 1968). Also, auxins such as IAA and NAA induced anomalous abscission when applied to the bases of excised apple pedicels, and both auxins and cytokinins induced anomalous abscission in excised pear pedicels (Pierik, 1971, 1973). Such results give emphasis to the view that levels and gradients of endogenous regulators are among the critical factors controlling abscission.

The assistance of Alice B. Addicott in the preparation of figures and literature citations is gratefully acknowledged.

REFERENCES

Abeles, F. B. (1967). *Physiologia Pl.* **20**, 442–454.
Abeles, F. B. (1973). "Ethylene in Plant Biology", Academic Press, New York and London.
Addicott, F. T. (1965). *In* "Encyclopedia of Plant Physiology" (W. Ruhland, ed.) 15/2, 1094–1126. Springer-Verlag, Berlin.
Addicott, F. T. (1970). *Biol. Rev.* **45**, 485–524.
Addicott, F. T. and Lynch, R. S. (1957). *Adv. Agron.* **9**, 67–93.
Addicott, F. T. and Lyon, J. L. (1969). *A. Rev. Pl. Physiol.* **20**, 139–164.
Addicott, F. T. and Lyon, J. L. (1973). *In* "Shedding of Plant Parts" (T. T. Kozlowski, ed.), pp. 85–124. Academic Press, New York and London.
Anderson, I. C., Greer, H. A. L. and Tanner, J. W. (1965). *In* "Genes to Genus" (F. A. Greer and T. J. Army, eds), pp. 103–115. International Minerals and Chemicals Corp., Skokie, Illinois.
Ashton, F. M. and Crafts, A. S. (1973). "Mode of Action of Herbicides", Wiley-Interscience, New York.
Audus, L. J. (1959). "Plant Growth Substances", Leonard Hill, London.
Avery, G. S., Jr., Burkholder, P. R. and Creighton, H. B. (1937). *Am. J. Bot.* **24**, 553–557.
Avery, G. S., Jr. and Johnson, E. B. (1947). "Hormones and Horticulture", McGraw-Hill, New York.
Batjer, L. P. (1954). *In* "Plant Regulators in Agriculture" (H. B. Tukey, ed.), pp. 117–131. Wiley, New York.
Batjer, L. P. and Thompson, A. H. (1948). *Proc. Am. Soc. hort. Sci.* **51**, 77–80.
Batjer, L. P. and Thomson, B. J. (1961). *Proc. Am. Soc. hort. Sci.* **77**, 1–8.
Batjer, L. P. and Williams, M. W. (1966). *Proc. Am. Soc. hort. Sci.* **88**, 76–79.
Beal, J. M. and Whiting, A. G. (1945). *Bot. Gaz.* **106**, 420–431.
Ben-Yehoshua, S. and Eaks, I. L. (1969). *J. Am. Soc. hort. Sci.* **94**, 292–298.
Beutel, J., Gerdts, M., LaRue, J. and Carlson, C. (1969). *Calif. Agric.* **23**(1), 6–7.
Beyer, E. M., Jr. (1973). *Pl. Physiol., Baltimore*, **52**, 1–5.
Beyer, E. M., Jr. and Morgan, P. W. (1971). *Pl. Physiol., Baltimore*, **48**, 208–212.
Biggs, R. H. and Leopold, A. C. (1957). *Pl. Physiol., Lancaster*, **32**, 626–632.
Bornman, C. H., Addicott, F. T., Lyon, J. L. and Smith, O. E. (1968). *Am. J. Bot.* **55**, 369–375.
Brian, P. W., Petty, J. H. P. and Richmond, P. T. (1959). *Nature, Lond.* **183**, 58–59.
Brown, H. S. and Addicott, F. T. (1950). *Am. J. Bot.* **37**, 650–656.
Buchanan, D. W. and Biggs, R. H. (1969). *J. Am. Soc. hort. Sci.* **94**, 327–329.

Bukovac, M. J., Zucconi, F., Larsen, R. P. and Kesner, C. D. (1969). *J. Am. Soc. hort. Sci.* **94,** 226–230.
Burkholder, C. L. and McCown, M. (1941). *Proc. Am. Soc. hort. Sci.* **38,** 117–120.
Carns, H. R. (1951). *Oxygen, respiration and other critical factors in abscission.* Doct. dissert., Univ. California, Los Angeles.
Carns, H. R. (1966). *A. Rev. Pl. Physiol.* **17,** 295–314.
Carns, H. R., Addicott, F. T. and Lynch, R. S. (1951). *Pl. Physiol., Lancaster,* **26,** 629–630.
Carns, H. R., Addicott, F. T., Baker, K. C. and Wilson, R. K. (1961). *In* "Plant Growth Regulation" (R. M. Klein, ed.), pp. 559–565. Iowa State Univ. Press, Ames.
Carpenter, W. J. (1956). *Proc. Am. Soc. hort. Sci.* **67,** 539–544.
Chandler, W. H. (1951). "Deciduous Orchards", Lea and Febiger, Philadelphia.
Chaney, W. R. and Leopold, A. C. (1972). *Can. J. Forest Res.* **2,** 492–495.
Cooper, W. C. and Henry, W. H. (1973). *In* "Shedding of Plant Parts" (T. T. Kozlowski, ed.), pp. 476–524. Academic Press, New York and London.
Cracker, L. E. and Abeles, F. B. (1969). *Pl. Physiol., Lancaster,* **44,** 1144–1149.
Crafts, A. S. and Crisp, C. E. (1971). "Phloem Transport in Plants", Freeman, San Francisco.
Crane, J. C. (1969). *HortSci.* **4,** 8–11.
Crowe, A. D. (1965). *Proc. Am. Soc. hort. Sci.* **86,** 23–27.
Eames, A. J. and MacDaniels, L. H. (1947). "An Introduction to Plant Anatomy", McGraw-Hill, New York.
Edgerton, L. J. (1971). *HortSci.* **6,** 378–382.
Edgerton, L. J. (1973). *In* "Shedding of Plant Parts" (T. T. Kozlowski, ed.), pp. 435–474. Academic Press, New York and London.
Edgerton, L. J. and Blanpied, G. D. (1970). *J. Am. Soc. hort. Sci.* **95,** 664–666.
Edgerton, L. J. and Hatch, A. H. (1972). *J. Am. Soc. hort. Sci.* **97,** 112–115.
Edgerton, L. J. and Hoffman, M. B. (1953). *Proc. Am. Soc. hort. Sci.* **62,** 159–166.
Elliott, F. C. (1969). "Cotton Defoliation Guide". Bull. No. L-145. Texas Agricultural Extension Service, College Station, Texas.
Emsweller, S. L. (1954). *In* "Plant Regulators in Agriculture" (H. B. Tukey, ed.), pp. 161–169. Wiley, New York.
Facteau, T. J., Hendershott, C. H. and Biggs, R. H. (1968). *Proc. Am. Soc. hort. Sci.* **92,** 195–202.
Gardner, F. E., Marth, P. C. and Batjer, L. P. (1939). *Proc. Am. Soc. hort. Sci.* **37,** 415–428.
Goldsmith, M. H. M. (1968). *In* "Biochemistry and Physiology of Plant Growth Substances" (F. Wightman and G. Setterfield, eds), pp. 1037–1050. Runge Press, Ottawa.
Goldsmith, M. H. M. (1969). *In* "The Physiology of Plant Growth and Development" (M. B. Wilkins, ed.), pp. 125–162. McGraw-Hill, New York.
Hackett, W. P., Sachs, R. M. and Debie, J. (1972). *Calif. Agric.* **26**(8), 12–13.
Hall, W. C. and Lane, H. C. (1952). *Pl. Physiol., Lancaster,* **27,** 754–768.
Hall, W. C. and Morgan, P. W. (1964). *In* "Régulateurs Naturels de la Croissance Végétale" (J. P. Nitsch, ed.), pp. 727–745. Centre National de la Recherche Scientifique, Paris.
Hall, W. C., Truchelut, G. B., Leinweber, C. L. and Herrero, F. A. (1957). *Physiologia Pl.* **10,** 306–317.
Hartmann, H. T. (1955). *Bot. Gaz.* **117,** 24–28.
Hartmann, H. T., Fadl, M. and Whisler, J. (1967). *Calif. Agric.* **21**(7), 5–7.
Hartmann, H. T., Heslop, A. J. and Whisler, J. (1968). *Calif. Agric.* **22**(7), 14–16.
Hartmann, H. T., El-Hamady, M. and Whisler, J. (1972). *J. Am. Soc. hort. Sci.* **97,** 781–785.
Henry, E. W. and Jensen, T. E. (1973). *J. Cell Sci.* **13,** 591–601.
Henry, E. W., Valdovinos, J. G. and Jensen, T. E. (1974). *Pl. Physiol., Baltimore,* **54,** 192–196.
Hield, H. Z., Burns, R. M. and Coggins, C. W., Jr. (1964). "Pre-harvest use of 2,4-D on citrus". Calif. Agric. Expt. Stn, Circ. 528, 1–10.
Hield, H. Z., Coggins, C. W., Jr. and Garber, M. J. (1958). *Calif. Agric.* **12**(5), 9–11.
Hield, H. Z., Coggins, C. W., Jr., Knapp, J. C. F. and Burns, R. M. (1966). *Calif. Citrogr.* **51,** 312–343.
Hield, H. Z., Palmer, R. L. and Lewis, L. N. (1968). *Calif. Citrogr.* **53,** 368–392.
Hield, H. Z., Palmer, R. L. and Lewis, L. N. (1969). *Calif. Citrogr.* **54,** 292–324.

Hildebrand, E. M. (1944). *Proc. Am. Soc. hort. Sci.* **45**, 53–58.
Hoover, M. (1972). "Defoliation and Other Harvest-aid Practices". Division of Agricultural Sciences, Univ. of California, Berkeley.
Jackson, M. B. and Osborne, D. J. (1972). *J. exp. Bot.* **23**, 849–862.
Jensen, T. E. and Valdovinos, J. G. (1968). *Planta*, **83**, 303–313.
Johnson, P. W. (1955). *Proc. 9th annu. Cotton Defoliation Conf.* 24–25.
Katterman, F. R. H. and Hall, W. C. (1961). *Pl. Physiol., Lancaster,* **36**, 816–819.
Kozlowski, T. T. (1960). *Proc. 17th N. Cent. Weed Control Conf.* 1–10.
Kozlowski, T. T. (1971). "Growth and Development of Trees", Vol. 2. Academic Press, New York and London.
Kozlowski, T. T. (1973). *In* "Shedding of Plant Parts" (T. T. Kozlowski, ed.), pp. 1–44. Academic Press, New York and London.
Krezdorn, A. H. (1969). *In* "Proceedings First International Citrus Symposium" (H. D. Chapman, ed.). Vol. III, pp. 1113–1119. Univ. California, Riverside.
Laibach, F. (1933). *Ber. dt. bot. Ges.* **51**, 386–392.
Larsen, F. E. (1967). *Proc. int. Pl. Propagators Soc. ann. Meeting* 157–172.
LaRue, C. D. (1936). *Proc. natn. Acad. Sci. U.S.A.* **22**, 254–259.
Leinweber, C. L. and Hall, W. C. (1959). *Bot. Gaz.* **120**, 183–186.
Licitis-Lindbergs, R. (1956). *Phytomorphology,* **6**, 151–167.
Love, R. M. (1963). *Calif. Agric.* **17**(10), 2–3.
Luckwill, L. C. (1953). *J. hort. Sci.* **28**, 25–40.
Luckwill, L. C. and Lloyd-Jones, C. P. (1962). *J. hort. Sci.* **37**, 190–206.
McConnell, W. P. and Kenerson, L. (1964). *J. For.* **62**, 463–466.
McIntyre, G. I. (1973). *Can. J. Bot.* **51**, 293–299.
Martin, G. C. (1973). *In* "Symposium on Growth Regulators in Fruit Production" (S. J. Wellensiek, ed.), pp. 345–352. Technical Communications International Society for Horticultural Science No. 34, Vol. 1, The Hague.
Martin, G. C. and Griggs, W. H. (1970). *HortSci.* **5**, 258–259.
Miller, C. S., Wilkes, L. H., Thaxton, E. L. and Hubbard, J. L. (1971). *Bull. Texas agric. Exp. Stn.* MP-1010. College Station.
Millington, W. F. and Chaney, W. R. (1973). *In* "Shedding of Plant Parts" (T. T. Kozlowski, ed.), pp. 149–204. Academic Press, New York and London.
Morgan, P. W. and Durham, J. I. (1972). *Pl. Physiol., Baltimore,* **50**, 313–318.
Morgan, P. W. and Durham, J. I. (1973). *Planta,* **110**, 91–93.
Morgan, P. W., Ketring, D. L., Beyer, E. M., Jr. and Lipe, J. A. (1972). *In* "Plant Growth Substances 1970" (D. J. Carr, ed.), pp. 502–509. Springer-Verlag, Berlin.
Mullendore, G. P., Thomas, R. O. and Cathey, G. W. (1972). "Guide for Use of Defoliants and Desiccants". Inf. Sheet 529. Crops Res. Div., USDA, Stoneville, Miss.
Mullett, J. H. (1966). *J. Aust. Inst. agric. Sci.* **32**, 218–219.
Osborne, D. J. (1968). *Nature, Lond.* **219**, 564–567.
Osborne, D. J. (1973). *In* "Shedding of Plant Parts" (T. T. Kozlowski, ed.), pp. 125–147. Academic Press, New York and London.
Osborne, D. J. and Moss, S. E. (1963). *Nature, Lond.* **200**, 1299–1301.
Pederson, M. W., Bohart, G. E., Marble, V. L. and Klostermeyer, E. C. (1972). *In* "Alfalfa Science and Technology" (C. H. Hanson, ed.), Monogr. 15, pp. 689–720. Am. Soc. Agron., Madison.
Pierik, R. L. M. (1971). *Naturwissenschaften.* **58**, 568–569.
Pierik, R. L. M. (1973). *Acta Hort.* **34**, 299–309.
Raphael, H. J., Panshin, A. J. and Day, M. W. (1954). *Q. Bull. Mich. St. Univ. agric. Exp. Stn.* **37**, 230–240.
Rappaport, L. (1957). *Pl. Physiol., Lancaster,* **32**, 440–444.
Reid, D. M. and Carr, D. J. (1967). *Planta,* **73**, 1–11.
Roberts, A. N. and Ticknor, R. L. (1970). *Am. hort. Mag.* **49**, 301–314.
Romberger, J. A. (1963). "Meristems, Growth, and Development in Woody Plants". U.S. Department of Agriculture, Forest Service Tech. Bull. 1293.
Sacher, J. A. (1973). *A. Rev. Pl. Physiol.* **24**, 197–224.

Shindy, W. W., Lyon, J. L., Gauer, W. O., Crosby, D. G. and Addicott, F. T. (1973). *Pl. Cell Physiol.* **14,** 169–176.
Stahly, E. A. and Williams, M. W. (1972). *J. Am. Soc. hort. Sci.* **97,** 724–726.
Stembridge, G. E. and Gambrell, C. E., Jr. (1972). *J. Am. Soc. hort. Sci.* **97,** 708–711.
Sutcliffe, J. F., Arch, P. D., Leggett, P. A., Phillips, B. J. and Sexton, R. (1969). *Abstracts, XIth int. bot. Congr.,* Seattle, U.S.A., p. 213.
Sweet, G. B. (1973). *In* "Shedding of Plant Parts" (T. T. Kozlowski, ed.), pp. 341–382. Academic Press, New York and London.
Swets, W. A. and Addicott, F. T. (1955). *Proc. Am. Soc. hort. Sci.* **65,** 291–295.
Teubner, F. G. and Murneek, A. E. (1955). *Bull. Mo. agric. Exp. Stn.* 590.
Thompson, A. H. (1957). *Bull. Md. agric. Exp. Stn.* A-88.
Thompson, A. H. and Rogers, B. L. (1972). *J. Am. Soc. hort. Sci.* **97,** 644–647.
Tukey, H. B. (ed.) (1954). "Plant Regulators in Agriculture", Wiley, New York.
Valdovinos, J. G. and Jensen, T. E. (1968). *Planta,* **83,** 295–302.
Valdovinos, J. G. and Muir, R. M. (1965). *Pl. Physiol., Lancaster,* **40,** 335–340.
Veen, H. (1966). *Acta bot. neer.* **15,** 419–433.
Walhood, V. T. (1957). *Proc. 12th annu. Beltwide Cotton Defoliation and Physiol. Conf.,* 1957, 24–31.
Warne, L. G. G. (1947). *Jl R. hort. Soc.* **72,** 193–195.
Wasscher, J. (1947). *Meded. Dir. Tuinb.* Oct., 547–555.
Weaver, R. J. (1972). "Plant Growth Substances in Agriculture", Freeman, San Francisco.
Weaver, R. J. and Pool, R. M. (1969). *J. Am. Soc. hort. Sci.* **94,** 474–478.
Webster, B. D. (1968). *Pl. Physiol., Lancaster,* **43,** 1512–1544.
Webster, B. D. and Leopold, A. C. (1972). *Bot. Gaz.* **133,** 292–298.
Weintraub, R. L., Brown, J. W., Nickerson, J. C. and Taylor, K. N. (1952). *Bot. Gaz.* **113,** 348–362.
Wertheim, S. J. (1973). *Scientia Hort.* **1,** 85–105.
Wester, H. V. and Marth, P. C. (1950). *Science, N.Y.* **111,** 611.
Whiting, A. G. and Murray, M. A. (1948). *Bot. Gaz.* **109,** 447–473.
Wilcox, H. E., Czabator, F. J., Girolami, G., Moreland, D. E. and Smith, R. F. (1956). *Tech. Publs N.Y. St. Coll. For.* 77.
de Wilde, R. C. (1971). *HortSci.* **6,** 364–370.
Williams, M. W. and Batjer, L. P. (1964). *Proc. Am. Soc. hort. Sci.* **85,** 1–10.
Wilson, W. C. and Coppock, G. E. (1969). *In* "Proceedings First International Citrus Symposium" (H. D. Chapman, ed.). Vol. III, pp. 1125–1134. Univ. California, Riverside.
Wittwer, S. H. (1954). *In* "Plant Regulators in Agriculture" (H. B. Tukey, ed.), pp. 62–80. Wiley, New York.
Worley, C. L. and Grogan, R. G. (1941). *J. Tenn. Acad. Sci.* **16,** 326–328.
Wright, S. T. C. and Hiron, R. W. P. (1972). *In* "Plant Growth Substances 1970" (D. J. Carr, ed.), pp. 291–298. Springer-Verlag, Berlin.
Wright, S. T. C. (1969). *Planta,* **86,** 10–20.
Zucconi, F., Stösser, R. and Bukovac, M. J. (1969). *BioScience* **19,** 815–817.

CHAPTER 7

EFFECTS ON SOLUTE TRANSPORT AND PLANT CONSTITUENTS

F. M. ASHTON AND D. E. BAYER

Department of Botany, University of California at Davis, Davis, California, USA

I. Introduction

The changes in plant constituents occurring after an application of a herbicide to a plant must be considered as a dynamic sequence of events. The major factors which determine the compositional status of plants are those associated with: (1) the herbicide; (2) the plant; and (3) the environment.

The results are determined by the concentration of the herbicide at the subcellular site of action. This concentration is influenced by penetration or uptake and translocation of the herbicide by the plant. The structure of the soil and the availability of nutrients and moisture also contribute to the dynamic nature of the processes influencing the effect of herbicides on plant constituents.

The variety of effects resulting from the kind of herbicide, its concentration, the species and age of the plant, and the interval between treatment and harvest all have a bearing on the plant response. This suggests why one observes the wide range of effects reported in the literature. The final results are determined not only by the nature of action of herbicide and the response of the particular plant species under investigation, but may be related also to a differential effect between a growth stimulatory concentration of the herbicide and a herbicidal concentration. It is well within reason to expect that sublethal rates of application of some of these herbicides may be used to control the metabolism of crop plants to obtain a desired chemical composition (see Chapters 15–17).

It cannot be over-emphasized that the results of herbicide treatment must be carefully evaluated, keeping in mind the dosage of herbicide, type of application, the species and portion analysed, and the interval from application to harvest. While it is difficult to evaluate all these factors from the literature the examples cited in the following discussion have been selected as representative examples of what might be expected from a given application of a certain herbicide.

II. Effects on the Transport of Solutes in the Plant

In general there are five major types of effects that herbicides have on transport that can be documented from the literature. These are (1) physical obstruction of the phloem caused by excessive unorganized cell division, (2) blockage as a result of callose formation, (3) localized injury and death of tissue, (4) stimulation or inhibition of metabolic sinks, and (5) influence on transpiration.

Some herbicides including dichlobenil and endothal affect assimilate transport by inducing the formation of callose which prevents the movement of assimilates. Maestri (1967) studied the effects of endothal on translocation in beans and cucumbers and showed that callose formation was particularly evident at veinlet endings and as pit callose in vein parenchyma. Excessive deposits of callose in the veins of cucumber leaves treated with endothal may be observed in Fig. 1. His work also showed a damaging effect on the plasma membrane. Leonard and Glenn (1968) reported that endothal prevented vein loading and thought that the site of action was located in the border parenchyma cells. If the movement of solutes into the phloem is an active process, a mechanism of this type would logically be associated with membranes of the parenchyma cells along the fine veins of the leaf. Blockage of this nature would inhibit the movement of solutes into the phloem in areas treated with endothal.

Glenn (1971) studying the effect of dichlobenil on transport in cotton found excessive deposits of callose on the sieve plates. Heavy deposits of pit callose were also noted in mesophyll and vein parenchyma cells. He concluded that movement through the phloem was blocked by plugging of the sieve elements.

Physical blockage of assimilate movement has been reported for several herbicides, in particular the phenoxy herbicides. Studies have shown that 2,4-D has two distinct effects on phloem tissue. These result in an initial inhibition of transport which increases with time. The early effect appears to be an early maturation and ageing of the sieve elements which in turn is followed by the second effect, a disruption and plugging of the vascular tissue. This disruption and plugging of the vascular tissue has been considered as a primary cause of death in 2,4-D-treated plants. MacLeod (1964) has suggested that blockage of transport as a result of MCPA treatment results in an accumulation of assimilates in the leaves of treated plants to levels that become toxic.

Akobundu et al. (1970) reported that phloem tissue in the basal regions of leaves of purple nutsedge (Cyperus rotundus) leaf sheaths was destroyed by dichlobenil treatment. This resulted in an accumulation of assimilates in the leaf since transport was prevented from occurring across the injured zone. Their studies were not designed to determine the presence of callose but disruption similar to that reported by Glenn (1971) could be found in the injured phloem tissue. They concluded that disruption only occurred in phloem tissue at a certain stage of development which suggests that possibly callose formation may have been involved in the early stages of transport blockage.

MH was shown to cause obliteration and plugging of sieve elements reducing translocation and thus causing sucrose and starch to build-up in the leaves. Crafts (1961) suggested that MH may result in disturbed metabolism and death as evidence by disruption of the vascular tissues.

Contact herbicides, although not generally thought of as blocking transport, may function in this capacity. Severe contact injury to foliage will prevent movement of assimilates out of the injured tissue.

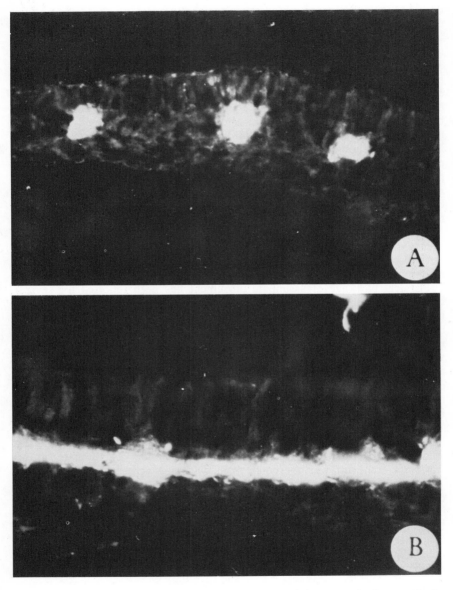

Fig. 1. Callose formation induced by 0·1 M endothal applied to cucumber leaves, 5 h after spraying. Cross sections were made in a cryostat. The sections were examined in an ultra-violet microscope, using a blue source filter and a yellow barrier filter. A. Cross section of three veins loaded with callose, ×100. B. Longitudinal view of a single vein also loaded with callose, ×250. Maestri, 1967.

The translocation of photosynthate and other phloem-mobile compounds is largely controlled by source-sink balances. Herbicides have been shown to alter these metabolic sources and sinks, and thus shift the normal source-sink balance. For example, herbicides that inhibit photosynthesis greatly reduce the major source in leaves and consequently reduce the translocation of all phloem-mobile compounds from the leaves, whereas, the auxin-type herbicides induce proliferated growth in the apical meristem and reversion of certain apparently mature tissues (i.e. hypocotyl) to a meristematic state. In general low concentrations of these compounds promote metabolism and growth whereas high concentrations are inhibitory. The stimulated state creates an active metabolic sink which increases the symplastic transport into the area. This results in a greater amount of phloem-mobile compounds in the stimulated tissues with concurrent reductions in the other tissues. However, if a normally active metabolic sink is inhibited by a herbicide, the amount of phloem-mobile compounds flowing into the inhibited area would be reduced.

Apoplastic or xylem-mobile compounds are largely influenced by transpiration rates. Some herbicides have been reported to influence transpiration. While rapid water loss may increase accumulation of minerals it may cause wilting of the plant and an ultimate reduction in both synthesis and respiratory processes. Conversely, herbicides that reduce transpiration will reduce the accumulation of compounds which are mobile in the xylem stream. This may be an important factor in movement and accumulation of herbicides as well as minerals, providing the herbicide is dependent on the transpiration stream for movement from the point of uptake to the site of action, generally from roots to the foliage of plants.

III. CARBOHYDRATE CONSTITUENTS

The relationship between the level of carbohydrates in plants and the susceptibility of weeds to herbicidal treatments was recognized over 30 years ago (Bakke *et al.* 1939). Interpretations of the results obtained are complicated by the large number of interacting variables that contribute to the observed response. Herbicides that affect synthetic processes in plants influence carbohydrate composition and utilization. Response at the cellular level depends on the concentration of the herbicide at the site of action. It is essential that the results of herbicide treatment be evaluated with these factors in mind.

A. ALIPHATICS

In general, results have shown total carbohydrates are reduced by herbicidal treatments with dalapon or TCA (Balandina and Gashnikova, 1969; McWhorter, 1961). Balandina and Gashnikova (1969), studying the effect of dalapon and TCA on cotton, reported that carbohydrate metabolism was reduced but dalapon caused less disruption than TCA. McWhorter (1961), using dalapon and TCA for the control of Johnson grass (*Sorghum halepense*), noted that the herbicide treatments providing the best control caused a general reduction of glucose with a corresponding increase in sucrose content. Dalapon in particular favoured sucrose production at the expense of glucose. However, the response reverted back to normal as the herbicidal effect diminished.

B. ARSENICALS

Duble and Holt (1970) found that repeated applications of the amine salt of methanearsonic acid (MAA) accelerated starch hydrolysis in tubers of purple nutsedge. The disappearance of starch was correlated with As content in the tubers. They suggested the carbohydrate fraction of the food reserves was preferentially utilized.

C. BIPYRIDYLIUMS

When *Chlorella vulgaris* grown in the dark was treated with concentrations of diquat that were toxic in the light, no significant change in the respiration quotient was observed. However, when *C. vulgaris* was kept in the dark for some hours before treatment with diquat, respiration increased two to three fold (Stokes and Turner, 1971). Pre-illumination or treating with glucose caused a greater stimulation in the breakdown of starch and most intermediate substances of carbohydrate metabolism.

Alexander and Montalvo-Zapata (1970) demonstrated that paraquat caused a reduction in sugar content in sugar-cane plants that were illuminated. The reduction was approximately identical to untreated plants kept in the dark. The reduction caused by paraquat was irreversible. They suggest that a free radical of paraquat is produced by photoreduction. However, the number of free radicles produced by photosynthesis would be greater than those produced by respiration.

D. DINITROANILINES

Lantican *et al.* (1969) reported that rice seedlings treated with 50 parts/10^6 of trifluralin for 24 h reduced both glucose and fructose content to undetectable levels. When trifluralin at 0·84 kg/ha was applied primarily to the hypocotyledonary tissue of sugar beet, no reduction of sucrose content of the roots were noted (Schweizer, 1970). However, in the area of aberrant roots, the purity of the sugar (sucrose/reducing sugar ratio) was lower than in normal tissue from untreated plants.

E. NITRILES

Dichlobenil at 0·2, 2, and 10 parts/10^6 caused an accumulation of assimilates in tubers and leaves of purple nutsedge (Akobundu *et al.* 1970). They suggested that this accumulation may have been due in part to an interference with the translocation process. An increase in respiration followed the accumulation in dichlobenil-treated tubers.

F. PHENOXY ACIDS

Extensive literature relating to this topic makes it impossible to cover it in detail in this chapter. An extensive review on the biochemical and metabolic changes in plants induced by chlorophenoxy herbicides has been written by Penner and Ashton (1966). It is apparent that the observations reported are influenced by the concentration of the herbicide used as well as the plant tissue analysed.

Early in the developmental use of 2,4-D, Tukey *et al.* (1945) proposed that a possible mechanism contributing to the herbicidal action of 2,4-D was the increased respiration that depletes food reserves. However, Smith *et al.* (1947) concluded that reduction of carbohydrates, whether by diminished production or increased utilization, was not the principal cause of death. He found, working with field bindweed (*Convolvulus arvensis*), that sugars increased in leaves, stems, and rhizomes in the first 10 days following treatment with 2,4-D and that the starch-dextrin fraction decreased to one-third the level of that in the control tissue.

Sub-lethal rates of 2,4-D may significantly alter the carbohydrate composition depending on stage of development. Payne *et al.* (1953) applied 2,4-D at 0·5 lb/acre (0·56 kg/ha) to Red McClure potato plants at early bloom stage and increased the specific gravity of the tubers, indicating an increase in the starch content. Nylund (1956) was not able to reproduce the effect with Pontiac potatoes. When Payne and Fults (1955) sprayed potato plants at an earlier stage a significant decrease in reducing sugars was observed, while an application made at a late stage caused a significant increase. Sucrose levels were not affected. Low reducing sugar content is desired for light colour in deep-fat-fried potato chips.

Wain *et al.* (1964) reported that 2,4-D at 10^{-5} M caused a greater reduction in carbohydrate content of storage tissue than did 3,5-D at 10^{-5} M. In artichoke tissue, 2,4-D reduced total carbohydrate content 70% while 3,5-D reduced it only 25%. An increase in hydrolase and invertase activity and an extensive hydrolysis of oligosaccharides was noted (Rutherford *et al.* 1969; Flood *et al.* 1970). Increased water uptake produced by 2,4-D treatment was associated with the hydrolysis of oligosaccharides to simpler sugars.

Bourke *et al.* (1962) concluded there was no relation between phytotoxicity of several phenoxyacetic acid compounds and their effect on glucose metabolism. In *Vicia faba* leaves treated with MCPA MacLeod (1964) found a 10-fold increase of sugar over the levels in untreated leaves and suggested that this was due to a partial interruption of the translocation flow from the leaves and that toxicity may result from a build-up of metabolites to toxic levels.

Coble and Slife (1971), studying physiological changes in the roots of honeyvine milkweed (*Ampelamus albidus*) following a foliar application of 2,4-D amine at 0·56 kg/ha found starch reserves in the roots were reduced 22 and 36% two and three days after treatment, respectively. No increase in soluble-sugar levels were observed and they suggested this was because sugars were utilized in the growth process. They were able to measure a 30-fold increase in cellulase activity three days after treatment. This increase was considered large enough to cause malfunction and disruption of the root system. Grant and Fuller (1971) found 2,4-D at 10^{-5} M decreased α-cellulose content of the cell wall fraction from *Vicia faba* root tips. Other work showed that 2,4-D increased hemicellulase activity but had no effect on cellulase and amylase activity (Tanimoto and Masuda, 1968). Coble and Slife (1971) found that the starch decrease, the protein increase, the alpha-amylase and nucleotide exudation, and the increase in cellulase activity induced in roots by a foliar application of 2,4-D coincided with tissue softening and initiation of decay. They suggested that 2,4-D had many distinct reactions involved in its herbicidal action.

From studies on root growth, Kim and Bidwell (1967) concluded that the main effect of the auxins IAA, and 2,4-D on sugar metabolism was not on uptake but rather on the metabolism of glucose. They suggested auxins have a specific effect on the formation of certain amino acids from glucose.

Mostafa and Fang (1971) reported that the *in vitro* effect of 2,4-D at 10^{-4} M on tissues of pea and maize plants fed 1- and 6-^{14}C-labelled glucose was a preferential release of the C-1 carbon of glucose as CO_2. The glucuronic acid pathway was augmented in maize stem and was inhibited in maize roots. The pentose phosphate pathway was affected in an opposite pattern. Increased amount of glucose catabolized via the pentose phosphate cycle was noted in etiolated cereal seedlings treated with 2,4-D at 10^{-3} M for 12 h (Black and Humphreys, 1962).

A reduction in the ratio between anabolic and catabolic utilization of glucose in stems of bean treated with 2,4-D was observed by Fang, 1962. This reduction was not detected in stem tissue of 2,4-D-treated cucumber, pea, tomato, or maize plants. The C_6/C_1 ratio in treated bean tissue increased with increasing 2,4-D concentrations. He indicated the reduction in amount of glucose catabolized via the pentose phosphate pathway was not due to either a limitation of enzymes or an inhibition of the enzymatic process.

The probable sites where 2,4-D attacks the basic processes of metabolism are discussed fully in Chapter 15. Precise information is still lacking on many aspects of these actions particularly in relation to internal concentrations of 2,4-D in the plant tissues.

Some workers suggest that the actions of some herbicides in certain combinations are complementary. Kalinin *et al.* (1968) have suggested that the action mechanism of the triazines and 2,4-D are complementary. Simazine and atrazine suppress the synthesis of carbohydrates and 2,4-D promotes their consumption.

G. THIOCARBAMATES

From field studies for the control of Johnson grass, Balandina and Gashnikova (1969) reported that EPTC reduced carbohydrate metabolism in cotton leaves.

H. TRIAZINES

Many reports have indicated that the triazine herbicides reduce the amount of glucose, fructose, and/or sucrose in plants; however, this is one of the first effects one would expect from a photosynthesis inhibitor.

The herbicides atrazine, chlorazine, propazine, and simazine were found to increase nitrogenous substances in maize and oats at the expense of sugars (Lukin *et al.* 1969). Reduction in fructose was followed by a reduction in sucrose content. Ploszynski *et al.* (1969) reported that the herbicides atratone, atrazine, prometone, prometryne, propazine, and simazine decreased the saccharide content of the aerial parts of *Echinochloa crus-galli, Sinapis alba,* and *Lepidium sativum* while the glucose content of *S. alba* and *L. sativum* was increased. Paramonova (1971), observed that prometryne reduced respiration but the lethal effect suggested a disruption of sugar synthesis. An increase in mono- over di-saccharides reflected a significant level of hydrolytic over synthetic processes.

Chodova and Zemanek (1971) demonstrated that eight days after treatment of roots of *Cirsium arvense* plants with simazine, the foliage contained less but the roots contained more carbohydrates. They also observed a reduction in respiration. They suggested that the reduction in respiration was related to a decrease in respiratory substrates since simazine is known to block photosynthesis.

J. UREAS

Several investigators have reported that the urea herbicides reduce sugar content in treated plants; however, as was indicated in section on triazine herbicides, this is one of the first effects one would expect from a photosynthetic inhibitor.

Figure 2 illustrates the carbohydrate-depletion pattern found in Johnson grass rhizomes planted in soil treated with diuron or bromacil. Growth

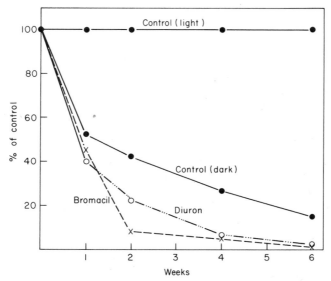

Fig. 2. Effect of bromacil and diuron on total carbohydrates of Johnson grass rhizomes (adapted from Opile, 1969).

from rhizomes planted and left in the dark depleted the total carbohydrate level at approximately the same rate as rhizomes treated with either bromacil or diuron. This general pattern in plants is representative of the carbohydrate-depletion patterns resulting from treatment with photosynthesis inhibitors.

Alimova and Nuriidinov (1969) reported that diuron apparently suppressed respiration in cotton by reducing the amount of hexosephosphate in the leaves. Paramonova (1971) reported that linuron reduced respiration and noted an increase of mono- over di-saccharides which reflects a significantly greater activity of hydrolytic as compared with synthetic processes.

K. UNCLASSIFIED HERBICIDES

1. *Bromacil*

The uracil herbicides have been reported to be photosynthetic inhibitors and therefore to reduce carbohydrate levels in treated plants. Bromacil and isocil at 1 and 4 lb/acre (1·1 to 4·5 kg/ha) was shown to decrease total carbohydrates of quackgrass following treatment (Swann and Buchholtz, 1966). After seven weeks, total carbohydrate content dropped from 37 to 12% on a dry weight basis. Carbohydrates of Johnson grass are also reduced by bromacil (Opile, 1969), Fig. 2.

2. *Picloram*

Guenthner (1970), showed that picloram induced an accumulation of carbohydrates in the phloem and proliferating tissue region of tomato and in the phloem of the stem of *Cirsium arvense*. Picloram at the rate of 50, 100, 200, and 500 parts/10^6 increased the amount of carbohydrates by 170 to 430% and reduced sugars in the exudate from maize seedlings by 190 to 270% (Lai and Semeniuk, 1970). The increase was positively correlated with the concentration of picloram. They further showed that in the presence of *Rhizoctonia solani* and *Pythium arrhenomanes* there was no increase in carbohydrates, presumably, because the pathogens used part of the exuded carbohydrates for their growth. The results suggest that increased carbohydrate exudation may account for the increased root damage on picloram-treated plants from soil-borne root pathogens.

3. *Pyrazon*

Pyrazon has been reported to decrease the glucose and sucrose contents of wheat seedlings (Reyes and de Capriles, 1970). The decrease was greater with increase in temperature.

Mashtakov and Kudryavtsev (1969), growing both diploid and tetraploid sugar-beet plants in pyrazon concentrations of 10^{-3} to 10^{-2} M for 48 h, found four phenolic compounds (one a phenolic glycoside) in the cotyledons. The phenolic content increased with increasing sensitivity to pyrazon whereas the simple carbohydrate content of the seedlings decreased with increasing sensitivity. They suggested the effect of pyrazon may depend to a great extent on the amount and qualitative composition of the phenolic compounds.

IV. Lipid Constituents

Lipids occur in all parts of the plant as part of the essential structures of all living cells. However, they also occur in higher concentrations in the fruit and seeds of plants as a reserve food. Most higher plants contain fat as a reserve food, often to the exclusion of starch. A notable exception to this are the cereals. Lipids as well as carbohydrates should be considered as part of a dynamic equilibrium food reserve system.

A. ALIPHATICS

The deposition of wax on leaf surfaces has been found to be affected by dalapon and TCA. Dewey *et al.* (1956), in attempting to enhance herbicide uptake, pretreated pea plants with dalapon and TCA and then followed these pretreatments with an application of dinoseb. They found that as little as 1·5 lb/acre (1·7 kg/ha) of either dalapon or TCA increased the wettability and permeability of the pea plants. Several other weed species showed the same response. They found that the amount of dinoseb required to control the weeds could be reduced and that the degree of selectivity between the peas and weeds remained approximately the same. The total amount of dinoseb required to control the weeds was reduced considerably. Pfeiffer *et al.* (1957), extending this information, demonstrated the relationship between the wax content of leaves and increasing rates of TCA (Table 1). Pfeiffer *et al.* (1957) and Juniper and Bradley (1959) found the same effect occurs with other herbicides such as MCPA, following a TCA pretreatment. They suggest that normal rates of application of TCA alters wax formation by leaves even though growth and yield is unaffected. When the total wax content was reduced spray retention increased. A higher rate of transpiration has been observed in TCA-treated plants (Corbadzijska, 1962). The reduction of wax could be responsible for a modified cuticle that would result in a greater loss of water which may lead to excessive wilting or death of the plant.

Structure of surface waxes as affected by TCA and other herbicides have been illustrated by Juniper (1959) and Still *et al.* (1970). At higher application rates of the herbicide, the crystalline nature of the surface wax disappeared and

TABLE I

PERCENTAGE DECREASE OF WAX CONTENT OF LEAVES DUE TO TCA SOIL-TREATMENT BEFORE SOWING (Pfeiffer *et al.* 1957)

Plant species	% Decrease of wax (lb/acre TCA)				Normal wax-content % of d.wt.
	0	*2·5*	*5·0*	*10·0*	
Peas	0	16	20	35	1·2
Kale	0	23	32	58	1·2
Annual nettle	0	5	14	18	1·1

the surface was smooth, often apparently devoid of wax or scattered with small wax flakes. Contact angles were determined on these leaves and they were shown to be more wettable than untreated leaves. However, Pfeiffer *et al.* (1959) suggested there was more involved than just wetting and that some other process associated with penetration was equally important.

B. AMIDES

Penner and Meggitt (1969), studying the effect of propachlor on seed lipids, noted that the linoleic acid content was increased and the stearic acid content decreased. Differences in fatty acid composition could not be correlated with yield or injury as a result of the herbicide treatment.

C. ARSENICALS

Arsenic reacts readily with dithiol compounds such as reduced lipoic acid forming a stable ring structure and the reaction is not reversible (Dixon and Webb, 1958). This reaction of energy transfer to ATP has been suggested as an explanation for membrane degradation (loss of turgor), injury, and death.

Duble and Holt (1970), studying the effect of the amine salt of MAA on the food reserves in tubers of purple nutsedge, found there was little or no effect on the fat content following repeated applications at three-week intervals. They suggest that the carbohydrates were utilized in preference to fats and proteins.

D. BENZOICS

Chloramben has been reported to stimulate the incorporation of ^{14}C from malonic acid-$[2-^{14}C]$ into lipids (Mann and Pu, 1968). Promotion of lipid synthesis was also demonstrated by other auxin-like herbicides in hemp sesbania (*Sesbania exaltata*). However, dinoben caused essentially complete inhibition of lipid synthesis at the concentrations used (1 to 20 mg/l). Other non-auxin herbicides were inhibitory to lipid synthesis.

E. CARBAMATES

Ivens and Blackman (1949), studying the effect of ethyl phenylcarbamate on mitosis in root cells, proposed that the carbamate esters combine with the lipid component of the spindle protein causing an intramolecular precipitation, collapse, and disintegration of the spindle.

F. DINITROANILINES

The oil or fatty acid composition of soybean seeds was not affected by a 0·75 lb/acre (0·84 kg/ha) treatment of trifluralin (Johnson and Jellum, 1969). Although Penner and Meggitt (1969, 1970) found no changes in total oil content of soybean seed after treatment they did observe an effect on the

TABLE II

EFFECT OF INCORPORATED HERBICIDE TREATMENTS ON SOY-BEAN OIL CONTENT, FATTY ACID COMPOSITION, SEED YIELD, AND PLANT INJURY AVERAGED OVER TWO CULTIVARS, CHIPPEWA AND HAROSOY (From Penner and Meggitt, 1970)

Herbicide	Rate kg/ha	% oil	Fatty acid composition, % of total[1]					Yield, kg/ha	Injury[2]
			16:0	18:0	18:1	18:2	18:3		
Control		20·1	11·5	4·2 a[3]	23·1	49·7 bcd	11·4	2,112	0·0 c
Trifluralin	0·84	20·0	11·3	3·6 a	20·5	53·2 ab	11·4	2,368	2·8 a
Trifluralin	1·12	20·1	10·4	3·0 b	20·2	54·5 a	11·9	2,428	1·8 abc
Nitralin	1·69	20·3	11·5	4·9 a	23·9	46·7 d	12·0	2,448	0·5 bc
Vernolate	2·24	20·3	10·6	3·8 a	22·6	51·0 abcd	11·9	2,226	1·8 abc
Vernolate	3·36	20·5	11·0	4·4 a	23·4	48·2 cd	13·1	2,152	2·5 ab

[1] Palmitic acid (16:0), stearic acid (18:0), oleic acid (18:1), linoleic acid (18:2), and linoleic acid (18:3).

[2] Injury ratings are on a 0 to 10 scale, where 0 is no injury and 10 is complete kill.

[3] Values within columns with similar letters are not significantly different at the 5% level.

quality of the oil. Linoleic acid content was increased and stearic acid was reduced (see Table II).

Hilton and Christiansen (1972) hypothesized that seedling tolerance to trifluralin is determined by the amount of endogenous lipid present and accumulating the herbicide. In laboratory studies, they showed that many lipids reduced the phytotoxic action of trifluralin. In further studies, they were able to demonstrate a positive correlation between the lipid content of seedlings and tolerance to trifluralin.

G. PHENOXY ACIDS

The influence of lipids on the absorption of 2,4-D has been demonstrated by Smith (1972). Using the Schulman liquid-liquid membrane model, 2,4-D absorption was shown to be more rapid from the hydrophilic phase to the lipophilic phase in the presence of the polar lipids (lecithin and monogalactosyl dilinolenate).

Absorption was reduced in the presence of the less polar triglycerides (tristearate, trioleate, and trilinolenate). The potato tuber parenchymatous tissue absorbed more 2,4-D in the presence of lecithin. Brian (1958), using the Langmuir trough technique, showed that less MCPA was absorbed in the presence of lecithin. The cationic charges on the protein were neutralized by the negative phosphate group of lecithin. The stearic factors on the anions of the protein presumably counteracted the positive charge on the quaternary nitrogen atom of lecithin. Thus, he concluded that plants containing less lecithin are less susceptible to MCPA than plants containing a higher proportion of lecithin or other phospholipids because they can absorb more anions.

The lipid content of several plants have been shown to be affected by 2,4-D and MCPA (Sell *et al.* 1949; Dunham, 1951; Epps, 1953; Kent and Hutchinson, 1957; Sell *et al.* 1957; Zoschke, 1957). An application of 2,4-D increased the oil content of the kernel of oats (Kent and Hutchinson, 1957); they reported that this increase was enhanced by increasing the rate of 2,4-D. Sell *et al.* (1949), found a slightly higher content of unsaponifiable material and fatty acids in the ether extract as well as an increase in the total lipid content of 2,4-D treated plants. However, an application of 2,4-D was shown to reduce the ether extractable lipids in all parts of the plant except the stem (Sell *et al.* 1957). Zoschke (1957), showed that an application of 2,4-D reduced the lipid content of sugar-beets.

Dunham (1951) measured a reduction in oil content as well as an adverse effect on the iodine number of flax seed oil following an application of 2,4-D of 4 to 24 oz/acre (0·28 to 1·7 kg/ha). A maximum reduction in oil content of 2·37% occurred when the application was made during the pre-bud and late-bud stages. A reduction of oil content in cotton seed from plants treated with rates of 2,4-D greater than 0·01 lb/acre (11·2 g/ha) has been reported by Epps (1953).

The lipase activity of *Ricinus* was greatly reduced by concentrations of 2,4-D in the order of 0·01% (Hagen *et al.* 1949; Ravazzoni and Valerio, 1956). Kunert (1959), observed that 2,4-D inhibited lipase activity in *Aspergillus niger*. However, he reported that MCPA increased lipase activity in the same species.

Mashtokov *et al.* (1967a), using as their criterion the increased extractability of chlorophyll in petroleum ether with time, found that both 2,4-D and MCPA weakened the chlorophyll-protein-lipoid complex in maize plants when applied at the 5- to 6-leaf stage. When flax was treated at the 10- to 12-inch-tall (25–30 cm) stage the greatest weakening of the complex in sensitive flax plants occurred with 2,4-D while the greatest weakening in the resistant flax plants occurred with MCPA.

Alfalfa treated with 2,4-DB formed appreciable quantities of 6-(2,4-dichlorophenoxy)caproic and 10-(2,4-dichlorophenoxy)decanoic acids and also limited amounts of 4-(2,4-dichlorophenoxy)crotonic acid and 2,4-D (Linscott *et al.* 1968). When the methyl ester of 4-(2,4-dichloro-phenoxy)crotonic acid was applied it was reduced to the methyl ester of 2,4-D and 2,4-DB. The 2,4-DB was then used for synthesis of the long-chained fatty acids mentioned above. A portion of the 2,4-D formed was also believed to enter into the fatty acid chain-lengthening process. They thought that the synthesis occurred in the epidermal cells and was associated with wax production. Thus, β-oxidation of alfalfa was reduced as the 2,4-DB was used as a substrate in fatty acid synthesis. Linscott and Hagin (1970) was also able to show that 2-carbon units could be added to the aliphatic side chain of 2,4-D, 4-(2,4-dichlorophenoxy)crotonic acid and 4-(2,4-dichlorophenoxy-β-hydroxy)butyric acid in legumes. Bach (1961) proposed that a lengthening of the side chain analogous to the above mentioned fatty acid biosynthesis could take place if 2,4-D-acetyl-CoA would react with CoA, CO_2, and ATP. Mono-

and di-carboxylic acids with potential sites for hydroxyl or keto linkages might be formed.

Hagin *et al.* (1970), observed the formation of α-(2,4-dichloro-phenoxy)propionic acid (2,4-DP) following the treatment of several grass species with 2,4-D. The 3-(2,4-DP) was considered a major metabolite and was located primarily in the epidermal cells. They could not explain the significance of the 3-(2,4-DP) but considered it to be involved in wax biosynthesis of these plants.

Mitochondria from soybean hypocotyls treated with 2,4-D were larger and showed enhanced incorporation of phosphate into phospholipids (Baxter and Hanson, 1968). It has been suggested that 2,4-D can act as a stimulator of phospholipid synthesis as evidenced by the apparent increase in mitochondrial size and the general stimulation of membrane synthesis.

H. THIOCARBAMATES

Gentner (1966), found that EPTC inhibited the formation and deposition of surface wax on cabbage leaves. He was able to correlate increased retention of spray droplets with reduction of surface wax. Transpiration was shown to increase in direct proportion to increased dosage. Wilkinson and Hardcastle (1970), showed that with the increased rates of EPTC application the thickness of both the upper and lower cuticle of the leaflet were decreased 15 and 20%, respectively.

Diallate reduced epicuticular lipids between 50 to 80% depending on how it was applied to the plant (Still *et al.* 1970). The ratio between the wax lipid components remained unchanged although the total quantity of surface lipid components (hydrocarbons, esters, secondary alcohols, and fatty acids) were reduced; primary alcohols were inhibited to a greater extent, i.e. to approximately one-fifth of that found in the untreated plants. Diallate may interfere with the biosynthesis of a precursor essential in the function of the elongation-decarboxylation pathway of lipid synthesis. If the block was at the elongation-decarboxylation step the resulting *n*-alkane would have been of shorter chain length than C_{31}, however, this was not what was found.

J. TRIAZINES

When isolated chloroplasts of spinach were treated with atrazine or simazine at 0, 10^{-6}, 10^{-5} or 10^{-4} M there was an increase in the concentration of unsaturated fatty acids (Smith and Wilkinson, 1973). Saturated fatty acids remained unchanged. The overall response was an increase in the total free fatty acid content.

K. TRIAZOLES

Wort and Loughman (1961), studying the incorporation of ^{32}P in amitrole-treated plants, observed that phospholipids of both root and shoot of barley plants decreased within 24 h after treatment. They suggest the effect of amitrole on phosphate metabolism occurs in the incorporation of phosphate into the components of the acid-insoluble fraction (nucleic acids, phospholipids, phosphoproteins).

L. UNCLASSIFIED HERBICIDES

1. *Pyridazinone (Sandoz 6706)*

When pyridazinone was applied to mustard (*Brassica juncea* (L.) Cos.) and barley, it increased the amount of fatty acids in the nonpolar lipids of the chloroplasts and reduced their concentration in the galactolipids (Hilton *et al.* 1971). Very little effect was found on the total content of bound fatty acids. The amount of linolenic acid relative to linoleic acid was reduced in the barley leaves. The addition of unsaturated fatty acids and their methyl esters overcame the inhibition of chlorophyll formation in pyridazinone-treated plants. The protective action was explained by the herbicide being absorbed into the lipid phase, thus limiting the availability of the herbicide.

V. NITROGENOUS CONSTITUENTS

Herbicides affect the level of protein and other nitrogenous constituents of higher plants in several ways, including absorption, translocation, and metabolism. The mechanisms of absorption and translocation of nitrogenous compounds probably vary only in minor details from those of herbicides. These are covered in Chapters 12 and 13 therefore, are not treated here. The effect of herbicides on protein and nucleic acid metabolism are discussed in Chapter 17. The herbicide-induced modifications of the above processes result in qualitative and quantitative shifts in the composition of proteins and other nitrogenous constituents of plant tissues.

A. ALIPHATICS

Mashtakov and Moshchuk (1967) reported that TCA increased the amount of asparagine and glutamine in a resistant variety of *Lupinus luteus*. They also noted a slight increase in protein and β-alanine and no change in the amount of free ammonia. However, in a sensitive variety, they reported a decrease in amides and β-alanine with an increase in free ammonia and protein. Similar results were also reported for dalapon (Mashtakov *et al.* 1967b). They concluded that the action of these herbicides was an inhibition of the enzymes which are involved in the conversion of ammonia to amides, thereby allowing the accumulation of toxic levels of free ammonia. Andersen *et al.* (1962) observed that dalapon caused an increase in the degradation of protein to amino acids and an increase in amides. Further breakdown of the free amino-acids with the liberation of ammonia was indicated. The amides appeared to act in ammonia detoxication by serving as storage sites for the released ammonia. After some period of time the amide and amino-acids returned to normal levels in the resistant sugar beets, but this did not occur in the susceptible yellow foxtail (*Setaria glauca*). Using dalapon-susceptible and resistant genotypes of Bermuda grass (*Cynodon dactylon*), Sistrunk (1969) found that the susceptible genotypes tended to contain higher concentrations of certain amino-acids, particularly asparagine and proline, than did the tolerant genotypes. He postulated that dalapon induced protein degradation thus resulting in ammonia toxicity and that selectivity is due to the ability of genotypes either to detoxify ammonia or to prevent its formation. Jain *et al.* (1966) reported an increase in radioactive aspartic acid, glutamic acid, asparagine, and glutamine following [^{14}C]-glucose feeding of dalapon-treated plants. Dalapon and TCA increased the synthesis of nitrogen compounds in the leaves of cotton and Johnson grass, whereas dalapon reduced the nitrogenous content of Johnson grass rhizomes (Balandina and Gashnikova, 1969). Mashtakov and Moshchuk (1967) reported that TCA inhibited proteinase activity especially in the resistant variety. However, Ashton *et al.* (1968) did not find any inhibition of proteinase activity in squash seeds grown in 10^{-3} M dalapon.

TCA is the classical compound used by the biochemist for precipitating proteins in the test tube; however, concentrations of about 5% are usually used for this purpose and this is entirely out of the physiological range of herbicidal action. Dalapon and TCA were reported to be protein precipitants by Redemann and Hamaker (1957) and concentrations as low as about 200 parts/10^6 produced a visible precipitate from egg-yolk and egg-white protein. This is still a relatively high concentration. Foy (1969) notes that halogenated acetates and propionates are theoretically able to alkylate the sulphydryl or amino groups in enzymes. Hydrogen bonding of dalapon to *N*-methylacetamide, a model protein, has been reported (Kemp *et al.* 1969). We have seen from the previous discussion that apparently many enzyme reactions are influenced by these herbicides, but none in such a specific way that would explain its mechanism of action. Compounds which actually bring about precipitation of proteins at 200 parts/10^6 must produce some conformational changes at a much lower concentration, perhaps in the physiological range. If such conformational changes do indeed occur, they could explain many of the results which have been reported.

B. AMIDES

Several amide-type herbicides have been shown to influence nucleic acid and/or protein metabolism and therefore should have some effect on nitrogenous constituents of a plant.

Alachlor was reported to increase the RNA and DNA content of root tissue and increase protein synthesis in excised hypocotyl and root sections of cucumber (Edmondson, 1969). CDAA does not appear to inhibit protein or RNA synthesis *per se* (Moreland *et al.* 1969) but does seem to inhibit the development of the activity of certain hormone-induced hydrolytic enzymes (Ashton *et al.* 1968; Moreland *et al.* 1969; Smith and Jaworski, unpublished).

Duke (1967) found that propachlor inhibited protein synthesis in both root and shoot sections and that it occurred just prior to the cessation of growth. Propachlor also inhibited the GA-induced development of phytase (Penner, 1970). Propanil is bound to proteins (Camper and Moreland, 1967); it appears to inhibit RNA and protein synthesis (Moreland *et al.* 1969) and reduces GA-induced α-amylase activity in half-seeds of barley (Moreland *et al.* 1969). However, the research of Baker and Pizzalato (1968) suggests that propanil may influence both the synthesis and breakdown of DNA and RNA in maize root tips; the actual effect depends on the section of the root tip examined. Nodes at the tip of quackgrass rhizomes, examined two to three weeks after the plant was sprayed with pronamide, were swollen and necrotic and had higher amounts of DNA, RNA, and protein than the control plants (Smith *et al.* 1971). Diphenamid has been reported to inhibit RNA synthesis in oat roots completely (Briquet and Wiaux, 1967); however, this is difficult to reconcile with several reports that diphenamid does not inhibit protein synthesis (Ashton and Crafts, 1973). Root tips of wheat germinated in diphenamid solutions showed an increase in total nitrogen with a subsequent decrease in acid-insoluble nitrogen (Yaklich, 1970).

C. AMITROLE

Burt and Muzik (1970) treated three upper leaves of a resistant and a susceptible ecotype of *Cirsium arvense* with amitrole and analysed the leaves, stem, and roots for several alcohol-soluble nitrogenous compounds. These were aspartic acid, glutamic acid, serine, glycine, asparagine, threonine, alanine, glutamine, lysine, arginine, proline, valine, leucine, phenylalanine, tyrosine, α-aminobutyric acid and pipecolic acid. In the leaves, one day after treatment most of these compounds decreased markedly but this was followed by an increase at five days. The changes within the stems were less consistent; perhaps there was a slight decrease at five days. In the roots, a substantial decrease had occurred at five days. The changes noted above are for the susceptible ecotype; the resistant ecotype did not necessarily show these same changes (see Fig. 3).

The protein content of light-grown amitrole-treated plants decreases, with a compensating increase in free amino-acids (McWhorter, 1963; Bartels and Wolf, 1965). This would suggest an increase in protein hydrolysis and/or a decrease in protein synthesis. Amitrole has been shown to have no effect on the activity of proteinases or dipeptidases isolated from squash cotyledons or little if any effect on their normal increase during germination (Ashton *et al.* 1968; Tsay and Ashton, 1971). Bartels and Wolf (1965) reported that wheat seedlings germinated in a petri dish containing a 10^{-4} M solution of amitrole for three days inhibited the incorporation of ^{14}C-labelled glycine into protein in the shoot by about 50%. However, Mann *et al.* (1965) and Moreland *et al.* (1969) concluded that amitrole did not inhibit the incorporation of ^{14}C-leucine into protein in stem or coleoptile sections. Brown and Carter (1968) also were unable to find a direct effect of amitrole on protein synthesis. These results

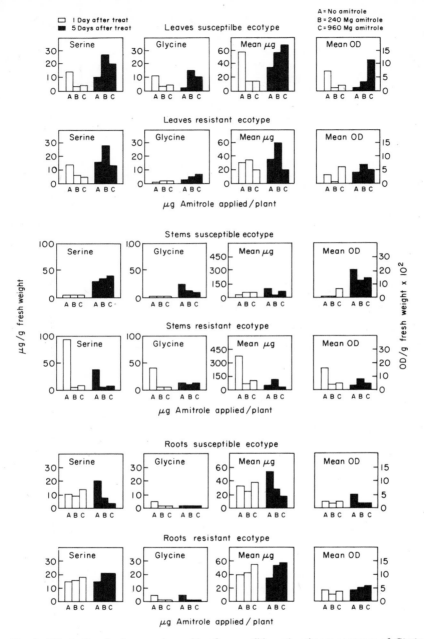

Fig. 3. Effect of amitrole on amino acids of susceptible and resistant ecotypes of *Cirsium arvense*. Mean μg = mean of aspartic acid, glutamic acid, alanine, valine, leucines, and γ-amino butyric acid. Mean OD = mean of asparagine, threonine, lysine, arginine, proline, phenylalanine, tyrosine, pipecolic acid, and some unknown ninhydrin reactive compounds (Burt and Muzik, 1970).

would seem to indicate that amitrole does not inhibit the protein synthesis reactions *per se* but rather inhibits the formation of a critical component necessary for protein synthesis.

Sund *et al.* (1960) found that purines would alleviate the growth-inhibitory effect of amitrole in bean and tomato. Castelfranco *et al.* (1963) showed that riboflavin would reduce the effects of amitrole in maize. However, Naylor (1964) reported that purines would not reverse chlorosis and growth inhibition caused by amitrole in higher plants. Bartels and Wolf (1965) suggested that amitrole may interfere with purine synthesis and thereby affect the nucleic-acid content of the plants. Consequently, a lower level of RNA would occur which would decrease the plants' ability to synthesize protein, and result in the accumulation of amino acids. Light-grown, amitrole-treated wheat seedlings were found to lack detectable quantities of chloroplast DNA whereas treated, dark-grown plants contained plastid DNA (Bartels and Hyde, 1970). They suggested that amitrole affects the accumulation of chloroplast DNA by inhibiting the formation of chloroplast membranes, enzymes, and pigments.

In micro-organisms amitrole inhibits the growth by competitive inhibition of imidazoleglycerol phosphate dehydratase (IGP dehydratase), an enzyme of histidine biosynthesis (Hilton *et al.* 1965; Klopotonski and Wiater, 1965). In higher plants, however, McWhorter and Hilton (1967) and Hilton (1969) suggested that histidine biosynthesis could not account for the herbicidal action of amitrole. In two recent papers, Wiater *et al.* (1971a, 1971b) demonstrated a histidine-biosynthetic pathway in barley and oats comparable to that in micro-organisms. *In vivo,* the plant IGP-dehydratase activity was more strongly inhibited by amitrole than the same enzyme from micro-organisms. Therefore, the inhibition of histidine biosynthesis may be a contributing factor to the herbicidal action of amitrole in young seedlings (those having exhausted seed reserves); but it is not by itself sufficient to explain all of the effects of amitrole (Hilton, 1972). The contribution of this inhibition to herbicidal action is perhaps masked by additional sites of inhibition which are equally sensitive.

D. ARSENICALS

Duble *et al.* (1968) found that the amine salt of MAA caused the accumulation of amino acids in roots of Johnson grass and suggested that the amine salt of MAA may block protein synthesis or some other biosynthetic pathway. Similar results were obtained from tops of Johnson grass and similar conclusions were reached by Scherl and Frans (1969). However, they also reported that increases in amino-acid content of cotton, a tolerant species, did not occur. Later studies by Duble and Holt (1970) using purple nutsedge, showed that the amine salt of MAA caused an increase in protein content of leaf tissue, with little effect on the protein content of tubers.

E. BENZOICS

Chen *et al.* (1972), using roots from wheat and cucumber seedlings germinated in the herbicide solutions, studied nucleic-acid and protein changes induced by dicamba, as well as 2,4-D, 2,4,5-T, and picloram. All four herbicides increased the DNA and protein content of roots at four days; the increase was greatest in cucumber. With increasing herbicide concentrations, there were progressive decreases in the RNA content in wheat but increases in cucumber. When the protein levels of both species were compared on a per unit RNA basis, the protein/RNA ratio was higher than the control for wheat but lower for cucumber. They suggested that differential alteration of RNA species and interference with protein synthesis was the basis for selectivity between these two species.

F. PHENOXY ACIDS

The literature in this topic is very voluminous; it is not possible to cover it exhaustively here. Much of this early research was reviewed by Penner and Ashton (1966). The chlorophenoxy herbicides affect the nitrogenous content of several tissues quite markedly by inducing metabolic sinks (see Section I of this chapter).

As early as 1947 Smith *et al.* (1947) observed that the total nitrogen in 2,4-D-treated field bindweed decreased in the leaves and increased in the stems, roots, and underground rhizomes (Table III). In 1949 Wort (1949) reported that in buckwheat plants treated with 1,000 parts/10^6 of the sodium salt of 2,4-D, the total nitrogen in the stems and roots increased and that of the leaves decreased

TABLE III

CHANGES IN COMPOSITION OF FIELD BINDWEED FOLLOWING TREATMENT WITH 2,4-D. VALUES ARE PERCENTAGE OF DRY WEIGHT (Smith *et al.* 1947)

| | | | Nitrogen | |
| | | Total Control (C) % | Treated (T) % | T/C |
Day	Tissue			
0	Leaf	4·78	—	—
2		4·57	4·80	1·05
4		5·08	4·41	0·87
7		4·76	4·11	0·86
10		5·00	2·31	0·66
0	Stem	2·25	—	—
2		2·23	2·52	1·13
4		2·50	2·72	1·09
7		2·45	3·19	1·30
10		2·53	3·40	1·35
0	Rhizome-Root	1·50	—	—
2		1·79	1·60	0·90
4		1·73	1·97	1·14
7		1·87	2·13	1·14
10		1·78	2·48	1·39

four to eight days after the 2,4-D application. Torchinskaya (1958) reported a loss of nitrogen from the leaves to the roots and stems of a number of weeds in which he followed the effects of 2,4-D applications.

Numerous investigators have attempted to determine the effect of the phenoxy herbicide on the levels of amino-acids and proteins present in plant tissues. Considerable care must be taken in interpreting the results in terms of metabolism since, in addition to the comments made in Section I of this chapter, it has been shown that uprooted bean plants left to die on the greenhouse bench undergo the same rapid decreases in protein and amino-acid contents of roots as do 2,4-D-treated plants (Muzik and Lawrence, 1958, 1959; Lawrence and Muzik, 1962). Furthermore, MacLeod (1964) suggested that the five-fold increase in the concentration of leucine, isoleucine, and valine of *Vicia faba* leaves after MCPA treatment or merely detaching the leaves was due to the interrupting of the translocation of metabolites from the leaves.

There is no question that auxins, including 2,4-D, alter nucleic acid metabolism (Hanson and Slife, 1969) and thus protein synthesis (see Chapter 17 for details). The report of Key *et al.* (1966) clearly showed that in tissues where growth is suppressed by 2,4-D nucleic acid metabolism is also suppressed and in tissues where growth is accelerated by 2,4-D nucleic acid metabolism is likewise accelerated. 2,4-D also affects the type of proteins found in pea roots (Norris and Morris, 1969). Several reports cited by Penner and Ashton (1966) suggest that 2,4-D not only induces changes in amino-acid and protein levels by affecting protein synthesis but also influences protein degradation.

Table IV gives data from a representative experiment which shows shifts of nitrogenous components induced by 2,4-D (Cardenas *et al.* 1968). A lethal dose of 2,4-D was applied to the midrib at the base of the first true leaf of three-week-old cocklebur (*Xanthium* sp.) plants. The plants were harvested two and seven days after treatment and analysed for total nitrogen, protein,

TABLE IV

EFFECT OF 2,4-D ON THE CONTENT OF TOTAL NITROGEN, PROTEIN, RNA, AND DNA OF VARIOUS PARTS OF THREE-WEEK-OLD COCKLEBUR PLANTS (*Xanthium* sp.) TWO AND SEVEN DAYS AFTER A LETHAL DOSE WAS APPLIED ON THE MIDRIB AT THE BASE OF THE FIRST TRUE LEAF, EXPRESSED AS % OF CONTROL (adapted from Cardenas *et al.* 1968)

Days	Total N		Protein		RNA		DNA	
	2	7	2	7	2	7	2	7
Apex	195	348	201	482	196	407	108	203
Stem	228	149	203	174	352	170	185	166
Tap root	143	116	176	108	271	127	333	149
Lateral roots	91	29	84	38	100	20	100	38
Opposite leaf	75	54	94	64	80	56	103	70
Alternate leaves	57	10	47	12	39	8	50	79
Cotyledons	83	45	83	50	86	56	—[1]	—[1]

[1] Too low to measure.

RNA, and DNA. At two days after treatment with 2,4-D, there were substantial increases in all four of these nitrogenous constituents in the apex, stem, and tap root with corresponding decrease in the other plant parts. However, at seven days after treatment, the values of the stem and tap root as well as the other plant part decreased (relative to day-2) with even larger increases in the apex. In addition, both photosynthesis and ion absorption were initially stimulated but declined sharply after the first day. Translocation to leaves and roots was drastically reduced in favour of the proliferating axis. The biochemical basis for these responses appears to be with aberrant nucleic-acid metabolism.

G. TRIAZINES

Nitrogen metabolism changes induced by the triazine herbicides are influenced by the amount of herbicide applied, the amount and form of nitrogen fertilizer supplied to the plant, environmental growth conditions and the species of plant; in certain cases they depend on whether the data are calculated on a percent by weight or per plant basis. This problem has been investigated rather thoroughly since some of the research has suggested that subtoxic levels of the triazine herbicides increase growth and the amount of nitrogen and/or protein in the plants. The effect of subtoxic levels are covered in Chapter 11, volume 2 and therefore will not be discussed in detail here.

In 1957, Bartley reported that simazine increased the growth and green colour of leaves of maize. Numerous papers have subsequently reported increased greening of leaves following the use of the triazine herbicides for weed control. Most of these reports have been from crops with some tolerance to these herbicides, suggesting that low levels of the triazine herbicides may be responsible. Gast and Grob (1960) observed that atrazine or simazine increased the protein content of maize. Many subsequent studies have shown the triazine herbicides increase the *percentage* of several nitrogen fractions in plants. Although some studies have demonstrated that there is actually an increase in the amount of the nitrogen components on a *per plant* basis others have not. Frequently the increase in the *percent* of nitrogen fractions has been attributed to a decrease in dry weight rather than a real increase in nitrogen.

A limited number of reports indicate that the triazine herbicides may also influence the amount of nucleic acids in plant cells or tissues. In measuring RNA synthesis, Moreland et al. (1969) used both orotic acid-[6-^{14}C] and ATP-[8-^{14}C] incorporation and found that by ATP incorporation, RNA synthesis was inhibited by atrazine, but, with the orotic acid method RNA synthesis was augmented by atrazine. However, since leucine-^{14}C incorporation into these same tissues was not significantly altered by atrazine, its effect on RNA synthesis may not be particularly meaningful to nitrogen metabolism in general. In experiments with maize, Gräser (1970) and Gräser and Tillich (1970) reported that simazine increased the concentrations of RNA and the inhibition of nucleic acid synthesis by 5-fluorouracil was nullified. Singh et al. (1969) reported that simazine increased the levels of DNA, RNA, and protein in maize.

Berüter and Temperli (1970) reported that prometryne inhibited nucleic acid synthesis prior to its inhibition of photosynthetic CO_2 fixation and suggested that nucleic acid metabolism may be the primary site of action rather than the generally accepted concept that it is in the photosynthetic apparatus.

H. UNCLASSIFIED HERBICIDES

1. *Picloram*

Picloram at low concentrations has been reported to reduce the soluble protein content of above-ground parts of monocots (maize, sorghum, rice, wheat) and sunflower but little change occurred in cowpeas and cotton (Baur *et al.* 1970). Picloram also effects the types of protein found in pea roots (Norris and Morris, 1969). Seeds of cucumber and wheat, germinated in picloram and harvested at four days, showed progressive increase in DNA and protein content of the roots (on a per plant basis) up to 10 parts/10^6; however, at 100 parts/10^6 they both decreased (Chen *et al.* 1972). In contrast, RNA progressively decreased up to 100 parts/10^6 of picloram in wheat and progressively increased in cucumber up to 10 parts/10^6 and decreased at 100 parts/10^6. In general 2,4-D, 2,4,5-T, and dicamba induced similar responses. All four herbicides inhibited root growth in both species at concentrations greater than $0 \cdot 1$ parts/10^6. At inhibitory concentrations, swelling and inhibition of secondary root formation occurred in cucumber seedlings but not in wheat. They suggested that a differential alteration of RNA species and an interference with protein synthesis were the basis of selectivity. Malhotra and Hanson (1970) sprayed picloram on seedlings of resistant species (barley, wheat, maize) as well as susceptible species (soybean, cucumber) and looked for changes in nucleic acid metabolism 24 h later. The total RNA and DNA content correlated inversely with herbicide resistance, resistant species being low in nucleic acids and the susceptible ones high. The increase in the nucleic acids of the sensitive species after treatment appeared to be associated with lower levels of ribonuclease and deoxyribonuclease. The inability of the resistant species to synthesize more RNA may be due to the high levels of nucleases.

2. *DCPA*

DCPA has been reported to induce increases in total nitrogen nitrate, aminoacids, and protein in stolons of Bermuda grass (Fullerton *et al.* 1970).

3. *Dicamba*

Chen *et al.* (1972) showed that dicamba increased DNA and protein contents of germinating cucumber and wheat seedling up to 10 parts/10^6 but at 100 parts/10^6 there was a decrease. However, RNA increases in cucumber up to 10 parts/10^6 whereas it progressively decreases in wheat.

4. *Uracils*

Hewitt and Notten (1966) found that bromacil and isocil inhibit the induction of nitrate reductase by nitrate and molybdenum in leaf tissues. These herbicides were more inhibitory than several anti-metabolites previously used. Inhibition was severe at a concentration of 1 µg/ml in the infiltrating solutions.

5. *EPTC*

EPTC applied to the soil at 15 and 20 kg/ha increased the synthesis of nitrogen compounds in the leaves of cotton and Johnson grass and reduced the nitrogenous content of Johnson grass rhizomes (Balandina and Gashnikova, 1969).

VI. MINERAL CONSTITUENTS

There appears to be a general agreement that the interpretation of data from studies on the influence of a herbicide on the mineral composition of a plant is very difficult (Wort, 1964a; Penner and Ashton, 1966; Adams and Espinoza, 1969). The final results are influenced by the several factors discussed in the Introduction and Section I of this chapter. However, the rate of growth is particularly important in regard to mineral nutrients. For example, if the total uptake of ions is not affected by the herbicide but growth is stimulated or inhibited the concentrations of minerals in the plant is diluted or concentrated, respectively.

A. ALIPHATICS

The uptake of ^{32}P by excised roots of maize plants was inhibited 50% by a 12 h treatment with 5×10^{-3} M dalapon (Ingle and Rodgers, 1961). *Chlorella vulgaris* treated with a concentration series up to 7.2×10^{-2} M of NaTCA showed a three-fold decrease in acid-insoluble phosphate and a corresponding increase in acid-soluble phosphate (Hassall, 1961). Cell division was retarded at concentrations above 1.0×10^{-2} M. The uptake of phosphorus, nitrogen, and glucose was depressed at concentrations above 6.0×10^{-2} M. He suggested that NaTCA interferes with nucleic acid metabolism.

B. AMIDES

A study of the effect of diphenamid on the uptake and distribution of macronutrient elements in cabbage grown in culture solution has been conducted (Nashed and Ilnicki, 1968b). Diphenamid decreased the uptake of Ca, Mg, K, and P, in that order. Shoots showed a reduction of these four elements, especially Ca. Roots showed a marked increase in Ca content and a marked reduction in Mg content relative to control plants.

C. BIPYRIDYLIUMS

The influence of Cu^{2+} and CN^- ions on paraquat toxicity to *Chlorella pyrenoidosa* was studied by Tsay *et al.* (1970). Cu^{2+} inhibited the phytotoxicity of paraquat whereas CN^- increased paraquat phytotoxicity with highly significant mutually antonistic interactions occurring between Cu^{2+} (2 parts/10^6) and CN^- (16 parts/10^6) at paraquat concentrations of 0·5 to 2·0 parts/10^6.

D. CARBAMATES

Chlorpropham was one of several herbicides investigated for their effect on nitrate levels of 14 plant species in regard to nitrate poisoning of livestock (Frank and Grigsby, 1957). The plants were analysed 1, 4, and 14 days after the herbicide treatment. The nitrate content of two species was increased, was decreased in four species, and in eight species no significant change was found due.

E. DINITROANILINES

Kust and Struckmeyer (1971) found that trifluralin induced nitrogen deficiency symptoms in soybeans in the field and these plants had fewer nitrogen-fixing bacterial nodules. Greenhouse experiments confirmed that trifluralin inhibited nodule formation. They suggested that trifluralin also inhibits utilization of cotyledonary reserves and redistribution of organic and mineral constituents of unifoliolate leaves because of various anatomical abnormalities induced. Supplying nitrogen to the treated plants did not overcome the inhibition of growth.

F. NITRILES

A comprehensive study on the effect of dichlobenil and boron deficiency on morphology, root histology, mitosis in roots, translocation, and pH of tissue homogenates strongly suggests that dichlobenil acts on the same basic process that is effected by a boron deficiency (Milborrow, 1964).

G. PHENOLS

Dinoseb altered the nitrate content of 14 weed species as follows: one was increased, six were decreased, and six were not changed (Frank and Grigsby, 1957). However, the nitrate content of *Solanum dulcamara* was unchanged after one day, increased after 4 days, and decreased after 14 days. The accumulation of phosphorus in leaf discs was inhibited by dinoseb and this was thought to be associated with its inhibition of oxidative phosphorylation (Wojtaszek *et al.* 1966).

PCP reduced transpiration and uptake of NH_4, K, Si, and particularly P by rice and barnyard grass grown in culture solution (Yoshida and Takahashi, 1967).

The chlorophenoxy herbicides have been extensively studied in regard to inorganic elements. Three articles have been written about this subject as well as other aspects of these herbicides (Wort, 1961, 1964; Penner and Ashton, 1966). In general, the mobilization and redistribution of minerals following chlorophenoxy-herbicide treatment appears to follow a pattern similar to that observed for amino-acids and sugars. The mineral content of the leaves decreases following herbicide treatment and that of the stem increases. The uptake of various ions may also be influenced by these herbicides; but as Wort (1961) noted, 2,4-D either inhibits or has little effect on the absorption of some 14 ions by plants. Rarely is ion uptake enhanced by 2,4-D treatment.

Several reports have shown that sub-lethal (stimulatory) amounts of 2,4-D applied to leaves of plants in combination with minor elements induce greater productivity (see reviews by Wort, 1961, 1964b). Also see Chapter 11, volume 2.

Considerable research has been conducted on the effect of sub-toxic levels of the triazine herbicides on nitrogen metabolism particularly in regard to nitrates, nitrate reductase, and proteins; however, this is covered in Chapter 11, volume 2 and will be omitted here except where essential for the discussion.

1. NPK Interaction

Many investigators have reported the interaction of fertilizers or mineral elements and the triazine herbicides. In general, fertilization with NPK increased triazine toxicity and the triazines increased the uptake of several elements; however, in some experiments the results were not in agreement with this generalization. Voitekhova (1968) treated barley with atrazine and found that intermediate levels of NPK induced the most injury. When N, P, and K were applied separately, N caused the greatest injury followed by K; P had little effect. Bagaev and Kalimullina (1968) reported that NPK increased the toxicity of propazine to carrots. Propazine increased the N and P content and reduced the K content when the carrot plant was young, but at harvest NPK was increased in the leaves and decreased in the roots relative to controls. Using maize treated with atrazine, Sosnovya (1971) found that applications of NPK increased NPK uptake. Protein N and nucleotide P content was increased but the phosphorus sugar-ester content of the grain was decreased. Zhukov (1968) also studied NPK-atrazine interaction in maize with particular attention to organic phosphorus compounds. At the 4 to 5 leaf stage; sugar-phosphorus, lipid phosphorus, and protein content increased but nucleotide phosphorus content decreased. At tasselling time there was a reduction in protein-phosphorus content. However, at harvest protein phosphorus was increased along with lipid phosphorus and nucleinic phosphorus while sugar phosphorus and nucleotide phosphorus was reduced.

Vicia faba plants were grown in culture solution and treated with 0·25, 0·5, 1·0, and 2·0 parts/10^6 of simazine, 10 times these levels of trietazine, as well as, 100 times these levels of chlorazine (Zinckenko *et al.* 1966). Simazine and trietazine induced progressive chlorosis and increased the P and K contents and also the K : Ca ratio of leaves. Chloroazine induced a darker green colour of the leaves and growth was not reduced as much as with simazine or trietazine. Chlorazine had no effect on N or P levels of leaves but reduced the K : Ca ratio. The K : Ca ratio of roots was decreased by all treatments. The difference in symptoms between chlorazine and simazine or trietazine suggest that even though the chlorazine was applied at a much higher concentration, these levels were still sub-lethal. Perhaps even lower levels of simazine or trietazine could induce the chlorazine type symptoms as well as inorganic element shifts similar to chlorazine.

Atrazine (3 mg) applied to a leaf of sunflower or mustard (*Sinapis alba*) and subsequently treated with ^{15}N and ^{32}P via sand culture, reduced the uptake of ^{15}N and ^{32}P into the aerial parts of the plants and influenced their distribution in plant organs (Chesalin and Timofeeva, 1968). Control plants had most of the ^{15}N and ^{32}P in the top leaves while the atrazine-treated plants had most of these elements in the stems. Atrazine caused the radioactivity of acid- and alcohol-soluble fractions to increase whereas that of RNA and DNA fractions decreased following the ^{32}P application to mustard. Atrazine caused the radioactivity of the proteins of leaves, stems, and roots to be decreased following ^{15}N application to sunflower.

2. *N Interaction*

The addition of simazine (1·5 parts/10^6) and N (84 mg) to a pot of N-deficient soil increased the dry matter and N content of maize to 27% and 29·5%, respectively, 29 days after planting (Freney, 1965). Using culture solutions, Freney found that high concentrations (0·3 to 3·0 parts/10^6) of simazine reduced dry matter, but a low concentration (0·06 parts/10^6) increased yield by 35·9% and increased N, P, K, and Mg uptake. He concluded that simazine increased plant growth by a direct effect on plant metabolism and not as a result of any interaction with the soil. Simazine, at 0·5 to 2 kg/ha, with low and high rates of N fertilizer reduced the total N and protein N contents of maize and rye (Frieske, 1969). However, simazine increased the nitrate accumulation of these two species at the high rate of N application. Simazine also increased the K content significantly and slightly increased the Ca, Mg, and P content of the plants.

3. *P Interaction*

The data from a series of field experiments with simazine and atrazine indicated that weed control was better on soil with high than with low phosphate fertility. Subsequently, Adams and associates investigated the interaction of P with atrazine or simazine primarily using soybean as the test plant (Adams and Otto, 1964; Adams, 1965; Adams and Espinoza, 1969). In general, increased amounts of P increased the phytotoxic action of these triazines (Table V). Either added P or simazine alone did not influence inorganic ion uptake; however when both were present Fe, K, Mn, and Si content of shoots were increased while Ca, Mg, B, Sr, and Mo decreased. At relatively high concentrations of atrazine (0.4 parts/10^6 in various soil types) the concentration of nearly all elements in the plant increased; however, this may have been associated with the inhibited rate of growth. In similar studies on the atrazine-P interaction in soybeans, Sun (1969) found that atrazine increased the shoot content of P, Ca, Mn, and Mg. Interaction between atrazine and P lowered the Mn content and resulted in a larger accumulation of K and B in the shoots. Mn deficiency and excess P reduced the photosynthetic capacity and increased the inhibitory effect of atrazine on photosynthesis. Mn appeared to protect the plants against atrazine injury.

The uptake of ^{32}P by oats was markedly inhibited by 0.25 to 0.50 parts/10^6

TABLE V

MINERAL COMPOSITION OF SOYBEAN SHOOTS GROWN ON FORMAN CLAY LOAM SOIL WITH THREE ATRAZINE AND FIVE PHOSPHORUS TREATMENTS (adapted from Adams and Espinoza, 1969)

$Ca(H_2PO_4)_2$ applied parts P/10^6	Growth dry wt., g/pot	P^1	K^1	Ca^1	Mg^1	Fe^2	Mn^2	B^2	Zn^2	Mo^2	Si^2	Ba^2
No atrazine												
0	2·19	0·22	1·24	1·42	0·43	49	46	49	26	0·35	0·06	20
50	5·04	0·22	1·08	1·68	0·47	46	46	46	22	0·36	0·06	22
100	5·00	0·34	1·11	1·69	0·48	46	42	46	22	0·40	0·08	24
150	4·85	0·32	1·01	1·51	0·44	45	40	45	16	0·33	0·08	22
200	5·04	0·41	1·12	1·68	0·45	44	50	44	20	0·44	0·13	23
0.2 parts/10^6 atrazine												
0	4·06	0·23	1·16	1·29	0·37	48	44	48	24	0·22	0·07	16
50	4·59	0·30	1·25	1·45	0·33	46	34	46	30	0·22	0·08	12
100	4·43	0·35	1·42	1·78	0·50	52	48	52	26	0·41	0·12	26
150	3·89	0·46	1·69	1·74	0·52	46	48	46	27	0·42	0·10	25
200	4·13	0·50	1·42	1·80	0·51	45	46	45	24	0·41	0·12	26
0.4 parts/10^6 atrazine												
0	2·70	0·41	2·12	1·80	0·56	64	63	64	40	0·60	0·12	30
50	2·42	0·45	2·19	1·90	0·60	58	54	58	37	0·67	0·17	32
100	2·14	0·55	2·58	1·89	0·55	58	40	58	37	0·68	0·20	36
150	1·83	0·61	2·81	1·88	0·56	60	42	60	38	0·69	0·20	37
200	1·43	0·64	2·76	1·97	0·52	56	50	56	36	0·72	0·14	38

[1] Values in percentage.
[2] Values in parts/10^6.

of simazine in soil (Zurawski *et al.* 1965), whereas, Dhillon *et al.* (1967) reported that ^{32}P uptake from culture solution was increased at 5 and 10 parts/10^6 of simazine but inhibited at 15 and 20 parts/10^6. Furthermore, translocation of ^{32}P seemed to be stimulated by all four concentrations of simazine.

Rudgers (1970) reported that atrazine had no effect of P-induced Zn deficiency symptoms of maize but did increase the P and Zn concentrations in leaves.

4. *K Interaction*

Leaves of oat plants grown in soil treated with simazine were treated with ^{42}K; 4, 8, and 10 days after emergence (Plozynski *et al.* 1968). Simazine inhibited the foliar uptake of ^{42}K at all application times. The authors attributed this to the interference of simazine with photosynthesis.

5. *Mg Interaction*

Mg-deficient tomato plants grown under continuous light were more susceptible to atrazine applied four weeks after planting than were plants with normal levels of Mg. However, under a 12-h photoperiod Mg levels had little or no effect on atrazine toxicity. Since Mg deficiency and continuous light each cause a decrease in chlorophyll, it was suggested that the influence of Mg on atrazine toxicity may be due to its effect on chlorophyll content rather than a direct effect on atrazine action.

K. AMITROLE

Since amitrole acts as a chelating agent, Menoret (1957) investigated the effect of several elements on amitrole toxicity to tomatoes. Both amitrole and the elements under study were applied to roots via culture solution. The addition of Cu, Ni, and Co reduced the toxicity of amitrole but Fe, Ca, and Mg had no effect. He concluded that the action of amitrole is associated with its chelation of Fe. However, this concept is not widely accepted (Carter, 1969; Ashton and Crafts, 1973).

McWhorter (1963) reported that the chlorotic tissue of maize induced by amitrole contained increased concentrations of Mn, NH_4, and total N. Protein content was usually reduced and amino-acids increased; concentrations of these and free NH_3 was influenced by light period. Herbicide toxicity was increased by ammonia fertilization as compared to nitrate fertilization.

Several studies have been conducted on P-amitrole interaction. Herrett and Linck (1968) compared the effects of amitrole on *Cirsium arvense* leaf discs from normal and P-deficient plants. With P-deficient leaf discs respiration was markedly increased and chlorosis was much more pronounced, relative to normal leaf discs. Studies on the influence of P on amitrole toxicity to groundnuts (in culture solution) and wheat seedlings (on agar) were conducted by Vega (1967). The P concentration of the culture medium which produced maximal growth also produced maximal amitrole injury. Wort and Laughman (1961) reported that amitrole diminished root uptake of ^{32}P, but the proportion of ^{32}P absorbed was higher in shoots of treated plants than those of control plants. The distribution of ^{32}P between the acid-insoluble fraction (nucleic acids, phospholipids, and phosphoproteins) and acid-soluble fraction (nucleotides, sugar-P, and inorganic-P) was essentially unchanged up to 4 h but at 24 h amitrole decreased ^{32}P incorporation into the acid-soluble fraction of both roots and shoots. They suggested that amitrole action involved nucleic acid synthesis. Suong (1961) proposed that amitrole inhibits photosynthesis by reducing the regeneration of ribulose phosphate.

L. UREAS

The uptake of ^{32}P by oats was reported to be markedly inhibited by 0·25 to 0·50 parts/10^6 of monuron in soil (Zurawski et al. 1965). However, when linuron was applied to roots via culture solution it did not effect P uptake of maize or crabgrass but did increase P uptake 30 to 50% in soybean (Nashed and Ilnicki, 1968a). Hogue (1968) found that when linuron was applied to leaves of tomato and parsnip, ^{32}P uptake by roots was markedly increased and the content of ^{32}P in each plant part examined (roots, stem and petioles, and leaves, particularly the leaves) was higher than the controls.

Leaves of oat plants grown in soil treated with monuron were treated with ^{42}K 3, 8, and 10 days after emergence (Plozynski et al. 1968). Monuron inhibited the foliar uptake of ^{42}K at all application times. Linuron applied via culture solution also inhibited root uptake of K in soybean very markedly but had only a slight inhibitory affect on K uptake by roots of maize and crabgrass (Nashed and Ilnicki, 1968a).

Hogue (1968) reported that when linuron was applied to leaves of tomato and parsnip, ^{45}Ca uptake and translocation by roots was inhibited. However, when linuron applied to roots via culture solution, Ca uptake by roots was markedly increased in maize, soybean, and crabgrass (Nashed and Ilnicki, 1968a).

Nashed and Ilnicki (1968a) investigated the effect of linuron applied to roots via culture solution on root uptake of SO_4, NO_3, and Mg (as well as P, K, and Ca mentioned above). Linuron increased the uptake of SO_4 three- to six-fold in soybean, maize, and crabgrass. NO_3 uptake was generally inhibited at the lower linuron dosages but tended to increase somewhat above control levels at higher dosages. Mg uptake increased markedly in crabgrass, increased slightly in maize, and decreased markedly in soybeans as the concentration of linuron increased. They suggested that changes in ion uptake in herbicide treated plants may be symptomatic of basic changes in metabolism or of changes in the permeability characteristics of plant membranes. The latter possibility is suggested by the actual loss of Mg from linuron-treated soybeans.

REFERENCES

Adams, R. S. (1965). *Weeds,* **13,** 113–116.
Adams, R. S. and Espinoza, W. G. (1969). *J. agric. Fd Chem.* **17,** 818–822.
Adams, R. S. and Otto, H. J. (1964). *Abst. Weed Sci. Soc. Am.* pp. 73–74.
Akobundu, I. O., Bayer, D. E. and Leonard, O. A. (1970). *Weed Sci.* **18,** 403–417.
Alexander, A. G. and Montalvo-Zapata, R. (1970). *J. Agric. Univ. P. Rico,* **54,** 264–296.
Alimova, F. R. and Nuriidinov, A. I. (1969). *Uzbek, biol. Zh.* **13,** 30–32.
Andersen, R. N., Behrens, R. and Linck, A. J. (1962). *Weeds,* **10,** 4–9.
Ashton, F. M. and Crafts, A. S. (1973). "Mode of Action of Herbicides", John Wiley and Sons, New York, 504 pp.
Ashton, F. M., Penner, D. and Hoffman, S. (1968). *Weed Sci.* **16,** 169–171.
Bach, M. K. (1961). *Pl. Physiol., Lancaster,* **36,** 558–565.
Bagaev, V. B. and Kalimullina, Kh. K. (1968). *Dokl. TSKhA,* **133,** 281–287.
Baker, J. B. and Pizzalato, T. D. (1968). *Pl. Physiol., Lancaster,* Suppl. **43,** 3.
Bakke, A. L., Goessler, W. G. and Loomis, W. E. (1939). *Res. Bull. Iowa agric. Exp. Stn.* No. 254.
Balandina, I. D. and Gashnikova, V. V. (1969). *Khimiya sel' Khoz.* **7,** 291–293.
Bartels, P. G. and Hyde, A. (1970). *Pl. Physiol., Baltimore,* **46,** 825–830.
Bartels, P. G. and Wolf, F. T. (1965). *Physiologia Pl.* **18,** 805–812.
Bartley, C. E. (1957). *Agric. Chem.* **12,** 34–36, 113–115.
Baur, J. R., Bovey, R. W. and Benedict, C. R. (1970). *Agron. J.* **62,** 627–630.
Baxter, R. and Hanson, J. B. (1968). *Planta,* **82,** 246–260.
Berüter, J. and Temperli, A. T. (1970). *Experientia,* **26,** 600–601.
Black, C. C. and Humphreys, T. E. (1962). *Pl. Physiol., Lancaster,* **37,** 66–73.
Bourke, J. B., Butts, J. S. and Fang, S. C. (1962). *Pl. Physiol., Lancaster,* **37,** 233–237.
Brian, R. C. (1958). *Pl. Physiol., Lancaster,* **33,** 431–439.
Briquet, M. V. and Wiaux, A. L. (1967). *Meded. Rijksfac. LandbWet. Gent.* **32,** 1040–1049.
Brown, T. C. and Carter, M. C. (1968). *Weed Sci.* **16,** 222–226.
Burt, G. W. and Muzik, T. J. (1970). *Physiologia Pl.* **23,** 498–504.
Camper, N. D. and Moreland, D. E. (1967). *Abstr. Weed Sci. Soc. Am.* pp. 66–67.
Cardenas, J., Slife, F. W., Hanson, J. B. and Butler, H. (1968). *Weed Sci.* **16,** 96–100.
Carter, M. C. (1969). *In* "Degradation of Herbicides" (P. C. Kearney and D. D. Haufman, eds), pp. 187–206. Marcel Dekker, Inc., New York.
Castelfranco, P., Oppenheim, A. and Yamaguchi, S. (1963). *Weeds,* **11,** 111–115.
Chen, L. G., Switzer, C. M. and Fletcher, R. A. (1972). *Weed Sci.* **20,** 53–55.
Chesalin, G. A. and Timofeeva, A. A. (1968). *Agrokhimiya,* **5,** 108–113.
Chodová, D. and Zemánek, J. (1971). *Biologia Pl.* **13,** 234–242.
Coble, H. D. and Slife, F. W. (1971). *Weed Sci.* **19,** 1–3.
Corbadzijska, B. (1962). *Nava. Trudy viss. selskostop. Inst. 'V. Kolarov.' Agron. Fak.* **11,** 87–98.
Crafts, A. S. (1961). "The Chemistry and Mode of Action of Herbicides", Interscience Publishers, New York, N.Y., 269 pp.
Dewey, O. R., Gregory, P. and Pfeiffer, R. K. (1956). *Proc. Br. Weed Control Conf.* 3rd Meeting, pp. 313–326.
Dhillon, P. S., Byrnes, W. R. and Merritt, C. (1967). *Weeds,* **15,** 339–343.
Dixon, M. and Webb, E. C. (1958). "Enzymes", Longman Group Ltd., Harlow, England, 782 pp.
Duble, R. L. and Holt, E. C. (1970). *Weed Sci.* **18,** 174–179.
Duble, R. L., Holt, E. C. and McBee, G. G. (1968). *Weed Sci.* **16,** 421–424.
Duke, W. B. (1967). *An investigation of the mode of action of N-isopropyl alphachloroacetanilide.* Ph.D. Dissertation, Univ. of Illinois, 124 pp.
Dunham, R. S. (1951). *In* "Plant Growth Substances" (F. Skoug, ed.), pp. 195–206. University of Wisconsin Press, Madison.
Edmondson, J. B. (1969). *Effects of 2-chloro-2',6'-diethyl-N-(methoxymethyl)acetanilide on cucumber seedling growth.* Ph.D. Dissertation, Univ. of Illinois, 51 pp.

Epps, E. A., Jr. (1953). *J. agric. Fd Chem.* **1,** 1009–1010.

Fang, S. C. (1962). *Pl. Physiol., Lancaster,* **27,** (Suppl), 24.

Flood, A. E., Rutherford, P. P. and Weston, E. W. (1970). *Phytochemistry,* **9,** 2431–2437.

Foy, C. L. (1969). *In* "Degradation of Herbicides" (P. C. Kearney and D. D. Haufman, eds), pp. 207–253. Marcel Dekker, Inc., New York.

Frank, P. A. and Grigsby, B. H. (1957). *Weeds,* **5,** 206–217.

Freney, J. R. (1965). *Aust. J. agric. Res.* **16,** 257–263.

Frieske, S. (1969). *Biul. Inst. Ochr. Roślin,* **45,** 23–32.

Fullerton, T. M., Murdock, C. L., Spooner, A. E. and Frans, R. E. (1970). *Weed Sci.* **18,** 711–714.

Gast, A. and Grob, M. (1960). *Pest Technol.* **3,** 68–73.

Gentner, W. A. (1966). *Weed Sci.* **14,** 27–30.

Glenn, R. K. (1971). *Some studies of dichlobenil on cotton, including its effect on assimilate transport.* M.S. Thesis, University of California, Davis, p. 65.

Grant, M. E. and Fuller, K. W. (1971). *J. exp. Bot.* **22,** 49–59.

Gräser, H. (1970). *TagBer. dt. Akad. LandwWiss. Berl.* **109,** Vorträge eins internationalen Symposiums 21–31 May 1968, 97–117.

Gräser, H. and Tillich, B. (1970). *TagBer. dt. Akad. LandwWiss. Berl.* **109:** Vorträge eines internationalen Symposium 26–31 May 1968, 83–95.

Guenthner, H. R. (1970). *Translocation, anatomical and histochemical effects of picloram.* Ph.D. Dissertation, Washington State Univ., p. 82.

Hagen, C. E., Glagett, C. D. and Helgeson, E. A. (1949). *Science, N.Y.,* **110,** 116–117.

Hagin, R. D., Linscott, D. L. and Dawson, J. E. (1970). *J. agric. Fd Chem.* **18,** 848–850.

Hanson, J. B. and Slife, F. W. (1969). *Residue Rev.* **25,** 59–67.

Hassall, K. A. (1961). *Physiologia Pl.* **14,** 140–149.

Herrett, R. A. and Linck, A. J. (1968). *Abstr. Weed Soc. Am.* pp. 36–37.

Hewitt, E. J. and Notten, B. A. (1966). *Biochem. J.* **101,** 39–40.

Hilton, J. L. (1969). *J. agric. Fd Chem.* **17,** 182–198.

Hilton, J. L. (1972). Personal communication.

Hilton, J. L. and Christiansen, M. N. (1972). *Weed Sci.* **20,** 290–294.

Hilton, J. L., Kearney, P. C. and Ames, B. N. (1965). *Archs Biochem. Biophys.* **112,** 544–547.

Hilton, J. L., St. John, J. B., Christiansen, M. N. and Norris, K. H. (1971). *Pl. Physiol., Baltimore,* **48,** 171–177.

Hogue, E. J. (1968). *Weed Sci.* **16,** 185–187.

Ingle, M. and Rodgers, B. J. (1961). *Weeds,* **9,** 264–272.

Ivens, G. W. and Blackman, G. E. (1949). *Symp. Soc. exp. Biol.* **3,** 266–282.

Jain, M. L., Kurtz, E. B. and Hamilton, K. C. (1966). *Weeds,* **14,** 259–262.

Johnson, B. J. and Jellum, M. D. (1969). *Agron. J.* **61,** 379–380.

Juniper, B. E. (1959). *New Phytol.* **58,** 1–4.

Juniper, B. E. and Bradley, D. E. (1959). *J. Ultrastruct. Res.* **2,** 16–27.

Kalinin, F. L., Merezhinskii, Yu. G. and Volovik, O. I. (1968). *Fiziol.-biokhim. Osnovy Pitan. Rasten., Kiev,* **4,** 20–27.

Kemp, T. R., Stoltz, L. P., Herron, N. W. and Smith, W. T. (1969). *Weed Sci.* **17,** 444–446.

Kent, N. L. and Hutchinson, J. B. (1957). *Ann. appl. Biol.* **45,** 481–488.

Key, J. L., Lin, C. Y., Gifford, E. M. and Dengler, R. (1966). *Bot. Gaz.* **127,** 87–94.

Kim, W. K. and Bidwell, R. G. S. (1967). *Can. J. Bot.* **45,** 1751–1760.

Klopotonski, T. and Wiater, A. (1965). *Archs Biochem. Biophys.* **112,** 562–566.

Kunert, G. (1959). *Naturwissenschaften,* **46,** 603.

Kust, C. A. and Struckmeyer, B. E. (1971). *Weed Sci.* **19,** 147–152.

Lai, M. T. and Semeniuk, G. (1970). *Phytopathology,* **60,** 563–564.

Lantican, B. P., Zamora, P. M., Robles, R. P. and Talatala, R. L. (1969). *Philipp. Agric.* **52,** 553–565.

Lawrence, J. M. and Muzik, T. J. (1962). *NW. Sci.* **36,** 39.

Leonard, O. A. and Glenn, R. K. (1968). *Weed Sci.* **16,** 352–356.

Linscott, D. L. and Hagin, R. D. (1970). *Weed Sci.* **18,** 197–203.

Linscott, D. L., Hagin, R. D. and Dawson, J. E. (1968). *J. agric. Fd Chem.* **16,** 844–848.

Lukin, V. V., Petunova, A. A. and Saburova, P. V. (1969). *Byull. vses. nauchno-issled. Inst. Zashch. Rast.* 59–63.

MacLeod, D. G. (1964). *Weed Res.* **4**, 275–282.

Maestri, M. (1967). *Structural and functional effects of endothal on plants.* Ph.D. Dissertation, Univ. of Calif., Davis, p. 100.

Malhotra, S. S. and Hanson, J. B. (1970). *Weed Sci.* **18**, 1–4.

Mann, J. D., Jordan, L. S. and Day, B. E. (1965). *Pl. Physiol., Lancaster,* **40**, 840–843.

Mann, J. D. and Pu, M. (1968). *Weed Sci.* **16**, 197–198.

Mashtakov, S. M., Deeva, V. P. and Volynets, A. P. (1967a). *Nauka Tekh., Minsk.* 78–93.

Mashtakov, S. M., Deeva, V. P. and Volynets, A. P. (1967b). *In* "Physiological Effects of Herbicides on Varieties of Crop Plants" (S. M. Mashtakov, ed.), pp. 47–49. Nauka Tekhnika, Minsk.

Mashtakov, S. M. and Kudryavtsev, G. P. (1969). *Soviet Pl. Physiol.* **16**, 593–598.

Mashtakov, S. M. and Moshchuk, P. A. (1967). *Agrokhimiya* **9**, 80–89.

McWhorter, C. G. (1961). *Weeds,* **9**, 563–568.

McWhorter, C. G. (1963). *Pl. Physiol., Lancaster,* **16**, 31–39.

McWhorter, C. G. and Hilton, J. L. (1967). *Physiologia Pl.* **20**, 30–40.

Menoret, Y. (1957). *Soc. Franc. de Physiol. Veg. B.* **3**, 131.

Milborrow, B. V. (1964). *J. exp. Bot.* **15**, 515–523.

Moreland, D. E., Malhotra, G. S., Gruenhagen, R. D. and Shokraii, E. H. (1969). *Weed Sci.* **17**, 556–563.

Mostafa, I. Y. and Fang, S. C. (1971). *Weed Sci.* **19**, 248–253.

Muzik, T. J. and Lawrence, J. M. (1958). *Proc. 16th Western Weed Control Conf.* p. 31.

Muzik, T. J. and Lawrence, J. M. (1959). *Nature, Lond.* **183**, 482.

Nashed, R. B. and Ilnicki, R. D. (1968a). *Weed Sci.* **16**, 188–192.

Nashed, R. B. and Ilnicki, R. D. (1968b). *Proc. 22nd N.E. Weed Control Conf.* p. 500.

Naylor, A. W. (1964). *J. agric. Fd Chem.* **12**, 21–25.

Norris, L. A. and Morris, R. O. (1969). *Res. Prog. Rept. Western Soc. Weed Sci.* pp. 103–104.

Nylund, R. E. (1956). *Am. Potato J.* **33**, 145–154.

Opile, W. R. (1969). *Depletion of carbohydrates in storage organs of four perennial weeds.* M.S. Thesis, Univ. of Calif., Davis, 96 pp.

Paramonova, T. V. (1971). *Khimiya v sel's skom Khozyaistve* **9**, 610–611.

Payne, M. G. and Fults, J. L. (1955). *Am. Potato J.* **32**, 144–149.

Payne, M. G., Fults, J. L., Hays, R. J. and Livingston, C. H. (1953). *Am. Potato J.* **30**, 46–49.

Penner, D. (1970). *Weed Sci.* **18**, 360–364.

Penner, P. and Ashton, F. M. (1966). *Residue Rev.* **14**, 39–113.

Penner, D. and Meggitt, W. F. (1969). *Proc. 24th N. cent. Weed Control Conf.* 81.

Penner, D. and Meggitt, W. F. (1970). *Crop Sci.* **10**, 553–554.

Pfeiffer, R. K., Dewey, O. R. and Brunskill, R. T. (1959). *Proc. 4th int. Congr. Crop Protection* (Hamburg, Germany, Sept. 8–15, 1957), pp. 523–525.

Płoszyński, M., Swietochowski, B. and Zurawski, H. (1969). *Roczn. Nauk roln. (Ser. A.)* **95**, 401–415.

Ploszynski, M., Zurawski, H. and Bors, J. (1968). *Pam. Pulaw.* **31**, 113–121.

Ravazzoni, C. and Valerio, R. (1956). *G. Biochim.* **5**, 37–44.

Redemann, C. T. and Hamaker, J. W. (1957). *Weeds,* **3**, 387–388.

Reyes, E. H. and de Capriles, R. L. (1970). *Agron. trop. Nogent.* **20**, 257–266.

Rudgers, L. A. (1970). *Proc. Soil. Sci. Soc. Am.* **34**, 240–244.

Rutherford, P. P., Weston, E. W. and Flood, A. E. (1969). *Phytochemistry,* **8**, 1859–1866.

Scherl, M. M. and Frans, R. E. (1969). *Weed Sci.* **17**, 421–427.

Schweizer, E. E. (1970). *Weed Sci.* **18**, 131–134.

Sell, H. M., Hamner, C. L., Rebstock, T. L., Weller, L. E., Miller, A. F. and Fukui, H. N. (1957). *Q. Bull. Mich. St. Univ., agric. Exp. Stn.* **40**, 44–50.

Sell, H. M., Luecke, R. W., Taylor, B. M. and Hamner, C. L. (1949). *Pl. Physiol., Lancaster,* **24**, 295–299.

Singh, O. S., Sinka, B. K. and Madan, S. K. (1969). *Indian J. Weed Sci.* **1**, 115–118.

Sistrunk, J. W. (1969). *Intra-species tolerance of bermudagrass,* Cynodon dactylon, *to dalapon (2,2-dichloro-propionic acid sodium salt).* Ph.D. Dissertation, Kansas State Univ., 78 pp.

Smith, A. E. (1972). *Weed Sci.* **20**, 46–48.

Smith, A. E. and Wilkinson, R. E. (1973). *Weed Sci.* **21**, 57–60.

Smith, F. G., Hamner, C. L. and Carlson, R. F. (1947). *Pl. Physiol., Lancaster,* **22**, 58–65.

Smith, L. W., Peterson, R. L. and Horton, R. F. (1971). *Weed Sci.* **19**, 174–177.

Sosnovaya, O. N. (1971). *Fiz. Biokhim. kul't. Rast.* **3**, 44–48.

Still, G. G., Davis, D. G. and Zander, G. L. (1970). *Pl. Physiol., Baltimore,* **46**, 307–314.

Stokes, D. M. and Turner, J. S. (1971). *Aust. J. biol. Sci.* **24**, 433–447.

Sun, C.-N. (1969). *Physiological effects of the phosphorus-atrazine (2-chloro-4-ethylamino-6-isopropylamino sym-triazine) interaction in soybean plants* (Glycine max). Ph.D. Dissertation, Univ. of Minnesota, 130 pp.

Sund, K. A., Putala, E. C. and Little, H. N. (1960). *J. agric. Fd Chem.* **8**, 210–212.

Suong, N.-T. (1961). *C.r. hebd. Séanc. Acad. Sci. Paris,* **252**, 1996–1998.

Swann, C. W. and Buchholtz, K. P. (1966). *Weeds,* **14**, 103–105.

Tanimoto, E. and Masuda, Y. (1968). *Physiologia Pl.* **21**, 820–826.

Torchinskaya, V. M. (1958). *Biol. Zbirn, L'vov. Derzhav. Univ. im. Ivana Franka,* **8**, 141.

Tsay, R. C. and Ashton, F. M. (1971). *Weed Sci.* **19**, 682–684.

Tsay, S.-F., Lee, J.-M. and Lynd, J. Q. (1970). *Weed Sci.* **18**, 596–598.

Tukey, H. B., Hammer, C. L. and Imhofe, B. (1945). *Bot. Gaz.* **107**, 62–73.

Vega, M. R. (1967). *Philipp. Agric.* **51**, 349–396.

Voitekhova, V. A. (1968). *Agrokhimiya,* **5**, 94–98.

Wain, R. L., Rutherford, P P., Weston, E. W. and Griffiths, C. M. (1964). *Nature, Lond.* **203**, 504–506.

Wiater, A., Krajensha-Grynkiewiez, K. and Klopotonski, T. (1971a). *Acta biochim. pol.* **18**, 299–307.

Wiater, A., Klopotonski, T. and Bagdasarian, G. (1971b). *Acta biochim. pol.* **18**, 309–314.

Wilkinson, R. E. and Hardcastle, W. S. (1970). *Weed Sci.* **18**, 125–128.

Wojtaszek, T., Cherry, J. H. and Warren, G. F. (1966). *Pl. Physiol., Lancaster,* **41**, 34–38.

Wort, D. J. (1949). *Am. J. Bot.* **36**, 673–676.

Wort, D. J. (1961). *In* "Encyclopedia of Plant Physiology, Vol. XIV" (W. Ruhland, ed.), pp. 1110–1136. Springer-Verlag, Berlin.

Wort, D. J. (1964a). *In* "The Physiology and Biochemistry of Herbicides" (L. J. Audus, ed.), pp. 291–334. Academic Press, London and New York.

Wort, D. J. (1964b). *In* "The Physiology and Biochemistry of Herbicides" (L. J. Audus, ed.), pp. 335–342. Academic Press, London and New York.

Wort, D. J. and Loughman, B. C. (1961). *Can. J. Bot.* **39**, 339–351.

Yaklich, R. W. (1970). *The metabolism of the herbicide (N,N-dimethyl-2,2-diphenylacetamide by corn root.* Ph.D. Dissertation, West Virginia Univ., 114 pp.

Yoshida, T. and Takahashi, J. (1967). *J. Sci. Soil Manure, Tokyo,* **38**, 342–344.

Zhukov, Yu. P. (1968). *Dokl. ISKhA,* **133**, 343–347.

Zinchenko, V. A., Fursenko, L. S. and Osinskya, T. V. (1966). *Agrokhimiya,* **9**, 103–109.

Zoschke, M. (1957). *Kühn-Arch.* **71**, 305–383.

Zurawski, H., Baranowski, R., Bors, J. and Ploszynski, M. (1965). *Symposium on Use of Isotopes in Weed Res.* p. 6.

CHAPTER 8

EFFECTS ON ETHYLENE PHYSIOLOGY

P. W. MORGAN

Department of Plant Sciences, Texas A&M University
College Station, Texas 77843, USA

I. Introduction

The predecessors of modern herbicides were linked to ethylene several ways in the early plant hormone literature. The connections all centred on the auxins. Although the auxins were first responsible for some of the interest in ethylene, they were also responsible later for the marked decline of that interest. Study of the auxins, meanwhile, resulted in the discovery of the phenoxy herbicides. These synthetic auxins were destined to play a key role in the development of modern chemical weed control.

Scientists at the Boyce Thompson Institute for Plant Research, in contrast to many others, were investigating both the newly discovered auxins and the highly phytoactive, unsaturated, hydrocarbon gases. They quickly recognized that the auxins and ethylene had many similar effects when applied to plants (Crocker *et al.* 1935). Using the leaf epinasty bioassay, they determined that "heteroauxin", indol-3yl-acetic acid (IAA), promoted the synthesis of ethylene by plant tissue (Zimmerman and Wilcoxon, 1935). Their straightforward proposal, that many of the apparent actions of auxins are actually mediated by ethylene produced in the plant, was ahead of its time. That concept, along with the proposal that ethylene is a plant hormone (Crocker *et al.* 1935), was dismissed in an unconvincing yet very effective manner (Michener, 1938). Apparently the intense interest in auxin left neither time nor sympathy for study of an apparent detractor.

As time passed, an additional connection between ethylene and auxin was recognized. Generally, the same groups of plants are sensitive to both auxins and ethylene (Hall *et al.* 1957; Hamner and Tukey, 1944). This selectivity of auxins and ethylene, typified by the herbicide-auxin 2,4-D, was one of the important properties that made the phenoxy herbicides so useful. It is well known in both cases that the selectivity is quantitative and not qualitative.

The similarity of plant responses to auxins and ethylene and their generally similar selectivity eventually led to a reinvestigation of the relationship. At the time of the reinvestigation the most obvious relationship to explain the similar responses was for the exogenous material (auxin in agricultural applications) to promote the synthesis of an endogenous substance (ethylene) which then produced a response. This would explain why exogenous ethylene produces the same response. Auxin-induced ethylene was associated with intact plant responses such as epinasty, apical bending, and leaf curling (Fig. 1 and Morgan and Hall, 1961, 1962, 1964; Hall and Morgan, 1964). The phenomenon of auxin-induced ethylene synthesis was promptly verified in other laboratories and other species (Abeles and Rubinstein, 1964; Burg and Burg, 1966a) and extended to a wide variety of specific responses.

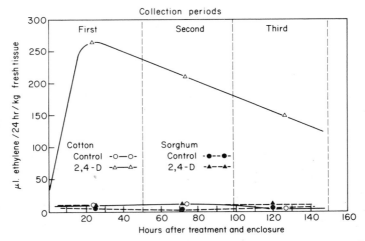

Fig. 1. Effect of 2,4-D on the synthesis of ethylene (µl/kg fresh wt. 24 h) by pre-flowering cotton and grain sorghum plants. Each curve is the average of two experiments (redrawn from Morgan and Hall, 1962).

This review deals primarily with relationships between herbicides and ethylene. The reader is directed to many excellent reviews on the broader field of ethylene metabolism and physiology (Burg, 1962; Pratt and Goeschl, 1969; Spencer, 1969; Mapson, 1969; Yang and Baur, 1969; Abeles, 1972, 1973; and others cited therein). Reviews on agricultural uses of ethylene physiology and ethylene substrates are also available (de Wilde, 1971 and Morgan, 1972).

II. Effects of Ethylene

Since auxinic herbicides and other auxins will induce ethylene synthesis in a variety of plant species, it is important to catalogue the range of effects that ethylene can have on plants.

A. PLANT RESPONSES TO ETHYLENE

The list of plant responses to ethylene is lengthy and many of the effects result from applications of auxins (Table I and Morgan, 1973). The number of recognized similar effects of ethylene and auxin has increased since the association was first reported (Crocker et al. 1935; Crocker, 1948). Expression of the association is obviously qualified by species, age, concentrations, environment, and other factors. Nevertheless, it is apparent that ethylene produces a wide range of plant responses that are both positive (growth promotion, dormancy release, and fruit ripening) and negative (growth inhibition, chlorophyll destruction, abscission). The visible symptoms and biochemical effects of auxins are often indistinguishable from those of ethylene (Fig. 2, Table II). Ethylene and auxins generally have different effects on abscission and dehiscence, seed and bud dormancy, and apical dominance release. Ethylene usually promotes these processes while auxin generally retards or maintains them (Table I). These relationships make ethylene physiology important in herbicide research and use.

B. METABOLIC AND CELLULAR EFFECTS OF ETHYLENE

Stimulation of respiration is one well known and widely studied effect of ethylene (Harvey, 1915; Denny, 1924; Herrero and Hall, 1960; Reid and Pratt, 1972). Ethylene is known to effect a large number of enzymes (see review in Abeles, 1972). Increases in protease and invertase activity were noted in the early work of Regeimbol and Harvey (1927). Among other enzyme effects are increased activity of α-amylase (Herrero and Hall, 1960), peroxidase (Herrero and Hall, 1960; Stahmann et al. 1966; Imaseki et al. 1968a), IAA-oxidase (Hall and Morgan, 1964), cellulase (Horton and Osborne, 1967; Abeles, 1969), and polyphenol oxidase (Herrero and Hall, 1960; Stahmann et al. 1966; Imaseki et al. 1968a).

Increases in protein synthesis due to ethylene treatment have been noted frequently (Abeles, 1972). Both RNA and protein levels as well as the synthetic capacity of chromatin were decreased in soybean seedlings in

TABLE I

EFFECTS OF APPLIED ETHYLENE ON PLANTS AND PLANT PARTS COMPARED TO SIMILAR AND DISSIMILAR EFFECTS OF AUXINS[1]

Similar Plant Responses to Ethylene and Auxins

GROWTH INHIBITION (Elmer, 1932; Knight and Crocker, 1913; Neljubow, 1901)
GROWTH PROMOTION (Imaseki and Pjon, 1970; Ku *et al.* 1970)
GEOTROPISM MODIFICATION (Knight and Crocker, 1913; Neljubow, 1901)
TISSUE PROLIFERATION (Harvey and Rose, 1915; Knight and Crocker, 1913)
ROOT AND ROOT HAIR INITIATION (Zimmerman and Wilcoxon, 1935)
LEAF EPINASTY (Doubt, 1917)
LEAF MOVEMENT INHIBITION (Crocker, 1948; Crocker *et al.* 1932; Doubt, 1917)
FORMATIVE GROWTH AND HOOK FORMATION (Goeschl *et al.* 1967; Hall *et al.* 1957; Kang *et al.* 1967)
CHLOROPHYLL DESTRUCTION (Harvey, 1925; Zimmerman *et al.* 1931)
ANTHOCYANIN SYNTHESIS PROMOTION (Hall *et al.* 1957)
ANTHOCYANIN SYNTHESIS INHIBITION (Morgan and Powell, 1970)
FLOWER INITIATION (Rodriguez, 1932; Traub *et al.* 1939)
FLOWER INHIBITION (Abeles, 1967*b*)
FLOWER SEX SHIFTS (Minina, 1952; Robinson *et al.* 1968)
FRUIT GROWTH STIMULATION (Galil, 1968; Maxie and Crane, 1968)
FRUIT DEGREENING (Denny, 1924)
FRUIT RIPENING (Hansen, 1942; Rosa, 1928; Wolfe, 1931)
RESPIRATORY CHANGES (Denny, 1924; Harvey, 1915)
STORAGE PRODUCT HYDROLYSIS (Harvey, 1915)
LATEX AND GUM EXUDATION AND GUTTATION PROMOTION (Abraham *et al.* 1969; Doubt, 1917; Harvey, 1915)

Dissimilar Plant Responses to Ethylene and Auxins

ABSCISSION AND DEHISCENCE (Doubt, 1917; Sorber, 1934)
SEED AND BUD DORMANCY RELEASE (Vacha and Harvey, 1927)
APICAL DOMINANCE RELEASE (Hall *et al.* 1957)

[1] References are for ethylene effects. For more complete discussion of and references to auxin effects see discussion in Section III, A and B of this review and Crocker *et al.* 1935; Crocker, 1948; Hall and Morgan, 1964; Morgan, 1967 and Pratt and Goeschl, 1969. For more detailed discussion of natural roles of ethylene, which are the same as some effects of the gas, see reviews listed in Introduction.

parallel with growth inhibition of a specific tissue (Table II and Holm *et al.* 1970). Similarly, ethylene inhibits DNA synthesis in tissues where growth is inhibited (Holm *et al.* 1970; Apelbaum and Burg, 1972*a*). In petioles, ethylene promoted RNA and protein synthesis (Abeles and Holm, 1966); the promotion of RNA synthesis was localized near the separation layer (Osborne, 1968).

Recent work has detailed the effect of ethylene at the cellular level in etiolated pea tissue where its ability to promote lateral cell expansion and inhibit elongation (Neljubow, 1901) had been long employed in the triple response bioassay for ethylene (Knight and Crocker, 1913). Ethylene inhibits cell division within a few hours (Apelbaum and Burg, 1972*a*). Elongation is

CONTROL   2,4-D   C₂H₄

Fig. 2. Similar effects of 2,4-D, picloram and ethylene on mesquite seedlings. (Upper) Plants treated with spray application of 10^{-5} M 2,4-D or picloram or 20 μl ethylene per litre air. Photograph taken after plants were in the dark for 2 h. Note absence of leaflet closing on treated plants in contrast to control. (Lower) Effect of 2,4-D and 100 μl of ethylene/l air on soybean hypocotyls. The two $\frac{1}{2}$ day-old seedlings were treated with 10^{-3} M 2,4-D and 100 μl of ethylene/l air. Photograph taken two days after treatment. Note swelling in subapical tissue in both ethylene and 2,4-D treated (from Holm and Abeles, 1968, by permission of the authors and publishers).

TABLE II

CHANGES IN RNA, DNA AND PROTEIN IN THE APICAL, ELONGATING, AND BASAL SECTIONS OF HYPOCOTYLS OF INTACT SOYBEAN SEEDLINGS IN RESPONSE TO 2,4-D AND ETHYLENE

Treatment	Fresh wt. in mg 10 Sect.			µg RNA 10 Sect.			µg DNA 10 Sect.			µg Protein 10 Sect.		
	Apical	Elong.	Basal	Apical	Elong.	Basal	Apical	Elong.	Basal	Apical	Elong.	Basal
A. 24 h												
Control	0·16	0·40	0·32	446	414	169	36	27	14	4,368	3,510	1,300
10^{-3} M 2,4-D	0·13	0·63	0·59	434	474	297	33	40	21	3,406	4,238	2,392
100 parts/10^6	0·16	0·57	0·56	434	595	245	32	39	27	3,692	4,862	2,730
B. 48 h												
Control	0·27	0·74	0·37	466	390	137	43	33	25	4,566	3,822	1,343
10^{-3} M 2,4-D	0·17	1·19	0·59	434	1,475	647	37	95	53	3,822	7,653	4,113
100 parts/10^6 ethylene	0·19	1·21	0·50	418	1,238	434	30	77	39	3,177	6,409	3,298

From: Holm, R. E. and F. B. Abeles (1968).

inhibited while isodiametric cell expansion is promoted (Burg and Burg, 1966a). Cellular expansion in sub-apical stem segments, though at a reduced rate, is actually prolonged in duration by appropriate levels of ethylene (Apelbaum and Burg, 1972b). While elongation is inhibited and swelling occurs, the cellulose microfibrillar orientation of the wall appears changed (Apelbaum and Burg, 1971; Eisinger and Burg, 1972).

C. EFFECTS OF ETHYLENE ON OTHER HORMONES

During the 1930s several researchers considered the possibility that ethylene might achieve some of its effects by modifying the activities of the only hormone then recognized—auxin. Michener (1938) concluded that ethylene does not influence the production or transport of auxin. Promotion of auxin destruction was not ruled out by Michener; ethylene appeared to increase the sensitivity of plants to auxin. However, the overall significance of ethylene in plant growth and development was minimized in the paper. van der Laan (1934) and Borgström (1939) both argued for an action of ethylene on lateral auxin transport. von Guttenberg and Steinmetz (1947) obtained evidence indicating that ethylene modifies polar auxin transport. But, as time passed, attention to ethylene declined. It was neglected in detailed treatments of plant growth substances as well as general works on plant physiology. By the mid-1950s, ethylene's endogenous roles and its effect on auxin were still not clearly known.

Several ethylene effects on auxin physiology are now well documented. Ethylene was shown to reduce basipetal or polar auxin transport in plants exposed to the gas (Morgan and Gausman, 1966). This effect of ethylene was not recognized in some experiments because excised segments, with rapidly declining transport capacity, rather than intact plants, were exposed to ethylene (Abeles, 1966b; Burg and Burg, 1966a). After this difficulty was recognized, the phenomenon was promptly verified and characterized (Fig. 3 and Burg and Burg, 1967b; Morgan et al. 1968). It occurs in a wide variety of species (Morgan et al. 1968). Evidence indicates that inhibition of auxin transport by endogenous ethylene is involved in natural leaf abscission as well as abscission induced by exogenous ethylene (Beyer and Morgan, 1971; Beyer, 1973). It is possible, in some cases, to demonstrate ethylene-mediated reduction of auxin-transport capacity in excised plant segments, but the effect is more pronounced in the intact system (Morgan and Gausman, 1966; Osborne and Mullins, 1969). Auxin-transport capacity declines and ethylene synthesis rises before natural abscission occurs (Beyer and Morgan, 1971). Exogenous ethylene will cause a reduction in auxin-transport capacity and will promote leaf abscission from intact plants (Beyer and Morgan, 1971); yet, exogenous auxin prevents the effect of exogenous ethylene on auxin transport (Osborne and Mullins, 1969; Beyer, 1973). Recent studies with intact plants reveal that exogenous ethylene must be applied to both the leaf blade and petiole to induce abscission; yet, auxin-transport capacity is reduced but not eliminated by both treatments (Beyer, 1975). Presumably reduction of auxin

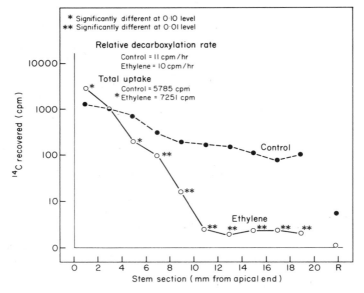

Fig. 3. Effect of ethylene on auxin transport. A 15 h ethylene pretreatment period modified the distribution, total uptake and the relative decarboxylation rate of NAA-1-^{14}C following 4 h of transport. Each datum is the average of two experiments with three replications per experiment and 10 stem sections per replication (Beyer and Morgan, 1969).

transport, as well as auxin supply, are involved in abscission. In addition to the association with abscission, reduction of auxin-transport capacity is the means by which ethylene inhibits growth in the subhook region of the intact, etiolated pea seedling (Apelbaum and Burg, 1972b). In tissues in which ethylene *promotes* elongation growth, the gas also *promotes* auxin transport (Musgrave and Walters, 1973).

Several lines of evidence indicate that exposure of plants to ethylene causes a reduction in auxin levels (see reviews by Burg and Burg, 1968a; Abeles, 1972; and Apelbaum and Burg, 1972b). This could be accomplished by promotion of auxin destruction or inhibition of synthesis. Ethylene has been clearly shown to increase the IAA-oxidase activity in extracts of fumigated tissue (Hall and Morgan, 1964). Since much evidence indicates that the enzymatic portion of IAA-oxidase is peroxidase (Fowler and Morgan, 1972; Hoyle, 1972), the numerous observations of ethylene-mediated increases in *in vitro* peroxidase activity (Herrero and Hall, 1960; Stahmann *et al.* 1966; Imaseki *et al.* 1968a and others reviewed by Abeles, 1972 and Morgan and Fowler, 1972) suggest that ethylene has the potential to promote IAA destruction. Ethylene clearly increases auxin destruction in transport experiments with some species (Table III and Morgan *et al.* 1968) and just as clearly fails to in other species (Burg and Burg, 1967b; Morgan *et al.* 1968). Interpretation of the results of these transport experiments, as well as ethylene

TABLE III

THE EFFECT OF A 3- AND 15-H ETHYLENE PRETREATMENT PERIOD ON THE DRY WEIGHT, CROSS-SECTIONAL AREA AND THE RATE OF DECARBOXYLATION OF IAA-1-^{14}C IN COTTON STEM SECTIONS[1]

Treatment	Treatment period (h)	Relative decarboxylation rate[2] (cpm/h)	Surface area (mm²)	Dry wt. (mg)	cpm/h/mg dry wt.	cpm/h/mm² surface area	Total uptake
Control	3	556	33·33	53·60	10·37	16·68	918
Ethylene	3	1,035[3]	38·20	48·97	21·14[3]	27·09[3]	946
Control	15	688	35·74	49·73	13·84	19·25	1,594
Ethylene	15	2,432[3]	33·20	56·20	43·27[3]	73·25[3]	1,522

[1] Stem sections were 22 mm in length. The apical 2 mm were excised prior to transport and used to determine the cross-sectional area of the absorbing surface. The remaining 20 mm sections were used to determine the relative rates of decarboxylation and the dry weights of the transport tissue. Each treatment period included 3 replications with 10 stem sections per replication. Cross-sectional areas were determined for each stem section while dry weights were determined for each replication. Each datum represents the average per replication.
[2] Cpm/h designates the average number of cpm of ^{14}C recovered as $^{14}CO_2$ per hour of transport.
[3] Statistically significant at the 0·05 level.

From: Beyer, E. M., Jr. and P. W. Morgan (1969).

effects on conjugation and other metabolic modifications of auxin, is complicated by the cut surface through which the IAA must pass. However, since ethylene consistently increases *in vitro* peroxidase activity, decarboxylation or conjugation of IAA must be considered as possibly one of the major aspects of ethylene action. Verification of this possibility awaits further study.

In *Coleus,* ethylene does not appear to promote auxin destruction, but it reduced the capacity of tissue homogenates to decarboxylate tryptophan (Valdovinos *et al.* 1967). This effect of ethylene on tryptophan breakdown was suggested to indicate a reduction in auxin synthesis. These results, in addition to an apparent promotion of IAA conjugation by ethylene, have been verified and extended (Ernest and Valdovinos, 1971).

Ethylene has been shown to have a rapid effect on lateral auxin transport (van der Laan, 1934; Burg and Burg, 1966a). Not only does ethylene inhibit lateral auxin transport, but recent studies indicate that with saturation levels of ethylene, lateral auxin transport is *reversed* in geotropically stimulated stem sections (Kang and Burg, 1973). No ethylene effect occurs on phototropically induced lateral transport (Kang and Burg, 1973); thus, the geotropic effect of ethylene may be intimately associated with the perception of gravity.

The possibility that ethylene modifies the synthesis, distribution or inactivation of hormones other than auxins has not been extensively studied. Ethylene does initiate the synthesis of ABA in rose petals, and the ABA subsequently inhibits ethylene synthesis (Mayak and Halevy, 1972). Ethylene-mediated fluctuations in levels of GA and cytokinins have not been reported; however, Lieberman and Kunishi (1972) have introduced the interesting concept that ethylene serves as a modulator of activity of the growth substances (cytokinins, auxins and gibberellins). Thus they find that promotive effects of auxins and cytokinins on ethylene synthesis are more than additive and the expected reduction in auxin levels have been observed from ethylene treatment. Lieberman and Kunishi's (1972) model of ethylene function emphasizes the need to determine the effect of the gas on cytokinins and gibberellins. There are numerous reports of ethylene modifying an effect of another hormone or the reverse, but such interactions could result from a level of action other than effects on hormone levels. This topic is developed in more detail elsewhere (Scott and Leopold, 1967; Lieberman and Kunishi, 1972; Abeles, 1973).

III. AUXINIC HERBICIDES AND ETHYLENE

A. AUXIN-INDUCED ETHYLENE SYNTHESIS

In addition to 2,4-D (Fig. 1), other auxinic herbicides promote ethylene synthesis including: 2,4,5-T (Maxie and Crane, 1967); picloram (Baur and Morgan, 1969); 2,5-dichlorophenoxyacetic acid (2,5-D) (Rubinstein and Abeles, 1965) and dicamba (Scott and Norris, 1970). IAA (Morgan and Hall, 1964) and naphth-1yl-acetic acid (NAA) (Abeles and Rubinstein, 1964) promote ethylene synthesis while the analogues indol-3yl-propionic acid (IPA) and indol-3yl-butyric acid (IBA), with less auxin activity, have little or no effect (Abeles and Rubinstein, 1964). Phenoxyacetic acid (PAA), several chloro-substituted phenoxyacetic acids (2-CPA; 4-CPA; 2,6-D; 2,4,6-T) and 2,4,5-trichlorophenoxyisobutyric acid, auxin analogues with little or no auxin activity, had essentially little effect on ethylene production by bean petiole explants (Fig. 4). This finding was verified later with cotton explants; however,

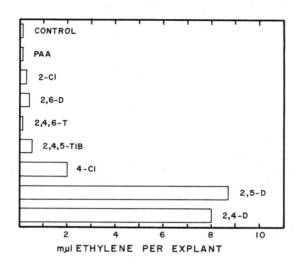

Fig. 4. Effect of various phenoxyacetic acids at 10^{-4} M on ethylene evolution by bean petiole explants. Explants were placed in plain agar 48 h before the start of the experiment (from Rubinstein and Abeles, 1965, by permission of the authors and publishers).

there was some variation in the relative effect of two of the weak auxin analogues (2-CPA and 2,4,5-trichlorophenoxyisobutyric acid) (Abeles, 1967a). The growth and germination inhibitor coumarin, when applied to etiolated bean hypocotyl hooks at relatively high concentrations, promotes ethylene synthesis, primarily in the hook, and simultaneously promotes growth of the basal, straight portion of the hypocotyl (Morgan and Powell, 1970). The growth promotion is not due to ethylene.

Ethylene synthesis induced by IAA and NAA in excised sections or explants was characterized by Abeles and Rubinstein (1964) and by Burg and Burg (1966a). The threshold concentration for measurable induction was between 10^{-6} and 10^{-5} M with stimulation continuing through 10^{-3} M IAA (Fig. 5). The time course revealed that peak ethylene synthesis occurred within 4 to 8 h. Protein synthesis inhibitors partially or completely block auxin-induced ethylene synthesis (Abeles, 1966a), and there is a lag of a few hours or less between treatment and response (Abeles, 1966a; Burg and Burg, 1966a). Upon closer examination the lag in ethylene production appeared to be less than 1 h or possibly non-existent (Chadwick and Burg, 1967; 1970). Steen and Chadwick (1973) have subsequently shown that there is an immediate

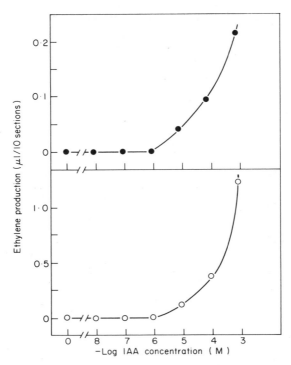

Fig. 5. (Upper) Effects of IAA on C_2H_4 production 1 cm subapical etiolated pea epicotyl sections during 18 h. (Redrawn from Burg and Burg, 1966a). (Lower) Effects of IAA on C_2H_4 production during 18 h by 1 cm subapical hypocotyl sections from light grown sunflowers. (Redrawn from Burg and Burg, 1968a by permission of the authors.)

promotion of ethylene synthesis at a low IAA concentration (1 nM) which is insensitive to cycloheximide and the additional promotion of ethylene synthesis at higher IAA levels is not evident until 2 to 4 h after treatment. This latter response is cycloheximide sensitive; so, there is apparently a direct effect of IAA on ethylene production at low concentrations. The low-concentration effect is different from the effect at higher auxin concentrations which appears to involve synthesis of new protein. Other data indicate that auxin promotes formation of a short-lived RNA which is involved in the synthesis of a very labile protein which controls the rate of ethylene synthesis (Kang *et al.* 1971; Sakai and Imaseki, 1971). The effect of IAA on ethylene production is much less lasting than that of 2,4-D (Fig. 6), and the release of ethylene parallels the level of free auxin (Kang *et al.* 1971). After an initial surge of uptake, free IAA is depleted by conjugation and decarboxylation while 2,4-D levels remain more stable (Kang *et al.* 1971). Peroxidase apparently does not regulate ethylene synthesis (Kang *et al.* 1971; Fowler and Morgan, 1972) as was suggested by results of studies of model systems (Yang, 1967).

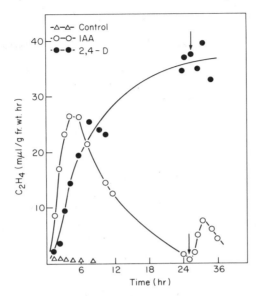

Fig. 6. Effect of 10 μM IAA or 10 μM 2,4-D on ethylene production by subapical pea stem sections. The arrows indicate when sections were transferred to fresh auxin solutions (Kang *et al.* 1971).

B. RESPONSES CAUSED BY ETHYLENE IN AUXIN-TREATED PLANTS

To avoid confusion, the phenomenon of auxin-induced ethylene synthesis should be viewed with the understanding that auxin can have at least three different effects on growth: (1) growth promotion at low levels that do not measurably promote ethylene synthesis, (2) promotion of ethylene synthesis at intermediate auxin levels where ethylene inhibits growth, (3) direct toxic or herbicidal inhibition of growth at high auxin levels where ethylene is not involved. These different relationships have been uncovered by many studies since the pioneering work of the Boyce Thompson group, but they have been considered simultaneously and in depth recently by Apelbaum and Burg (1972a and b).

The Boyce Thompson group viewed auxin-induced ethylene rather than auxin itself as the active agent in many responses, but they called attention especially to root and root hair initiation (Zimmerman and Wilcoxon, 1935). Hansen (1942) demonstrated that promotion of ethylene synthesis paralleled auxin-induced fruit ripening. Morgan and Hall (1962) connected auxin-induced ethylene to epinasty, bending and curling of apex and leaves, and other intact plant responses. The role of ethylene in epinasty has been verified in a convincing manner (Maxie and Crane, 1967; Stewart and Freebairn, 1969). Burg and Burg (1966a) showed that auxin-induced ethylene is the cause of the inhibition of growth of stem sections exposed to supra-optimal levels of auxin (see also Burg and Burg, 1968a). The group at Ft. Detrick determined that ethylene was the active agent in auxin-promoted abscission of explant petioles (Abeles and Rubinstein, 1964; Rubinstein and Abeles, 1965; Abeles, 1967a); the same was demonstrated for intact leaves (Hallaway and Osborne, 1969; Baur and Morgan, 1969; Morgan and Baur, 1970). Ethylene produced in response to auxin treatment has been implicated in a wide variety of other responses including, promotion of flowering in pineapple (Burg and Burg, 1966b); inhibition of growth of pea roots (Chadwick and Burg, 1967; 1970); inhibition of pea bud growth (Burg and Burg, 1968b); geotropic curvature (Abeles and Rubinstein, 1964; Chadwick and Burg, 1967); inhibition of leaf movement (Baur and Morgan, 1969; Morgan and Baur, 1970); tissue swelling or proliferation (Burg and Burg, 1966a; Apelbaum and Burg, 1972b); fruit growth stimulation (Maxie and Crane, 1968; Galil, 1968) and flower sex shifts in cucurbits (Shannon and de la Guardia, 1969); anthocyanin synthesis inhibition (Morgan and Powell, 1970); flower petal fading and abscission (Hall and Forsyth, 1967; Burg and Dijkman, 1967); inhibition of flowering in *Xanthium* (Abeles, 1967b); latex flow promotion in *Hevea* (D'Auzac and Ribaillier, 1969); etiolated seedling hook closing (Goeschl *et al.* 1967; Kang *et al.* 1967; Kang and Ray, 1969; Morgan and Powell, 1970).

Rigorous proof that an action of auxin, mimicked by ethylene, is due to auxin-induced ethylene synthesis involves more than measuring ethylene released by auxin-treated and control tissues. Abeles (1973) has summarized the five criteria which have been developed as the investigation of this regulatory phenomenon has progressed. The criteria are: (1) kinetics of auxin-induced ethylene synthesis should be such that timing and amount of ethylene allow a cause and effect relationship, (2) ethylene should produce the same effect more rapidly than auxin, (3) saturating levels of ethylene should reduce, mask or remove the response to auxin and additional effects of auxin must be due to it rather than ethylene, (4) reduction of the internal concentration of ethylene in auxin-treated tissue by vacuum or flushing with air should reduce or delay the response. (If the response occurs only when the tissue is enclosed, its *in vivo* significance must be questioned!), (5) CO_2 is usually a competitive inhibitor of ethylene action, and it should delay or reduce the effects of auxins (Burg and Burg, 1967a). It should be noted that there are systems in which CO_2 is not a competitive inhibitor but rather is toxic or ineffective (Abeles, 1968), slightly promotive, or essential for the ethylene effect (Wegm *et al.* 1972; Ketring and Morgan, 1972). Another type of proof is to block auxin-induced ethylene synthesis and show that auxin is then inactive in the system but that ethylene is active. This has been done with heat and epinasty (Stewart and Freebairn, 1969).

C. ROLE OF ETHYLENE IN THE HERBICIDAL ACTION OF AUXINS

Does ethylene produced by plants after treatment with auxinic herbicides contribute to the herbicidal effects? Auxins, such as 2,4-D, and ethylene are generally selective to the same groups of plants (Hall *et al.* 1957). Further, 2,4-D promotes ethylene synthesis much more in a susceptible species (cotton) than in a resistant species (grain sorghum) (Morgan and Hall, 1962; Hall and Morgan, 1964). Thus, some relationship was proposed to exist between herbicidal action and promotion of ethylene synthesis. One study supports the proposal (Burg and Burg, 1968a) and others do not (Holm and Abeles, 1968; Abeles, 1968).

No one has suggested that ethylene is the agent by which herbicides kill plants. Were this true one would expect ethephon and ethylene to be more effective herbicides and to produce longer lasting damage. Few plants die from rather drastic exposures to ethylene (10 species and varieties out of 114 tested in 2, 5 or 10 µl ethylene/l air for 10 days; Heck and Pires, 1962) and recovery after exposure is usually rapid. Nevertheless, several interesting relationships exist. One is that analogues inactive as auxins (and consequently less effective as herbicides than 2,4-D, etc.) are also inactive as ethylene synthesis promoters (Fig. 4). [This statement must be viewed in the context that a wide variety of environmental and chemical stresses will promote ethylene synthesis (see following sections); thus, non-specific wound or stress ethylene can be expected from applications of high concentrations of almost any compound.] In etiolated pea and sunflower sections, auxin-induced ethylene formation accounts for the growth inhibition of supra-optimal levels of auxin (Burg and Burg, 1968a). In light-grown pea and sunflower stem sections, elongation is not inhibited in concentrations of IAA up to 10^{-3} M; thus, the relationship of a toxic or herbicidal level of auxin was not tested. In contrast, in two "resistant" monocots, *Avena* and *Zea,* IAA causes little or no growth inhibition via induced ethylene formation (Burg and Burg, 1968a). A significant inhibition of both growth and ethylene production does occur in the monocots treated with 10^{-3} M IAA and both responses are absent in the dicots. At this high concentration, growth inhibition is partially, but not completely, reproduced by benzoic acid and acetic acid, when the maximum promotive level of IAA is also applied (10^{-4} M). It seems clear in this study (Burg and Burg, 1968a) that the response of the dicots and monocots to IAA-induced ethylene is substantially different.

That herbicidal activity of 2,4-D is not due to ethylene production is indicated by the fact that 50 µl ethylene/l air or 10^{-3} M 2,4-D have equal but not additive effects on growth (i.e. both elongation and weight increases), of the hook region of etiolated peas (Apelbaum and Burg, 1972a) (10^{-4} M 2,4-D gave similar results). However, 10^{-3} M 2,4-D inhibits cell division in the hook slightly more than 10^{-4} M 2,4-D or saturating levels of ethylene. The inhibition of cell division (over that produced by ethylene) is even more strongly expressed in root tips by 10^{-4} M 2,4-D (10^{-3} M 2,4-D kills the tissue). In the subhook zone of peas, the herbicidal action of 2,4-D is distinguished from the effect of ethylene in two ways (Apelbaum and Burg, 1972b). Elongation is inhibited modestly and swelling is promoted by 10^{-4} M 2,4-D. When these plants are grown under hypobaric conditions to remove auxin-induced ethylene, elongation growth is not effected but swelling is prevented. A higher concentration of 2,4-D (10^{-3} M) inhibits both elongation and fresh weight increase maximally and no additional effect is achieved by adding ethylene.

Ethylene has the same effect on fresh weight increase but not on elongation. Thus, in both cell division and cell elongations, 2,4-D has toxic effects which exceed those produced by ethylene. This separation of some of the effects of auxinic herbicides from ethylene was also observed by Scott and Norris (1970) where 2,4-D, but not ethylene, promoted cell divisions in pea root pericycles.

While auxin (2,4-D) applied to intact cotton and grain sorghum plants promoted ethylene synthesis in both species, maximum observed production rates were 8·1 µl/kg h for cotton and 0·5 µl/kg h for grain sorghum (Morgan and Hall, 1962). Abeles (1968) stated that 2,4-D-induced ethylene production rates were 45 µl/g 48 h for light-grown soybean versus 5·5 µl/g 48 h for corn. Growth inhibition of etiolated soybean stem sections in supra-optimal levels of auxin result from induced ethylene production (Holm and Abeles, 1968). One of the clearest demonstrations that ethylene causes auxin herbicide symptoms is an experiment where 2,4-D caused tissue swelling which was prevented by putting plants under vacuum (Apelbaum and Burg, 1972b). The vacuum system did not alter the growth of control peas, but removed a volatile responsible for swelling. Thus, in all species studied, the quantitative effect of auxin on ethylene synthesis is larger in plants where induced ethylene will cause some growth inhibition. In conclusion, there is no evidence that auxin-induced ethylene itself kills plants. On the other hand, herbicide responses which precede kill, such as leaf and stem epinasty and leaf abscission, are accounted for by ethylene (Morgan and Hall, 1962; Maxie and Crane, 1967; Hallaway and Osborne, 1969; Baur and Morgan, 1969; Morgan and Baur, 1970). Thus, there still seems to be an undefined association between sensitivity to auxin-induced ethylene production and sensitivity to auxinic herbicides (selective action).

D. CRITICAL EVALUATION OF ROLES OF AUXIN-INDUCED ETHYLENE

Some criticisms of the concept of ethylene mediation of "auxin effects" have been published (Andreae et al. 1968; Muir and Richter, 1972). That the inhibition of pea-root growth by IAA is due to ethylene (Chadwick and Burg, 1967) was disputed by Andreae et al. (1968). One reason for the initial disagreement was that Chadwick and Burg (1967 and 1970) worked primarily at auxin levels below 10^{-5} M IAA while Andreae et al. (1968) worked mostly above that level. Actually all three papers agree that auxin-induced ethylene does not explain the growth inhibition of roots (toxic or herbicidal effects) observed at all IAA concentrations above 10^{-5} M. While the variety of evidence appears more than adequate to support the hypothesis in both root sections and roots of intact plants (Chadwick and Burg, 1970), there may be some need to analyze the criticisms further. Andreae et al. (1968) and Muir and Richter (1972) found that growth inhibition by IAA is reversed by removal of IAA but not so with ethylene. Recovery was tested after a 16 h treatment, a test considered valid because the growth rate of the auxin-treated tissues recovered. In fact, the treatments were IAA plus ethylene (made endogenously) versus ethylene (supplied exogenously). Auxin prevents some responses to ethylene (defoliation, Hall, 1952; Hall and Morgan, 1964; inhibition of polar auxin transport, Osborne and Mullins, 1969; Beyer, 1973), but auxin does not appear to prevent the effect of ethylene on growth (see Fig. 5, Chadwick and Burg, 1970). Thus, auxin may allow ethylene-mediated growth inhibition in one treatment, but prevent senescence promotion by ethylene. Recovery from ethylene-mediated growth inhibition is easily demonstrated at shorter exposures, but the degree of recovery begins to decline after 16 h (Fig. 6, Chadwick and Burg, 1970), probably a reflection of the well known senescence-promoting effect of ethylene (Abeles *et al.* 1971). The failure of applied ethylene to inhibit growth as rapidly as IAA (Andreae *et al.* 1968) is probably accounted for by two considerations. First, the direct herbicidal effect of auxins discussed previously occurs rapidly giving the impression that the time course of auxin effects are earlier and of greater magnitude than those of ethylene (Chadwick and Burg, 1970; Apelbaum and Burg, 1972*a* and *b*). A different impression would probably come from tests at auxin levels where all growth inhibition is due to ethylene. Second, sections were placed in a medium in which IAA was already dissolved or in a hormone-free medium. Ethylene gas was then injected into the head space in the latter treatment, an arrangement which may have allowed the IAA to enter the sections more rapidly than the ethylene which had to equilibrate with the medium. Muir and Richter (1972) did not find increased ethylene synthesis at the lowest auxin concentration that gave growth inhibition. In fact, they did not observe any ethylene production from root or stem segments at IAA levels of 10^{-7} and 10^{-6} M for 18 h. This situation raises doubts about the ability of their measurement system to sense low levels of ethylene, because Chadwick and Burg (1970) found a family of separate curves for ethylene production from 10^{-7} to 10^{-4} M IAA and control. Differences were detected at hourly intervals, and complete separation of all of the IAA curves from the control was evident at the 1 and 2 h observations. The CO_2 level of 1% used by Muir and Richter (1972) was too low to inhibit ethylene action; 10 to 15% CO_2 is necessary to inhibit competitively responses to one to several μl ethylene per litre air (Burg and Burg, 1967*a*). This analysis does not deal with every criticism of the concept that induced ethylene causes some of the growth inhibition and other responses produced by supra-optimal levels of auxin. Many of the disagreements may result from an impression that ethylene is claimed to account for everything that auxin does. In the pea root system, ethylene is pretty clearly responsible for most of the growth inhibition at low concentrations and little of it in the toxic range.

IV. NON-AUXINIC REGULATION OF ETHYLENE SYNTHESIS

A. STRESS ETHYLENE

That physical wounding promotes synthesis of ethylene, a common emanation from dead or dying cells, has been known from the work of Williamson (1950). Promotion of ethylene release by cutting fruits, stem sections, explants, and other tissue is now a universally recognized phenomenon (see Burg, 1962). It is the basis for the ancient slashing technique for ripening figs (Galil, 1968). The so-called wound ethylene might be better termed stress ethylene (Abeles, 1973), because wounding is only one of a number of stresses which promote release of the gas. Chilling temperatures (Cooper et al. 1969), water stress (McMichael et al. 1972), γ-radiation (Shah and Maxie, 1965), physical resistance to growth (Goeschl et al. 1966), insects (King and Lane, 1969), and diseases (Williamson, 1950; Weise and DeVay, 1970) promote ethylene synthesis. Disturbing the normal geotropic orientation of tissue promotes ethylene synthesis (Denny, 1936) along with brief heating (Paterson, 1973).

Chemical treatments which cause tissue damage, some of them very non-specific, promote ethylene synthesis. Examples include: endothal (Hall, 1952; Abeles and Abeles, 1972); potassium iodide (Rubinstein and Abeles, 1965); cycloheximide (Cooper et al. 1968); Cu^{2+} (Cooper et al. 1968; Abeles and Abeles, 1972); Fe^{3+} (Cooper et al. 1968); ozone (Abeles and Abeles, 1972) and phenol derivatives (Fuchs, 1970). It should be noted, however, that excessive chemical treatments which cause rapid desiccation actually reduce ethylene release (Hall, 1952); therefore, even stress ethylene is of metabolic origin and requires living cells for its synthesis. This conclusion is supported by the observation that ethylene synthesis induced by chemical stress utilizes [U-^{14}C]-methionine as a substrate and synthesis is inhibited by cycloheximide (Abeles and Abeles, 1972). Apparently, stress ethylene arises from biochemical pathways similar to those of endogenous ethylene.

B. INDUCTION OF ETHYLENE SYNTHESIS BY GROWTH SUBSTANCES AND OTHER COMPOUNDS

In addition to auxins and chemical stress agents, many other substances promote ethylene synthesis. Many of these substances, in contrast to chemical-stress agents, are effective at relatively low concentrations. The list includes: gibberellins (Abeles and Rubinstein, 1964; Lewis et al. 1968; Cooper et al. 1968; Ketring and Morgan, 1970), cytokinins (Abeles et al. 1967; Fuchs and Lieberman, 1968; Burg and Burg, 1968a; Radin and Loomis, 1969; Ketring and Morgan, 1971), abscisic acid (Abeles, 1967a; Cracker and Abeles, 1969), malformin (Curtis, 1969), amino-acids (Lieberman et al. 1966; Rubinstein and Abeles, 1965), ascorbic acid (Cooper et al. 1968), Co^{2+} (Kang et al. 1967), propionic acid (Lieberman and Kunishi, 1969) and coumarin (Morgan and Powell, 1970). This list may not be complete, but it does indicate the wide range of substances which will promote ethylene synthesis. There are some qualifications that must be kept in mind. For example, abscisic acid promotes ethylene synthesis when applied to ageing or abscising tissue (Abeles, 1967a; Cracker and Abeles, 1969), but it inhibits or prevents ethylene synthesis when used to maintain peanut seed dormancy (Ketring and Morgan, 1972), to inhibit pea seedling growth (Gertman and Fuchs, 1972), or to promote rose petal senescence (Mayak and Halevy, 1972). In a like manner cycloheximide applied to leaves or fruit at injury-producing levels *promotes* ethylene synthesis (Cooper et al. 1968), but when tissue sections are placed in a solution of the substance, cycloheximide *prevents* auxin-induced ethylene production (Abeles, 1966a).

C. NON-AUXINIC HERBICIDES AND DEFOLIANTS

Since 2,4-D and other auxin herbicides have such a marked effect on ethylene production, it is surprising that no comprehensive survey has been made to determine whether representative members of other classes of herbicides also have this property. As stated before, a range of compounds which produce chemical stress also promote ethylene synthesis. The list includes several defoliants which are also used as contact herbicides. In addition to endothal (Hall, 1952), mono-iodoacetamide, TCA, and sodium ethyl mercurithiosalicylate are all active at low concentrations (Imaseki *et al.* 1968b). But higher concentrations, which rapidly killed the tissue, stopped ethylene production (Nakagaki *et al.* 1970), as noted earlier for endothal (Hall, 1952).

V. MANIPULATION OF ETHYLENE PHYSIOLOGY

Since ethylene produces a wide variety of responses (Table I), and it is a by-product of treatment of plants with a wide variety of substances including auxinic and contact herbicides, ethylene physiology in general has many practical applications. Agricultural use of ethylene is made more attractive since effects of the gas can be manipulated in a wide variety of ways. As has been developed in detail elsewhere (Morgan, 1972, 1973), one can: (1) apply ethylene gas, (2) apply ethylene substrate (ethephon or similar compounds), (3) promote ethylene synthesis, (4) inhibit ethylene synthesis, (5) increase sensitivity to ethylene, (6) reduce sensitivity to ethylene, or (7) remove ethylene. Since some of these options are related to the use of herbicides, they deserve a more detailed treatment here.

A. ETHYLENE-PRODUCING CHEMICALS

Since ethylene, synthesized in response to herbicides such as 2,4-D, has regulatory effects on the treated plants, it follows that ethylene generated in plant tissue by other means should also be an effective plant growth regulator (Morgan, 1967). Such a material is available in the form of 2-chloroethylphosphonic acid (ethephon, CEPA, Ethrel™) (Fig. 7), a substance rapidly liberating ethylene in plant tissue or basic solutions (Cooke and Randall, 1968). Other substances liberating ethylene are β-hydroxyl-ethylhydrazine (BOH) (Palmer *et al.* 1967), monoethylsulphate (Kumamoto *et al.* 1969), and ethylpropylphosphate (Niagara 10637) (Dollwet and Kumamoto, 1970) (Fig. 7).

In theory, the ethylene substrates should be as effective as, but more convenient than, ethylene since they are applied in liquid form. The extensive literature on ethephon has been reviewed by de Wilde (1971) and summarized in two technical bulletins of Amchem Products, Inc., Ambler, Pennsylvania. Ethephon, marketed as Ethrel, has label or experimental-label registration in

ETHYLENE-PRODUCING CHEMICALS

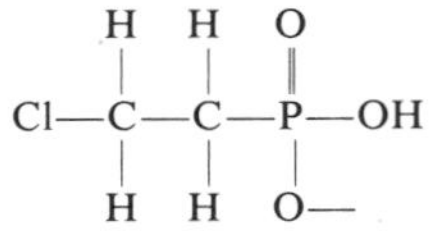

2-Chloroethylphosphonic acid
(Ethephon)

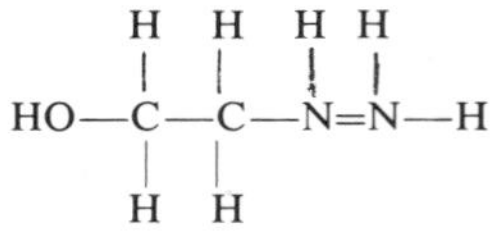

2-Hydroxyethylhydrazine
(BOH)

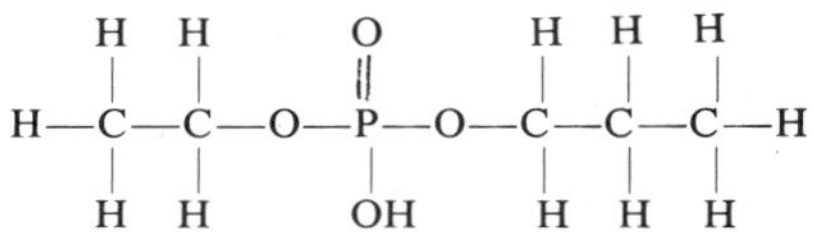

Ethylpropylphosphate

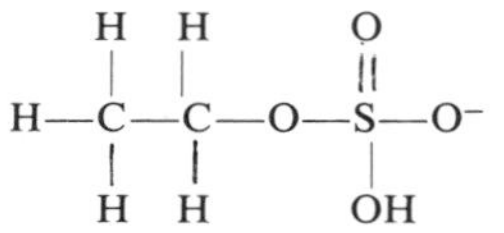

Monoethylsulphate

Fig. 7. Ethylene-releasing chemicals that are members of the model system $Y-CH_2-CH_2-X$ which yields ethylene from redox-coupled, cleavage reactions (redrawn from Kumamoto *et al.* 1969, by permission of the authors and the American Chemical Society).

the USA for harvest-aid use (promotion of ripening, earlier harvest, easier release) on walnuts, filberts, apples, tomatoes, cherries, cantaloupe, and pineapple and for induction of flowering in pineapple. Ethrel is marketed to increase latex yield from rubber trees. A companion formulation, Florel, is marketed for use in hybrid-seed production in cucumbers and squash and floral induction in ornamental bromeliads.

Ethylene and ethephon break dormancy of seed, tubers, bulbs, rhizomes and buds (Table I, Pratt and Goeschl, 1969 and de Wilde, 1971). Several workers have developed the concept that ethephon might be used to release dormancy in reproductive structures of undesirable plants followed by treatment with a herbicide. Mesquite, a woody shrub, escapes kill by auxinic herbicides via dormant basal stem buds that grow after the plant top has been killed. Ethephon releases buds in mesquite which should make them more susceptible to kill (Morgan *et al.* 1969). Ethephon has shown some ability to promote growth of underground organs in Johnson grass (C. A. Beasley, personal communication, 1967), quackgrass (Duke *et al.* 1970), and nutsedge (Jackson *et al.* 1971). Egley and Dale (1970) found that seed dormancy of the corn parasite, witchweed (*Striga lutea* Lour.), is broken by ethephon resulting in a uniform stand of seedlings that are relatively easily controlled. While none of these findings have yet led to a commercial practice, it would seem that ethephon and other methods to manipulate ethylene physiology have considerable potential in weed control. This potential depends on the willingness of researchers and users to accept the inconvenience and expense of a system that requires two separate applications. If simultaneous applications (ethephon and herbicides) would increase kill, then auxinic herbicides should be as effective alone because they promote ethylene synthesis.

B. INCREASING SENSITIVITY TO ETHYLENE

In some specialized cases, it has been possible to make plants much more sensitive to ethylene and/or ethephon. Since ethylene-induced and natural leaf abscission appear to involve a reduction of auxin transport capacity by ethylene (Beyer and Morgan, 1971; Beyer, 1973; Beyer, 1975), other auxin-transport inhibitors should promote abscission. This increase in the abscission action of ethylene or ethephon by auxin-transport inhibitors has been demonstrated (Beyer, 1972 and DPX-1840-DuPont Information Bulletin, March 1971; Morgan and Durham, 1972; 1973). Another effect of 3,3a-dihydro-2-(*p*-methoxyphenyl)-8H-pyrazolo-(5,1a)-isoindol-8-one (DPX-1840) was release of apical dominance (DPX-1840-DuPont Information Bulletin, March, 1971) (Beyer, 1972; Morgan and Durham, 1973). Morey (1974) has extended examination of this action of the auxin transport inhibitor DPX-1840 to honey mesquite; he found that it modified xylem development and caused the formation of almost fibreless wood composed of parenchyma cells. There was an extensive release of basal stem buds of mesquite thus making the plant more susceptible to total kill by herbicides. Since ethephon produces bud release similar to DPX-1840 in honey mesquite (Morgan *et al.* 1969), and ethephon and DPX-1840 are synergistic (Morgan and Durham, 1972), these compounds used together may modify the growth habit of many perennial weeds making them more susceptible to herbicidal kill.

Another method to increase sensitivity to ethylene is the recently observed synergism of endothal and ethephon in bean and woody plant leaf abscission (Sterrett *et al.* 1973). Since endothal causes stress-ethylene production, its use with ethephon may prolong the duration of elevated ethylene levels in leaves. The two substances were used at levels at which neither alone would cause abscission, and there was an indication the endothal inhibits translocation of ethephon out of the leaves (Sterrett *et al.* 1973). Continued study of leaf abscission in cotton has led to the discovery that gibberellic acid promotes the rate or extent of ethylene-induced leaf abscission (Fig. 8). GA was more effective than naptalam, one of the transport inhibitors, and significant

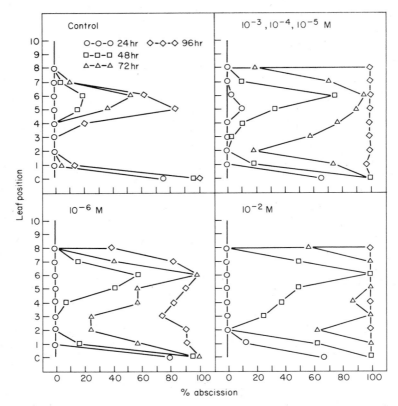

Fig. 8. Effect of various concentrations of GA₃ on cotton leaf abscission by position from 24 day-old plants induced by 1·56 μl ethylene per litre air over a 4-day period (Morgan and Durham, 1975). Leaves numbered from the base upward beginning with the cotyledonary leaves (c).

promotion of ethylene-induced abscission came at GA concentrations as low as 10^{-6} M (Morgan and Durham, 1975).

VI. Summary

Ethylene physiology is of interest to individuals studying or using herbicides for several reasons. A wide variety of responses produced by auxinic herbicides are mimicked by exposure to ethylene or ethylene substrates. A few responses of plants to ethylene are opposite to those of the auxinic herbicides. In some circumstances ethylene modifies auxin physiology and conversely, auxin modification of ethylene physiology, also occurs. Not only are there many ethylene effects and ethylene-auxin interactions, but there are also several ways that ethylene physiology can be manipulated. Thus, an understanding of ethylene physiology can be useful or perhaps essential in herbicide and growth regulator research and development.

Ethylene is apparently a universal product of plant tissue under environmental and chemical stress, a fact that could be useful in efforts to manipulate or temporarily alleviate the undesirable effects of stresses.

Manipulation of ethylene physiology in conjunction with herbicide uses can now be realized. Several ethylene-producing chemicals are known, one of which is available commercially. Several growth regulators promote synthesis of ethylene. In some cases it may be possible to prevent an undesirable effect of internal ethylene, for example defoliation, by application of small amounts of auxinic herbicides. Some aspects of sensitivity to ethylene can be enhanced, namely by application of auxin transport inhibitors, gibberellic acid or the defoliant-herbicide, endothal. A weed-control strategy of increasing sensitivity or susceptibility of target species by modification of metabolism, growth form or dormancy is theoretically possible (see Chapter 5). Modification of ethylene physiology could be an important tool in developing that strategy into a commercially useful practice.

Ethylene is certainly not the dominant component of the plant hormone complex; in fact, growth control by the balance of several substances is receiving increased attention. However, ethylene should be recognized as an important part of the hormone complex.

REFERENCES

Abeles, A. L. and Abeles, F. B. (1972). *Pl. Physiol., Baltimore,* **50,** 496–498.
Abeles, F. B. (1966a). *Pl. Physiol., Lancaster,* **41,** 585–588.
Abeles, F. B. (1966b). *Pl. Physiol., Lancaster,* **41,** 946–948.
Abeles, F. B. (1967a). *Physiologia Pl.* **20,** 442–452.
Abeles, F. B. (1967b). *Pl. Physiol., Lancaster,* **42,** 608–609.
Abeles, F. B. (1968). *Weed Sci.* **16,** 498–500.
Abeles, F. B. (1969). *Pl. Physiol., Lancaster,* **44,** 447–452.
Abeles, F. B. (1972). *A. Rev. Pl. Physiol.* **23,** 259–292.
Abeles, F. B. (1973). "Ethylene in Plant Biology", Academic Press, New York and London.
Abeles, F. B. and Holm, R. E. (1966). *Pl. Physiol., Lancaster,* **41,** 1337–1342.
Abeles, F. B. and Rubinstein, B. (1964). *Pl. Physiol., Lancaster,* **39,** 963–969.
Abeles, F. B., Holm, R. E. and Grahagan, H. E. (1967). *Pl. Physiol., Lancaster,* **42,** 1351–1357.
Abeles, F. B., Craker, L. E. and Leather, G. R. (1971). *Pl. Physiol., Lancaster,* **47,** 7–9.
Abraham, P. D., Wycherley, P. R. and Pakianathan, S. W. (1969). *J. Rubb. Res. Inst. Malaya,* **20,** 291.
Andreae, W. A., Venis, M. A., Jursic, F. and Dumas, T. (1968). *Pl. Physiol., Lancaster,* **43,** 1375–1379.
Apelbaum, A. and Burg, S. P. (1971). *Pl. Physiol., Lancaster,* **48,** 648–652.
Apelbaum, A. and Burg, S. P. (1972a). *Pl. Physiol., Baltimore,* **50,** 117–124.
Apelbaum, A. and Burg, S. P. (1972b). *Pl. Physiol., Baltimore,* **50,** 125–131.
Baur, J. R. and Morgan, P. W. (1969). *Pl. Physiol., Lancaster,* **44,** 831–838.
Beyer, E. M., Jr. (1972). *Pl. Physiol., Baltimore,* **50,** 322–327.
Beyer, E. M., Jr. (1973). *Pl. Physiol., Baltimore,* **52,** 1–5.
Beyer, E. M., Jr. (1975). *Pl. Physiol., Baltimore,* **55,** 322–327.
Beyer, E. M., Jr. and Morgan, P. W. (1969). *Pl. Cell Physiol., Tokyo,* **10,** 787–799.
Beyer, E. M., Jr. and Morgan, P. W. (1971). *Pl. Physiol., Baltimore,* **48,** 208–212.

278 P. W. MORGAN

Borgström, G. (1939). *K. Fysiograf. Sallsk. Lund Forh.* **9**, 135–174.
Burg, S. P. (1962). *Pl. Physiol., Lancaster,* **13**, 265–302.
Burg, S. P. and Burg, E. A. (1966a). *Proc. natn. Acad. Sci. USA* **55**, 262–269.
Burg, S. P. and Burg, E. A. (1966b). *Science, N.Y.* **152**, 1269.
Burg, S. P. and Burg, E. A. (1967a). *Pl. Physiol., Lancaster,* **42**, 144–152.
Burg, S. P. and Burg, E. A. (1967b). *Pl. Physiol., Lancaster,* **42**, 1224–1228.
Burg, S. P. and Burg, E. A. (1968a). *In* "Biochemistry and Physiology of Plant Growth Substances" (F. Wightman and G. Setterfield, eds), pp. 1275–1294. Runge Press, Ottawa.
Burg, S. P. and Burg, E. A. (1968b). *Pl. Physiol., Lancaster,* **43**, 1069–1074.
Burg, S. P. and Dijkman, M. J. (1967). *Pl. Physiol., Lancaster,* **42**, 1648–1650.
Chadwick, A. V. and Burg, S. P. (1967). *Pl. Physiol., Lancaster,* **42**, 415–420.
Chadwick, A. V. and Burg, S. P. (1970). *Pl. Physiol., Lancaster,* **45**, 192–200.
Cooke, A. R. and Randall, D. I. (1968). *Nature, Lond.* **218**, 974–975.
Cooper, W. C., Rasmussen, G. K., Rodgers, B. J., Reece, P. C. and Henry, W. H. (1968). *Pl. Physiol., Lancaster,* **43**, 1560–1576.
Cooper, W. C., Rasmussen, G. K. and Waldon, E. S. (1969). *Pl. Physiol., Lancaster,* **44**, 1194–1196.
Cracker, L. E. and Abeles, F. B. (1969). *Pl. Physiol., Lancaster,* **44**, 1144–1149.
Crocker, W. (1948). "Growth of Plants", Reinhold Publ. Corp., New York, pp. 139–171.
Crocker, W., Zimmerman, P. W. and Hitchcock, A. E. (1932). *Contrib. Boyce Thompson Inst. Pl. Res.* **4**, 177–218.
Crocker, W., Zimmerman, P. W. and Hitchcock, A. E. (1935). *Contrib. Boyce Thompson Inst. Pl. Res.* **7**, 231–248.
Curtis, R. W. (1969). *Pl. Cell Physiol., Tokyo,* **10**, 909–916.
D'Auzac, J. and Ribaillier, D. (1969). *C.r. Séanc. hebd. Acad. Sci. Paris,* **268**, 3046–3049.
Denny, F. E. (1924). *Bot. Gaz.* **77**, 322–329.
Denny, F. E. (1936). *Contrib. Boyce Thompson Inst. Pl. Res.* **7**, 341–347.
de Wilde, R. C. (1971). *HortSci.* **6**, 364–370.
Dollwet, H. H. A. and Kumamoto, J. (1970). *Pl. Physiol., Lancaster,* **46**, 786–789.
Doubt, S. L. (1917). *Bot. Gaz.* **63**, 209–224.
Duke, W. B., Boldt, P. F. and Smith, C. S. (1970). *Proc. NEast. Weed Control Conf.* **24**, 369–374.
Eisinger, W. R. and Burg, S. P. (1972). *Pl. Physiol., Baltimore,* **50**, 510–517.
Egley, G. H. and Dale, J. E. (1970). *Weed Sci.* **18**, 586–589.
Elmer, O. H. (1932). *Science, N.Y.* **75**, 193–194.
Ernest, L. C. and Valdovinos, J. G. (1971). *Pl. Physiol., Baltimore,* **48**, 402–406.
Fowler, J. L. and Morgan, P. W. (1972). *Pl. Physiol., Baltimore,* **49**, 555–559.
Fuchs, Y. (1970). *Pl. Physiol., Baltimore,* **45**, 533–534.
Fuchs, Y. and Lieberman, M. (1968). *Pl. Physiol., Lancaster,* **43**, 2029–2036.
Galil, J. (1968). *J. econ. Bot.* **22**, 178–190.
Gertman, E. and Fuchs, Y. (1972). *Pl. Physiol., Baltimore,* **50**, 194–195.
Goeschl, J. D., Rappaport, L. and Pratt, H. K. (1966). *Pl. Physiol., Lancaster,* **41**, 877–884.
Goeschl, J. D., Pratt, H. K. and Bonner, B. A. (1967). *Pl. Physiol., Lancaster,* **42**, 1077–1080.
Hamner, C. L. and Tukey, H. B. (1944). *Science, N.Y.* **100**, 154–155.
Hall, W. C. (1952). *Bot. Gaz.* **113**, 310–322.
Hall, W. C. and Morgan, P. W. (1964). *In* "Régulateurs Naturels de la Croissance Végétale" (J. P. Nitsch, ed.), pp. 727–745. C.N.R.S., Paris.
Hall, W. C., Truchelut, G. B., Leinweber, C. L. and Herrero, F. A. (1957). *Physiologia Pl.* **10**, 306–317.
Hall, I. V. and Forsyth, F. R. (1967). *Can. J. Bot.* **45**, 1163–1166.
Hallaway, M. and Osborne, D. J. (1969). *Science, N.Y.* **163**, 1067–1068.
Hansen, E. (1942). *Bot. Gaz.* **103**, 543–558.
Harvey, E. M. (1915). *Bot. Gaz.* **56**, 439–442.
Harvey, E. M. and Rose, R. C. (1915). *Bot. Gaz.* **60**, 27–44.
Harvey, R. B. (1925). *Bull. Minn. agric. Exp. Stn.* No. **222**, p. 20.
Heck, W. W. and Pires, E. G. (1962). *Misc. Publs Tex. agric. Exp. Stn.* **613**, p. 12.

Herrero, F. A. and Hall, W. C. (1960). *Physiologia Pl.* **13**, 736–750.
Holm, R. E. and Abeles, F. B. (1968). *Planta,* **78,** 293–304.
Holm, R. E., O'Brien, T. J., Key, J. L. and Cherry, J. H. (1970). *Pl. Physiol., Baltimore,* **45,** 41–45.
Horton, R. F. and Osborne, D. J. (1967). *Nature, Lond.* **214,** 1086–1088.
Hoyle, M. C. (1972). *Pl. Physiol., Baltimore,* **50,** 15–18.
Imaseki, H. and Pjon, C. (1970). *Pl. Cell Physiol., Tokyo,* **11,** 827–829.
Imaseki, H., Uchiyama, M. and Uritani, I. (1968*a*). *Agric. biol. Chem.* **32,** 387–389.
Imaseki, H., Uritani, I. and Stahmann, M. A. (1968*b*). *Pl. Cell Physiol., Tokyo,* **9,** 757–768.
Jackson, E. K., Jangaard, N. O. and James, A. L. (1971). *Pl. Physiol., Baltimore,* **47,** (Suppl.) 15.
Kang, B. G. and Burg, S. P. (1973). *Abstr. 8th int. Conference Plant Growth Substances, Tokyo.* Science Council of Japan. Abs. No. 66.
Kang, B. G. and Ray, P. M. (1969). *Planta,* **87,** 206–216.
Kang, B. G., Yocum, C. S., Burg, S. P. and Ray, P. M. (1967). *Science, N.Y.* **156,** 958–959.
Kang, B. G., Newcomb, W. and Burg, S. P. (1971). *Pl. Physiol., Baltimore,* **47,** 504–509.
Ketring, D. L. and Morgan, P. W. (1970). *Pl. Physiol., Baltimore,* **45,** 268–273.
Ketring, D. L. and Morgan, P. W. (1971). *Pl. Physiol., Baltimore,* **47,** 488–492.
Ketring, D. L. and Morgan, P. W. (1972). *Pl. Physiol., Baltimore,* **50,** 382–387.
King, E. E. and Lane, H. C. (1969). *Pl. Physiol., Lancaster,* **4,** 903–906.
Knight, L. I. and Crocker, W. (1913). *Bot. Gaz.* **55,** 337–371.
Ku, H. S., Suge, H., Rappaport, L. and Pratt, H. K. (1970). *Planta,* **90,** 333–339.
Kumamoto, J., Dollwet, H. H. A. and Lyons, J. M. (1969). *J. Am. chem. Soc.* **91,** 1207–1210.
Lewis, L. N., Palmer, R. L. and Hield, H. Z. (1968). *In* "Biochemistry and Physiology of Plant Growth Substances" (F. Wightman and G. Setterfield, eds), pp. 1303–1313. Runge Press, Ottawa.
Lieberman, M. and Kunishi, A. T. (1969). *Pl. Physiol., Lancaster,* **44,** 1446–1450.
Lieberman, M. and Kunishi, A. T. (1972). *In* "Plant Growth Substances 1970" (D. J. Carr, ed.), pp. 549–560. Springer Verlag, Berlin.
Lieberman, M., Kunishi, A., Mapson, L. W. and Wardale, D. A. (1966). *Pl. Physiol., Lancaster,* **41,** 376–382.
Mapson, L. W. (1969). *Biol. Rev.* **44,** 155–187.
Maxie, E. C. and Crane, J. C. (1967). *Science, N.Y.* **155,** 1548–1550.
Maxie, E. C. and Crane, J. C. (1968). *Proc. Am. Soc. hort. Sci.* **92,** 255–267.
Mayak, S. and Halevy, A. H. (1972). *Pl. Physiol., Baltimore,* **50,** 341–346.
McMichael, B. L., Jordan, W. R. and Powell, R. D. (1972). *Pl. Physiol., Baltimore,* **49,** 658–660.
Michener, H. D. (1938). *Am. J. Bot.* **25,** 711–720.
Minina, E. G. (1952). *Moscow Acad. Sci. USSR* 1537–1647.
Morey, P. R. (1974). *Weed Sci.* **22,** 6–10.
Morgan, P. W. (1967). *Proc. 21st Cotton Defoliation and Physiology Conference* **21,** 150–155. (National Cotton Council, Memphis, Tenn., USA.)
Morgan, P. W. (1972). *Misc. Publs Tex. agric. Exp. Stn.* **1018,** p. 12.
Morgan, P. W. (1973). *Acta Horticulturae,* **34,** 41–54.
Morgan, P. W. and Baur, J. R. (1970). *Pl. Physiol., Baltimore,* **46,** 655–659.
Morgan, P. W. and Durham, J. I. (1972). *Pl. Physiol., Baltimore,* **50,** 313–318.
Morgan, P. W. and Durham, J. I. (1973). *Planta,* **110,** 91–93.
Morgan, P. W. and Durham, J. I. (1975). *Pl. Physiol., Baltimore,* **55,** 308–311.
Morgan, P. W. and Fowler, J. L. (1972). *Pl. Cell Physiol., Tokyo,* **13,** 727–736.
Morgan, P. W. and Gausman, H. W. (1966). *Pl. Physiol., Lancaster,* **41,** 45–52.
Morgan, P. W. and Hall, W. C. (1961). *Proc. Cotton Defoliation and Physiol. Conf.* (National Cotton Council, Memphis, Tenn., USA) **15,** 20–22.
Morgan, P. W. and Hall, W. C. (1962). *Physiologia Pl.* **15,** 420–427.
Morgan, P. W. and Hall, W. C. (1964). *Nature, Lond.* **201,** 99.
Morgan, P. W. and Powell, R. D. (1970). *Pl. Physiol., Baltimore,* **45,** 553–557.
Morgan, P. W., Beyer, E., Jr. and Gausman, H. W. (1968). *In* "Biochemistry and Physiology of

Plant Growth Substances" (F. Wightman and G. Setterfield, eds), pp. 1255–1273. Runge Press, Ottawa.

Morgan, P. W., Meyer, R. E. and Merkle, M. G. (1969). *Weed Sci.* **17,** 353–355.

Muir, R. M. and Richter, E. W. (1972). *In* "Plant Growth Substances 1970" (D. J. Carr, ed.), pp. 518–525. Springer Verlag, Berlin.

Musgrave, A. and Walters, J. (1973). *New Phytol.* **72,** 783–789.

Nakagaki, Y., Hirai, T. and Stahmann, M. A. (1970). *Virology,* **40,** 1–9.

Neljubow, D. (1901). *Beih. bot. Zbl.* **10,** 128–139.

Osborne, D. J. (1968). *In* "Biochemistry and Physiology of Plant Growth Substances" (F. Wightman and G. Setterfield, eds), pp. 815–840. Runge Press, Ottawa.

Osborne, D. J. and Mullins, M. G. (1969). *New Phytol.* **68,** 977–991.

Palmer, R. L., Lewis, L. N., Hield, H. Z. and Kumamoto, J. (1967). *Nature, Lond.* **216,** 1216–1217.

Paterson, D. R. (1973). *Abstr. annu. Meet. Am. Soc. hort. Sci.* **70,** 32.

Pratt, H. K. and Goeschl, J. D. (1969). *A. Rev. Pl. Physiol.* **20,** 541–584.

Radin, J. W. and Loomis, R. S. (1969). *Pl. Physiol., Lancaster,* **44,** 1584–1589.

Regeimbal, L. O. and Harvey, R. B. (1927). *J. Am. chem. Soc.* **49,** 1117–1118.

Reid, M. S. and Pratt, H. K. (1972). *Pl. Physiol., Baltimore,* **49,** 252–255.

Robinson, R. W., Shannon, S. and de la Guardia, M. D. (1968). *BioScience,* **19,** 141–142.

Rodriguez, A. G. (1932). *J. Dep. Agric. P. Rico,* **16,** 5–18.

Rosa, J. T. (1928). *Hilgardia,* **3,** 421–443.

Rubinstein, B. and Abeles, F. B. (1965). *Bot. Gaz.* **126,** 255–259.

Sakai, S. and Imaseki, H. (1971). *Pl. Cell Physiol., Tokyo,* **12,** 349–359.

Scott, P. C. and Leopold, A. C. (1967). *Pl. Physiol., Lancaster,* **42,** 1021–1022.

Scott, P. C. and Norris, L. A. (1970). *Nature, Lond.* **227,** 1366–1367.

Shah, J. and Maxie, E. C. (1965). *Physiologia Pl.* **18,** 1115–1120.

Shannon, S. and de la Guardia, M. D. (1969). *Nature, Lond.* **223,** 186.

Sorber, D. G. (1934). *Diamond Walnut News,* June 1934, pp. 3–4.

Spencer, Mary. (1969). *In* "Fortschritte der Chemie organisches Naturstoffe" (L. Zechmeister, ed.), Vol. **27,** pp. 31–80. Springer Verlag, Wien, New York.

Stahmann, M. A., Clare, B. G. and Woodbury, W. (1966). *Pl. Physiol., Lancaster,* **41,** 1505–1512.

Steen, D. A. and Chadwick, A. V. (1973). *Pl. Physiol., Baltimore,* **52,** 171–173.

Sterrett, J. P., Leather, G. R. and Tozer, W. E. (1973). *Pl. Physiol., Baltimore,* **51,** (Suppl) 29.

Stewart, E. R. and Freebairn, H. T. (1969). *Pl. Physiol., Lancaster,* **44,** 955–958.

Traub, H. P., Cooper, W. C. and Reece, P. C. (1939). *Proc. Am. Soc. hort. Sci.* **37,** 521–525.

Vacha, G. A. and Harvey, R. B. (1927). *Pl. Physiol., Lancaster,* **2,** 187–192.

Valdovinos, J. G., Ernest, L. C. and Henry, E. W. (1967). *Pl. Physiol., Lancaster,* **42,** 1803–1806.

van der Laan, P. A. (1934). *Recl. Trav. bot. neerl.* **31,** 691–742.

von Guttenberg, H. and Steinmetz, E. (1947). *Pharmazie,* **2,** 17–21.

Wegm, F. B., Smith, O. E. and Kumamoto, J. (1972). *Pl. Physiol., Baltimore,* **49,** 869–872.

Weise, M. V. and DeVay, J. E. (1970). *Pl. Physiol., Baltimore,* **45,** 304–309.

Williamson, C. E. (1950). *Phytopathology,* **40,** 205–208.

Wolfe, H. S. (1931). *Bot. Gaz.* **92,** 337–366.

Yang, S. F. (1967). *Archs. Biochem. Biophys.* **112,** 481–487.

Yang, S. F. and Baur, A. H. (1969). *Qualitas Pl. Mater. veg.* **19,** 201.

Zimmerman, P. W. and Wilcoxon, F. (1935). *Contrib. Boyce Thompson Inst. Pl. Res.* **7,** 209–229.

Zimmerman, P. W., Hitchcock, A. E. and Crocker, W. (1931). *Contrib. Boyce Thompson Inst. Pl. Res.* **3,** 459–481.

CHAPTER 9

EFFECTS ON PLANT CELL MEMBRANE STRUCTURE AND FUNCTION

R. S. MORROD

Imperial Chemical Industries Ltd, Plant Protection Division, Jealott's Hill Research Station, Bracknell, Berkshire, England

I. INTRODUCTION

The last decade has seen a tremendous, worldwide effort to unravel the secrets of biological membranes, with the result that some aspects of their structure and function are now well understood and current research is moving at an exciting pace. However, most of this work has been concerned with mammalian cells and information about higher plant membranes is only just beginning to catch up, after lagging behind for some time. For this reason, membrane publications in the herbicide field are extremely limited and have been restricted to relatively simple, mode-of-action studies. Nevertheless, several herbicides are claimed to produce toxicity via a direct or indirect effect on the cell membranes, whose vital role in plant physiology has been recognized for many years. Unfortunately, these observations are so diverse that it is very difficult to inter-relate them and in most cases, the data are too limited for any conclusions to be drawn about the molecular basis of activity.

However, an attempt has been made to group observations together in a meaningful way and to relate them wherever possible to our current understanding of biomembranes. Only results primarily concerned with the plasmalemma and the tonoplast have been included in this chapter, but references to the membranes of organelles, such as chloroplasts and mitochondria, can be found in Chapters 15 and 16.

II. STRUCTURE AND FUNCTION OF CELL MEMBRANES

It is not the purpose of this chapter to review the structure and function of biological membranes, but a short summary of recent developments in this field may help to put the effects of herbicides into perspective. Many of the first clues to the existence of membranes came from experiments with plant cells and a fascinating account of these early probings was presented at a symposium in 1962 by Homer W. Smith (Smith, 1962). More recently, Finean (1972) has traced how the current ideas of biological membrane structure have developed from the continuous bilayer theory (Danielli and Davson, 1935), the unit membrane concept (Robertson, 1959) and the particulate lipoprotein model (Vanderkooi and Green, 1970). It is now generally acknowledged that most of the phospholipids are in a bilayer configuration, but globular membrane proteins are at least partly submerged in this phase and in some cases may span the whole membrane (Fig. 1). In this way, both ionic and apolar interactions between the proteins and the phospholipids play an important part in membrane stability. This type of structure seems to be more capable of fulfilling the physical and biochemical requirements of a membrane, than some of the earlier models. For a more detailed discussion of membrane structure, the reader is referred to proceedings edited by Green (1972) and a review by Oseroff et al. (1973).

One of the most important discoveries of recent years is the recognition that biological membranes are extremely fluid. Spectroscopic probes added to artificial, phospholipid membranes (liposomes), have revealed a flexibility gradient along the hydrocarbon chains, increasing towards the terminal methyl

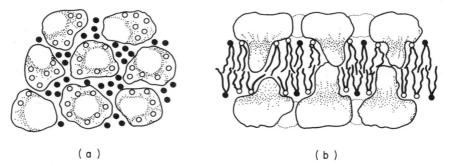

(a)   (b)

Fig. 1. A recent model for the structure of biomembranes showing the arrangement of proteins and phospholipids (a) surface view, (b) cross-section (after Finean, 1972).

group, which depends upon the composition of the membrane (Lee *et al.* 1973). Therefore, the interior of the membrane is similar to a hydrocarbon oil, but the extent of this molecular motion decreases dramatically when the temperature is reduced below the liquid-crystalline transition of the membrane. Below this point, the membrane is unable to fulfil all its functional roles, but some systems in nature are able to overcome the effect. For example, chill-resistant plants appear to synthesize more unsaturated fatty acids, as a means of keeping their membranes operative, below the temperature at which normal plant membranes undergo a phase transition to a more rigid form. Fluidity is also conferred by the lateral movement of phospholipids, which may diffuse at rates up to 1 μm/s (Traüble and Sackmann, 1972). Some membrane proteins may also be mobile, but their configuration and function are much more difficult to elucidate. The above, taken together with the biochemical turnover of components, illustrates the very dynamic nature of biomembranes, which almost certainly plays an important role in their many functions (McConnell, 1970).

Studies of the bioenergetics of plant membranes have been mainly concerned with chloroplasts and mitochondria, but recent methods for isolating plasma membranes may produce a change of emphasis. The greatest difficulty in pinpointing the site of a membrane process may depend on being able to separate the plasmalemma from the tonoplast. Nevertheless, examples of the facilitated transport of solutes and electrogenic ion pumping have been identified in plant cell membranes, but as yet, these are not as well defined as their counterparts in mammalian systems, such as the erythrocyte. It is partly because of this lack of basic information that the effects of herbicides, on energetically driven transport processes, are so poorly understood.

III. MEMBRANE DISORGANIZATION

A. DIRECT

A complete disorganization of membrane structure probably occurs only when a herbicide is able to dissociate the stabilizing, polar or hydrophobic bonds between the component proteins and phospholipids. Examples of both types of reaction are cited in the literature, but no detailed studies have been made to elucidate the molecular events involved.

The action of aromatic hydrocarbons and oils on plant tissue is so rapid that biochemical mechanisms have been ruled out by investigators. Currier (1951) showed that short exposures of barley, carrot or tomato plants to the vapour of benzene, toluene, xylene or trimethyl benzene produced a rapid darkening of the leaf tips, which he thought indicated the leakage of cell sap. The darkening quickly spread to the rest of the plant and there was a loss of turgor, accompanied by the odour that is normally associated with macerated tissue. In most cases, a short exposure to the hydrocarbon vapour produced death within 24 h, but lower concentrations of benzene ($2 \cdot 2 \times 10^{-4}$ mol benzene per litre of air) induced only a chronic response in barley. However, acute symptoms rapidly developed when the concentration of benzene was increased only slightly to 3.2×10^{-4} M. Discussing this phenomenon, Van Overbeek and Blondeau (1954) suggested that the chronic and acute symptoms were really various degrees of the same process, which may be regulated by the amount of chemical actually reaching the active site. This is thought to be the plasma membrane (Currier and Peoples, 1954; Dallyn and Sweet, 1951) which may be partially protected by the retention of these hydrophobic compounds in the cuticular wax. When the hydrocarbon does dissolve in the membrane the fatty acid chains of the phospholipids will be forced apart. Initially, this may only increase the permeability, but as the concentration of hydrocarbon increases in the membrane, the bilayer configuration could be completely' dissociated. Bulky hydrocarbons probably disorganize the phospholipids most effectively and this may be the reason why aromatic compounds are more toxic than the aliphatics, even though the latter have much higher partition coefficients. Similarly, Leonard and Harris (1950) found that heptene-2, which has a bent configuration, induced very rapid discolouration and death in 7-day-old cotton hypocotyls, whereas the straight chain n-heptane produced no visible injury. However, the effect of one double bond became less significant as the length of the chain increased beyond 10 carbon atoms. These observations can be correlated with the natural variations found in phospholipid chains, such as unsaturation or the inclusion of a cyclopropane ring, which can increase the permeability of a membrane by creating local molecular disorder. Besides affecting lipid-lipid interactions, the hydrocarbons probably disrupt hydrophobic bonding between lipids and the apolar regions of membrane proteins. The Lipid Globular Protein Mosaic Model proposed by Singer (1971), stresses the importance of these interactions for proteins that lie within the fluid, lipid bilayer. Therefore, the dissociation of these bonds may not only produce physical instability, but also lead to the degradation of membrane-bound enzymes.

The toxicity of an oil preparation is often enhanced by the oxidation of olefins to aliphatic acids. When an oxidized oil is washed with alkaline water, these acids are removed and the toxicity of the oil returns to its original level. Van Overbeek and Blondeau (1954) ascribed the effect to the undissociated form of the acid, which, particularly for short chain aliphatic acids, is very active against the membranes of beet-root tissue.

Several other herbicides exhibit a contact action in the field, which can be attributed to a physical interference with membrane stability, similar to that of the hydrocarbon oils. Endothal has no formative effects and its toxication consists of rapid penetration, desiccation and browning of the foliage. These effects have been attributed to membrane disorders, induced by ionic interactions with the compound, producing ion leakage, modified tonoplast plasmolysis and water loss. Veinal callose formation, which is typical of endothal phytotoxicity, has also been claimed to result from membrane effects, specifically from a leaky plasmalemma (Maestri, 1967). A similar burning, contact action observed with nitrofen, has been attributed to hydrophobic interactions with the cell membranes. The increased permeability of beet-root sections induced by nitrofen was enhanced by dimethylsulphoxide, but abolished by 0·2 M sucrose or mannitol, which may stabilize the weakened membranes osmotically. The lipid-nitrofen conjugates extracted by Hawton and Stobbe (1971) may play a role in the mechanism of action. The phenolic herbicides can act as uncouplers of oxidative phosphorylation, but at high concentrations, under favourable conditions, the loss of turgor is so rapid that they may be physically interfering with the structural organization of the membrane. Similarly, acrolein is a general plant-cell toxicant of high reactivity, related to its unsaturation, which is reported to destroy the integrity of plant cell membranes and to react with various enzyme systems (Ashton and Crafts, 1973).

Sodium arsenite is another toxic, contact herbicide, which causes rapid wilting by loss of turgor. This effect may be produced by the loss of metabolic energy associated with the uncoupling action of trivalent arsenic (van Overbeek, 1964). However, the pentavalent arsenical, cacodylic acid, can also uncouple oxidative phosphorylation, but does not induce a rapid loss of turgor (Sachs and Michael, 1971). Therefore, the action of sodium arsenite may be more intimately connected with its reactivity to protein sulphydryl groups;

$$2R-SH + O = As-R^1 \rightleftharpoons \begin{matrix} RS \\ \diagdown \\ \diagup \\ RS \end{matrix} As-R^1 + H_2O \tag{1}$$

In this way, membrane-bound enzymes, such as ATPase, may be inhibited and lead to a change in the permeability of the membrane (Spoerl, 1969; Sutherland et al. 1967). In addition, a 10^{-3} M solution of the sulphydryl reagent, p-chloromercuribenzoate, has been shown to stop protoplasmic streaming completely within a few minutes (Abe, 1959) and this may represent another potent inhibitory mechanism of sodium arsenite.

When high concentrations of 2,4-D are applied to leaves, the growth effects typical of lower concentrations are superseded by a contact, burning effect, which has been associated with membrane disruption. This was corroborated by White and Hemphill (1972), who used the electron microscope to study the sensitivity of old and young leaves of tobacco to 2,4-D. A solution of approximately 2 mM 2,4-D (Na salt) injected into mesophyll tissue, did not appear to affect the ultrastructure of an expanding tobacco leaf. In contrast, the tonoplast of a mature leaf was breached after only 30 min and the plasmalemma disintegrated shortly afterwards. This effect was attributed to the interaction of 2,4-D with membrane proteins, which may play different roles in old and young cells. A similar breakdown of membrane structure was observed within 4 h of treating Phaseolus vulgaris with $1·6 \times 10^{-3}$ M 2,4-D (Hallam, 1970). This effect occurred only when the tissue was illuminated and was not produced by other mono- and di-chloro substituted phenoxyacetic acids. Bachelard and Ayling (1971) also showed that 2,4-D was the membrane-active component of Tordon 50D (2,4-D + picloram), producing chloroplast and cytoplasmic membrane breakdown within 4 to 8 h in leaves of *Pinus radiata*. The changes observed in the electron microscope were correlated with a large decrease in the resistance ratio of the treated tissue.

Little is known about the mode of action of TCA and dalapon, but it is likely that both affect the conformation of vital membrane and cytoplasmic proteins. TCA at 100 kg/ha killed *Cynodon dactylon,* by reducing protoplasmic streaming and increasing the concentration of the cell sap (Corbadzijska, 1962). Other herbicides which are claimed to have some effect on membrane structure are bromacil (Ashton *et al.* 1969), bromoxynil, ioxynil and fluorodifen (Ashton and Crafts, 1973) and atrazine (Ashton *et al.* 1963).

B. INDIRECT

Biochemical breakdown of membrane structure should include a discussion of the deleterious effects of herbicides on the turnover of membrane proteins and lipids, but in the absence of such reports, examples are cited of membrane damage which appear to be produced via a biochemical reaction.

The mechanism of action of the bipyridylium herbicides, paraquat and diquat, is interesting because it depends upon the perversion of a normal biochemical process, to produce a chemical toxicant. Many experiments have been performed to deduce the requirements for paraquat activity (Mees, 1960; Merkle *et al.* 1965; Dodge, 1971), but recently, Harris (1970) has completed a detailed study of the sequence of events leading to cell death. He concluded that the primary effect was to divert normal photosynthetic electron flow into the production of some toxic substance(s), probably hydrogen peroxide, which has been shown to damage the membranes of plant cells (Siegel and Halpern, 1965). Electron microscopy showed that 6 h after the application of 10^{-4} M paraquat to flax cotyledons, the tonoplast membrane was damaged at points where it was in close contact with the chloroplasts. Simultaneously, there was a major increase in the level of malondialdehyde, which is a product of lipid peroxidation. Peroxides can react with unsaturated fatty acids to cause chain cleavage or the formation of hydroxyl compounds, both of which would destroy the amphipathic nature of the phospholipids. As a result of these changes a large efflux of K^+ from the treated leaves was observed (Fig. 2). Harris also feels, that this loss of compartmentalization would release hydrolytic enzymes from the vacuole into the cytoplasm, so producing a secondary effect. A similar mechanism of action has been reported for diquat in *Chlorella vulgaris* (Stokes *et al.* 1970).

Reports of the time taken for paraquat and diquat to induce membrane damage vary considerably. This is probably related to the time the compound takes to reach its site of action and the degree of adsorption occurring *en route*. Leaf wilting is the first sign of paraquat action in the field and under the favourable conditions of bright sunlight, this can occur after only 30 min (R. C. Brian, personal communication). However, the report by Baur *et al.* (1969), that paraquat damages the plasmalemma of mesquite cells after only 5 min is atypical of the plant response under laboratory conditions. I have

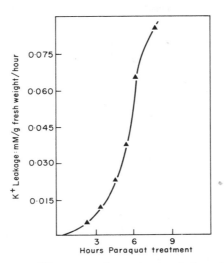

Fig. 2. Membrane permeability measured as potassium leakage from slices of paraquat-treated flax cotyledons (Harris and Dodge, 1972).

found (unpublished results) that up to 2 h are required before the effects of paraquat on membrane integrity become apparent. This conclusion was reached by using epidermis-free leaf discs, in which there is no cuticular barrier to penetration of the mesophyll tissue (Morrod, 1974). Epidermis-free discs were cut from tobacco leaves and exposed to a $7 \cdot 5 \times 10^{-3}$ M solution of paraquat for 6 h, under light and dark conditions. At hourly intervals, pairs of tissue discs were removed from the paraquat solutions and transferred to an isotonic buffer for 1 h in darkness. An increase in the amount of paraquat released from light-treated discs was observed after 2 h and this corresponded with the onset of wilting (Fig. 3). After 6 h in contact with paraquat, the discs were completely flaccid and almost all of the herbicide could be recovered. This type of efflux behaviour agrees with the observations in whole plants, that is, after the initial contact effect a flush of paraquat moves basipetally, probably by reverse xylem flow (Baldwin, 1963; Slade and Bell, 1966).

A much more dramatic loss of membrane integrity was observed when naked plant protoplasts were exposed to paraquat. Boulware and Camper (1972) found that 10^{-3} M to 10^{-6} M paraquat completely destroyed a population of isolated, tomato fruit protoplasts within 30 min. This rapid effect may be attributed to the intimate contact between the herbicide and the cell in the absence of extracellular components, plus the recognized instability of protoplasts in a hypertonic medium (Ruesink, 1971). In comparison, isolated peanut cells (with a cell wall) did not exhibit membrane damage even after a 2-h exposure to 10^{-5} M diquat (Davis and Shimabukuro, 1973).

Propanil inhibits several biochemical processes, the Hill reaction (Good, 1961), respiration (Hofstra and Switzer, 1968) and RNA and protein synthesis

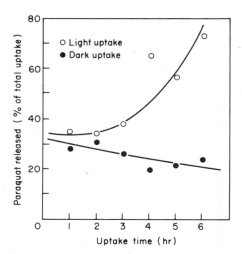

Fig. 3. Release of ^{14}C paraquat in 1 h from epidermis-free tobacco leaf discs after exposure to the herbicide under light or dark conditions (Morrod, unpublished).

(Moreland *et al.* 1969). However, effects on the cell and chloroplast membranes may also contribute to the phytotoxic effect (Hofstra and Switzer, 1965). Spots with a water-soaked appearance developed only on sprayed plants after three or four days and this observation corresponded with an increase in the permeability of beet-root tissue. This correlation suggests that propanil induces membrane damage and this conclusion was supported by the breakdown of treated chloroplasts *in vitro*. Considering the time taken for the effect to emerge, it seems probable that membrane breakdown results from a reduction in biochemical energy.

IV. PERMEABILITY

The plasmalemma and the tonoplast membranes effectively protect the cytoplasm, where most of the delicately balanced, biochemical reactions of the cell occur. The vacuole has been ascribed a lytic or lysosomal character (Matile, 1968; Villiers, 1971) and therefore it is particularly important that the tonoplast continues to act as a permeability barrier. The semi-permeable properties of the biological membrane are thought to be conferred mainly by a close-packed sheet of phospholipids, which form a hydrophobic barrier. Water-soluble macro-molecules are generally unable to penetrate this layer (unless pinocytosis occurs), but smaller solutes may diffuse across the barrier at a rate which depends on the structure of the molecule and the composition of the membrane. The factors which dictate permeability have been mainly deduced from experiments with animal and algal cells, but the same principles almost certainly apply to plant cell membranes. For many molecules, permeability can be summarized by the following simple equation:

$$\text{Flux} = \frac{KD}{l} \qquad (2)$$

where K is the partition coefficient between the membrane and the aqueous phase, D is the diffusion coefficient in the membrane and l is the thickness of the membrane. A small number of organic molecules cross the membrane more rapidly than the equation predicts and this has led many investigators to believe that pores exist in the membrane. The partition coefficient term above, describes the ability of the molecule to enter the hydrophobic interior of the membrane, so that compounds with a high partition coefncient tend to penetrate easily. On the other hand, a compound that is highly polar or forms hydrogen bonds with water, does not easily enter the membrane and permeability to them is low. It is worth noting that inorganic ions, amino acids and sugars have an intrinsically rather low penetrability for this reason. However, plant and animal cells have developed membrane transport systems, capable of speeding up the penetration of metabolically important molecules (Shtarkshall et al. 1970; Maretzki and Thom, 1972). These translocators, which are thought to be membrane proteins, impose a strict, steric selectivity, with the result that most exogenous compounds, such as herbicides, cross the cell membrane by simple diffusion. Many herbicides are rather lipophilic and will probably have little difficulty in penetrating plant cells. Therefore, although it is often stated that the membrane is there to exclude toxic substances, it is really only effective as a barrier to hydrophilic compounds.

The rate of diffusion of a compound through the interior of the membrane has been found to depend on the size and shape of the permeant and the close packing of the phospholipid molecules (Diamond and Wright, 1969; De Gier et al. 1972). The latter is illustrated by the effect of unsaturated or bulky acyl chains, which increase permeability, compared with the condensing effect of certain sterols, which can reduce the flux. As a result, membranes with different chemical composition may transport the same compound at different rates. This phenomenon has been illustrated by Hingson and Diamond (1972), who found that the permeation of a given nonelectrolyte varied considerably through several animal, epithelial membranes. If similar differences exist between plant species, the effect may form an important part of a physical basis for herbicidal selectivity. The effect of herbicides on plant cell membrane permeability is usually monitored by examining the flux of endogenous solutes. Therefore, examples quoted in this section are intimately associated with other effects described throughout this chapter.

The mode of action of dichlobenil seems to be connected with a change in membrane permeability, which may be induced by a dual mechanism. Price (1969) found that a lag period of 8 to 12 h occurred before 10^{-4} M dichlobenil was able to produce a significant increase in the efflux of betacyanin from beetroot discs and even after 24 h, only six times more dye was present in the treated solution compared with a control. Although dichlobenil has a high lipid solubility (Massini, 1961), the speed and magnitude of these changes do not suggest a strong biophysical effect, such as that produced by the phytotoxic oils. However, the increase in permeability was partially reversed by Ca^{2+} ions, which are known to increase the ordering of membrane phospholipids (Butler et al. 1970). A similar, but instantaneous, increase in permeability was observed when the beet-root tissue was treated with 10^{-4} M dinitrophenol (DNP) and so the lag phase observed with dichlobenil suggests that a protracted biochemical initiation may be operating. Both dichlobenil (10^{-5} M) and DNP (10^{-4} M) increased the respiration of beet-root tissue, but the herbicide failed to uncouple oxidative phosphorylation in isolated cucumber mitochondria. However, the related compound 3-hydroxy-2,6-dichlorobenzonitrile was as effective as DNP in this respect. This compound is a major metabolite of dichlobenil in mammals, where it produces a toxic effect by uncoupling oxidative phosphorylation. Similarly, the 3-hydroxy- and 4-hydroxy-derivatives of dichlobenil form more than 95% of the primary plant metabolites and are capable of producing phytotoxicity independently (Verloop and Nimmo, 1969). As a result of the action of dichlobenil and/or its metabolites on the tonoplast, phenols have been observed to leak from the vacuole and form phytotoxic polymers in the cytoplasm (Price, 1969).

The permeability changes induced by 2,4-D vary with concentration, in the same way as the anatomical response of plants to the compound. Low concentrations of 2,4-D (10^{-6} M), which promote longitudinal growth, stimulate the efflux of nucleic acids from pea roots, but higher concentrations ($> 10^{-4}$ M) inhibit this flux (Strube and Fellenberg, 1972). This bimodal effect has also been reported for the auxin, indol-3yl-acetic acid (Masuda, 1953). Jenner (1963, 1965) has described a specific effect of 2,4-D on glucose permeability. When segments, cut from the mesocotyl of etiolated oat seedlings, were immersed in water, organic substances leaked out slowly, but the addition of 10 parts/10^6 2,4-D (5×10^{-5} M) increased the efflux of the sugars, glucose and fructose equally. However, 100 parts/10^6 2,4-D promoted the outflow of sugar more than amino acids and glucose more than fructose. Other metabolic inhibitors and poisons also accelerated the efflux of the sugars, but the composition of the external medium was always the same as that of the cells. Wedding et al. (1959) also noticed that 2,4-D specifically modified the permeability of Chlorella pyrenoidosa to glucose, mannitol and undissociated phosphate, without affecting the flux of inorganic ions. However, the concentration of 2,4-D which inhibits nonelectrolyte permeation, is the same as that found to increase the electrical resistance of Nitella membranes by Bennett and Rideal (1954). A decrease in membrane permeability has also been reported for dicamba by Magalhâes and Ashton (1969). They measured the conductivity of a solution containing leaf sections, cut from plants of purple nutsedge (Cyperus rotundus L), five days after spraying with dicamba. All the concentrations used decreased the conductivity of the medium when expressed as a percentage of the total possible conductivity, observed after heating the tissue to 80°C for 30 min (Table I). The lowest concentration induced only a marginal decrease, which is in agreement with the observation by Reid and Hurtt (1970), that 10^{-4} M and 10^{-5} M dicamba do not affect the permeability of beet-root cells. Magalhâes and Ashton attributed the effect of dicamba to an increase in the resistance of the membrane, thus decreasing the release of ions into the bathing medium. Unfortunately, membrane resistance was not directly measured and after five days in contact with the herbicide, the effect may have been more directly associated with the other important plant interactions attributed to dicamba (Quimby, 1967; Moreland *et al.* 1969; Arnold and Nalewaja, 1971).

TABLE I

RELEASE OF ELECTROLYTES FROM LEAVES OF *Cyperus rotundus* L. FOLLOWING TREATMENT WITH DICAMBA[1]

| Concentration of dicamba | Time after excision (min) | | | | | Regression equation |
	5	30	60	120	180	
Control	2·17	3·00	3·99	5·97	7·95	$y = 2·01 + 0·03302x$
10^{-2} M	1·87	2·45	3·14	4·53	5·92	$y = 1·75 + 0·02317x$
10^{-3} M	1·70	2·32	3·08	4·59	6·10	$y = 1·57 + 0·02576x$
10^{-4} M	1·67	2·47	3·44	5·36	7·29	$y = 1·51 + 0·03212x$

[1] Measured by conductivity and expressed as a % of the total possible electrolyte release (Magalhâes and Ashton, 1969).

The permeability of the *Chlorella pyrenoidosa* membrane was studied by following the efflux of ^{14}C-intracellular material from cells which had been allowed to incorporate $^{14}CO_2$ photosynthetically (Sikka *et al.* 1973). It was observed that the efflux increased upon treatment with low concentrations (3–30 μM) of 2,3-dichloro-1,4-naphthoquinone (dichlone). The rate of loss as well as the total loss of ^{14}C increased with an increase in the concentration of dichlone. In the dichlone-treated cells, the leakage was observed within 1 min of the addition of the chemical and the effect on cell permeability was irreversible. Cells exposed to dichlone in the light or under anaerobic conditions released significantly greater amounts of ^{14}C-material than cells treated in the dark or under aerobic conditions (Fig. 4). The aqueous alcohol-soluble fraction of the cell was the source of the released material. The proportion of the alcohol-soluble ^{14}C that leaked out of the cell varied with the time of ^{14}C-assimilation prior to treatment with dichlone. In the dichlone-treated cells, practically all the sucrose, alanine, serine, and glycine leaked out of the cell, whereas glutamic and aspartic acids were lost only partially. Essentially no ^{14}C-lipids were lost from the cells during dichlone treatment. The extreme rapidity of the effect of dichlone on permeability and the low concentrations at which it acted suggest that the cell membrane may be a primary site of action. The authors attributed the effect of dichlone to a rapid reaction with the sulphydryl and amino groups of membrane proteins, rather than to a free radical mechanism, which would have shown a dependence on oxygen the reverse of that actually observed.

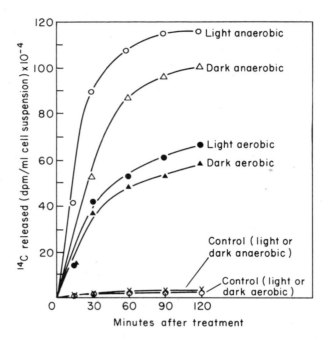

Fig. 4. Release of intracellular materials upon treatment of *Chlorella* with dichlone in light and dark, under aerobic and anerobic conditions. *Chlorella* cells were allowed to incorporate $^{14}CO_2$ for 30 min. ^{14}C-labelled cells were centrifuged, washed and resuspended in fresh medium. These cells were incubated with 30 μM dichlone at 25°C under the conditions indicated. Samples of cell suspension were removed at intervals, filtered and the release of ^{14}C-material was measured by radioactive counting of the filtrate. One ml of cell suspension contained $2 \cdot 48 \times 10^6$ dpm prior to quinone treatment (Sikka *et al.* 1973).

V. Ion Transport

This section is concerned only with the effects of herbicides on ion accumulation in cells, rather than the overall uptake and translocation of ions through the plant, which is described in Chapter 7. Many of the established details of ion transport in plants have been determined from studies with giant algal cells (MacRobbie, 1970; Spanswick, 1972; Vredenberg and Tonk, 1973). Used as large single cells, the *Characean* species are ideal for studying the relationship of ion fluxes to the existing membrane potential, measured with microelectrodes. On the other hand, higher plant tissue presents technical difficulties, because the additional diffusion steps through the cuticle and the apoplast may obscure the true membrane kinetics. However, many careful studies have been made and the reader is directed to a monograph on ion transport for more information (Anderson, 1973). Active ion transport occurs in both root and leaf cells (Pitman and Saddler, 1967; Higinbotham *et al.*

1970) with the result that a high cellular K^+/Na^+ ratio usually exists combined with a high total salt content. This distribution of ions across the membrane is produced by a complicated interdependence of cation and anion fluxes. However, it appears that Na^+ is actively extruded (Etherton, 1967), whilst K^+ and Cl^- are pumped into the cells (Pitman and Saddler, 1967; Gerson and Poole, 1972). Kinetically, two ion transport mechanisms have been identified, a high affinity system in the plasmalemma (Epstein and Rains, 1965) and a low affinity system in either the plasmalemma (Epstein, 1972) or the tonoplast (Laties, 1969). Ion-dependent, membrane-bound ATPase has been located in both root (Hodges *et al.* 1972) and shoot cells (Lai and Thompson, 1971), and has been implicated in the pumping mechanism. However, these plant ATPases seem to differ from those found in mammalian cells, in that, with the exception of one report (Cram, 1968), they are insensitive to ouabain. The energy supply for ion transport has been investigated by the use of various metabolic inhibitors and it is clear that ion transport is coupled to respiration (Polya and Atkinson, 1969) and the partial reactions of photosynthesis (Lüttge *et al.* 1971). Also, van Steveninck and van Steveninck (1972) have recently found a link between protein and nucleic-acid synthesis and the development of ion uptake mechanisms in beet-root tissue. Therefore, herbicides could influence ion transport in root or shoot directly, by physico-chemical interactions with the cell membrane, or indirectly, by affecting oxidative phosphorylation, photosynthesis and possibly, nucleic-acid and protein synthesis. Since these are the major biochemical mechanisms attributed to herbicides (Moreland, 1967), one would expect changes in ion transport to go hand in hand with the onset of phytotoxicity. Circadian rhythms (Bünning and Moser, 1972), stomatal opening (Allaway and Hsiao, 1973) and cell elongation (Rayle and Johnson, 1973) are related to ion flux, so it seems likely that indirect effects of herbicides on these mechanisms may contribute quite considerably to the overall phytotoxic response. On the other hand, the osmotic stability of plant cells is not as dependent on ion pumping as are many mammalian cells (Richardson and Neergaard, 1972). Unfortunately, there is little detailed information about the effects of herbicides on the complex ion-transport mechanisms described above.

It is acknowledged that phenolic herbicides can uncouple oxidative phosphorylation (Ashton and Crafts, 1973) and so affect active ion transport by depressing the level of ATP in the cell. The method by which mitochondrial energy is transformed into ion pumping has not been elucidated, but recently the activity of a membrane-bound ATPase has been correlated with the amount of ion transported (Leonard and Hanson, 1972*b*). Besides the phenolic herbicides, dicamba and dichlobenil may also act in this way, after being metabolized to hydroxyl compounds. Wojtaszek (1966) found that the level of ATP in plant cells is regulated by the availability of inorganic phosphorus, the uptake of which requires in turn, the energy stored in ATP. Therefore, he observed that plants which were resistant to dinoseb accumulated much more ^{32}P and generated more ATP than susceptible species. Similarly, Matlib and Kirkwood (1970) found that MCPA and MCPB inhibited phosphate uptake in bean roots, by uncoupling or inhibiting oxidative phosphorylation. However, phenols are often rapidly detoxified by conjugation with glucose or other organic molecules and therefore it is worth considering that part of their effect may be initiated at the plasmalemma. Glass (1973), in a study of the effect of substituted phenols on phosphate uptake by barley roots, concluded that these compounds affect the permeability of the plasmalemma to phosphate, but not by denaturing specific membrane carriers or uncoupling oxidative phosphorylation. Amongst a series of hydroxylated benzoic acids, the inhibition of phosphate uptake was positively correlated with the partition coefficient of the molecule and so the author concluded that the effect was due to an increasing concentration of phenol in the membrane. However, this correlation may also refer to the ability of the molecule to transport a proton through the hydrophobic part of the membrane, which according to the chemiosmotic theory (Mitchell, 1961) is the basis of uncoupling in mitochondria. This effect of phenols could also operate at the plasmalemma, where the existence of proton gradients have been observed (Lucas and Smith, 1973; Hager et al. 1971) and may play an important role in the overall scheme of inorganic ion transport.

It has already been seen that the auxin herbicides, depending on their concentration, markedly affect the permeability of plant cell membranes and this is reflected in the effect of 2,4-D on ion transport. In the whole plant, low concentrations of 2,4-D stimulate ion absorption, particularly in the roots following leaf application (Cooke, 1957), but as the concentration of 2,4-D rises, inhibition begins (Cardenas et al. 1966). This effect is well illustrated by

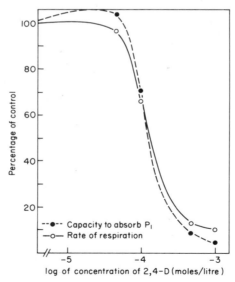

Fig. 5. The effect of 2,4-D treatment (18 h) on respiration and the capacity of beetroot discs to absorb phosphate (Palmer, 1966).

Fig. 5, which shows how respiration and the development of an enhanced phosphate absorption in washed root tissue are increasingly inhibited by 2,4-D. Palmer and Blackman (1964) also found that this herbicide increased the rate of release of labelled phosphate from preloaded pea epicotyls, which may indicate an increase in membrane permeability or a decrease in the metabolic energy required to maintain a high intracellular concentration of phosphate. The relevance of these results to the killing effect of 2,4-D is not clear, because 2,4,6-trichlorophenoxyacetic acid which is an antiauxin also inhibited phosphate uptake. Other reports suggest that 2,4-D does not inhibit the uptake of a particular ion; on the contrary, the response depends upon the species (Blackman, 1955) and the combination of ions in contact with the treated cells (Swenson and Burström, 1960).

The herbicidal activity of 2,4-D has been ascribed to changes in RNA and protein synthesis (Ashton and Crafts, 1973) and this biochemical response may be initiated from the plasma membrane. Hardin *et al.* (1972) claim that a transcriptional factor is released from isolated soybean membranes by $0 \cdot 1 \, \mu M$ 2,4-D, whereas the inactive compound 3,5-D does not produce this response. The authors feel that binding of 2,4-D to a receptor site associated with the plasma membrane, may simultaneously initiate the rapid growth response and the delayed nuclear changes characteristic of the compound. This hypothesis is supported by the observation that 2,4-D (but not 3,5-D) can initiate a proton efflux from *Avena* coleoptiles (Rayle and Johnson, 1973), which is now thought to be the direct stimulus for cell elongation (Hager *et al.* 1971). A further association between 2,4-D, ion absorption and protein synthesis has been reported by Leonard and Hanson (1972*a*). They found that 1×10^{-4} M 2,4-D prevents the development of enhanced phosphate absorption in washed beet-root tissue, almost as effectively as the protein-synthesis inhibitor, cycloheximide. These effects must form part of the great number of structural and biochemical modifications involved in 2,4-D activity, but their contribution will largely depend on the effective concentration of the compound *in vivo*.

Other herbicides which have been reported to inhibit phosphate uptake are dalapon (Ingle and Rogers, 1961), simazine and monuron (Zurawski *et al.* 1965), propanil (Hofstra and Switzer, 1965), ioxynil (Kerr and Wain, 1964) and TCA (Hassell, 1961). In addition, 10^{-5} M DNOC produces a 50% decrease in growth, water uptake and salt uptake, but at higher concentrations the inhibition of salt uptake predominates (Riepma, 1958).

A special form of ion transport appears to be responsible for stomatal movement in the leaf epidermis. Several workers have found that the estimated uptake of K^+ and an organic anion by the guard cells can account for a major proportion of the observed osmotic changes during stomatal opening (Pallas and Wright, 1973; Allaway and Hsiao, 1973). Therefore, reports that 2,4-D (Bradbury and Ennis, 1952) and atrazine (Wills *et al.* 1963) induce stomatal closure and that 2,3,6-TBA stimulates stomatal opening (Mason, 1960), may be due to the direct or indirect effects of these herbicides on K^+ transport at the guard cell membrane.

VI. MISCELLANEOUS

A. SURFACTANTS

An aspect of herbicide application which is not often considered is the contribution of the surfactant to the overall development of phytotoxicity. In most cases, a surfactant alone is unlikely to kill the plant, but often causes marked scorching of the leaves, and this may play an important part in the initiation of the true herbicidal effect. The extent of plant damage by formulating agents is most clearly recognized by plant pathologists and entomologists, particularly those seeking bacterial control with cationic compounds. Surfactant phytotoxicity is usually attributed to cell-membrane damage, probably through forcing apart the fatty-acid chains or solubilizing the hydrophobic regions of the membrane. However, Mirimanoff (1953) concluded that the toxicity of surfactants bore no relationship to their surface activity or wetting ability, but was dependent upon the chemical structure and on the nature of the plant cell.

Cations are acknowledged to be the most phytotoxic group of surfactants, a fact which led indirectly to the discovery of paraquat. As an example, bio-electric potential measurements and microscopical examinations showed that cationic detergents kill *Nitella* more effectively than anionic compounds, by rupturing the plasmalemma in a manner analogous to the haemolytic action of detergents (Anon, 1957). In the same report, it was stated that 10^{-3} M dodecyl trimethylammonium bromide was highly toxic to pea sections, whereas the same concentration of sodium dodecyl sulphate had no effect. However, the anionic surfactants are not without effect, sodium *n*-dodecylbenzene sulphonate (3 mM) was found to induce drastic changes in the cell components and membranes of *Phalaenopsis* protocorms after five days (Healey *et al.* 1971). Similarly, decenylsuccinic acid decreased the water permeability of *Allium cepa* cells at low concentration without apparent injury, but was lethal at higher concentrations (Lee *et al.* 1972).

Non-ionic surfactants are widely used in pesticide formulation and are thought of as "milder detergents"; nevertheless, they too can adversely affect cell metabolism (Parr and Norman, 1964). For non-ionics such as Tween-20 and Triton X-100, the degree of phytotoxicity decreases as the polyoxy-ethylene (hydrophillic) content increases (Buchanan and Stainforth, 1966). The effect of Triton X-100 on plant cells has been studied in detail by Haapala (1970). She found that the inhibition of protoplasmic streaming in *Nitellopsis* and the onset of sugar and ion leakage from beet-root cells, coincided with the formation of surfactant micelles, which may interact with the plasmalemma. However, the observation that solute uptake was stimulated below the critical micelle concentration may be related to the report by Brierley *et al.* (1972) that low concentrations of Triton X-100 can act as an ionophore for potassium in mitochondria. Millaway (1969) found that Tween-20 can also increase the efflux of betacyanin from beet root discs and inhibit the uptake of Rb$^+$ ions at a concentration of only 0·01%, but in this case, the cells later recovered.

Several authors find that the damaging effect of surfactants can be alleviated by cooling, possibly because of the greater stability of membranes below the phospholipid transition temperature, which appears to be 10° to 12°C for some plant membranes (Raison and McMurchie, 1972).

B. WATER PERMEABILITY

The major factors which regulate water permeability in plant cells have recently been reviewed by Kuiper (1972). He points out that the osmotic permeability coefficient (Lp), which lies between 5 and 100×10^{-4} cm/s, is reduced by cooling and affected by changes in the composition and configuration of membrane phospholipids. The evidence for active transport of water is equivocal, but the uptake and flow of water through roots appears to be metabolically mediated and related to protein synthesis. However, water movement in the whole plant is dictated by root- and leaf-cell turgor pressure and the rate of transpiration. For this reason, Krihning (1965a) felt that the decrease in water uptake in cut shoots of *Lycopersicon esculentum* exposed to 5×10^{-4} M 2,4-D, was due to a general reduction in transpiration, rather than any deleterious effects on plasmolytic processes. Nevertheless, the same author (Krihning, 1965b) reported that 2,4-D increased the time required for deplasmolysis of epidermal cells of *Solanum tuberosum* L. Simazine had the opposite effect, but several other triazines prolonged deplasmolysis, by increasing the plasma viscosity and consequently decreasing water permeability (Auerbach *et al.* 1968). A completely different response to 2,4-D has been observed with treated chichory root discs (Flood *et al.* 1970; Rutherford *et al.* 1966). A 10^{-5} M solution of 2,4-D caused a very large increase in water uptake, which did not occur with 3,5-D (Fig. 6). This effect

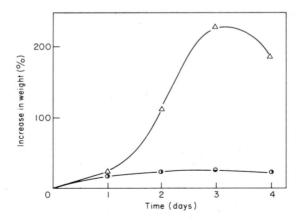

Fig. 6. The water uptake induced in discs of chicory root at 25°C by treatment with water or 10^{-5} M solutions of either 2,4-D or 3,5-D, for varying periods of time. △, 10^{-5} M 2,4-D; ●, 10^{-5} M 3,5-D; ○, water (Flood *et al.* 1970).

seemed to be due, at least in part, to an increase in the osmotic pressure of the cells, brought about by extensive hydrolysis of oligosaccharides, but at the same time there was also a marked stimulation of invertase activity, protein synthesis and respiration. However, van Overbeek (1944) found that the sap of tissues treated with growth substances possessed a lower osmotic concentration than that of non-treated tissues and concluded that the enhancement of water uptake was due to a decrease in wall pressure only or to an increase in non-osmotic water uptake or both.

C. ELECTRICAL EFFECTS

There are very few reports of the direct effect of herbicides on the electrical membrane potential or the surface charge of plant cells, but it can be assumed that changes take place when ion transport is modified. Higinbotham (1973) has recently produced an excellent review of the electropotentials of plant cells, in which he quotes that plant cells have a resting potential of about 100 mV, inside negative. Part of this membrane potential appears to be metabolically produced, because it can be inhibited by cyanide or DNP (Higinbotham et al. 1970). The surface charge of the plasmalemma is also negative (-25 mV), but can be reduced or even made positive when cells are exposed to a very small amount of a basic polypeptide (Grout et al. 1973). If a herbicide had a similar effect, the surface-charge resistance to the transport of inorganic ions may be altered.

Gunar et al. (1968) used microelectrodes to follow the electrophysiological response of *Nitella flexilis* to chemical stimuli. 2,4-D, 2,4-D amine, 2,4-D butyl ester, dalapon (1%), IAA (3%) and GA_3 (0·01%), when applied as droplets to one end of the cell, all made the resting potential negative. The depolarized section acted like a stimulant, producing a series of extensive, rhythmic, action potentials along the protoplasm, which for 2,4-D occurred at a speed of 4·4 cm/s and an amplitude of 67 mV. These changes in the biopotentials were attributed to the effect of the compounds on the protoplasmic membrane, thereby affecting the functional efficiency of the cells and their adaptive responses. In a thorough biophysical study, Bennett and Rideal (1954) also found that phenoxy compounds affect the electrical properties of *Nitella* cells. 2,4-D and MCPA increased the membrane resistance by more than three-fold, but only at pH 3·8 where both molecules are largely undissociated. The greatest resistance change occurred between 1 and 6×10^{-4} M 2,4-D, which corresponds to the concentration range of the herbicide that inhibits growth, ion transport and permeability.

D. LIPID BIOSYNTHESIS

The biochemical synthesis and degradation of membrane components occurs continuously and plays an important role in the stability of the membrane structure. Unfortunately, those seeking the mechanism of action of herbicides appear to have ignored this important pathway. An exception are Mann and

Pu (1968), who tested the effect of 30 herbicides on the incorporation of ^{14}C-malonic acid into lipid by excised hypocotyls of hemp sesbania *(Sesbania exaltata)*. The total chloroform-methanol extract was counted and no doubt contained several radioactively labelled compounds, formed by a variety of metabolic routes. Nevertheless, several herbicides caused more than 25% inhibition or stimulation of lipid synthesis (Table II). The stimulation shown by 2,4-D and 2,4,5-T decreased with increasing concentration (about 5×10^{-6} M to 1×10^{-4} M), whereas picloram and amiben produced a constant effect in the same concentration range.

TABLE II

HERBICIDES CAUSING MORE THAN 25% INHIBITION OR STIMULATION OF LIPID BIOSYNTHESIS IN
SESBANIA HEMP HYPOCOTYLS

Chemical name	Common name	Effect at 20 parts/10⁶
3-amino-2,5-dichlorobenzoic acid	amiben	stimulation
2-chloro-*N,N*-diallylacetamide	CDAA	inhibition
2-chloroallyldiethyldithiocarbamate	CDEC	inhibition
2,6-dichlorobenzonitrile	dichlobenil	inhibition
3-nitro-2,5-dichlorobenzoic acid	dinoben	inhibition
7-oxabicyclo (2.2.1) heptane-2,3-dicarboxylic acid	endothal	inhibition
3,5-diiodo-4-hydroxybenzonitrile	ioxynil	inhibition
pentachlorophenol	PCP	inhibition
4-amino-3,5,6-trichloropicolinic acid	picloram	stimulation
2,4-dichlorophenoxyacetic acid	2,4-D	stimulation
2,4,5-trichlorophenoxyacetic acid	2,4,5-T	stimulation

(Mann and Pu, 1968).

St. John and Hilton (1973) have also investigated the effect of herbicides on lipid metabolism. In a well executed series of experiments, they showed that dinoseb and MBR 8251 (1,1,1-trifluoro-4'-(phenylsulphonyl)-methanesulphono-*o*-toluidine) will inhibit the enzymic synthesis of glycerides *in vitro*. Furthermore, dinoseb and MBR 8251 caused marked alterations in the lipid levels of four-day-old wheat plants *(Triticum aestivum* L., "Mediterranean" (C.I. 5303)), whereas dinitrophenol (DNP) was only slightly effective (Table III). The reduction in polar lipids was due to a substantial decrease in all the component fatty acids of polar lipids. These changes might be expected to lead to an alteration in membrane structure and function, because the polar glycolipids and phospholipids are the main lipid constituents of the chloroplast, mitochondrial and cell membranes of plants. This proved to be so and 10^{-4} M dinoseb was found to increase the permeability of intact wheat roots to ions and amino acids within 2 h of application. However, MBR 8251 was without effect over the 6-h period tested. This result suggests that dinoseb may have a direct effect on contact with membranes, but in addition, the authors felt that, both herbicides are capable of altering membrane structure and function through the inhibition of polar lipid synthesis.

TABLE III

LIPID LEVELS IN FOUR-DAY-OLD WHEAT SEEDLINGS TREATED WITH HERBICIDES

Treatment[1]	Free fatty acids	Neutral lipids	Polar lipids
	(μg/g dry wt)	(μg/g dry wt)	(μg/g dry wt)
Control	199	3,515	10,127
DNP	293	2,479	8,466
MBR 8251	632	2,323	5,501
Dinoseb	402	2,113	2,800

[1] Lipids were extracted from the first leaf and coleoptile as described by St. John and Hilton (1973); chemical concentrations were; DNP, 5×10^{-4} M; MBR 8251, 10^{-4} M; dinoseb, 5×10^{-5} M.

E. PHYTOTOXINS

The future is likely to see a much greater exploitation of naturally occurring phytotoxic agents as general herbicides. Certainly, the devastation that is wrought by fireblight (*Erwinia amylovora*) suggests that the extraction and identification of toxins could be worthwhile. One such phytotoxin, helminthosporal, can damage plant cell membranes and it is suggested that this may be an important factor in the susceptibility of cereal tissue to an attack by *Bipolaris sorokiniana*, the fungus which produces the toxin. White and Taniguchi (1972) found that a 1 mM solution of helminthosporal increased the apparent free space in barley roots and initiated a large efflux of betacyanin by attacking the cell membranes of beet-root tissue. A similar toxin produced by *Helminthosporium maydis* caused partial depolarization of the electrical potential difference in corn roots within 5 min and inhibited the electrogenic influx of K^+ in stomatal guard cells (Mertz and Arntzen, 1973). Furthermore, Strobel (1973) claims to have isolated a membrane protein from sugar cane, which specifically binds helminthosporoside, the host-specific toxin produced by *Helminthosporium sacchari*. Taken together, these observations clearly indicate that several naturally occurring phytotoxins exert their effect by altering the permeability of the plasma membrane.

VII. CONCLUSIONS

The elucidation of the mode of action of a herbicide is a very difficult exercise and only a relatively few complete biochemical mechanisms have been resolved. Most herbicides have an opportunity to affect membrane function, when contact is first established with the cell, during movement away from the point of application. However, if there is an effect, it may be overlooked, because the biochemist often has to resort to a cell-free system, to detect the effect of a herbicide on a particular biochemical pathway, such as the Hill Reaction, oxidative phosphorylation or protein synthesis. This is understandable, since the inter-relationship between many biochemical events is rather poorly understood, which is particularly true of plant cell membrane phenomena in relation to other cell processes.

Amongst the papers reviewed in this chapter there are a great many interesting observations and conclusions, but there is also a great deal which has been left unsaid and uninvestigated. Many authors seem to be quite content to attribute herbicidal activity to an observed permeability change, without considering the mechanism by which the molecule produces this effect or whether the magnitude of the change would be sufficient to induce cell death. The work on paraquat mode of action by Harris and Dodge (1972) is an exception in this respect, because the authors have tried to present a balanced picture of the sequence of events by combining biochemistry and electron microscopy. Similarly, Currier (Currier, 1951; Currier and Peoples, 1954) showed great thoughtfulness in his appraisal of the molecular factors involved in the attack of the phytotoxic hydrocarbons on membrane structure. Generally, the experimental methods described are rather unsophisticated. The simple betacyanin method, introduced in 1949 by Veldstra and Booij, continues to be the most popular test, to the detriment of progress. Frequently, a herbicide, which increases the efflux of betacyanin, is claimed to work by producing "a loss of membrane integrity", however marginal the effect and with little reference to the concentration required. Many chemicals are exposed to beetroot tissue only, when it would often be more appropriate to measure the efflux of endogenous material from leaf sections. In any case, permeability changes revealed by this sort of test give only a superficial indication of the action of a herbicide and must be correlated with other effects on the cell. Fuller use could be made of isolated cells, protoplasts, membrane vesicles and artificial membranes to pinpoint the biochemical and bio-physical reactions to a chemical. Particularly in conjunction with the new spectroscopic techniques for probing molecular events in membranes. Only by understanding the molecular interactions of current toxicants can we hope to move towards designing a herbicide for a specific target.

One has only to scan the current membrane literature to realize that the contribution from the plant sciences is very small indeed. Nevertheless, progress is being made and we can expect a greater understanding of membrane structure to emerge in the near future, followed later by a visualization of the role of membrane turnover and bioenergetic function. As this knowledge becomes available, the involvement of membranes in the overall scheme of cell function is likely to be increasingly appreciated and the claim by Hardin *et al.* (1972) that 2,4-D and IAA initiate their effects from a membrane receptor site seems to point the way in this respect.

REFERENCES

Abe, S. (1959). *Kagaku,* **29**, 361–362.
Allaway, W. G. and Hsiao, T. C. (1973). *Aust. J. biol. Sci.* **26**, 309–318.
Anderson, W. P. (1973). "Ion Transport in Plants", Academic Press, London.
Anon (1957). *9th Rep. Commonw. Sci. industr. Res. Org. Australia,* p. 27 (iii).

Arnold, W. E. and Nalewaja, J. D. (1971). *Weed Sci.* **19**, 301–305.
Ashton, F. M. and Crafts, A. S. (1973). "Mode of Action of Herbicides", John Wiley and Sons, New York.
Ashton, F. M., Cutter, E. G. and Huffstutter, D. (1969). *Weed Res.* **9**, 198–204.
Ashton, F. M., Gifford, E. M., Jr. and Bisalputra, T. (1963). *Bot. Gaz.* **124**, 336–343.
Auerbach, S., Prüfer, P. and Weise, G. (1968). *Biol. Zbl.* **87**, 425–438.
Bachelard, E. P. and Ayling, R. D. (1971). *Weed Res.* **11**, 31–36.
Baldwin, B. C. (1963). *Nature, Lond.* **198**, 872–873.
Baur, J. R., Bovey, R. W., Baur, P. S. and El-Seify, Z. (1969). *Weed Res.* **9**, 81–85.
Bennett, M. C. and Rideal, E. (1954). *Proc. R. Soc.* **B124**, 483–496.
Blackman, G. E. (1955). *In* "The Chemistry and Mode of Action of Plant Growth Substances", pp. 253–259. Butterworths, London.
Boulware, M. A. and Camper, N. D. (1972). *Physiologia Pl.* **26**, 313–317.
Bradbury, D. and Ennis, W. B., Jr. (1952). *Am. J. Bot.* **39**, 324–328.
Brierley, G. P., Jurkowitz, M., Merola, A. J. and Scott, K. M. (1972). *Archs Biochem. Biophys.* **152**, 744–754.
Buchanan, G. A. and Staniforth, D. W. (1966). *Abstr. Meet. Weed Soc. Am.*, pp. 44–45.
Bünning, E. and Moser, I. (1972). *Proc. natn. Acad. Sci. USA* **69**, 2732–2733.
Butler, K. W., Dugas, H., Smith, I. C. P. and Schneider, H. (1970). *Biochem. Biophys. Res. Commun.* **40**, 770–776.
Cardenas, J., Slife, F. W. and Hanson, J. B. (1966). *Abst. Meet. Weed Soc. Am.* pp. 43–44.
Cooke, A. R. (1957). *Weeds,* **5**, 25–28.
Corbadzijska, B. (1962). *Nauchni. Trud. vissh. selskostop. Inst. Vasil Kolarov.,* **11**, 87–98.
Cram, W. J. (1968). *J. exp. Bot.* **19**, 611–616.
Currier, H. B. (1951). *Hilgardia,* **20**, 383–406.
Currier, H. B. and Peoples, S. A. (1954). *Hilgardia,* **23**, 155–173.
Dallyn, S. L. and Sweet, R. D. (1951). *Proc. Am. Soc. hort. Sci.* **57**, 347–354.
Danielli, J. F. and Davson, H. (1935). *J. cell. comp. Physiol.* **5**, 495–508.
Davis, D. G. and Shimabukaro, R. H. (1973). *Abstr. Weed Sci. Soc. Am.* p. 73.
Diamond, J. M. and Wright, E. M. (1969). *A. Rev. Physiol.* **31**, 581–645.
Dodge, A. D. (1971). *Endeavour,* **30**, 130–135.
Epstein, E. (1972). *In* "Mineral Nutrition of Plants: Principles and Perspectives". John Wiley Inc., New York.
Epstein, E. and Rains, D. W. (1965). *Proc. natn. Acad. Sci. USA* **53**, 1320–1324.
Etherton, B. (1967). *Pl. Physiol., Lancaster,* **42**, 685–690.
Finean, J. B. (1972). *Sub-Cell. Biochem.* **1**, 363–373.
Flood, A. E., Rutherford, P. P. and Weston, E. W. (1970). *Phytochemistry,* **9**, 2431–2437.
Gerson, D. E. and Poole, R. J. (1972). *Pl. Physiol., Lancaster,* **50**, 603–607.
Gier, J. de, Haest, C. W. M., van der Neut-Kok, E. C. M., Mandersloot, J. G. and van Deenan, L. L. M. (1972). *Proc. 8th FEBS Meeting, Amsterdam,* **28**, part II.
Glass, A. D. M. (1973). *Pl. Physiol., Lancaster,* **51**, 1037–1041.
Good, N. E. (1961). *Pl. Physiol., Lancaster,* **36**, 788–803.
Green, D. E. (ed.) (1972). "Membrane Structure and Its Biological Applications". *Ann. N.Y. Acad. Sci.* **195.**
Grout, B. W. W., Willison, J. H. M. and Cocking, E. C. (1973). *J. Bioenerg.* **4**, 311–328.
Gunar, I. I., Tsareva, L. A. and Sinyukhin, A. M. (1968). *Izv. timiryazev. sel'-khoz. Akad.* **1**, 3–14.
Haapala, E. (1970). *Physiologia Pl.* **23**, 187–201.
Hager, A., Menzel, H. and Krauss, A. (1971). *Planta,* **100**, 47–75.
Hallam, N. D. (1970). *J. exp. Bot.* **21**, 1031–1038.
Hardin, J. W., Cherry, J. H., Morre, D. J. and Lembi, C. A. (1972). *Proc. natn. Acad. Sci. USA* **69**, 3146–3150.
Harris, N. (1970). "Studies on the mode of action of paraquat and diquat". Ph.D. Thesis, University of Bath, England.
Harris, N. and Dodge, A. D. (1972). *Planta,* **104**, 201–219.
Hassall, K. A. (1961). *Pl. Physiol., Lancaster,* **14**, 140–149.

Hawton, D. and Stobbe, E. H. (1971). *Weed Sci.* **19,** 555–558.
Healey, P. L., Ernst, R. and Arditti, J. (1971). *New Phytol.* **70,** 477–482.
Higinbotham, N. (1973). *A. Rev. Pl. Physiol.* **24,** 25–46.
Higinbotham, N., Graves, J. S. and Davis, R. F. (1970). *J. membrane Biol.* **3,** 210–222.
Hingson, D. J. and Diamond, J. M. (1972). *J. membrane Biol.* **10,** 93–135.
Hodges, T. K., Leonard, R. T., Bracker, C. E. and Keenan, T. W. (1972). *Proc. natn. Acad. Sci. USA* **69,** 3307–3311.
Hofstra, G. and Switzer, C. M. (1965). *Pl. Physiol., Lancaster,* **40S,** xiv.
Hofstra, G. and Switzer, C. M. (1968). *Weed Sci.* **16,** 23–28.
Ingle, M. and Rogers, B. J. (1961). *Weeds,* **9,** 264–272.
Jenner, C. F. (1963). *Rep. Waite agric. Res. Inst.* pp. 36–37.
Jenner, C. F. (1965). *Rep. Waite agric. Res. Inst.* p. 76.
Kerr, M. W. and Wain, R. L. (1964). *Ann. appl. Biol.* **54,** 441–446.
Krihning, J. (1965a). *Phytopath. Z.* **52,** 372–394.
Krihning, J. (1965b). *Phytopath. Z.* **52,** 241–268.
Kuiper, P. J. C. (1972). *A. Rev. Pl. Physiol.* **23,** 157–172.
Lai, Y. F. and Thompson, J. E. (1971). *Biochim. Biophys. Acta,* **233,** 84–90.
Laties, G. G. (1969). *A. Rev. Pl. Physiol.* **20,** 89–116.
Lee, A. G., Birdsall, N. J. M. and Metcalfe, J. C. (1973). *Chem. Br.* **9,** 116–123.
Lee, O. Y., Stadelmann, E. J. and Weiser, C. J. (1972). *Pl. Physiol., Lancaster,* **50,** 608–615.
Leonard, R. T. and Hanson, J. B. (1972a). *Pl. Physiol., Lancaster,* **49,** 430–435.
Leonard, R. T. and Hanson, J. B. (1972b). *Pl. Physiol., Lancaster,* **49,** 436–440.
Leonard, O. A. and Harris, V. C. (1950). *Proc. sth. Weed Control Conf.*
Lucas, W. J. and Smith, F. A. (1973). *J. exp. Bot.* **24,** 1–14.
Lüttge, U., Ball, E., von Willert, K. (1971). *Z. PflPhysiol.* **65S,** 326–335.
MacRobbie, E. A. C. (1970). *Q. Rev. Biophys.* **3,** 251–294.
Maestri, M. (1967). "Structural and functional effects of endothal on plants". Ph.D. Thesis, Univ. California Davis, California, USA.
Magalhâes, A. C. and Ashton, F. M. (1969). *Weed Res.* **9,** 48–52.
Mann, J. D. and Pu, M. (1968). *Weed Sci.* **16,** 197–198.
Maretzki, A. and Thom, M. (1972). *Pl. Physiol., Lancaster,* **49,** 177–182.
Mason, G. W. (1960). "The absorption, translocation and metabolism of 2,3,6-trichlorobenzoic acid in plants". Ph.D. Thesis, University of California Davis, California, USA.
Massini, P. (1961). *Weed Res.* **1,** 142–146.
Masuda, Y. (1953). *Bot. Mag., Tokyo,* **66,** 256–261.
Matile, P. (1968). *Planta,* **79,** 181–196.
Matlib, M. A. and Kirkwood, R. C. (1970). *Pestic. Sci.* **1,** 193–196.
McConnell, H. M. (1970). *In* "The Neurosciences: Second Study Program" (Schmitt, F. O., ed.), pp. 697–706. Rockefeller University Press, New York.
Mees, G. C. (1960). *Ann. appl. Biol.* **48,** 601–612.
Merkle, M. G., Leinweber, C. L. and Bovey, R. W. (1965). *Pl. Physiol., Lancaster,* **40,** 832–835.
Mertz, S. M. and Arntzen, C. J. (1973). *Pl. Physiol., Baltimore,* **51S,** 16.
Millaway, R. M. (1969). "Surfactant effects on cell permeability of *Beta vulgaris* L. root tissue". Thesis, Iowa State University, USA.
Mirimanoff, A. (1953). *Protoplasma,* **42,** 250–260.
Mitchell, P. (1961). *Nature, Lond.* **191,** 144–148.
Moreland, D. E. (1967). *A. Rev. Pl. Physiol.* **18,** 365–386.
Moreland, D. E., Malhotra, S. S., Gruenhagen, R. D. and Shokraii, E. H. (1969). *Weed Sci.* **17,** 556–562.
Morrod, R. S. (1974). *J. exp. Bot.* **25,** 521–533.
Oseroff, A. R., Robbins, P. W. and Burger, M. M. (1973). *A. Rev. Physiol.* **42,** 647–682.
Overbeek, J. van (1944). *Am. J. Bot.* **31,** 265–269.
Overbeek, J. van (1964). *In* "The Physiology and Biochemistry of Herbicides" (L. J. Audus, ed.), pp. 387–400. Academic Press, London.

Overbeek, J. van and Blondeau, R. (1954). *Weeds,* **3,** 55–65.
Pallas, J. E., Jr. and Wright, B. G. (1973). *Pl. Physiol., Baltimore,* **51,** 588–590.
Palmer, J. M. (1966). *Pl. Physiol., Lancaster,* **41,** 1173–1178.
Palmer, J. M. and Blackman, G. E. (1964). *Nature, Lond.* **203,** 526–527.
Parr, J. F. and Norman, A. G. (1964). *Pl. Physiol., Lancaster,* **39,** 502–507.
Pitman, M. G. and Saddler, H. D. W. (1967). *Proc. natn. Acad. Sci., USA* **57,** 44–49.
Polya, G. M. and Atkinson, M. R. (1969). *Aust. J. biol. Sci.* **22,** 573–584.
Price, H. C. (1969). "The toxicity, distribution and mode of action of dichlobenil (2,6-dichloro-benzonitrile) in plants". Ph.D. Thesis, Michigan State University, USA.
Quimby, P. C. (1967). "Studies relating to the selectivity of dicamba for wild buckwheat (*Polygonum convolvulus* L.) *vs.* Selkirk wheat (*Tritieum aestivum*) and a possible mode of action". Ph.D. Thesis, N. Dakota State University, USA.
Raison, J. K. and McMurchie, E. J. (1972). *Abstr. Commun. Meet. Fed. Eur. Biochem. Soc.* **8,** 132.
Rayle, D. L. and Johnson, K. D. (1973). *Pl. Physiol., Baltimore,* **51S,** 2.
Reid, C. P. P. and Hurtt, W. (1970). *Physiologia Pl.* **23,** 124–130.
Richardson, I. W. and Neergaard, E. B. (1972). "Physics for Biology and Medicine", p. 153. John Wiley and Sons Ltd., London.
Riepma, P. (1958). *Proc. 10th int. Symp. Phytopharm., Meded. LandbHogesch. Gent* **23,** pp. 950–958.
Robertson, J. D. (1959). *Biochem. Symp.* No. 16, p. 3.
Ruesink, A. W. (1971). *Meth. Enzym.* **23,** 197–209.
Rutherford, P. P., Griffiths, C. M. and Wain, R. L. (1966). *Ann. appl. Biol.* **58,** 467–476.
Sachs, R. M. and Michael, J. L. (1971). *Weed Sci.* **19,** 558–564.
Shtarkshall, R. A., Reinhold, L. and Harel, H. (1970). *J. exp. Bot.* **21,** 915–925.
Siegel, S. M. and Halpern, L. A. (1965). *Pl. Physiol., Lancaster,* **40,** 792.
Sikka, H. C., Saxena, J. and Zweig, G. (1973). *Pl. Physiol., Baltimore,* **51,** 363–367.
Singer, S. J. (1971). *In* "Structure and Function of Biological Membranes" (L. I. Rothfield, ed.), p. 145. Academic Press, New York.
Slade, P. and Bell, E. G. (1966). *Weed Res.* **6,** 267–274.
Smith, H. W. (1962). *Circulation,* **26,** 987–1012.
Spanswick, R. M. (1972). *Biochim. biophys. Acta,* **288,** 73–89.
Spoerl, E. (1969). *J. membrane Biol.* **1,** 468–478.
St. John, J. B. and Hilton, J. L. (1973). *Weed Sci.* **21,** 477–480.
Steveninck, R. F. M. van and Steveninck, M. E. van (1972). *Physiologia Pl.* **27,** 407–411.
Stokes, D. M., Turner, J. S. and Markus, K. (1970). *Aust. J. biol. Sci.* **23,** 265–274.
Strobel, G. A. (1973). *J. biol. Chem.* **248,** 1321–1328.
Strube, U. and Fellenberg, G. (1972). *Planta,* **108,** 59–66.
Sutherland, R., Rothstein, M. A. and Weed, R. I. (1967). *J. Cell Physiol.* **69,** 185–198.
Swenson, G. and Burström, H. G. (1960). *Physiologia Pl.* **13,** 846–855.
Traüble, H. and Sackmann, E. (1972). *J. Am. chem. Soc.* **94,** 4499–4510.
Vanderkooi, G. and Green, D. E. (1970). *Proc. natn. Acad. Sci. USA* **66,** 615–621.
Veldstra, H. and Booij, H. L. (1949). *Biochim. biophys. Acta,* **3,** 287–312.
Verloop, A. and Nimmo, W. B. (1969). *Weed Res.* **9,** 357–370.
Villiers, T. A. (1971). *Nature, New Biol.* **233,** 57–58.
Vredenberg, W. J. and Tonk, W. J. M. (1973). *Biochim. biophys. Acta,* **298,** 354–368.
Wedding, R. T., Erickson, L. C. and Black, M. K. (1959). *Pl. Physiol., Lancaster,* **34,** 3–10.
White, J. A. and Hemphill, D. D. (1972). *Weed Sci.* **20,** 478–481.
White, G. A. and Taniguchi, E. (1972). *Can. J. Bot.* **50,** 1415–1420.
Wills, G. D., Davis, D. E. and Funderburk, H. H. (1963). *Weeds,* **11,** 253–255.
Wojtaszek, T. (1966). *Weeds,* **14,** 125–129.
Zurawski, H., Baranowski, R., Bors, J. and Ploszynski, M. (1965). *Symp. Use of Isotopes in Weed Res., Vienna,* 1965. *Int. Atom. En. Agency (IAEA Proc. Ser.).*

CHAPTER 10

EFFECTS IN RELATION TO WATER AND CARBON DIOXIDE EXCHANGE OF PLANTS

J. L. P. VAN OORSCHOT

Institute for Biological and Chemical Research on Field
Crops and Herbage, Wageningen, The Netherlands

I. INTRODUCTION

Many herbicides affect water and carbon dioxide exchange of plants. The interference of herbicides with photosynthesis and respiration at the cellular level is discussed in other chapters, but in this one the influence on intact plants is emphasized. This may help to broaden the interpretation of data obtained with isolated systems, because of the various factors involved in the accumulation of herbicides within the plant. Net photosynthesis is also a useful indication of the potential organic matter production of plants. Combining this discussion with that for water exchange is called for, since both carbon dioxide and water vapour diffuse through the stomata of terrestrial plants. The various aspects of the action of herbicides considered in relation to water and carbon dioxide exchange of plants under normal conditions are briefly reviewed in the following sections.

II. WATER UPTAKE AND TRANSPIRATION OF PLANTS

A. UPTAKE AND TRANSPORT OF WATER

Water easily diffuses through cell walls of the epidermis of the root and the root hairs, and moves quite freely in the intercellular spaces and the cell walls of the cortex, but encounters a resistance within the endodermis because of a suberized layer extending around each of the cells (Casparian strip). In consequence, water and ion transport may be forced across the endodermal cytoplasm, so that the latter may have a regulating influence on uptake. Root pressure is evident in the exudation of xylem sap from wounds and cut stems of non-transpiring plants, and is sometimes assumed to be due to active absorption of water. However, according to Slatyer (1967) there is "no real evidence of a phenomenon other than that of passive water movement following active ion transport into the xylem".

A driving force quantitatively much greater than root pressure is the suction in the xylem induced by transpiration in the leaves, and transmitted thence to the roots through the continuous liquid system in the xylem. The resistance to water flow within the vascular cylinder is small. A network of veins with ultimate branches of single vascular elements in the leaf laminae facilitates the water movement within the mesophyll tissue, of which both palisade and spongy parenchyma have a large surface area directly in contact with the substomatal spaces. As in the cortex water transport to the evaporating surface is mainly through cell walls.

The transport of water through the plant can be considered to occur along gradients of water potential from the soil to the atmosphere. With adequate water supply liquid-phase transport is the dominant process. The total flow is ultimately determined by the difference between vapour pressure at the evaporating surface and that in the ambient air. The water flow is regulated mainly by the stomata, but low root permeability to water may result indirectly in stomatal closure and decrease of transpiration rate (Slatyer, 1967).

B. STOMATAL PHYSIOLOGY

Stomata occur predominantly in the lower epidermis. Their size and frequency also vary within the same species, but since stomata tend to be smaller where they are more numerous, the total pore area per unit leaf area varies little (Meidner and Mansfield, 1968). Most guard cells have an elliptical shape, but are typical dumb-bell shaped in graminaceous stomata. The guard cells are in the relaxed state when the pore is closed. Increase of their turgor pressure above that of the surrounding epidermal cells opens the pore. Because of the thickened ventral walls along the pore but even more probably because of the orientation of the micellae in these walls the thin dorsal walls bulge into neighbouring epidermis cells (Meidner and Mansfield, 1968; Aylor *et al.* 1973).

Different hypotheses on the mechanism of stomatal movement have been proposed, but according to Meidner and Mansfield (1968) "none of them is entirely satisfactory", while Zelitch (1969) stated that "the mechanism cannot presently be given with precision". The old hypothesis that the change in turgor pressure is caused by the osmotic effect of the conversion of starch into sugar is sometimes linked with that concerned with the dark carboxylation reactions (formation of organic acids by incorporation of CO_2 into phosphoenolpyruvic acid) which would explain the closure of stomata at increasing CO_2 concentration. The closure of stomata by surface-active agents like alkenylsuccinic acids (Zelitch, 1964) supported another hypothesis suggesting that changes in membrane permeability of the guard cells might determine water flow from and into these cells, but it is not immediately clear how this would operate in the stomatal mechanism (Meidner and Mansfield, 1968).

There are indications that ATP is involved in stomatal opening by the promotion of active ion uptake (especially K^+) by the guard cells which increases their osmotic pressure (Zelitch, 1969). Zelitch (1965) suggested that ATP might be formed during the oxidation of glycollic acid. This oxidation is inhibited by α-hydroxysulphonates which also induce stomatal closure. High CO_2 concentration inhibits the synthesis of glycollic acid which would explain its effect on stomata. However, the interpretation of these results, especially of those with CO_2, was questioned by Meidner and Mansfield (1968). From the closure of stomata by abscisic acid (see Section V, B) it was assumed that the level of abscisic acid in the guard cells would regulate stomatal opening, but according to Hsiao (1973) further experimental evidence is necessary.

C. TRANSPIRATION

The diffusion of water during transpiration is usually described by:

$$T = \frac{e_l - e_a}{r_l + r_a} \tag{1}$$

in which T = transpiration (cm^3 water vapour s^{-1} cm^{-2}), e_l = water vapour pressure at evaporating leaf surface (mmHg), e_a = water vapour pressure in the air (mmHg), r_l = resistance to water vapour transfer in the leaf (s cm^{-1}) and r_a = resistance to water vapour transfer in the surrounding air layer (s cm^{-1}). The total resistance ($r_l + r_a$) is calculated by measuring T and e_a, and e_l determined from leaf temperature assuming water vapour pressure saturation. In an evaporation model under similar experimental conditions r_a can be determined, so that r_l may be found by subtracting r_a from the total resistance ($r_l + r_a$). The primary pathway of transpiration is through the stomata with a parallel subsidiary pathway via the cuticle. Under certain conditions the difference in resistance between the cuticular and stomatal pathway can be determined, so that r_s (stomatal resistance) is known as well.

The relative magnitudes of the various diffusion resistances are determined by various factors. r_a decreases appreciably with wind speed. The cuticular resistance varies with plant species (relatively low in shade plants, high in xerophytes), but is much higher than r_a. The resistance in the stomata is lower than in the cuticle; the minimum value of r_s (at fully open stomata) ranges between 0·5 and 4·0 s cm^{-1}.

Transpiration increases with light intensity, because of both increased stomatal opening and rise in leaf temperature (increase of e_l in Eq. 1). The increase in stomatal opening is due to the reduction in CO_2 concentration by photosynthesis (Slatyer, 1967) or to the supply of ATP via the metabolism of glycollic acid (Zelitch, 1969). Transpiration increases with lower, but decreases with higher ambient CO_2 concentration (e.g. Gaastra, 1959) because of the effect of CO_2 concentration in the intercellular spaces on stomatal opening, even in darkness (Meidner and Mansfield, 1968). Stomatal closure is the main cause of the transpiration decline as water stress in plants develops, but for the different mechanisms which may be involved see Slatyer (1967), Meidner and Mansfield (1968) and Hsiao (1973). Air humidity affects transpiration rate through its effect on e_a (Eq. 1). Increase in wind speed lowers r_a (Eq. 1), so that transpiration is enhanced, the degree of stomatal control of transpiration being higher than in still air. Apart from its effect on e_l (Eq. 1) transpiration also increases with temperature because of the effect on stomatal opening (Zelitch, 1971).

III. The Exchange of Carbon Dioxide

A. PHOTOSYNTHESIS

The various aspects of photosynthesis can be briefly summarized by the following simplified equations (see also Good and Izawa, 1973):

$$2H_2O + 2NADP^+ + nADP + nPi \xrightarrow[\text{energy}]{\text{light}} O_2 + 2NADPH + 2H^+ + nATP \qquad (2)$$

$$CO_2 + 2NADPH + 2H^+ + nATP \longrightarrow CH_2O + 2NADP^+ + H_2O + nADP + nPi \qquad (3)$$

$$\overline{CO_2 + H_2O \xrightarrow[\text{energy}]{\text{light}} CH_2O + O_2} \qquad + \quad (4)$$

Eq. (2) represents the photochemical processes in which radiation is absorbed by the photosynthetic apparatus of the chloroplasts in two light reactions, and converted into chemical energy (NADPH and ATP), which is used for biosynthesis in the dark reactions of Eq. (3). For details see Chapter 16.

In the biochemical reactions of Eq. (3) carbon dioxide is reduced to carbohydrate. There are two major pathways for CO_2 fixation. In the Calvin-Benson pathway carbon dioxide is bound to ribulose diphosphate from which two molecules of phosphoglyceric acid are produced (hence called C$_3$ plants). After reduction this yields the primary carbohydrate. The second pathway has been discovered in an increasing number of tropical grasses and some other plants (see e.g. Downton, 1971). Here, in the mesophyll cells CO_2 is bound to phosphoenolpyruvate yielding C_4-dicarboxylic acids (hence called C_4 plants), which are probably transported to the bundle sheath cells, and there incorporated into the Benson-Calvin pathway (see e.g. Black, 1973). Other differences between C_3 and C_4 plants will be discussed in later sections.

In intact plants one of the components of the overall reaction equation (Eq. 4) may be measured. A non-destructive method in higher plants usually involves the measurement of CO_2 uptake of plants or leaves enclosed in a chamber (Šesták et al. 1971). This uptake is a diffusion process. The binding of carbon dioxide in the biochemical processes (Eq. 3) creates a concentration gradient between the external CO_2 pool in the atmosphere of terrestrial plants (usually somewhat above 0·03%) and that in the chloroplasts which is considered to be close to zero.

The flux of carbon dioxide from the atmosphere to the chloroplasts is usually represented by the following formula, but various refinements have been made as well (see e.g. Šesták et al. 1971):

$$P = \frac{C_a - C_{chl}}{r_a + r_s + r_m} \tag{5}$$

in which P is photosynthesis (g cm^{-2} s^{-1}), C_a and C_{chl} are the CO_2 concentrations in the air and chloroplasts, respectively (g cm^{-3}), and r_a, r_s and r_m are the resistances to CO_2 diffusion in the boundary layer, the stomata and the mesophyll, respectively (s cm^{-1}). The total resistance ($r_a + r_s + r_m$) is calculated from P and C_a (C_{chl} is supposed to be zero), while its separate components may be computed when transpiration is measured simultaneously (Eq. 1). However, when calculated in this way, r_m is strictly not a diffusion resistance of the mesophyll, but includes also components of the photosynthetic process itself, so that the term residual resistance should be preferred (see Gifford and Musgrave, 1973). In general r_a is small, while r_m is usually higher than r_s in C_3 plants with open stomata, but r_s may be highest in C_4 plants (e.g. McPherson and Slatyer, 1973). It is not possible to go into the complicated determination of various resistances to CO_2 transfer caused by photorespiration (see e.g. Ludlow and Jarvis, 1971; Jarvis, 1971).

B. DARK RESPIRATION AND PHOTORESPIRATION

Dark respiration of intact plants is usually measured by oxygen uptake or carbon dioxide evolution. The respiratory quotient (CO_2 produced/O_2 taken up) is close to unity for most leaves. It has been assumed for a long time that this respiration continues in the light at the same rate, so that photosynthesis is diminished by a constant amount. However, there are indications that dark respiration is inhibited by light and replaced by the light-dependent photorespiration (Ludlow and Jarvis, 1971; Zelitch, 1971). Its presence is shown by the post-illumination outburst of CO_2, the carbon dioxide evolution in the light in CO_2 free air, and the increase of net photosynthesis at low oxygen concentrations; all methods probably underestimate the rate of photorespiration (Goldsworthy, 1970) which has been estimated to be at least 30 to 50% of the rate of net CO_2 uptake of leaves in the light (Black, 1973).

Photorespiration is present in C_3 plants, but difficult to detect in C_4 plants (Black, 1973). This difference is correlated with higher values of the CO_2 compensation point in C_3 plants (Jackson and Volk, 1970; Black, 1973). C_4 plants utilize water more efficiently, but the relationship to photorespiration is not yet clear (see also Section C).

The mechanism of photorespiration is not really known, but the hypothesis (Zelitch, 1971) is favoured that CO_2 arises from the oxidation of glycollic acid in the peroxisomes (Tolbert, 1971):

$$CH_2OH.COOH + O_2 \rightarrow HCOOH + H_2O + CO_2 \qquad (6)$$

It is thought that glycollic acid is an early product of photosynthesis, which would explain the linkage of photorespiration to photosynthesis. Both synthesis and oxidation of glycollic acid are stimulated by oxygen.

C. ENVIRONMENTAL FACTORS IN CO_2 EXCHANGE

Light intensity increases net photosynthesis in C_3 plants until saturation is attained. This light saturation is not observed in C_4 plants in which the maximum rate is considerably higher. The difference is ascribed to photorespiration in C_3 plants, and it has been suggested that in C_4 plants the bundle sheath cells in the mesophyll enable the refixation of photorespired CO_2 (Volk and Jackson, 1972; Black, 1973).

Temperature increases dark respiration, but photorespiration is enhanced also (Zelitch, 1971), while stomatal opening may also be increased (see Section II, C). Net photosynthesis of C_4 plants increases with temperature up to about $35°C$, but that of C_3 plants is rather indifferent to temperature, although it increased at higher CO_2 concentration (Gaastra, 1959). Zelitch (1971) thinks that this difference in response to temperature is due to photorespiration in C_3 plants.

Increase in the ambient CO_2 concentration decreases stomatal opening (see Section II, C), but the greater stomatal resistance is over-compensated by the larger concentration gradient of carbon dioxide (Eq. 5), so that net photosynthesis of some C_3 plants increased with CO_2 concentration up to about $1,000$ parts/10^6 at light saturation (Gaastra, 1959).

Water stress may decrease net photosynthesis by its effect on stomatal opening, but effects not connected with stomata may be involved as well (Hsiao, 1973). There are indications that stomata of C_4 plants are more responsive to water stress and other environmental conditions than those of C_3 plants which may be a factor in their greater efficiency of water use (Akita and Moss, 1972, 1973; Black, 1973). Water stress may also decrease dark respiration (Hsiao, 1973).

IV. Herbicides and Carbon Dioxide Exchange of Plants

A. EFFECTS ON DARK RESPIRATION

Interference of herbicides with mitochondrial function by the uncoupling of oxidative phosphorylation or otherwise (Chapter 15) could result in an effect on dark respiration of intact plants. The picture arising from various reports is rather conflicting as is illustrated in the examples given in Tables I and II.

Table I indicates the influence on intact plants or plant tissues of some herbicides which have been reported to inhibit oxidative phosphorylation (e.g. Lotlikar *et al.* 1968). The stimulation with dinoseb at lower concentrations corresponded to that reported for 2,4-dinitrophenol in intact plants or tissues (e.g. Kandler, 1958; Poskuta *et al.* 1967), but at higher concentration dark respiration decreased. With other herbicides either stimulation, no effect, or inhibition has been observed. Reviewing the chlorophenoxy herbicides Penner and Ashton (1966) listed various reports on the stimulation of respiration in plant tissues, but inhibition was reported in a number of other cases. There are indications that lower concentrations stimulate and higher ones inhibit dark respiration, a phenomenon not uncommon for uncouplers. However, Osborne and Allaway (1961) explained the stimulation of respiration by 2,4-D by movement of sugars into treated areas.

TABLE I

EFFECT OF SOME HERBICIDES ON DARK RESPIRATION (CO_2 EVOLUTION OR O_2 UPTAKE) OF INTACT PLANTS OR PLANT TISSUES

Herbicide	Plants or Tissues	Effect[1]	References
Dinoseb	Leaf discs	+, −	Wojtaszek *et al.* (1966)
MCPA	Seedlings	+, −	Mashtakov (1968); Chodová and Zemánek (1971)
2,4-D	Seedlings	+	Chesalin *et al.* (1968), Mashtakov (1968)
	Tomato leaves	O	Krihning (1965*b*)
Chlorpropham	Leaves	+	Eshel and Warren (1967)
	Pea seedlings	−	Wassink and van Elk (1961)
EPTC	Embryos	+	Ashton (1963)

[1] Stimulation (+), no effect (O), or inhibition (−).

The complexity is also demonstrated by the results with carbamates (Table I). Post-emergence sprays with chlorpropham stimulated leaf respiration, but respiration of cotton roots and germinating pea seedlings was decreased by the same herbicide. According to Wassink and van Elk (1961) this inhibition is only apparent, because chlorpropham inhibited the growth of seedling tissue, of which the respiration rate was higher than that of the cotyledons. Ashton (1963) arrived at a similar conclusion for EPTC. This herbicide stimulated the respiration of excised embryos when calculated on a fresh weight basis, but had no effect when expressed on an embryo basis, because it prevented normal increase in fresh weight. Stimulation of dark respiration has been reported for nitrofen (Pereira *et al.* 1971), and inhibition for some 2,6-dinitroanilines (Moreland *et al.* 1972).

Results with some photosynthesis-inhibiting herbicides in Table II indicate stimulation of respiration in a number of cases, especially in aquatic plants.

TABLE II

EFFECT OF SOME PHOTOSYNTHESIS-INHIBITING HERBICIDES ON DARK RESPIRATION (CO_2 EVOLUTION OR O_2 UPTAKE) OF INTACT PLANTS OR PLANT TISSUES

Herbicide	Plants or Tissues	Effect[1]	References
Diuron	Algae	+, O	Geoghean (1957); Orr *et al.* (1971)
	Intact plants	−	Nasyrova *et al.* (1968)
Linuron	Excised leaves	+	Kuratle (1968)
	Intact plants	−	Olech (1967*b*)
Simazine	*Elodea*	+	Roth (1958)
	Leaves	O	Krihning (1965*b*), Tieszen (1970)
	Leaves	−	Chodová and Zemánek (1971)
Pyrazone	Leaf discs	O	Frank and Switzer (1969)
	Leaves	−	Olech (1967*c*); Romanovskaya *et al.* (1968)
Propanil	Leaf tissue	−	Hofstra and Switzer (1968)
Diquat	Algae	+, −	van Rensen (1971)

[1] Stimulation (+), no effect (O), or inhibition (−).

Substituted ureas and triazines have also been reported to affect oxidative phosphorylation. No effect on dark respiration was reported in some others, but most authors observed inhibition. As pointed out by Olech (1967*b, c*) inhibition of dark respiration could result from inhibition of photosynthesis, especially in experiments of longer duration because of a gradually growing deficit in assimilates. In those cases where photosynthesis also was measured, it was much more depressed than dark respiration. The inhibition of dark respiration by propanil is probably of secondary nature, since this effect was much smaller than, and not as direct as the effect on photosynthesis (Hofstra and Switzer, 1968). The reported stimulation of dark respiration by diquat was followed by inhibition (van Rensen, 1971).

B. INFLUENCE OF INHIBITORS OF LIGHT REACTIONS ON
NET PHOTOSYNTHESIS

Current information on inhibitors of photosynthetic reactions indicates that various herbicides interfere with the light reactions of this process, but until now there is no evidence for a direct influence on dark reactions. However, in consequence of the inhibition of light reactions the reduction of carbon dioxide by ATP and NADPH in the dark reactions is suppressed, so that the internal CO_2 concentration is raised and the influx of this gas into the leaf is diminished

(Eq. 5). The interference with photosynthetic reactions is discussed in Chapter 16, so that it may suffice here to indicate their activity briefly.

The following groups have been reported to act in Photosystem II, or, for some of them, more precisely at its reducing side e.g. ureas, triazines, uracils, pyridazinones, benzimidazoles, triazinones, acylanilides, hydroxybenzonitriles, (bis)carbamates, dinoseb, thiazoles, pyriclor, imidazopyridines, alkylhydroxy-quinoline oxides, cycloaliphatic carboxamides and N-aryl-lactams. Some of these have also been reported to inhibit photophosphorylation. Bipyridylium herbicides and quinones act at the reducing side of Photosystem I. Uncoupling or inhibition of photophosphorylation has for example been reported for alkyldinitrophenols, phenylhydrazones, ethyloxazolidine-2,4-diones, 2,6-dinitroanilines, bromofenoxim and diphenylethers.

Büchel (1972) and Good and Izawa (1973) also discuss the various aspects of photosynthetic inhibition. There are a number of reports on some of these herbicides reducing the incorporation of CO_2 and the formation of carbohydrates, but for this aspect the discussion on carbohydrate metabolism in Chapter 7 should be studied.

In most studies on the influence of herbicides on photosynthesis only net CO_2 uptake or O_2 evolution is determined without accounting for the losses caused by photorespiration (see Section III, B). Measurement of the rate of the latter process is much more complicated, but the available information will be discussed in the next section (see Section IV, C). In intact higher plants net photosynthesis is usually determined by continuously measuring CO_2 uptake with an infra-red gas analyzer, momentary measurements being made by uptake of $^{14}CO_2$ over a short interval. Net photosynthesis of aquatic plants is mostly determined by measuring O_2 production by the Warburg technique or otherwise. The Warburg technique has also been used for studying effects of herbicides on photosynthesis of leaf tissues of terrestrial plants, but the gas exchange of such tissues could be a limiting factor in the interpretation of the results. When this method is used in studying the direct effect on photosynthesis by immersing leaf tissues in, or floating them on herbicide solution, some effects may be detected, but this does not take into account the various factors which may determine herbicide accumulation in the leaves under natural conditions.

There are numerous reports on strong inhibitions sometimes amounting to complete suppression of net photosynthesis in intact plants and isolated plant tissues by herbicides applied at low concentrations. In contrast to reports on dark respiration, stimulation has not been observed. A detailed discussion is not feasible because of the mass of data; a number of results from various groups are therefore collected in Tables III to VIII, but even then completeness is impossible. These tables include results with aquatic plants, with isolated plant tissues treated directly with herbicide solution, and with intact plants of which either the leaves or the roots were treated with herbicides.

Both leaf and root application of most ureas (Table III) decreased net photosynthesis. Various authors observed greater inhibition at higher concentration, and van Rensen *et al.* (1972) recorded a logarithmic relation

TABLE III

DECREASE IN NET PHOTOSYNTHESIS (CO_2 UPTAKE OR O_2 EVOLUTION) IN INTACT PLANTS OR PLANT TISSUES BY SOME SUBSTITUTED UREAS

Herbicide	Plants, Application[1] and References
Monuron	A: Gingras (1966); Nazarov and Kutyurin (1969); Funderburk and Lawrence (1964); S/L: Sasaki and Kozlowski (1967); N: Ashton et al. (1961); van Oorschot (1965)
Diuron	A: van Rensen (1971); Zweig et al. (1968); Orr et al. (1971); T: Eshel (1969b); Poskuta et al. (1967); Lüttge et al. (1971); S: Sedgley and Boersma (1969); N: van Oorschot (1964, 1968b, 1970)
Linuron	L: Olech (1967b); Hogue and Warren (1968); N: van Oorschot (1964)
Fluometuron	T: Eshel (1969b); Goren (1969); Rubin and Eshel (1971); N: van Oorschot (1964, 1968b)

[1] A = to aquatic plants in solution; T = to leaf tissues in solution; to soil (= S), nutrient solution (= N) or leaves (= L) of intact terrestrial plants.

between the concentration of diuron (and simeton) in the suspension and the inhibition of photosynthesis in *Scenedesmus*. Unpublished results of the present writer suggest that the relation between concentration in the nutrient solution and photosynthetic inhibition of intact bean (*Phaseolus vulgaris*) plants is more complex. Environmental factors also appear to affect the degree of inhibition. In short experiments with unicellular algae inhibition decreased with light intensity, but in prolonged experiments with higher plants the opposite effect was observed (see Section E). For a discussion on the environmental factors increasing the transpiration rate of plants and resulting in a greater inhibition of photosynthesis by root-applied herbicides the reader

TABLE IV

DECREASE IN NET PHOTOSYNTHESIS (CO_2 UPTAKE OR O_2 EVOLUTION) IN INTACT PLANTS OR PLANT TISSUES BY SOME TRIAZINES

Herbicide	Plants, Application[1] and References
Simazine	A: Roth (1958); Sutton et al. (1969); T: Krihning (1965b); N: Ashton et al. (1960); van Oorschot (1968b); Tieszen (1970)
Atrazine	A: Funderburk and Lawrence (1964); Bolhár-Nordenkampf (1970); T: Shimabukuro and Swanson (1969); Bolhár-Nordenkampf (1970); S/L: Olech (1967a); Sasaki and Kozlowski (1967); N: Couch and Davis (1966); Imbamba (1970); Pochinok (1971)
Simeton	A: van Rensen (1971); N/L: Ashton et al. (1960); van Oorschot (1966, 1970)
Terbutryne	N: van Oorschot (1968b); S: Figuerola and Furtick (1972)

[1] For meaning of symbols see Table III.

is referred to Section V, C. When linuron was applied to the underside of the leaves Olech (1967*b*) found a more rapid inhibition, as with leaf application at higher temperature and higher air humidity.

Olech (1967*a*) obtained similar results with leaf-applied atrazine (Table IV). Photosynthesis of rye growing at a high temperature was more decreased by root-applied simazine than that of plants growing at lower temperatures (Tieszen, 1970). In addition to the triazines listed in this table, root-applied metribuzin also decreased net photosynthesis (unpublished results).

TABLE V

DECREASE IN NET PHOTOSYNTHESIS (CO$_2$ UPTAKE OR O$_2$ EVOLUTION) IN INTACT PLANTS OR PLANT TISSUES BY SOME DIAZINES

Herbicide	Plants, Application[1] and References
Bromacil	L: Shriver and Bingham (1973); N: Couch and Davis (1966); van Oorschot (1968*b*)
Lenacil	N: van Oorschot (1965, 1968*b*)
Terbacil	L: Barrentine and Warren (1970); N: van Oorschot (1968*b*)
Pyrazone	T: Frank and Switzer (1969); S/L: Olech (1967*c*); N: van Oorschot (1965, 1968*b*); Romanovskaya et al. (1968); Hilton et al. (1969)
Metflurazone	N: Hilton et al. (1969)

[1] For meaning of symbols see Table III.

The range and activity observed with diazines (Table V) are similar to those of ureas and triazines. Acylanilides (Table VI) appear to act primarily via the leaves of various plants. Propanil especially caused a rapid inhibition of net photosynthesis (Hofstra and Switzer, 1968). Unpublished experiments have shown that a number of other acylamides (diphenamid, propachlor and pronamide) did not affect net photosynthesis.

TABLE VI

DECREASE IN NET PHOTOSYNTHESIS (CO$_2$ UPTAKE OR O$_2$ EVOLUTION) IN INTACT PLANTS BY SOME ACYLAMIDES

Herbicide	Plants, Application[1] and References
Propanil	L: Nakamura et al. (1968); Matsunaka (1969); Ivanova (1970); Hofstra and Switzer (1968)
Pentanochlor	L: Colby (1964); Nakamura and Matsunaka (1969)
Cypromid	L: Nakamura and Matsunaka (1969)

[1] For meaning of symbols see Table III.

Although under field conditions bipyridylium herbicides are active only when applied to leaves, they also appear to inhibit net photosynthesis when given to the roots (Table VII). This applies also to paraquat (unpublished results). Under these conditions the activity of diquat was similar to that of

TABLE VII
DECREASE IN NET PHOTOSYNTHESIS (CO_2 UPTAKE OR O_2 EVOLUTION) IN INTACT PLANTS BY SOME
BIPYRIDYLIUMS

Herbicide	Plants, Application[1] and References
Diquat	A: van Rensen (1969); Zweig et al. (1968); Funderburk and Lawrence (1964); L/N: van Oorschot (1966); Couch and Davis (1966)
Paraquat	A: Funderburk and Lawrence (1964)

[1] For meaning of symbols see Table III.

simeton (van Oorschot, 1966), but in a suspension of *Scenedesmus* diquat was less active (van Rensen, 1971). Simultaneous supply of diquat and a Hill reaction inhibitor resulted in additive inhibition of net photosynthesis of *Lemna* (Funderburk and Lawrence, 1964), *Phaseolus* (van Oorschot, 1966) and *Scenedesmus* (van Rensen, 1971), but prevented the development of symptoms in *Phaseolus* to a large extent.

TABLE VIII
DECREASE IN NET PHOTOSYNTHESIS (CO_2 UPTAKE OR O_2 EVOLUTION) IN INTACT PLANTS OR PLANT
TISSUES BY VARIOUS HERBICIDES

Herbicide	Plants, Application[1] and References
Phenmedipham	L: Arndt and Kötter (1968); van Oorschot (1968a)
Subst. phenols	A: Huang and Gloyna (1968); Desmoras and Jacquet (1964); T: Poskuta et al. (1967); N/L: van Oorschot (1974a)
Ioxynil	A: Desmoras and Jacquet (1964); N/L: van Oorschot (1974a)
2,6-Dinitroanilines	T: Moreland et al. (1972)
Phenylhydrazones	T: Lüttge et al. (1971); Jones and Osmond (1973)
Quinones	A: Zweig et al. (1968)
Thiazoles	L: Matsunaka and Nakamura (1972)
Petroleum oils	L: Helson and Minshall (1962)

[1] For meaning of symbols see Table III.

Table VIII lists a variety of herbicides and other compounds which have been reported to inhibit net photosynthesis. Unpublished experiments by the writer indicated inhibition also by the following herbicides applied at low concentrations: azolamid, bentazon, bromofenoxim, chlorflurazole, flumezin, karbutilate, methazole, nitrofen and oxadiazon.

In addition to the herbicides mentioned in Table III to VIII, others (e.g. carbamates and phenoxy herbicides) have been reported to decrease net photosynthesis of plants, but higher concentrations are needed. In experiments of longer duration leaf injury or other effects on the plant could indirectly cause a decrease in photosynthesis.

C. INHIBITION OF NET PHOTOSYNTHESIS AND PHOTORESPIRATION

Net photosynthesis is diminished to a considerable extent because of losses by photorespiration, especially in C_3 plants (see Section III, B). We may wonder whether the various herbicides reported in the previous section to inhibit net photosynthesis also affect photorespiration. Results of extended experiments may provide some answers to this question. For example, continuous exposure of the roots of some C_3 plants to linuron, simeton, diquat and isocil ultimately caused a rate of CO_2 evolution in the light which approached that of dark respiration (van Oorschot, 1964, 1966, 1974a). Such results suggest that these herbicides also inhibit photorespiration. However, after application of atrazine to some plants in the light, Bolhár-Nordenkampf (1970) observed a 25—70% higher CO_2 evolution than in darkness, which could indicate that photorespiration is inhibited less than photosynthesis. He and Lee et al. (1971) suggested that the tolerance of maize and other C_4 plants to atrazine might be associated with the different pathway of CO_2 incorporation. According to Imbamba (1970), however, the C_4 plant *Amaranthus hybridus* was equally sensitive to atrazine.

Direct measurements of the influence of herbicides on photorespiration are scanty. Downton and Tregunna (1968) studied photosynthesis and photorespiration of wheat and maize at normal and low oxygen concentrations in the ambient air (see Section III, B) as affected by diuron. They concluded

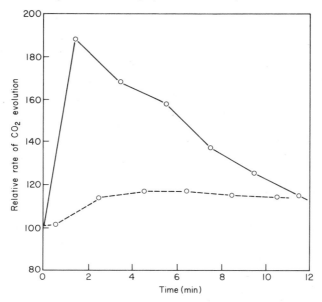

Fig. 1. Effect of atrazine (1.9×10^{-4} M, ——) on the post-illumination evolution of carbon dioxide from barley leaves (——=untreated). Steady state dark respiration = 100; leaves placed in darkness at time zero. (Reproduced with permission from Imbamba and Moss, 1971.)

that both photosynthesis and photorespiration were inhibited by diuron. Similar results were obtained with *Nitella flexilis,* a green alga, treated with diuron. Imbamba (1970) and Imbamba and Moss (1971) reported that atrazine suppressed photorespiration of barley leaves, measured as absence of post-illumination CO_2 outburst. This is illustrated in Fig. 1. It had no effect on the oxidation of glycollic acid in leaves of barley and *Amaranthus hybridus,* but prevented its synthesis in barley leaves. Gibbs (1971) found an inhibition of glycollic-acid formation by diuron in spinach chloroplasts. In all probability the present information suggests that herbicides which inhibit light reactions also suppress photorespiration.

The loss in potential productivity by photorespiration of C_3 plants induced a search for chemicals suppressing photorespiration (e.g. Corbett and Wright, 1971) in addition to other approaches. Zelitch (1973a, b) found an inhibition of glycollic acid synthesis in tobacco leaf discs by isonicotinic acid hydrazide, and observed an inhibition of its oxidation by α-hydroxy-2-pyridinemethanesulphonate (α-HPMS). The latter decreased photorespiration in short-term experiments, while net CO_2 uptake increased at 35°C but not at 25°C. The effects were not observed with maize. However, the interpretation of these results is difficult, since hydroxysulphonates have been reported also to inhibit photosynthetic CO_2 fixation in spinach chloroplasts and leaf segments (Murray and Bradbeer, 1971). Moreover, they affect stomatal aperture (see Section V, B).

D. INHIBITION OF NET PHOTOSYNTHESIS BY STOMATAL CLOSURE

Although the inhibition of net photosynthesis by a large variety of herbicides is usually attributed to their direct interference with light reactions (see Section B), net photosynthesis might also be affected by a change in stomatal aperture. Stomata constitute a resistance to CO_2 diffusion into the leaves (Eq. 5). Increase in CO_2 concentration within the leaves because of an inhibition of photosynthetic reactions may induce stomatal closure (see Section II, B). Therefore, the final effect of such herbicides might be due partly to stomatal closure. By measuring inhibition of net photosynthesis and transpiration by atrazine Imbamba and Moss (1971) determined the resistances to CO_2 diffusion (Eq. 5). They concluded that the inhibition of net photosynthesis was completely (barley) or mainly (maize) due to an inhibition of the photosynthetic reactions. The small effect on stomatal closure in maize is in agreement with reports that stomata of C_4 species were more responsive to changes in leaf water content, CO_2 concentration and light (Boyer, 1970; Akita and Moss, 1972, 1973). The fact that the degree of photosynthetic inhibition by various herbicides inhibiting Hill activity is much greater than that of transpiration in several C_3 plants (van Oorschot, 1965, 1970, 1974a) also indicates that stomatal closure is of secondary importance. However, the effect of other compounds might be different (see Section V, B).

There is evidence that plant hormones are affected when water deficits develop in plants. The abscisic acid content is especially important (Livne and

Vaadia, 1972), and stomatal closure has been induced by an exogenous supply of this hormone (e.g. Jones and Mansfield, 1970; Tal and Imber, 1972; Kriedemann *et al.* 1972). It is not surprsing therefore that net photosynthesis is inhibited by external application of abscisic acid to leaves or roots (Mittelheuser and van Steveninck, 1971; Poskuta *et al.* 1972). This photosynthetic inhibition was probably due to stomatal closure, because transpiration was more sensitive than photosynthesis and abscisic acid increased the stomatal diffusion resistance considerably (Table IX). On the contrary, cytokinins and gibberellic acid may increase stomatal opening in several plants (Livne and Vaadia, 1972) and stimulate net photosynthesis (Poskuta *et al.* 1972; Lester *et al.* 1972). High concentrations of auxins and of the related phenoxy herbicides cause stomatal closure and reduction in transpiration rate, which could also explain their reported inhibition of net photosynthesis.

TABLE IX

EFFECT OF 3.8×10^{-6} M ABSCISIC ACID ON THE RATE OF PHOTOSYNTHESIS AND STOMATAL RESISTANCE $(r_s,$ Eq. 1), OF DETACHED PRIMARY WHEAT LEAVES[1]. (REPRODUCED WITH PERMISSION FROM MITTELHEUSER AND VAN STEVENINCK, 1971).

Time (min)	Photosynthesis $(mg\ CO_2\ dm^{-2}\ h^{-1})$		Stomatal Resistance $(s\ cm^{-1})$	
	Control	Abscisic acid	Control	Abscisic acid
30	10.8 ± 0.3[2]	3.4 ± 0.6	2.8	15.7
40	7.0 ± 0.7	2.2 ± 0.1	3.5	34.5

[1] Different leaves used for both measurements.
[2] Standard error of the mean.

In dry climates considerable attention has been paid to the development of antitranspirants, and the effects of these products on net photosynthesis will be briefly discussed. Phenylmercuric acid (PMA) has been most intensively investigated. Earlier reports indicated that net photosynthesis was inhibited to a less extent than transpiration, but later studies (e.g. Bravdo, 1972; Sij *et al.* 1972) showed an equal or even greater inhibition of net photosynthesis. This might be related to the interference of PMA with chloroplast reactions (e.g. Siegenthaler and Packer, 1965). The same could apply to alkenylsuccinic acids which also inhibit transpiration and net photosynthesis (van Oorschot, 1974a). However, the relative effect of these compounds on net photosynthesis and transpiration is clearly different from that of photosynthesis-inhibiting herbicides (see Fig. 4 and Section V, B).

E. THE HERBICIDE EFFECT OF PHOTOSYNTHESIS INHIBITION

An obvious result of photosynthesis inhibition is the lack of carbohydrate formation leading to starvation of treated plants; but will this fully account for the herbicidal action of compounds inhibiting photosynthesis? There are, indeed, a number of reports indicating that exogenous supply of carbohydrates to treated plants delayed the toxic effects of some ureas (Gentner and Hilton, 1960; Davis, 1966), triazines (Moreland et al. 1959; Allen and Palmer, 1963; Hilton et al. 1969) and diazines (Hilton et al. 1964, 1969; Eshel, 1969a). Of course, the substitution of carbohydrates normally synthesized in older leaves by exogenous supply of carbohydrates to treated plants will decrease toxicity, but it could not prevent damage when the leaves matured (Hilton et al. 1964; Davis, 1966). Compared to untreated plants starving in darkness, herbicide-treated plants die much more rapidly in the light.

That other factors also are involved in the herbicidal action of these herbicides is indicated by the disintegration of chloroplasts in treated plants. In the light thylakoids swell, and then the lamellar system disorganizes, and rupture of the tonoplast and the chloroplast envelope follows. Such effects have been observed with atrazine (Ashton et al. 1963; Hill et al. 1968), bromacil (Ashton et al. 1969), pyrazone (Anderson and Schaelling, 1970), propanil (Hofstra and Switzer, 1968) and pyriclor (Geronimo and Herr, 1970). Light appeared to be necessary, for chloroplasts of atrazine-treated plants in darkness resembled those of untreated dark controls. Details of the effect of herbicides on ultrastructure of cells are to be found in Chapter 3 and in a survey of Anderson and Thomson (1973).

Another indication is the effect of light intensity. Various authors have observed that toxicity increased with light intensity, but in some studies this light effect could have been due at least partly to enhanced uptake and accumulation of the herbicide in the leaves during treatment (see Section V, C). However, other studies indicated that higher light intensity after treatment increased the toxicity of monuron, diuron, simeton, pentanochlor, ioxynil and pyrazone (Minshall, 1957; Colby and Warren, 1962; Carpenter et al. 1964; Romanovskaya and Straume, 1970; van Oorschot and van Leeuwen, 1974). Higher post-treatment illumination also resulted in greater reduction of net photosynthesis as indicated in Fig. 2. This is in contrast with the effect of light intensity on net photosynthesis in short experiments with unicellular algae, where greater inhibition was found at low light intensities (Gingras, 1966; van Rensen, 1971). As discussed below, secondary effects will be induced by longer exposure. The similarity of the damage induced by exposure to simeton and to CO_2 free air in the light (van Oorschot, 1974b) suggests that the photosynthetic apparatus is damaged when photosynthesis of illuminated plants is prevented.

Ashton et al. (1963) and Davis (1966) suggested that the acute toxicity is not due to starvation, but is a secondary result of photosynthesis inhibition. Both in isolated chloroplasts (Stanger and Appleby, 1972), in unicellular algae (Paromenskaya and Lyalin, 1968; Huang and Gloyna, 1968; Saakov, 1973a, b), and in leaves of higher plants (Zaki et al. 1967; Saakov, 1973a, b) the chlorophyll and carotene content decreased in the light when exposed to various photosynthesis inhibitors (including triazines, ureas, phenylhydrazones and substituted phenols). With diuron Stanger and Appleby (1972) found greater sensitivity of carotenoid pigments, which are supposed to protect

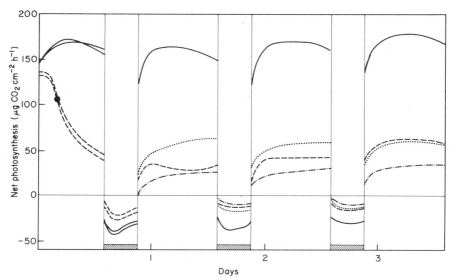

Fig. 2. Effect of post-treatment illumination (17 h/day) on net photosynthesis (μg CO_2 cm^{-2} h^{-1}) of *Phaseolus*. Root treatment with simeton from time zero to ●. All plants were exposed to 63 J m^{-2} s^{-1} radiation for 14 h; thereafter to low (26 J m^{-2} s^{-1}, · · · · · untreated, —·— simeton) or high light intensity (106 J m^{-2} s^{-1}, — untreated, –– simeton); periods of darkness are indicated by hatching. (Reproduced with permission from van Oorschot and van Leeuwen, 1974.)

chloroplasts against photo-oxidation (Krinsky, 1966). They proposed that the phytotoxicity of diuron is due to photo-oxidation. In non-herbicidal, photo-oxidative bleaching of *Cucumis* leaf discs at 1°C carotene was also the most sensitive pigment (van Hasselt, 1972).

Photo-oxidation as a result of inhibition of photosynthesis by herbicides or conditioned otherwise during illumination may explain the increased damage at higher light intensity and the greater photosynthetic inhibition shown in Fig. 2. Chloroplast degradation is probably related to photo-oxidation, since chloroplast swelling has been reported for some thermophylic plants under non-herbicidal conditions of photo-oxidation, viz. high light intensity and low temperature (Taylor and Craig, 1971).

Although bipyridylium herbicides interfere differently in photosynthesis (Chapter 16), their ultimate effect on plants is rather similar, but appears more rapidly. Toxicity of diquat increased with the post-treatment level of illumination (Mees, 1960), while bleaching of *Chlorella* and damage to plastids was observed (Stokes *et al.* 1970). Similar effects were found with flax cotyledons in paraquat solution (Harris and Dodge, 1972). The requirement of oxygen for the toxic action of bipyridylium herbicides is well documented (e.g. Mees, 1960; Merkle *et al.* 1965; van Rensen, 1969), and the formation of toxic peroxide radicals are held responsible for the rapid effect because of disruption of cell membranes (see Chapters 9 and 16).

F. HERBICIDE SELECTIVITY AND INHIBITION
OF NET PHOTOSYNTHESIS

Differences in inhibition of net photosynthesis between tolerant and susceptible plants could reflect an inherent tolerance to a particular herbicide. In such studies the use of intact plants is preferred because of the various factors involved in the accumulation of herbicides within the leaves, for it has been shown that the Hill activity of isolated chloroplasts of resistant species is as sensitive to simazine as that of susceptible species (Moreland and Hill, 1962), as with linuron (Kuratle, 1968), terbacil (Herholdt, 1968), pyrazone (Frank and Switzer, 1969), pentanochlor (Colby and Warren, 1965; Shirakawa, 1969) and phenmedipham (Arndt and Kötter, 1968).

An inhibition of net photosynthesis by foliage-applied herbicides, lower in tolerant than in susceptible species, has been reported for pyrazone (Olech, 1967c; Romanovskaya et al. 1968), linuron (Hogue and Warren, 1968; Olech, 1968) and phenmedipham (Arndt and Kötter, 1968; van Oorschot, 1968a). This may be due to a lower uptake or a higher detoxicification in tolerant plants. Recovery from inhibition of photosynthesis in tolerant plants observed after foliar application of the same herbicides and of pentanochlor (Colby, 1964), terbacil (Barrentine and Warren, 1970), propanil (Nakamura et al. 1968; Hofstra and Switzer, 1968; Ivanova, 1970) and 5-chloro-4-methyl-2-propionamide-1,3-thiazole (Matsunaka and Nakamura, 1972) could indicate inactivation when recovery is fast enough. However, dilution of the herbicide within the leaves by growth may also play a role, especially when recovery occurs slowly. The recovery from photosynthesis inhibition by phytotoxic petroleum oils in tolerant parsnips was considered to be due to physico-chemical effects (Helson and Minshall, 1962).

Root-applied herbicides have also been reported to induce different

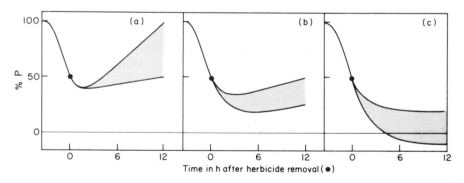

Fig. 3. Herbicide inactivation as indicated by recovery from inhibition of net photosynthesis (P in %). a. High inactivation (e.g. simazine-maize, pyrazone-sugar beet, terbacil-perppermint). b. Weak inactivation (e.g. monolinuron-beans, bromacil-peppermint, metoxuron-wheat). c. Absence of inactivation (e.g. simazine-strawberry, lenacil-sugar beet, bromacil-apple). (Schematical after van Oorschot, 1968b.)

inhibition of net photosynthesis in tolerant and susceptible plants, for example pyrazone (Olech, 1967c; Frank and Switzer, 1969) and linuron (van Oorschot, 1964; Olech, 1968). Kuratle's (1968) analysis showed that the distribution of linuron in tolerant carrots is more confined to the roots. Furthermore differences in transpiration rate between tolerant and susceptible plants could induce differences in photosynthesis inhibition (see Section V, C).

Recovery from inhibition of photosynthesis by root-applied herbicides in short-term experiments was interpreted as an indication of herbicide inactivation in tolerant plants (van Oorschot, 1965; Hilton et al. 1969). In this technique herbicide absorption from nutrient solution was allowed to continue until net photosynthesis had fallen to about 50% of its original, uninhibited rate. From that moment further uptake was prevented by the application of a herbicide-free nutrient solution, while the photosynthetic rate was measured during the subsequent period of 10–15 h. Such short-term experiments were realized by applying relatively high herbicide concentrations in the nutrient solution ($2 \cdot 10^{-5}$ M) under conditions favourable to herbicide accumulation in the leaves (high transpiration rate, Section V, C), so that the 50%-rate was attained within a few hours. During such experiments increase in leaf area and herbicide dilution by growth are of minor importance.

In this way recovery from inhibition of photosynthesis by various herbicides was determined (van Oorschot, 1968b). Figure 3 gives a schematic representation of various curves of relative net photosynthesis obtained over 12 h after herbicide removal, together with some examples in each of the three different classes. High herbicide inactivation (a), recovery of net photosynthesis above that at the moment of removal) evidently permits selective application on the seedbag of these crops. In case of weak herbicide inactivation (b), net photosynthesis after 12 h being higher than the minimum value, but lower than at removal) there will be more restrictions, e.g. soil type, depth of sowing, etc. No tendency to inactivation was found in the third group ((c), net photosynthesis not increased above the minimum value), so that the examples in this group probably represent cases of depth protection under field conditions, although it should be realized that absence of any recovery within 12 h does not exclude inactivation over a longer period.

The observed recovery indicates that inactivation occurs within the leaves. After brief treatment with atrazine Shimabukuro and Swanson (1969) found recovery of oxygen evolution of leaf discs of sorghum, but not of peas, together with degradation of this herbicide in sorghum. This degradation probably occurred outside the chloroplasts. They explained the recovery from photosynthesis inhibition by assuming equilibrium between reversibly-bound atrazine in the chloroplasts and that in the cytoplasm, so that degradation in the cytoplasm would diminish the concentration in the chloroplasts. This gradual "washing-out effect" of chloroplasts in leaves by inactivation outside the chloroplasts is comparable to the reversible inhibition of oxygen evolution in unicellular algae (van Rensen and van Steekelenburg, 1965; Zweig et al. 1968) and that of the Hill reaction in isolated chloroplasts (Moreland and Hill, 1962) by washing.

V. Herbicides in Relation to Water Uptake and Transpiration of Plants

A. INTERFERENCE WITH WATER UPTAKE

Water uptake into intact plants is closely related to transpiration rate (Section II), so that interference of herbicides with water uptake could be the result of changes in transpiration rate and *vice versa*. Therefore, the separate discussion of water uptake in this section, and that of transpiration in the next does not imply that the herbicides under discussion influence one or other process exclusively; it merely reflects the two ways in which the effect on water use of plants was determined.

Although increased water uptake into excised tissues has been observed, most of the reports on intact plants indicate a decrease in water uptake or transpiration by phenoxy herbicides (see survey by Penner and Ashton, 1966). Kozinka (1968) found that MCPA (10^{-3} M in nutrient solution) decreased water uptake of some plants more than the simultaneously determined transpiration rate, but in detached tomato shoots Krihning (1965a) observed hardly any difference between uptake and transpiration when they were treated with 2,4-D (10^{-4} M). Even if measured simultaneously it is difficult to assess which process is primarily affected. Some authors emphasize the effect on water uptake by the roots, but others relate the effect to closure of stomata in the leaves (Livne and Vaadia, 1972).

There are various reports that ureas and triazines reduce water uptake, both when applied to detached shoots in solution (e.g. Minshall, 1960; Sikka *et al.* 1964; Krihning, 1965a) and in the nutrient solution of intact plants (e.g. Smith and Buchholtz, 1964; Wills *et al.* 1963; Lindsay and Hartley, 1963; Shone and Wood, 1972). Pyrazone also reduced water uptake of intact plants (Frank and Switzer, 1969). A decrease in water uptake was also reported for some foliar-applied ureas, triazines and propanil (Smith and Buchholtz, 1964; Nakamura *et al.* 1968; Gill *et al.* 1972). In some cases closure of stomata was observed. This could result from inhibition of photosynthesis by these herbicides, and so the reduced water uptake is probably a secondary effect. However, root-applied dinoseb induced wilting, and Smith and Buchholtz (1964) thought its effect was primarily due to a reduction in water uptake by the roots. The effect of these herbicides on transpiration rate and stomatal opening is further discussed in the next section.

B. INFLUENCE ON TRANSPIRATION RATE AND STOMATAL OPENING

Before discussing herbicides we will briefly deal with the effect of other chemicals on transpiration rate. Some of them could have a direct effect on stomatal opening, for example phenylmercuric acid (PMA) (Zelitch, 1969). However, as indicated earlier (Section IV, D), its selectivity in diminishing transpiration to a greater extent than net photosynthesis is sometimes rather weak, since mesophyll photosynthesis may be inhibited as well (Squire and Jones, 1971). This could also suggest that PMA influences stomatal opening by increasing the internal CO_2 concentration (Mansfield, 1967) through inhibition of photosynthesis (Section II, B). Another chemical with antitranspirant properties is decenylsuccinic acid (DSA), but inhibition of photosynthesis has also been reported (Section IV, D). Stomatal closure was also brought about by α-hydroxy-2-pyridinemethanesulphonate (α-HPMS) (Zelitch, 1965), which interferes with the oxidation of glycollic acid. The latter is supposed to play a role in the opening of stomata (Section II, B) and in photorespiration (Section III, B). Presumably the action of α-HPMS is not very specific, since Moss (1968) found inhibition of CO_2 fixation and closure of stomata in tobacco leaves. The inhibition of stomatal aperture by various other chemicals observed by Mouravieff (1971, 1972) emphasizes the close connection between closure of stomata and inhibition of photosynthesis.

The role of some endogenous plant hormones is different. Gibberellic acid increased stomatal opening, and it has been suggested that the reduced transpiration rate commonly observed with chlormequat is related to its interference with the biosynthesis of gibberellic acid (Livne and Vaadia, 1972). Kinetin and other cytokinins also increased stomatal opening and stimulated transpiration rate in several plants, but not in others. It has been suggested that the effect is of a direct nature (Livne and Vaadia, 1972). Another plant hormone, abscisic acid, induced stomatal closure and reduced transpiration rate. Jones and Mansfield (1970) demonstrated that the effect was not reversed by CO_2-free air, and concluded that it was not the result of increased internal CO_2 concentration. This corresponds to the higher sensitivity of transpiration over that of photosynthesis (see Section IV, D). These results suggest that abscisic acid could be more efficient in saving water than the antitranspirants discussed before (Jones and Mansfield, 1972).

All the reports on the effect of photosynthesis-inhibiting herbicides on transpiration of plants indicate a reduction. The reduced water uptake with the herbicides mentioned in the previous section is probably also due to reduced transpiration. Furthermore, stomatal closure or decrease in transpiration rate have been reported for some ureas and triazines applied to the roots (Kuratle, 1968; Velev, 1969) or to the leaves (Walker and Zelitch, 1963; Dutta et al. 1973). A similar reduction was observed with leaf-applied bromoxynil (Friesen and Dew, 1967), propanil (Ivanova, 1970) and nitrofen (Pereira et al. 1971). Humburg (1970) investigated various herbicides for influence on transpiration rate and concluded that Hill activity inhibitors were most effective in reducing transpiration.

The close connection between transpiration and photosynthesis explains the relation observed between reductions in transpiration rate and dry weight accumulation of plants (Graham and Buchholtz, 1968; Płoszyński et al. 1969). The results of experiments in CO_2-free air (Allaway and Mansfield, 1967; Imbamba and Moss, 1971) suggest that the effects on transpiration rate are secondary. Closure of stomata by monuron was reversed, and that by atrazine did not occur in the absence of CO_2. This indicates that closure of stomata by these herbicides at normal CO_2 concentration is brought about by the accumulation of carbon dioxide due to inhibited photosynthesis. A similar result was obtained with PMA, but stomatal closure by 2,4-D could not be reversed by CO_2-free air (Mansfield, 1967), from which the conclusion was drawn that 2,4-D has a more direct effect on the guard cells.

Simultaneous measurement of transpiration and net photosynthesis may also indicate which process is primarily affected. For example, Fig. 4a shows that root-applied simeton decreased the net photosynthesis of beans (*Phaseolus vulgaris*) earlier and to a greater extent than transpiration, suggesting that the effect on transpiration is secondary. A smaller decrease in transpiration than in net photosynthesis was observed with various plants treated with ureas, triazines, diazines and bipyridylium herbicides (van Oorschot, 1964, 1965, 1966, 1970, 1974a; Pochinok, 1971).

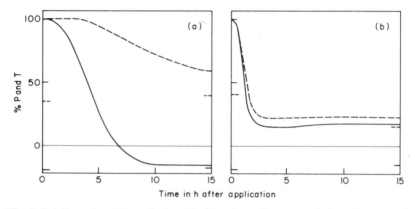

Fig. 4. Relative rates of transpiration (T,--) and net photosynthesis (P,—) of bean plants treated with 10^{-5} M simeton (a) or PMA (b) in the nutrient solution. Horizontal dashes on ordinates represent respiration and transpiration in darkness. (Reproduced with permission from van Oorschot, 1974a.)

A different situation was found with PMA (Fig. 4(b). Here, the effects on net photosynthesis and transpiration of *Phaseolus vulgaris* are rather similar. Such results were also obtained with high osmotic stress, DSA and root-applied ioxynil (van Oorschot, 1974a). However, the effects of leaf-applied ioxynil were comparable to those of simeton (Fig. 4(a). Nakamura *et al.* (1968) also reported an appreciable inhibition of both photosynthesis and transpiration with leaf-applied propanil, while unpublished experiments of the writer indicate a similar sensitivity of both processes when bean leaves were sprayed with nitrofen and fluorodifen. It is questionable whether the correlated inhibition of both processes always indicates that transpiration is primarily affected, for preliminary results also indicated similar effects of simeton and other photosynthesis inhibitors on maize, a C_4-plant of which the stomata are more responsive to changes in CO_2-concentration (see Section IV, D).

The phytotoxicity of all herbicides will ultimately lead to decreased transpiration rate, especially because of the causal relation between chlorophyll content and stomatal aperture (Shimshi, 1967), but a discussion of this is beyond the scope of this chapter. Finally, an opposite effect on transpiration rate should be mentioned. Gentner (1966) observed increased transpiration of cabbage leaves in which wax formation was inhibited by EPTC. Other herbicides interfering with wax formation, for example TCA and dalapon could act similarly.

C. WATER UPTAKE, TRANSPIRATION AND HERBICIDE ABSORPTION

Passively-absorbed herbicides (Chapter 11) are transported within the xylem (Chapter 12), so that factors influencing water uptake and transpiration may affect herbicide accumulation within the leaves. In particular sufficient water availability, high light intensity and temperature, and low air humidity enhance the transpiration rate of plants (Section II). The influence of these (and other) environmental factors on herbicide toxicity is fully reviewed in Volume 2, Chapter 7, but in this section these are only discussed in relation to water uptake and transpiration.

Variation in one or more factors influencing transpiration rate affected toxicity of root-absorbed herbicides. Minshall (1954, 1957) observed lower toxicity of monuron when applied during darkness or under low light. Differential toxicity of herbicides by variation in these factors has been reported in a number of cases in which a direct effect of light after treatment may have played a role, but where increased transpiration rate and herbicide uptake could partly account for the increased toxicity. Higher light intensity and temperature or lower air humidity increased the toxicity due to the inhibition of photosynthesis by atrazine (Ashton, 1965; Olech, 1967a), prometryne (Noda and Ibaraki, 1968), terbutryne (Houseworth and Tweedy, 1971; Figuerola and Furtick, 1972) and pyrazone (Olech, 1967c; Koren and Ashton, 1971). Similar results were found with monuron (Ashton, 1965), diuron (Ladonin, 1972), propazine (Ladonin and Chivikina, 1969) and methabenzthiazuron (Fedtke, 1973).

In other studies herbicide absorption was mostly determined with ^{14}C-labelled herbicides, although it should be realized that accumulation of radioactivity in plant parts does not necessarily correspond to that of the herbicide. Higher temperature during root absorption of the herbicides from nutrient solution increased their accumulation or that of radioactivity within the leaves or the shoots when plants were treated with chloroxuron (Geissbühler et al. 1963), linuron (Penner, 1971), atrazine (Wax and Behrens, 1965; Penner, 1971), pyrazone (Romanovskaya and Straume, 1970; Koren and Ashton, 1973) and propanil (Hodgson, 1971). Similar results were obtained for chloroxuron at high light intensity and low air humidity (Geissbühler et al. 1963) and for atrazine at low air humidity (Wax and Behrens, 1965). Some factors may complicate the results when plants are growing in soil (Roeth and Levy, 1971).

Sedgley and Boersma (1969) determined water uptake, and found an almost linear relation with the decrease in photosynthesis of diuron-treated wheat plants at different soil temperatures. The decrease per unit volume of water absorbed was lower at higher soil water stress. Van Oorschot (1970) measured transpiration rate and net photosynthesis of *Phaseolus* plants as affected by root treatments with some ureas, triazines and uracils under different conditions of transpiration obtained by variation in air humidity, light intensity and temperature. The results indicated that the degree of photosynthesis inhibition is determined by the transpiration rate of the plants. The lower degree of inhibition by diuron and lenacil at equal transpiration rates suggested decreased mobility in the plant. Unpublished experiments by the writer at enhanced CO_2 concentration showed smaller inhibitions of photosynthesis by some of these herbicides (lower transpiration rate by stomatal closure, see Section II, B).

Water uptake has been correlated with herbicide absorption in a number of studies. Sheets (1961) observed that the accumulation of radioactivity in oat and cotton leaves was dependent on water uptake from nutrient solution containing labelled simazine. Similarly, the simazine content of *Pinus silvestris* increased with water uptake at lower air humidity (Uhlig, 1968), and atrazine uptake in soybean increased with water uptake at higher root temperature (Vostral *et al.* 1970). Loss of radioactivity from a solution with labelled

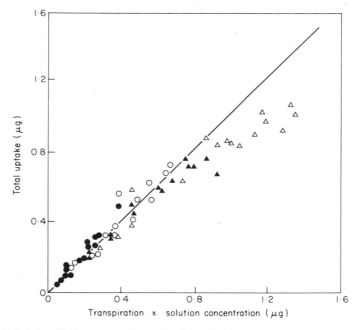

Fig. 5. Relationship between total uptake of atrazine from nutrient solution and supply by mass-flow. Exposure time: 3 days (●), 5 days (○), 7 days (▲) and 10 days (△). (Reproduced with permission from Walker, 1972.)

methazole increased in proportion to water uptake by cotton plants over seven weeks (Jones and Foy, 1972), but Moody *et al.* (1970) observed no clear relation between such losses from solutions with some herbicides and water uptake by soybean plants. Absorption of DNOC from nutrient solution by rye plants was higher than could be accounted for by water uptake, especially at low pH (Bruinsma, 1967).

Shone and Wood (1972) determined the accumulation of simazine within the shoots of barley under different conditions of water uptake, and found less accumulation when water uptake was decreased by lower root temperature, higher air humidity, lower light intensity, or the addition of metabolic inhibitors and mannitol to the nutrient solution. However, its concentration in the transpiration stream (amount of simazine per unit volume of absorbed water) was always lower than that in the nutrient solution, which could partly be attributed to retention of simazine in the roots.

Walker (1972) measured the radioactivity of wheat seedlings growing in nutrient solution with labelled atrazine at different concentration (0·012–0·06 mg/l), while water uptake was determined over several days. The straight line in Fig. 5 represents the expected herbicide uptake by mass flow supply, but also here actual uptake lags somewhat behind water uptake, especially over the longer periods. A greater discrepancy between expected and observed uptake was found with various soils in which processes of adsorption and breakdown may interfere. Similar results were obtained with linuron (Walker, 1973), although its translocation to the shoots of several plants was different (Walker and Featherstone, 1973). Diuron absorption by barley from nutrient solution was also related to water uptake (Moyer *et al.* 1972). In simulating a transpiration-induced flow in detopped tomato plants by a pressure cell, Perry (1973) found that the concentration of diuron and linuron in the exudate was unaffected by changes in the rate of sap flow. This supports the view that herbicide transport in the plant is proportional to transpiration rate.

This proportionality could be a factor in herbicide selectivity. For example, Uhlig (1968) observed more water and simazine uptake in the sensitive *Larix decidua,* while oats accumulated more radioactivity in the leaves than maize and *Amaranthus hybridus* from nutrient solution with labelled prometryne (Singh *et al.* 1972). Especially the lower transpiration rates of C_4 plants could be of importance in their tolerance to herbicides. Furthermore, herbicide accumulation and toxicity could be determined by weather and soil conditions, which influence the transpiration rate of plants.

REFERENCES

Akita, S. and Moss, D. N. (1972). *Crop Sci.* **12**, 789–793.
Akita, S. and Moss, D. N. (1973). *Crop Sci.* **13**, 234–237.
Allaway, W. G. and Mansfield, T. A. (1967). *New Phytol.* **66**, 57–63.
Allen, W. S. and Palmer, R. D. (1963). *Weeds*, **11**, 27–31.
Anderson, J. L. and Schaelling, J. P. (1970). *Weed Sci.* **18**, 455–459.
Anderson, J. L. and Thomson, W. T. (1973). *Residue Rev.* **47**, 167–189.

Arndt, F. and Kötter, C. (1968). *Weed Res.* **8,** 259–271.
Ashton, F. M. (1963). *Weeds,* **11,** 295–297.
Ashton, F. M. (1965). *Weeds,* **13,** 164–168.
Ashton, F. M., Cutter, E. G. and Huffstutter, D. (1969). *Weed Res.* **9,** 198–204.
Ashton, F. M., Gifford, E. M. and Bisalputra, T. (1963). *Bot. Gaz.* **124,** 336–343.
Ashton, F. M., Uribe, E. G. and Zweig, G. (1961). *Weeds,* **9,** 575–579.
Ashton, F. M., Zweig, G. and Mason, G. A. (1960). *Weeds,* **8,** 448–451.
Aylor, D. E., Parlange, J. Y. and Krikorian, A. D. (1973). *Am. J. Bot.* **60,** 163–171.
Barrentine, J. L. and Warren, G. F. (1970). *Weed Sci.* **18,** 373–377.
Black, C. C. (1973). *A. Rev. Pl. Physiol.* **24,** 253–286.
Bolhár-Nordenkampf, H. (1970). *Biochem. Physiol. Pfl. Jena,* **161,** 342–357.
Boyer, J. S. (1970). *Pl. Physiol., Baltimore,* **46,** 236–239.
Bravdo, B. (1972). *Physiologia Pl.* **26,** 152–156.
Bruinsma, J. (1967). *Acta bot. neerl.* **16,** 73–85.
Büchel, K. H. (1972). *Pestic. Sci.* **3,** 89–110.
Carpenter, K., Cottrell, H. J., De Silva, W. H., Heywood, B. J., Leeds, W. G., Rivett, K. F. and Soundy, M. L. (1964). *Weed Res.* **4,** 175–195.
Chesalin, G. A., Ladonin, V. F. and Spesivtsev, L. G. (1968). *Dokl. vses. Akad. sel.'-khoz. Nauk,* **1968**(3), 9–13.
Chodová, D. and Zemánek, J. (1971). *Biologia Pl.* **13,** 234–242.
Colby, S. R. (1964). *Diss. Abstr.* **25,** 715.
Colby, S. R. and Warren, G. F. (1962). *Weeds,* **10,** 308–310.
Colby, S. R. and Warren, G. F. (1965). *Weeds,* **13,** 257–263.
Corbett, J. R. and Wright, B. J. (1971). *Phytochemistry,* **10,** 2015–2024.
Couch, R. W. and Davis, D. E. (1966). *Weeds,* **14,** 251–255.
Davis, E. A. (1966). *Weeds,* **14,** 10–17.
Desmoras, J. and Jacquet, P. (1964). *Meded. LandbHogesch. Gent* **29,** 633–643.
Downton, W. J. S. (1971). *In* "Photosynthesis and Photorespiration" (M. D. Hatch, C. B. Osmond and R. O. Slatyer, eds), pp. 554–558. Wiley-Interscience, New York.
Downton, W. J. S. and Tregunna, E. B. (1968). *Pl. Physiol., Lancaster,* **43,** 923–929.
Dutta, T. R., Gupta, S. R. and Gupta, J. N. (1973). *Curr. Sci.* **42,** 243–244.
Eshel, Y. (1969a). *Weed Res.* **9,** 167–172.
Eshel, Y. (1969b). *Weed Sci.* **17,** 492–496.
Eshel, Y. and Warren, G. F. (1967). *Weeds,* **15,** 237–241.
Fedtke, C. (1973). *Pestic. Sci.* **4,** 653–664.
Figuerola, L. F. and Furtick, W. R. (1972). *Weed Sci.* **20,** 60–63.
Frank, R. and Switzer, C. M. (1969). *Weed Sci.* **17,** 344–348.
Friesen, H. A. and Dew, D. A. (1967). *Can. J. Pl. Sci.* **47,** 533–537.
Funderburk, H. H. and Lawrence, J. M. (1964). *Weeds,* **12,** 259–264.
Gaastra, P. (1959). *Meded. LandbHogesch. Wageningen,* **59**(13), 1–68.
Geissbühler, H., Haselbach, C., Aebi, H. and Ebner, L. (1963). *Weed Res.* **3,** 181–194.
Gentner, W. A. (1966). *Weeds,* **14,** 27–31.
Gentner, W. A. and Hilton, J. L. (1960). *Weeds,* **8,** 413–417.
Geoghean, M. J. (1957). *New Phytol.* **56,** 71–80.
Geronimo, J. and Herr, J. W. (1970). *Weed Sci.* **18,** 48–53.
Gibbs, M. (1971). *In* "Photosynthesis and Photorespiration" (M. D. Hatch, C. B. Osmond and R. O. Slatyer, eds), pp. 433–441. Wiley-Interscience, New York.
Gifford, R. M. and Musgrave, R. B. (1973). *Aust. J. biol. Sci.* **26,** 35–44.
Gill, H. S., Burley, J. W. A. and Bendixen, L. E. (1972). *Indian J. Weed Sci.* **4,** 1–7.
Gingras, G. (1966). *Physiol. vég.* **4,** 1–65.
Goldsworthy, A. (1970). *Bot. Rev.* **36,** 321–340.
Good, N. E. and Izawa, S. (1973). *In* "Metabolic Inhibitors" (R. M. Hochster, M. Kates and J. H. Quastel, eds), Vol. 4, pp. 179–214. Academic Press, New York and London.
Goren, R. (1969). *Weed Res.* **9,** 121–135.
Graham, J. C. and Buchholtz, K. P. (1968). *Weed Sci.* **16,** 389–392.
Harris, N. and Dodge, A. D. (1972). *Planta,* **104,** 201–209.

Helson, V. A. and Minshall, W. H. (1962). *Can. J. Bot.* **40**, 887–896.
Herholdt, J. A. (1968). *Diss. Abstr. B* **30**, 1978–1979.
Hill, E. R., Putala, E. C. and Vengris, J. (1968). *Weed Sci.* **16**, 377–380.
Hilton, J. L., Monaco, T. J., Moreland, D. E. and Gentner, W. A. (1964). *Weeds,* **12**, 129–131.
Hilton, J. L., Scharen, A. L., St. John, J. B., Moreland, D. E. and Norris, K. H. (1969). *Weed Sci.* **17**, 541–547.
Hodgson, R. H. (1971). *Weed Sci.* **19**, 501–507.
Hofstra, G. and Switzer, C. M. (1968). *Weed Sci.* **16**, 23–28.
Hogue, E. J. and Warren, G. F. (1968). *Weed Sci.* **16**, 51–54.
Houseworth, L. D. and Tweedy, B. G. (1971). *Weed Sci.* **19**, 732–735.
Hsiao, Th. C. (1973). *A. Rev. Pl. Physiol.* **24**, 519–570.
Huang, J. C. and Gloyna, E. F. (1968). *Wat. Res.* **2**, 347–366.
Humburg, N. E. (1970). *Diss. Abstr. B* **31**, 5759.
Imbamba, S. K. (1970). *Diss. Abstr. B* **31**, 1016.
Imbamba, S. K. and Moss, D. N. (1971). *Crop Sci.* **11**, 844–848.
Ivanova, E. A. (1970). *Khimiya sel.' Khoz.* **8**, 208–209.
Jackson, W. A. and Volk, R. J. (1970). *A. Rev. Pl. Physiol.* **21**, 385–432.
Jarvis, P. G. (1971). *In* "Plant Photosynthetic Production, Manual of Methods" (Z. Šesták, J. Čatský and P. G. Jarvis, eds), pp. 566–631. Junk, The Hague.
Jones, D. W. and Foy, C. L. (1972). *Weed Sci.* **20**, 116–120.
Jones, H. G. and Osmond, C. B. (1973). *Aust. J. biol. Sci.* **26**, 15–24.
Jones, R. J. and Mansfield, T. A. (1970). *J. exp. Bot.* **21**, 714–719.
Jones, R. J. and Mansfield, T. A. (1972). *Physiologia Pl.* **26**, 321–327.
Kandler, O. (1958). *Physiologia Pl.* **11**, 675–684.
Koren, E. and Ashton, F. M. (1971). *Weed Sci.* **19**, 587–592.
Koren, E. and Ashton, F. M. (1973). *Weed Sci.* **21**, 241–245.
Kozinka, V. (1968). *Biologia Pl.* **10**, 398–408.
Kriedemann, P. E., Loveys, B. R., Fuller, G. L. and Leopold, A. C. (1972). *Pl. Physiol., Baltimore,* **49**, 842–847.
Krihning, J. (1965a). *Phytopath. Z.* **52**, 372–394.
Krihning, J. (1965b). *Phytopath. Z.* **53**, 65–84.
Krinsky, N. (1966). *In* "Biochemistry of Chloroplasts" (T. W. Goodwin, ed.), Vol. 1, pp. 423–430. Academic Press, London and New York.
Kuratle, H. (1968). *Diss. Abstr.* **29**, 1825.
Ladonin, V. F. (1972). *Vest. sel.'-khoz. Nauki Mosk.* **1972**(11), 19–24.
Ladonin, V. F. and Chivikina, T. V. (1969). *Khimiya sel.' Khoz.* **7**, 690–693.
Lee, Ch.E., Lee, Y. H. and Hew, Ch.S. (1971). *Nanyang Univ. J.* **5**, 132–137.
Lester, D. C., Carter, O. G., Kelleher, F. M. and Laing, D. R. (1972). *Aust. J. agric. Res.* **23**, 205–213.
Lindsay, R. V. and Hartley, G. S. (1963). *Weed Res.* **3**, 195–204.
Livne, A. and Vaadia, Y. (1972). *In* "Water Deficits and Plant Growth" (T. T. Kozlowski, ed.), Vol. III, pp. 255–275. Academic Press, New York and London.
Lotlikar, P. D., Remmert, L. F. and Freed, V. H. (1968). *Weed Sci.* **16**, 161–165.
Ludlov, M. M. and Jarvis, P. G. (1971). *In* "Plant Photosynthetic Production, Manual of Methods" (Z. Šesták, J. Čatský and P. G. Jarvis, eds), pp. 294–315. Junk, The Hague.
Lüttge, U., Ball, E. and von Willert, K. (1971). *Z. PflPhysiol.* **65**, 326–335.
Mansfield, T. A. (1967). *New Phytol.* **66**, 325–330.
Mashtakov, S. M. (1968). *Sel.'-khoz. Biol.* **3**, 64–71.
Matsunaka, S. (1969). *Proc. 2nd Asian-Pacific Weed Contr. Interchange,* **1969**, 198–207.
Matsunaka, S. and Nakamura, H. (1972). *Weed Res. Japan,* **13**, 29–31.
McPherson, H. G. and Slatyer, R. O. (1973). *Aust. J. biol. Sci.* **26**, 329–339.
Mees, G. C. (1960). *Ann. appl. Biol.* **48**, 601–612.
Meidner, H. and Mansfield, T. A. (1968). "Physiology of Stomata", McGraw-Hill, London.
Merkle, M. G., Leinweber, C. L. and Bovey, R. W. (1965). *Pl. Physiol., Lancaster,* **40**, 832–835.
Minshall, W. H. (1954). *Can. J. Bot.* **32**, 795–798.

Minshall, W. H. (1957). *Weeds,* **5,** 29–33.
Minshall, W. H. (1960). *Can. J. Bot.* **38,** 201–216.
Mittelheuser, C. J. and van Steveninck, R. F. M. (1971). *Planta,* **97,** 83–86.
Moody, K., Kust, C. A. and Buchholtz, K. P. (1970). *Weed Sci.* **18,** 642–647.
Moreland, D. E., Farmer, F. S. and Hussey, G. G. (1972). *Pestic. Biochem. Physiol.* **2,** 354–363.
Moreland, D. E., Gentner, W. A., Hilton, J. L. and Hill, K. L. (1959). *Pl. Physiol., Lancaster,* **34,** 432–435.
Moreland, D. E. and Hill, K. L. (1962). *Weeds,* **10,** 229–236.
Moss, D. N. (1968). *Crop Sci.* **8,** 71–76.
Mouravieff, I. (1971). *Physiol. vég.* **9,** 109–118.
Mouravieff, I. (1972). *Physiol. vég.* **10,** 515–528.
Moyer, J. R., McKercher, R. B. and Hance, R. J. (1972). *Can. J. Pl. Sci.* **52,** 668–670.
Murray, D. R. and Bradbeer, J. W. (1971). *Phytochemistry,* **10,** 1999–2003.
Nakamura, H., Koizumi, J. and Matsunaka, S. (1968). *Weed Res. Japan,* **7,** 100–104.
Nakamura, H. and Matsunaka, S. (1969). *Weed Res. Japan,* **8,** 33–38.
Nasyrova, T., Mirkasimova, Kh. and Pazilova, S. (1968). *Uzbek. biol. Zh.* **12,** 23–26.
Nazarov, N. M. and Kutyurin, V. M. (1969). *Soviet Pl. Physiol.* **16,** 342–345.
Noda, K. and Ibaraki, K. (1968). *Weed Res. Japan,* **7,** 105–110.
Olech, K. (1967a). *Annls Univ. Mariae Curie-Sklodowska,* **22,** 149–166.
Olech, K. (1967b). *Annls Univ. Mariae Curie-Sklodowska,* **22,** 167–183.
Olech, K. (1967c). *Annls Univ. Mariae Curie-Sklodowska,* **22,** 185–195.
Olech, K. (1968). *Annls Univ. Mariae Curie-Sklodowska,* **23,** 193–200.
Orr, A. R., Kessler, J. E. and Tepaske, E. R. (1971). *Am. J. Bot.* **58,** 459–460.
Osborne, D. J. and Allaway, M. (1961). *In* "Plant Growth Regulation", pp. 329–338. Iowa State Univ. Press, Ames, Iowa.
Paromenskaya, I. N. and Lyalin, G. N. (1968). *Soviet Pl. Physiol.* **15,** 842–846.
Penner, D. (1971). *Weed Sci.* **19,** 571–576.
Penner, D. and Ashton, F. M. (1966). *Residue Rev.* **14,** 39–113.
Pereira, J. F., Splittstoesser, W. E. and Hopen, H. J. (1971). *Weed Sci.* **19,** 662–666.
Perry, M. W. (1973). *Weed Res.* **13,** 325–330.
Płoszyński, M., Świętochowski, B. and Żurawski, H. (1969). *Roczn. Nauk roln.* **95-A-3,** 401–415.
Pochinok, Kh.N. (1971). *Fiziol. Biokhim. Kul'turn. Rast.* **3,** 538–544.
Poskuta, J., Antoszewski, R. and Faltynowicz, M. (1972). *Photosynthetica,* **6,** 370–374.
Poskuta, J., Nelson, C. D. and Krotkov, G. (1967). *Pl. Physiol., Lancaster,* **42,** 1187–1190.
Roeth, F. W. and Levy, T. L. (1971). *Weed Sci.* **19,** 93–97.
Romanovskaya, O. I., Gubar, G., Irbe, I. and Straume, O. P. (1968). *Latvijas PSR Zinatnu Akademijas Vestis* **1968**(8), 90–97.
Romanovskaya, O. I. and Straume, O. P. (1970). *Latvijas PSR Zinatnu Akademijas Vestis,* **1970**(7), 18–22.
Roth, W. (1958). *Experientia,* **14,** 137–138.
Rubin, B. and Eshel, Y. (1971). *Weed Sci.* **19,** 592–594.
Saakov, V. S. (1973a). *Biochem. Physiol. Pfl. Jena,* **164,** 199–212.
Saakov, V. S. (1973b). *Biochem. Physiol. Pfl. Jena,* **164,** 213–227.
Sasaki, S. and Kozlowski, T. T. (1967). *Can. J. Bot.* **45,** 961–971.
Sedgley, R. H. and Boersma, L. (1969). *Weed Sci.* **17,** 304–306.
Šesták, Z., Čatsky, J. and Jarvis, P. G. (1971). "Plant Photosynthetic Production, Manual of Methods". Junk, The Hague.
Sheets, T. J. (1961). *Weeds,* **9,** 1–13.
Shimabukuro, R. H. and Swanson, H. R. (1969). *J. agric. Fd. Chem.* **17,** 199–207.
Shimshi, D. (1967). *New Phytol.* **66,** 455–461.
Shirakawa, N. (1969). *Weed Res. Japan,* **9,** 11–15.
Shone, M. G. T. and Wood, A. V. (1972). *J. exp. Bot.* **23,** 141–151.
Shriver, J. W. and Bingham, S. W. (1973). *Crop Sci.* **13,** 45–49.
Siegenthaler, P. A. and Packer, L. (1965). *Pl. Physiol., Lancaster,* **40,** 785–791.
Sij, J. W., Kanemasu, E. T. and Teare, I. D. (1972). *Crop Sci.* **12,** 733–735.

Sikka, H. C., Davis, D. E. and Funderburk, H. H. (1964). *Proc. sth. Weed Control Conf.* **17**, 340–350.
Singh, J. N., Basler, E. and Santelmann, P. W. (1972). *Pestic. Biochem. Physiol.* **2**, 143–152.
Slatyer, R. O. (1967). "Plant-Water Relationships", Academic Press, London and New York.
Smith, D. and Buchholtz, K. P. (1964). *Pl. Physiol., Lancaster,* **39**, 572–578.
Squire, G. R. and Jones, M. B. (1971). *J. exp. Bot.* **22**, 980–991.
Stanger, C. E. and Appleby, A. P. (1972). *Weed Sci.* **20**, 357–363.
Stokes, D. M., Turner, J. S. and Markus, K. (1970). *Aust. J. biol. Sci.* **23**, 265–274.
Sutton, D. L., Durham, D. A., Bingham, S. W. and Foy, C. L. (1969). *Weed Sci.* **17**, 56–59.
Tal, M. and Imber, D. (1972). *New Phytol.* **71**, 81–84.
Taylor, A. O. and Craig, A. S. (1971). *Pl. Physiol., Baltimore,* **47**, 719–725.
Tieszen, L. L. (1970). *Pl. Physiol., Baltimore,* **46**, 442–444.
Tolbert, N. E. (1971). *A. Rev. Pl. Physiol.* **22**, 45–74.
Uhlig, S. K. (1968). *Arch. PflSchutz,* **4**, 215–227.
Van Hasselt, Ph.R. (1972). *Acta bot. neerl.* **21**, 539–548.
Van Oorschot, J. L. P. (1964). *Meded. LandbHogesch. Gent,* **29**, 683–694.
Van Oorschot, J. L. P. (1965). *Weed Res.* **5**, 84–97.
Van Oorschot, J. L. P. (1966). *Jaarb. Inst. biol. scheik. Onderz. LandbGewass.* **1966**, 41–50.
Van Oorschot, J. L. P. (1968a). *I.I.R.B.* **3**, 132–142.
Van Oorschot, J. L. P. (1968b). *Proc. 9th Br. Weed Control Conf.* 624–632.
Van Oorschot, J. L. P. (1970). *Weed Res.* **10**, 230–242.
Van Oorschot, J. L. P. (1974a). *Acta bot. neerl.* **23**, 36–41.
Van Oorschot, J. L. P. (1974b). *Weed Res.* **14**, 75–79.
Van Oorschot, J. L. P. and van Leeuwen, P. H. (1974). *Weed Res.* **14**, 81–86.
Van Rensen, J. J. S. (1969). *Meded. LandbHogesch. Wageningen,* **69–14**, 1–11.
Van Rensen, J. J. S. (1971). *Meded. LandbHogesch. Wageningen,* **71–9**, 1–80.
Van Rensen, J. J. S., Justesen, S. H. and Tammes, P. M. L. (1972). *Acta bot. neerl.* **21**, 372–380.
Van Rensen, J. J. S. and van Steekelenburg, P. A. (1965). *Meded. LandbHogesch. Wageningen,* **65–13**, 1–8.
Velev, B. (1969). *Rast. Nauki,* **6**, 125–131.
Volk, R. J. and Jackson, W. A. (1972). *Pl. Physiol., Baltimore,* **49**, 218–223.
Vostral, H. J., Buchholtz, K. P. and Kust, C. A. (1970). *Weed Sci.* **18**, 115–117.
Walker, A. (1972). *Pestic. Sci.* **3**, 139–148.
Walker, A. (1973). *Pestic. Sci.* **4**, 665–675.
Walker, A. and Featherstone, R. M. (1973). *J. exp. Bot.* **24**, 450–458.
Walker, D. A. and Zelitch, I. (1963). *Pl. Physiol., Lancaster,* **38**, 390–396.
Wassink, E. C. and van Elk, B. C. M. (1961). *Meded. LandbHogesch. Wageningen* **61–17**, 1–14.
Wax, L. M. and Behrens, R. (1965). *Weeds,* **13**, 107–109.
Wills, G. D., Davis, D. E. and Funderburk, H. H. (1963). *Weeds,* **11**, 253–255.
Wojtazek, T., Cherry, J. H. and Warren, G. F. (1966). *Pl. Physiol., Lancaster,* **41**, 34–38.
Zaki, M. A., Taylor, H. F. and Wain, R. L. (1967). *Ann. appl. Biol.* **59**, 481–491.
Zelitch, I. (1964). *Science, N.Y.,* **143**, 692–693.
Zelitch, I. (1965). *Biol. Rev.* **40**, 463–482.
Zelitch, I. (1969). *A. Rev. Pl. Physiol.* **20**, 329–350.
Zelitch, I. (1971). "Photosynthesis, Photorespiration and Plant Productivity", Academic Press, New York and London.
Zelitch, I. (1973a). *Proc. natn. Acad. Sci. U.S.A.* **70**, 579–584.
Zelitch, I. (1973b). *Pl. Physiol., Baltimore,* **51**, 299–305.
Zweig, G., Hitt, J. E. and McMahon, R. (1968). *Weed Sci.* **16**, 69–73.

CHAPTER 11

HERBICIDE ENTRY INTO PLANTS

MARTIN J. BUKOVAC*

*Department of Horticulture, Michigan State University
East Lansing, Michigan, USA*

I. INTRODUCTION

The primary pathway for herbicide entry into plants will vary with the specific weed control problem in question. When herbicides are applied directly to the above-ground plant parts of a mixed crop and weed population, the leaves and stems serve as the primary intercepting and absorbing organs. In contrast, the root and emerging shoot are important pathways of entry when the herbicide is applied to or incorporated into the soil. In selective control of established woody plant species, however, the bark may serve as the site of entry.

Irrespective of the mode of application, the applied chemical must penetrate into the plant tissue before a biological response can be induced. Although marked differences may exist as to the relative significance of shoot versus root uptake, depending on the stage of plant development at the time of treatment, there are numerous similarities in movement of the applied chemical from the plant surface to the underlying living tissues.

Aerial plant parts (leaves and specialized structures, stems, flowers, fruits, etc.) are covered by a lipoidal, non-cellular, non-living membrane called the

---

* Dedicated to my Father.

cuticle. This membrane serves as the primary barrier to penetration of materials, particularly polar substances, into plants. Roots are also covered by a membrane similar in structure and composition to that found on above-ground plant parts, which no doubt serves a similar function.

Because of similarities in the outer covering of both shoots and roots, many factors, which influence the interaction of an applied chemical with the plant surface and the movement of the chemical across the cuticle during penetration into the plant, will, in many respects, be applicable to both entry into shoots and roots. Thus, in this chapter I will discuss the movement of chemicals, as applied in an aqueous spray, across the cuticular membrane and will use illustrative data generated primarily with leaf and fruit cuticles. When considering absorption of herbicides by roots, those factors affecting the availability of the chemical to the roots are discussed in Volume 2, Chapter 1, and the translocation of chemicals absorbed by roots and shoots is dealt with in Chapter 12 and will not be considered in this chapter.

II. Penetration Through the Shoot Epidermis

A. NATURE OF THE PLANT SURFACE

1. *Gross Morphology*

The gross morphology of the plant may play an important role in penetration by influencing the amount of the foliar-applied chemical intercepted and retained (Blackman *et al.* 1958). The stage of plant development, the shape and lamellar area of the leaf, the angle or orientation of the leaves in relation to the spray, and specialized features, which provide for localized accumulation, may all play an important role. The above factors will determine to a large extent, but not exclusively, the amount of spray intercepted and, thus, the quantity of applied chemical retained by the plant. For example, Davies *et al.* (1967) found a close relationship between leaf angle in barley (*Hordeum distichon*) and retention of ioxynil sprays as measured by a reduction in growth. Maximum retention occurred when leaves were positioned at 50° to 90° to the incidence of the spray (in this case vertical). Similarly, leaf orientation was found to be important for both easy- and difficult-to-wet soybean (*Glycine max*) cultivars (Ennis *et al.* 1952). Other gross morphological features, which may influence retention, would be expected to have a similar effect.

Gross morphology, thus, becomes an important factor in foliar penetration in so far as it may influence spray retention, which in turn determines the dose available for penetration.

2. *Specialized Structures*

Spray retention may also be influenced by specialized structures, e.g. trichomes, stomata, veins, etc. (Ennis *et al.* 1952; Challen, 1960, 1962; Furmidge, 1962; Sargent, 1966; Holloway, 1970). Trichomes are common features on many plant surfaces. They vary markedly in size, morphology, frequency, distribution and function. Challen (1962) showed that water retention was greater on leaves having an "open" than on those having a "closed" trichome pattern. The "open" pattern may enhance wetting due to capillary action, while the "closed" pattern could depress it by entrapment of air beneath the water droplets. Surfaces over veins where trichomes are often plentiful, over guard cells and around the bases of trichomes, often differ in wettability from areas over other epidermal cells. Such differential wettability not only leads to variation in retention over a given surface, but may be the basis for selective permeability often associated with specialized structures. A detailed discussion of factors affecting spray retention can be found in Volume 2, Chapter 8.

3. Cuticle

a. *Morphology.* The cuticle is a thin, continuous, non-cellular lipoidal membrane covering plant surfaces (Fig. 1a). It extends over specialized structures (Brongniart, 1834) and frequently covers the guard, mesophyll and epidermal cells adjoining substomatal chambers and intercellular air spaces (Fig. 1c, 1d) in leaves (Artz, 1933; Scott, 1950; Norris and Bukovac, 1968). On the cell wall side, the cuticle may project down between the anticlinal walls of epidermal cells (Priestley, 1943; Norris and Bukovac, 1968; Baker, 1970, 1971). Often in mature leaves and fruits entire epidermal cells may be encapsulated by cutinization of the cell wall.

Cutin is the chief structural component of the cuticle. Generally, a highly orientated layer of wax (referred to as cuticular or embedded) is found within the cutin matrix. This wax layer appears strongly birefringent (Fig. 1(b)) when viewed under plane-polarized light (Meyer, 1938; Roelofsen, 1952; Sitte and Rennier, 1963; Norris and Bukovac, 1968). Epicuticular wax is deposited on the outer surface, often in a specific and characteristic pattern. The quantity of epicuticular wax present varies widely; in a number of weed species it ranged from less than 10 to about 20 μg cm^{-2} (Baker and Bukovac, 1971).

Between the cutin matrix and the epidermal cell wall is a region rich in pectins (Anderson, 1928), which is believed to be continuous with the middle lamella of the anticlinal walls of the epidermal cells. Chemical (Holloway and Baker, 1968) or enzymatic (Skoss, 1955; Orgell, 1955) digestion of this pectic layer in many plants results in the separation of the cuticle from the leaf or fruit.

The cuticular components exterior to the pectin-rich region, i.e. cutin, cuticular and epicuticular waxes collectively, are frequently referred to as the cuticular membrane (see Esau, 1962), and I will adopt this terminology here. Occasionally, the demarcation between the cutin and the cellulose of the epidermal cell wall may not be precise, there being a gradual transition of one component into the other. The cuticle of the upper epidermis of a pear (*Pyrus communis*) leaf is illustrated diagrammatically in Fig. 2.

b. *Surface Fine-Structure.* The surface of the cuticle often follows the outlines of the epidermal cells beneath (Priestley, 1943). In addition, the cuticular surface may in itself have features or fine-structure which bear no relationship

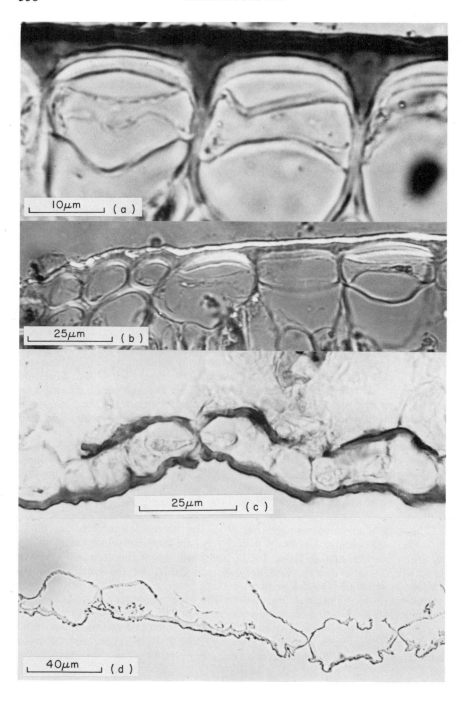

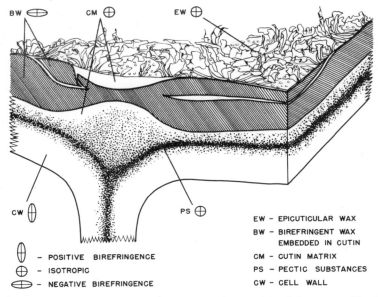

BW, CM, EW

CW, PS

- POSITIVE BIREFRINGENCE
- ISOTROPIC
- NEGATIVE BIREFRINGENCE

EW – EPICUTICULAR WAX
BW – BIREFRINGENT WAX
      EMBEDDED IN CUTIN
CM – CUTIN MATRIX
PS – PECTIC SUBSTANCES
CW – CELL WALL

Fig. 2. Schematic representation of the structure of the cuticle of the upper epidermis of a *Pyrus communis* leaf. Not necessarily drawn to scale. (After Norris and Bukovac, 1968.)

to the underlying cells. Most of the fine-structural characteristics of the cuticle, however, can be ascribed to the epicuticular wax. The surface features of leaves of some selected weed species are illustrated in Fig. 3.

Attempts were made as early as 1871 by deBary (Martin and Juniper, 1970) to group plants according to morphological characteristics of the epicuticular wax. More recently, Amelunxen *et al.* (1967) categorized surface wax features as (a) rodlet, (b) granular, (c) platelet, (d) layered or crust, (e) aggregate or (f) liquid or semifluid. The chemical composition which influences the crystallization pattern appears to be the important factor in determining the surface pattern. The use of epicuticular wax patterns in taxonomic classification of plants remains questionable, since multiple forms are frequently found side by side on the same organ, and the environment under which the wax is deposited has a marked effect on fine-structure (Juniper, 1960; Baker, 1972; Whitecross and Armstrong, 1972).

c. *Chemistry.* The chemistry of plant cuticles has been investigated extensively in recent years, and many of the constituents of the waxes and cutin have been

Fig. 1. Photomicrographs of the cuticle from *Pyrus communis* leaves in transverse section. (a). Upper epidermis showing cuticular pegs extending down between epidermal cells. (b). Birefringence of the cuticular wax layer viewed under plane-polarized light. (c). Cuticle extending through a stomatal pore and covering guard and epidermal cells adjacent to the substomatal cavity. (d). An isolated cuticular membrane from the abaxial surface showing extensive development of the "internal cuticle".

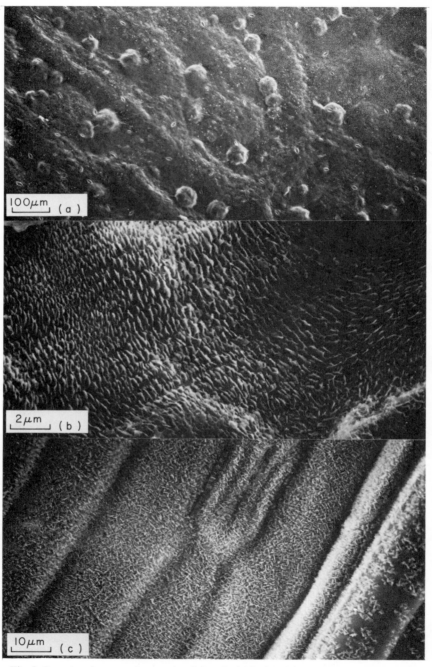

Fig. 3. Scanning electron micrographs of the adaxial leaf surfaces of *Amaranthus retroflexus* (A) and *Chenopodium album* (B) and the abaxial leaf surface of *Agropyron repens* (C).

elucidated with the aid of thin-layer and gas-liquid chromatography and mass spectrometry.

Although the epicuticular waxes of most plants contain a wide diversity of complex constituents, certain classes of compound are common to leaf waxes of many plants (Kreger, 1948; Mazliak, 1968; Kolattukudy, 1970; Martin and Juniper, 1970; Kolattukudy and Walton, 1973). In general, the epicuticular waxes are complex mixtures of long chain alkanes, alcohols (both primary and secondary), ketones, aldehydes, esters and fatty acids. Often each of the above classes is present as a homologous series with a predominant chain length of from C_{21} to C_{37}. The fatty acids, primary alcohols and aldehydes generally contain even numbers of carbon atoms, whereas the alkanes, ketones and secondary alcohols contain odd numbers. The acids and alcohols may be found free or combined as esters. Cyclic components, triterpenoids being most representative (e.g. ursolic and oleanolic acids), are frequently found in fruit cuticles (Kariyine and Hashimoto, 1959). Studies on the detailed chemistry of numerous leaf and fruit waxes have been reviewed recently by Martin and Juniper, 1970.

Although considerable progress has been made in elucidating the wax constituents, two points need to be stressed. Firstly, often no critical attempt has been made to separate the epicuticular from the cuticular waxes, and thus only limited information is available on the chemistry of the cuticular waxes *per se*. In dandelion (*Taraxacum officinale*), plantain (*Plantago major*) and forget-me-not (*Myosotis arvensis*), the cuticular waxes contained lesser quantities of hydrocarbons, esters and alcohols and higher proportions of polar constituents than did the epicuticular waxes (Baker and Bukovac, 1971). Secondly, little is known about the distribution and orientation of the constituents in the wax layers. The nature and distribution of the functional groups would be useful information.

Cutin, which makes up the structural matrix of the cuticle, is a polyester of long chain fatty and hydroxy fatty acids. It is not soluble in most organic solvents and is resistant to decay. The principal constituents are 10,16-dihydroxyhexadecanoic acid, 10,18-dihydroxyoctadecanoic acid, and 9,10,18-trihydroxyoctadecanoic acid (Matic, 1956; Baker and Martin, 1963; Eglington and Hunneman, 1968; Mazliak, 1968; Baker and Holloway, 1970; Holloway and Dees, 1971; Kolattukudy and Walton, 1973). The exact structure of the cutin polymer is not known; however, ester linkages between the hydroxyl and carboxyl groups dominate, and some peroxide and ether linkages may also be present (Heinen and Brand, 1963; Crisp, 1965; Kolattukudy and Walton, 1973).

Cutin behaves like a highly cross-linked, high-capacity ion-exchange resin of a weak organic-acid type (Schönherr and Bukovac, 1973). Three dissociable groups were observed in wax-free isolated tomato (*Lycopersicon esculentum*) fruit cutin in the pH range of 3 to 6, 6 to 9 and 9 to 12. Since cutin contains both polar and non-polar groups, it has both hydrophilic and hydrophobic properties (van Overbeek, 1956a). In water, cutin is believed to swell, separating the wax platelets and leading to increased permeability.

4. *Interaction of the Applied Chemical with the Plant Surface*

The interaction of a chemical spray with the plant surface is very complex and may be influenced by factors associated with both the surface and the spray solution. The epicuticular wax may influence wettability in two ways. Firstly, the fine-structure, particularly that of plant surfaces rich in crystalline wax, may result in a "micro" rough surface from which the spray droplets tend to be reflected (Fogg, 1948; Brunskill, 1956; Hartley and Brunskill, 1958; Juniper, 1959; Challen, 1960). The amount of wax present probably contributes little to wettability providing that some minimum level is present, i.e. sufficient to cover the cutin matrix adequately. Secondly, the chemistry of the epicuticular wax is important (Juniper, 1959; Silva Fernandes, 1965; Holloway, 1969a). The nature of the chemical groups exposed at the surface is critical (Fogg, 1948; Adam, 1958). The most difficult-to-wet leaf surfaces are rich in alkanes and those most easily wetted are rich in alcohols (Holloway, 1969a, 1969b, 1970). Other characteristics of the plant surface which influence spray retention will play a role.

Equally important, in many respects, are the characteristics of the spray solution, i.e. surface tension of the carrier, polarity, chemical structure of the toxicant and physical characteristics of the spray are some of the more important factors. Often several factors may be simultaneously involved and not easily separated. A discussion of the role of such factors in spray retention may be found in Volume 2, Chapter 8.

B. PENETRATION THROUGH THE CUTICLE

A wide variety of chemicals which differ in structure and polarity penetrate through the cuticle. The exact pathways involved are not known, but generally it is felt that the non-polar compounds follow a lipoidal route and polar compounds an aqueous route. The nature of such pathways remains conjectural and will be discussed in greater detail later. Specialized structures like stomata and trichomes may also play an important role.

Few studies have stressed the nature of penetration taking into account a specific pathway. A simplified overview of penetration into a cutinized plant tissue can be visualized as follows: (a) sorption into the cuticle, (b) movement across the cuticular membrane, (c) desorption into the apoplast and (d) uptake by the underlying cells. Generally, movement across the cuticular membrane is a physical process which may be affected directly by a number of factors, e.g. pH, particle size, cuticle thickness, etc., and indirectly by rate of uptake, transport and metabolism by the underlying tissue, in so far as these events may alter the concentration gradient across the cuticle.

1. *Binding and Sorption*

The surface of the plant is chiefly non-polar; however, there are some dissociable functional groups at the surface (Orgell, 1957). The half-dissociation point for the functional groups at the cuticular surface of pear leaves was estimated to be in the range of pH 2·8 to 3·0 (Bukovac and Norris, 1966). Therefore, at typical spray solution pH, they would be dissociated and the cuticular surface would carry a slight negative charge. The density of such groups does not appear to be great and electrostatic binding is probably not a critical factor in foliar penetration. The sorption or partitioning of the applied chemical into the cuticle would appear to be of much greater significance for entry of organic herbicides. Here, the polarity of the applied chemical appears critical. Non-polar compounds are more readily sorbed than polar compounds (Bukovac and Norris, 1966; Bukovac et al. 1971), e.g. 2,4-D is more readily sorbed as the non-dissociated molecule than as the anion, and progressive chlorination of phenoxyacetic acid results in a corresponding increase in lipid solubility and sorption (Bukovac et al. 1971). Ionic compounds also become associated with the cuticular surface, but generally to a lesser degree than non-dissociated molecules (Yamada et al. 1965, 1966).

2. Factors Affecting Movement Across the Cuticular Membrane

a. *Leaf Surface*. Both surfaces of leaves are readily penetrated by herbicides (Currier and Dybing, 1959; Foy, 1964; Sargent, 1965; Hull, 1970). The greater permeability of the lower leaf surface cannot be attributed to any one factor. Usually, the lower surface is rich in stomata and trichomes, which may play a role in penetration under some conditions (Hull, 1970). Of some 35 weed species studied, only three had a greater number of stomata in the upper than in the lower leaf surface (Ormrod and Renney, 1968). However, trichome-rich surfaces are more difficult to wet. Fine-structure of the surface wax frequently differs considerably between the two surfaces. The cuticle of the lower surface generally contains as much or more waxes than the upper, but the cuticular wax may not be as highly orientated (Norris and Bukovac, 1968). Unfortunately, critical studies have not been performed to establish if the cuticle, *per se,* on the lower epidermis is more permeable than that at the upper epidermis, or if the observed differences in permeability are related to secondary factors.

b. *Time-course*. The absorption of foliar-applied chemicals from spray droplets is initially rapid (Hull, 1970; Barrentine and Warren, 1970; Singh et al. 1972; Sands and Bachelard, 1973a). With increasing time after treatment, penetration decreases at an increasing rate (Luckwill and Lloyd-Jones, 1962). The progressive reduction in penetration with time has been associated with the rate at which the droplet evaporates (Luckwill and Lloyd-Jones, 1962). Penetration apparently continues from the residue on the leaf surface, since there is a slight positive slope to the absorption curve. Continued penetration from the residue is more pronounced if the chemical is hygroscopic or if a surfactant or humectant, e.g. glycerine (Hopp and Linder, 1946), is added to the spray solution. Rewetting of the residue, either experimentally or through the action of dew, may markedly enhance penetration (Bukovac, 1965).

When penetration is critically followed from a donor solution through an isolated cuticular membrane into a receiver solution, taking precaution to avoid large changes in concentrations, transfer across the cuticle is linear with time (Norris and Bukovac, 1969; Bukovac *et al.* 1971).

c. *Concentration.* When penetration is measured directly, it is generally linearly related to external herbicide concentration (Sargent and Blackman, 1962; Darlington and Cirulis, 1963; Sargent and Blackman, 1972). The relationship between concentration and penetration of non-contact type herbicides, when measured by plant response, may deviate from linearity. This is because at high concentrations of the herbicide, physiological changes may be induced in the uptake and transport processes, thus altering subsequent penetration (Hull, 1957).

d. *pH.* Differences of opinion still exist as to the role of hydrogen ion concentration on the penetration of foliar-applied herbicides. To assess the effect of pH fully one should separate the effects upon the penetrant from those on the plant system. The chemical nature of the herbicide is also a prime factor.

Hydrogen ion concentration plays a significant role in the penetration of weak organic acid type herbicides (Sargent, 1965). The undissociated molecule is more lipid soluble and penetrates more readily than the anion (Simon and Beevers, 1951; Sargent and Blackman, 1962). Penetration generally follows closely, but not absolutely, the dissociation curve as illustrated for naphth-1yl-acetic acid (NAA) in Fig. 4. Similar relationships have been shown for penetration of 2,4-D into bean (*Phaseolus vulgaris*) foliage (Sargent and

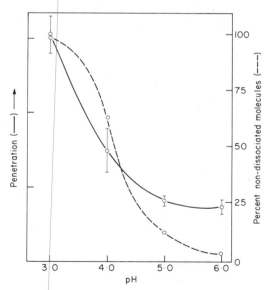

Fig. 4. Graph illustrating a generalized relationship between NAA penetration (————) into *Pyrus communis* leaves and the percent nondissociated molecules (- - - - - -) present as a function of pH.

Blackman, 1962) and into *Chlorella* (Wedding and Erickson, 1957) and for NAA into pear leaves (Greene and Bukovac, 1972). In most instances some penetration continues at pH values several units above the pK, and Wedding and Erickson (1957) proposed that the permeability of *Chlorella* cells to the 2,4-D anion is a constant fraction of their permeability to the undissociated molecule. For herbicides where the hydrogen ion concentration has little or no effect on dissociation (amides, esters, etc.), the pH of the spray solution has little or no effect on penetration (Crafts, 1961; Robertson *et al.* 1971; Greene and Bukovac, 1971).

The effect of pH appears primarily on the penetrant (Norris and Bukovac, 1971). Although there are dissociable groups at the cuticular surface, they do not pose an insurmountable obstacle to penetration. The dissociable groups within the cuticle, even in the case of isolated cuticles, appear to be little affected by the pH of the spray solution. This is undoubtedly because the spray solution does not penetrate sufficiently to influence these groups. Should adequate penetration occur so as to influence the functional groups exposed, an effect of pH on the permeability of the cuticle would be expected. pH may also indirectly influence penetration by modifying membrane potential or the metabolic activity of cells involved in the uptake-transport processes.

Orgell and Weintraub (1957) and Patten (1959) report enhanced penetration of 2,4-D at pH levels above the pK, particularly with ammonium and dihydrogen phosphate ions. This effect is difficult to explain; however, the inorganic ions may influence membrane swelling, thus altering permeability.

e. *Chemical Structure.* Molecular structure may affect penetration of an herbicide into plant tissue (van Overbeek, 1956b; Sargent, 1964; Saunders *et al.* 1965; Sargent *et al.* 1969; Buta and Steffens, 1971; Robertson *et al.* 1971). Penetration of phenoxyacetic acid into bean leaf tissue increased with progressive chlorination of the parent molecule (Sargent *et al.* 1969). On the other hand, chlorination of benzoic acid depressed penetration. Similar relationships held with regard to penetration of these same compounds through isolated tomato fruit cuticles (Bukovac *et al.* 1971).

Differences between the rates of penetration of these particular derivatives arise from differences in their lipid solubility. Chlorination of phenoxyacetic acid has little or no effect on the pK, whereas chlorination of benzoic acid markedly depresses the pK. Therefore, at a given pH value the chlorinated phenoxyacetic acid is more, and the chlorinated benzoic acid less lipid soluble than the corresponding parent acid. Chlorination of the benzoic acid ring of maleimides also resulted in an increase in lipid solubility of the parent molecule (van Overbeek *et al.* 1955). Darlington and Cirulis (1963) reported a poor correlation between partition coefficients between water and chloroform and penetration of such diverse compounds as sugars, chloracetate and valerate; however, penetration of α-chloroacetamides and N-substituted derivatives increased with an increase in chloroform solubility. Thus, in general, modification of molecular structure which results in increased lipid solubility will enhance foliar penetration.

f. *Surfactants.* Surfactants may enhance, depress, or have no effect on herbicide activity (Foy and Smith, 1969). The role of surfactants in modifying herbicide penetration is complex and not well understood. One can visualize surfactant effects on the spray solution, on spray plant surface interaction and on plant processes *per se*.

Probably the most frequently cited effect of surfactants is the lowering of the surface tension of the spray solution. Generally, as the concentration of the surfactant is increased, the surface tension is lowered to a point beyond which further addition of surfactant is without effect. This point, the critical micelle concentration (CMC), lies between 0·01 and 0·5% for most efficient surfactants (Osipow, 1964). A similar relationship has been observed between surfactant concentration and wetting of the plant surface (Becher and Becher, 1969). In contrast, herbicidal effectiveness (penetration) is often maximal at concentrations 10 times the CMC concentration or greater (Foy and Smith, 1965). Chemical interaction between the herbicide and the surfactant may also occur in the spray solution, in most instances resulting in reduced efficacy (Freed and Montgomery, 1958; Smith and Foy, 1967).

Lowering the surface tension improves wetting and consequently the area of contact between the applied herbicide and the leaf surface is increased. However, the relationship between the surfactant and wetting is often quite specific. A surfactant producing a given surface tension may improve wetting, hence retention, of a difficult-to-wet plant surface, while run-off from an easy-to-wet surface may be excessive and retention less than in the absence of a surfactant (Schönherr and Bukovac, 1972b). Similarly, the drying time and characteristics of the spray droplet could be considerably modified by the presence of a surfactant. Surfactants may modify the plant surface by solubilizing the waxes (Furmidge, 1959), or may interact with the cutin matrix, thus altering its charge characteristics and swelling properties and, hence, the membrane resistance to diffusion of a specific chemical. The effects of high concentrations (greater than 1%) are not well understood, and consequently offer intriguing areas of further study.

Effects of surfactants on plants have been extensively reviewed by Parr and Norman (1965). Of particular interest in herbicide uptake are the effects on the plant surface (Furmidge, 1959) and on membrane permeability (Kuiper, 1967). Tween 20 in the presence or absence of diquat enhanced the efflux of betanin from beetroot tissue discs (Sutton and Foy, 1971). Brian (1972) demonstrated an effect of nonylphenol and ethylene oxide condensates on paraquat uptake and transport. Greatest uptake was associated with lowest percentage transport. He concluded that this was not due to an interaction of uptake with transport but rather the effect of the surfactants on the two processes.

There is interest in facilitating herbicide entry by additives, not necessarily surfactants, which cause injury to the cuticle and epidermis (Turner, 1972), or by enhancing herbicide activity with ammonium and dihydrogen phosphate ions (Turner and Loader, 1972) in some as yet unknown way.

g. *Plant Factors and Status*. The stage of plant development, e.g. ratio of young to mature leaves, leaf stem ratio, may markedly influence spray retention (Davies *et al.* 1967). Mature fully expanded leaves generally absorb less than immature expanding leaves (Bukovac, 1965). Leaves damaged mechanically or by insects are more permeable than non-damaged leaves. Adequate moisture favours absorption, probably by maintaining the cuticle in a highly hydrated state (van Overbeek, 1956a), and provides for optimum plant function. Foliage on bean plants developing at low soil temperature (13°C) absorb less than those on plants developing at 18° or 24°C (Phillips and Bukovac, 1967). Generally, moisture stress reduces absorption (Currier and Dybing, 1959). The complexity of the effect of moisture stress has been discussed by Hull (1970). More recent data would suggest that moisture stress influences foliar absorption by its effect on transport (Pallas and Williams, 1962; Singh et al. 1972).

h. *Environmental Factors (Light, Temperature, Humidity) Affecting Absorption.* The effects of environmental factors on foliar absorption is very complex. Firstly, light, temperature and humidity may exert their effect during the absorption process, or they may influence plant development prior to absorption, resulting in either an increase or a decrease in sensitivity of the plant to a given herbicide dose (Hammerton, 1967). Further, simultaneous control of all three factors is extremely difficult, and thus the data reported are often confounded to varying degrees.

First, I shall emphasize the impact of light, temperature and humidity during the absorption process, and then discuss how they influence plant sensitivity.

(1) Light. Perhaps one of the first effects of light on penetration of some herbicides relates to persistence of the applied dose on the plant surface. Some organic compounds are rapidly photodegraded and, hence, dosage decreases rapidly with time. Care, therefore, must be exercised in interpreting studies where penetration is measured by loss of activity from the plant surface (Luckwill and Lloyd-Jones, 1962).

Penetration is enhanced by light, and relatively low intensities (5,000 to 15,000 lx) are adequate for maximal response (Sargent and Blackman, 1965; Greene and Bukovac, 1971). The complexity of the light effect on 2,4-D penetration is illustrated by the studies of Sargent and Blackman (1965) with *Phaseolus*. Penetration continued at a steady state in the dark. In light, penetration was slow and steady for about 4 h; thereafter, a striking increase in rate occurred. This surge in penetration was not observed at low 2,4-D concentrations (100 mg l⁻¹), at 1°C, or when penetration into the adaxial surface was followed. Furthermore, of several species studied, a surge was observed only in *Phaseolus* and *Beta vulgaris* (Sargent and Blackman, 1972). In contrast, Brian (1967) reported greater penetration of diquat and paraquat into tomato leaves in darkness than in light.

Light-enhanced uptake of 2,4-D (Sargent and Blackman, 1969) and NAA (Greene and Bukovac, 1972) can be negated by inhibitors of the Hill reaction indicating that it is dependent upon a supply of ATP. The light effect is probably not due to mass flow of the treating solution through stomata (Sargent and Blackman, 1962), although conflicting reports continue to appear (Sands and Bachelard, 1973b).

(2) Temperature. Foliar absorption of herbicides is temperature dependent,

and specific examples of temperature effects have been reviewed by Sargent (1965) and Hull (1970). The relatively high temperature coefficients obtained (1·5 to 3·0) suggested that the effect of temperature was on metabolic processes. However, as reviewed by van Overbeek (1956a), sharp increases in Q_{10} values can be associated with changes in the physical nature of the fatty substances in the plasma membrane.

We have demonstrated striking temperature effects directly on the permeability of the cuticular membrane by isolating the cuticle from the underlying living tissue (Norris and Bukovac, 1968). Greater penetration of NAA was obtained with an increase in temperature from 5° to 35°C, with Q_{10} values of 2·4 to 5·6. The temperature effect was completely reversible (Fig. 5) and was demonstrated in isolated cuticles freed of both the epicuticular and cuticular waxes (Norris and Bukovac, 1969). Of particular interest was a marked change in permeability of pear leaf cuticle in the range of 15° to 25°C.

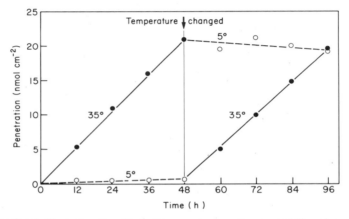

Fig. 5. Graph illustrating the effect of temperature on the permeability of enzymatically isolated *Pyrus communis* leaf cuticle to NAA. (After Norris and Bukovac, 1969.)

The high Q_{10} values observed for the penetration of NAA through isolated cuticle are thought to be related to the lipoidal nature of the cuticle. Sutcliffe (1962) pointed out that high temperature coefficients could be expected for lipid membranes because of the relatively high energy barrier that must be overcome for a molecule to diffuse through the membrane. The marked change in cuticular permeability between 15° and 25°C may be associated with a phase change in the cuticle, although we have no direct evidence.

Sands and Bachelard (1973a) found, in the light, increased penetration of picloram into *Eucalyptus viminalis* leaves with an increase in temperature, but only a slight increase or no effect in the dark. This observation is difficult to explain. Based on temperature effects on penetration through isolated cuticles, one would expect a similar effect of temperature in light and dark.

(3) Humidity. Foliar absorption of herbicides is generally favoured by high

relative humidity. High relative humidity increases the drying time of the spray droplets (Prasad *et al.* 1967), favours stomatal opening (Prasad *et al.* 1967), enhances transport (Pallas, 1960) and may increase the permeability of the cuticular membrane (van Overbeek, 1956*a*). Greater absorption of NAA occurred at high (100%) than at low (37%) relative humidity not only from the original spray droplet, but also from the spray residue after the aqueous phase evaporated (Luckwill and Lloyd-Jones, 1962). Greater quantities of 2,4-D and benzoic acid were absorbed and transported in bean plants at 70 to 74% r.h. than at 34 to 48% r.h. The increased absorption was correlated with degree of stomatal opening (Pallas, 1960). Similarly, dinitro-*o*-cresol and dalapon absorption was favoured by high relative humidity (Westwood *et al.* 1960; Prasad *et al.* 1967). Relative humidity was the most important environmental factor in the absorption of maleic hydrazide, three to five times more being absorbed at 100% r.h. than at 50% r.h. (Smith *et al.* 1959).

(4) Effect of environment prior to treatment. The environment under which the plant develops may markedly affect the absorption of a chemical subsequently applied to the foliage. Leaves expanding in full sunlight produce a heavier cuticle than those developing in shade (Skoss, 1955). Predisposing plants to high (20° to 30°C) temperature and humidity (70 to 100%) results in greater penetration than when predisposed to low temperature and humidity (Westwood and Batjer, 1958; Edgerton and Haesler, 1959; Donoho *et al.* 1961). Low root temperature (Phillips and Bukovac, 1967) and moisture stress (Skoss, 1955) reduce foliar absorption.

The above environmental factors markedly affect the quantity, fine-structure and chemistry of the epicuticular wax (Hull, 1958; Juniper, 1960; Baker, 1972; Whitecross and Armstrong, 1972). Modification of the epicuticular wax may alter wettability and retention as well as permeability *per se*. Under field conditions, light, temperature and humidity interact, and quantification of the relative significance of each is extremely difficult.

3. *Relative Role of Cuticular Components*

Of the three principal components of the cuticle, the epicuticular wax is the prime barrier to foliar penetration. Epicuticular wax, by nature of its fine-structure and chemistry, determines to a large extent the wetting properties and, thus, retention by the plant surface. Disruption of these waxes by light brushing has been shown to enhance wetting and water loss (Hall and Jones, 1961), and to increase the uptake of 3-chlorophenoxypropionic acid (Bukovac, 1965). For leaves exhibiting a contact angle greater than 90°, the epicuticular wax is the dominant factor governing wettability, and removal of these waxes with an organic solvent generally results in increased wettability (Holloway, 1969*b*). Notable exceptions are *Acer pseudoplatanus* and *Nymphaea* leaves (Fogg, 1948). In any case, the contact angle seldom exceeds 100°.

The relative effects of the epicuticular and cuticular waxes on permeability of *Lycopersicon esculentum* fruit cuticle is noted in Table I. 2,4-D penetration

Table I
EFFECT OF CUTICULAR COMPONENTS ON THE PENETRATION OF NAA AND 2,4-D THROUGH
ENZYMATICALLY ISOLATED *Pyrus communis* LEAF AND *Lycopersicon esculentum*
FRUIT CUTICLES (DATA AFTER BUKOVAC 1970)

	Pyrus leaf cuticle NAA	Lycopersicon fruit cuticle 2,4-D
	(% of control)	
Control (not dewaxed)	100	100
Epicuticular wax removed	—[1]	816
Epicuticular and cuticular waxes removed	1,185	951

[1] Not determined.

was increased approximately 8-fold after removal of epicuticular wax and only slightly more (9-fold) upon removal of both cuticular and epicuticular waxes. Sorption and penetration of NAA was 5- to 10-fold greater after removal of the surface wax from *Pyrus communis* leaf cuticle (Norris and Bukovac, 1972). A similar effect on permeability to *N*-isopropyl-α-chloroacetamide was observed for *Prunus armeniaca* leaves (Darlington and Barry, 1965).

Studies on the relative permeability of *Lycopersicon* fruit cuticle, with and without waxes, to derivatives of phenoxyacetic acid revealed that dewaxed cuticles were more permeable to the non-chlorinated or mono-chlorinated derivatives compared with the di- or tri-chlorinated derivatives, i.e. waxes presented a greater barrier to penetration of the more polar derivatives (Bukovac *et al.* 1971).

4. *Specific Pathways Through the Cuticle*

Crafts (1956) probably was first to suggest that there may be specific pathways through the cuticle. He based his suggestion on mounting evidence that lipid-soluble compounds penetrated the cuticle more readily than water soluble chemicals, plus the recognition that dissociated molecules of weak organic acids (polar in nature) also penetrated and induced a response. He proposed that lipid-soluble compounds followed a lipoidal route and water soluble compounds an aqueous pathway. To date little is known about the nature of any specific pathway through the cuticle which might accommodate foliar-applied chemicals.

It is generally felt that lipid-soluble compounds are sorbed and diffuse through the lipoidal components of the cuticle. Once on the cell-wall side, if sufficiently water soluble, they desorb from the cuticle into the cell wall. The pathway for water-soluble or ionic compounds is less clear. The cuticle is frequently described as being sponge-like and hydrated, suggesting that there may be an aqueous continuum between the plant and its environment.

Convincing data on cuticular transpiration provides strong evidence of water movement through the cuticle, but there is little direct evidence that foliar-applied compounds enter by these same routes. Roberts *et al.* (1948) using histochemical techniques, observed pectic strands in the cuticle of *Malus* leaves and suggested that they might serve as pathways for polar materials. Such structures have not been generally found in plant cuticles. Although the existence of pores or microchannels is still controversial, there is overwhelming evidence supporting the absence of such structures in the cuticle (see review by Hull, 1970).

Much has been written on the involvement of ectodesmata (also referred to as Gilson diffusion patterns, mercuric precipitates, crystalline mercurial patterns and ectocythodes) as pathways in foliar penetration (Hull, 1970; Franke, 1971a, 1971b). Evidence for their role in foliar absorption has been extensively reviewed by Franke (1961, 1967, 1970).

Ectodesmata were first thought to be protoplasmic extensions of epidermal cells into the outer cell wall, much like plasmadesmata except that they terminated at the cell wall cuticle interface (Schumacher, 1942; Schumacher and Lambertz, 1956; Schnepf, 1959; Sievers, 1959; Franke, 1962). Franke (1964b) later presented evidence that ectodesmata were not protoplasmic strands but rather well-defined cell wall structures rich in reducing substances. Crisp (1965) confirmed that these structures were not plasmic.

It should be pointed out that ectodesmata are demonstrated indirectly by mercury precipitates formed during a fixation process (Schnepf, 1959). From this precipitate conclusions on the distribution and morphology of ectodesmata have been drawn. Based on the above procedures, ectodesmata reportedly occur most commonly over anticlinal walls of epidermal cells, around guard cells, in stomatal ledges, at the base of trichomes and in outer walls of epidermal cells over major veins (Franke, 1961). They vary in size and shape and are always localized in the cell wall. It should be stressed that these features have not been shown to exist in the cuticle.

We have recently suggested that ectodesmata, as demonstrated by Gilson fixation, may not be definable cell-wall structures (Schönherr and Bukovac, 1970a, 1970b). The characteristic distribution pattern and number can be altered by brushing the leaf surface or by removing the epicuticular wax with chloroform. These new ("induced") ectodesmata are now present in areas not formerly rich in ectodesmata and are randomly distributed.

Convincing evidence that ectodesmata are not definable cell-wall structures comes from studies with *Allium cepa* bulb scale and leaf cuticles (Schönherr and Bukovac, 1970a). Isolated cuticles mounted on gelatin pretreated with ascorbic acid as a reductant were processed to demonstrate ectodesmata. The distribution of ectodesmata in the cell-wall-free cuticle-gelatin model was identical with that observed in the stripped epidermis (Fig. 6), indicating that their distribution is governed by the cuticle and not by structures in the cell wall. The pattern observed most likely reflects sites in the cuticle preferentially permeable to mercuric chloride. This is not to claim that protoplasmic extensions do not extend into the cell wall; however, evidence is convincing

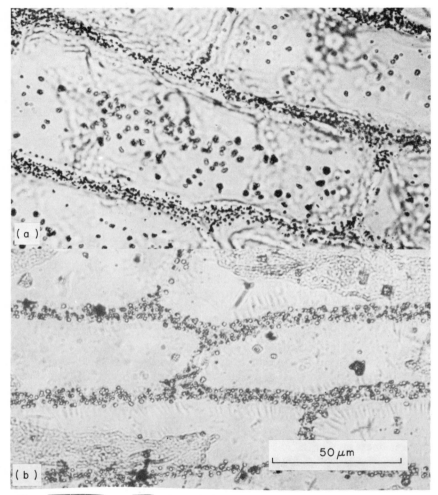

Fig. 6. Photomicrographs (surface view) illustrating the distribution of preferentially permeable areas in *Allium cepa* bulb scale cuticle to mercuric chloride. The dots represent mercury precipitates beneath the area of the cuticle permeable to mercuric chloride. (a). Distribution pattern in stripped inner epidermis. (b). Distribution pattern obtained with isolated cuticle of inner epidermis mounted on gelatin treated with ascorbic acid. Note identical distribution pattern as with stripped epidermis. (After Schönherr and Bukovac, 1970a.)

that the features demonstrated by Gilson fixation and currently termed ectodesmata are not plasmic in nature and are not definable cell-wall structures. What is critical for foliar absorption is that there are areas in the cuticle which are preferentially permeable to polar compounds. Such preferentially permeable areas could play a significant role in foliar penetration by serving as polar bridges across the chiefly lipoidal cuticular membrane.

1. *Trichomes*

In addition to influencing wetting and retention, trichomes are more directly involved in foliar absorption (Linskens *et al.* 1965). Leaf hairs, depending on their morphology, may provide a microclimate which can alter the drying time of aqueous sprays and thus the absorption pattern. Since in most instances leaf hairs are extensions of epidermal cells, an increased epidermal area is exposed to the spray solution. Those trichomes containing living protoplasm may be of particular significance.

Hallier (1868) established that glandular trichomes of *Pelargonium* absorbed the anthocyanin pigment of cherry juice applied to the leaf. Similarly indigo-sulphuric acid was taken up by trichomes of *Myosotis, Dianthus* and *Tradescantia* as well as *Pelargonium*. Uptake of fluorochrome dyes (Strugger, 1939) and berberine sulphate (Butterfass, 1956) has been associated with trichomes. Glandular hairs associated with major veins on *Phaseolus* leaves became labelled before other epidermal cells when the leaf was presented a solution of silver nitrate (Schönherr, 1969). Anticlinal walls and the area around the base of trichomes appeared preferentially permeable (Fig. 7(a), (b)). The greater permeability around the base or foot cells of trichomes may be related to delayed development of the cuticle in this region (Sifton, 1963).

2. *Stomata*

Absorption of foliar-applied chemicals from aqueous sprays is usually greater through the lower than upper leaf surface (Currier and Dybing, 1959; Sargent and Blackman, 1962; Hull, 1970). The greater permeability of the lower leaf surface has been attributed, in a general way, to the greater number of stomata present. However, the lower epidermis may have other characteristics, e.g. trichomes, morphology and chemistry of the cuticle, degree of wax present, etc., which may also have an effect on penetration. Nevertheless, good correlations exist in *Phaseolus* between the numbers of stomata present, but not necessarily the degree of stomatal opening, and penetration (Sargent and Blackman, 1962; Jyung *et al.* 1965).

Stomata appear to play a two-fold role in foliar penetration. Firstly, under certain conditions, the aqueous spray solution may move in mass through the stomatal pore and diffuse through the air space of the leaf (Ebeling, 1939; Turrell, 1947; Dybing and Currier, 1961; Schönherr and Bukovac, 1972a; Greene and Bukovac, 1974). Secondly, the cuticle over the guard and associated accessory cells may be more permeable and these structures, *per se,* serve as preferred sites of entry (Weber, 1930; Franke, 1964a; Neumann and Jacob, 1968).

Guard cells are known to be more permeable to urea, glycerine (Weber, 1930) and fluorochrome dyes (Butterfass, 1956) than the other epidermal cells, and participate in cuticular transpiration (Strugger, 1939; Maercker, 1965). Further, Butterfass (1956) demonstrated that the cuticular ledges were particularly permeable to berberine sulphate, and that the ledges of open

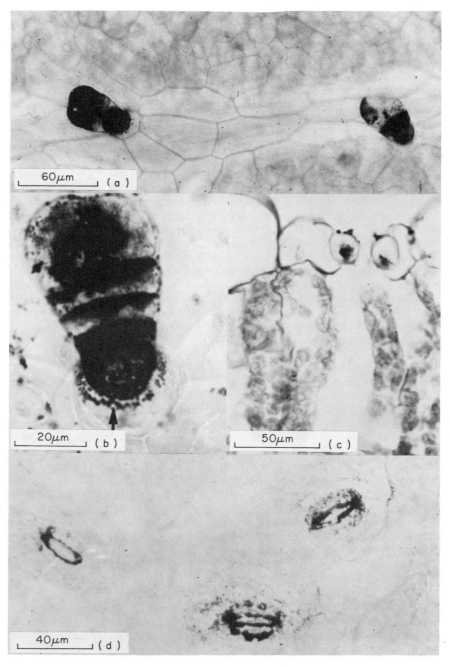

Fig. 7. Photomicrographs of preferential penetration of silver into trichomes ((a), (b)), cuticular ledges (c) and guard cells and anticlinal walls (d). (After Schönherr, 1969.)

stomata became labelled earlier than those of closed stomata. Schönherr (1969) made a similar observation with silver nitrate (Fig. 7(c), (d)). Other diverse molecules applied to leaves, such as aminobutyric acid (Neumann and Jacob, 1968) and sucrose (Franke, 1964a), become preferentially localized in the guard cells. Thus, stomata play a principal role in foliar absorption via the guard cells.

Evidence for mass movement of sprays into leaves through the stomatal pores is conflicting. While stomatal pore penetration has been reported for aqueous solutions (Skoss, 1955; Foy, 1964; Currier et al. 1964; Eddings and Brown, 1967), other workers (Weaver and DeRose, 1946; Turrell, 1947; Sargent and Blackman, 1962; Jyung et al. 1965) have concluded that mass movement of aqueous sprays does not occur in this manner.

The mechanism of stomatal pore penetration is complex. The surface tension of the liquid, the contact angle (wettability) produced by the liquid on the plant surface, the morphology and chemistry of the pore wall and whether or not pressure is involved must all be taken into account. Aqueous solutions with surface tensions approaching that of pure water will not penetrate the stomatal pore (Turrell, 1947; Adam, 1948; Schönherr and Bukovac, 1972a). Penetration can be achieved if the surface tension of the spray solution is equal to or less than the critical surface tension (Fox and Zisman, 1950) of the plant surface, i.e. the solution forms zero contact angle (complete wetting) on the plant surface. When complete wetting is not achieved, pore penetration can be expected only if the contact angle formed by the spray is less than the minimum wall angle of the stomatal pore or if pressure is applied (Schönherr and Bukovac, 1972a). The critical surface tension of most cuticular surfaces is ·in the neighbourhood of 30 dyne cm^{-1}. Few surfactants commonly used in commercial agriculture will reduce the surface tension below 30 dyne cm^{-1}.

One further consideration is important. Even after entry into the substomatal chamber, the chemical must transgress a cuticular membrane covering the guard, epidermal (Fig. 1(c), (d)) and mesophyll cells (Frey-Wyssling and Häusermann, 1941).

Stomatal penetration of aqueous sprays with commonly used surfactants appears minimal. Even with efficient surfactants, only 4 to 5% of the stomata of Pyrus communis leaves were infiltrated (Greene and Bukovac, 1974). However, because of complex interacting factors, direct evidence under field conditions is not available to assess the contribution of stomata quantitatively. Herbicides applied in organic solvents (Stalfet, 1916) and in oils (Turrell, 1947) of low surface tension may penetrate via stomatal pores.

III. Penetration Through the Stem (Bark)

The degree of penetration through the stem and bark will vary considerably depending on the growth characteristics and stage of plant development. Compared to the foliage, stems present a limited target area; however, there is an extensive and highly developed transport system beneath the surface capable of rapid export of the absorbed chemical. Unfortunately, only limited data exist which describe in depth the nature of shoot and bark penetration;

however, rather general concepts have emerged based mostly on performance following localized application to these structures.

Herbicides are readily absorbed by young stem tissue (Parker, 1966; Prendeville, 1968; Oliver *et al.* 1968; Nishimoto and Warren, 1971); however, marked differences may exist among species (O'Brien and Prendeville, 1972). In general, penetration into immature stem tissue is assumed to follow closely that discussed above for leaves.

Movement through the bark of woody plants is quite different. Although it is difficult to generalize, the following can be visualized for numerous woody plants. The periderm, a protective tissue replacing the epidermis after it dies and sloughs off, is composed of the phellogen, the phellem (cork) and the phelloderm (Esau, 1962). The cork tissue is composed of tightly packed cells devoid of intercellular spaces and, when mature, lacking in protoplasm. These cells contain tannins and are highly suberized. Other constituents commonly found in the periderm are fatty acids, lignin, cellulose and terpenes. Based on structure and composition, the periderm should exhibit low permeability to water and foliar-applied chemicals, particularly those polar in nature.

Any feature which may serve to by-pass the bark should be a potential pathway for penetration. Lenticels are essentially radial channels into or completely through the cork layer. The constituent cells are generally not suberized and readily absorb water. With continued secondary growth, radial cracks and fissures develop which may serve as portals of entry.

Even though the bark appears to present a formidable barrier to penetration, plant response to a wide range of chemicals, e.g. nutrient salts (Tukey *et al.* 1954), herbicides (Crafts, 1964) and complex antibiotics (Lemin *et al.* 1960) has been noted following bark application. To what extent these substances penetrated through the bark, *per se,* or through lenticels, pruning wounds, breaks or sites of injury is not clear. Harley *et al.* (1956) claimed that nutrient salts did not penetrate unbroken bark, but rather entered through natural openings, leaf scars and particularly through growth cracks.

In application of herbicides to woody plants, the bark is commonly accepted as a serious barrier. For acceptable performance the herbicide is usually prepared in a lipid-soluble form (e.g. ester), formulated in oil at a high (5 to 10%) concentration and sprayed liberally or painted on the bark. Aqueous sprays are not efficacious. An alternative practice is to by-pass the bark by introducing the herbicide through cuts (Leonard, 1957) or by injections (Shipman, 1973). In such cases the bark is mechanically penetrated and the herbicide is introduced into the transport system, water soluble formulations being most useful (Turner, 1973).

IV. Penetration through the Root

A. INTRODUCTION

Many herbicides are applied to the soil where the root serves as the primary absorbing organ. However, soil-applied herbicides may also enter seeds (Phillips *et al.* 1972; Scott and Phillips, 1973) and during seedling emergence or in some established plants may penetrate through the shoot (Parker, 1966; Prendeville, 1968; O'Brien and Prendeville, 1972). For the relative role of each in penetration, the reader is referred to the above-mentioned papers. In this chapter, I will discuss only the penetration of soil-applied herbicides into the root. The numerous interactions of the applied chemical in the soil environment and movement of the chemical to the root surface is stressed in Volume 2, Chapter 1. There are numerous reports documenting that herbicides of diverse molecular structure, size and solubility are readily absorbed by roots (Crafts and Yamaguchi, 1960). Unfortunately, few provide a quantitative assessment of the nature of the uptake process. Our concepts rely extensively on data generated for nutrient ion and water absorption.

B. WATER AND SOLUTE ABSORBING ZONES AND STRUCTURE

Roots present a large absorbing surface with zones preferentially permeable to water and solutes. Young roots are covered with a cuticular membrane (Scott *et al.* 1963). Even though older roots are highly suberized, some water and salt absorption takes place (Kramer, 1946).

In young plants the primary absorbing region of a root lies between 5 and 50 mm behind the root tip (Tanton and Crowdy, 1972). The root in the absorbing zone is characterized by a xylem sufficiently differentiated to be functional and an endodermis not sufficiently lignified to pose a serious barrier. Probably most of the water and solutes are absorbed in the same region.

Detailed studies with primary roots of *Zea mays* revealed that maximum uptake capacities for sodium decreased with increasing distance from the root tip (Eshel and Waisel, 1972). Both quantitative and qualitative differences in the sodium uptake mechanism were noted along the root.

In young roots an appreciable volume (7 to 10%) of the root is readily permeable to water and solutes, and this volume is frequently referred to as the "free space". This space is probably composed primarily of the intercellular air spaces and cell-wall volume of the cortex. Essentially, the "free space" is a continuum of the ambient environment. At the innermost region of the cortex, there is a distinct single layer of closely packed cells (endodermis) whose radial and transverse walls are partially lignified, and in some cases suberized (Esau, 1962). The endodermis constitutes a barrier to free movement of water and solutes to the stele.

C. FACTORS AFFECTING UPTAKE

1. *Time-course of Uptake*

In general, the time-course of root-uptake is characterized by an initial rapid phase for the first 30 min to 2 h followed by a second phase of slower but steady uptake. This generalized curve is typical for several species and compounds (Smith and Sheets, 1967; Isenee *et al.* 1971). Kinetic studies with *Lemna minor* show a similar pattern for dalapon (Prasad and Blackman, 1965a); however, for 2,3,5-tri-iodobenzoic acid (Blackman and Sargent, 1959) and 2,4-D (Blackman *et al.* 1959) the initial uptake was rapid with the subsequent rate falling off to zero and then becoming negative, i.e. a loss or efflux of the compounds from the roots. A similar egress after initial absorption has been noted for DNOC from winter rye (*Secale cereale*) roots by Bruinsma (1967), and for other herbicides (Moody *et al.* 1970).

2. *Temperature, pH and Concentration*

Root uptake is temperature dependent, being depressed upon lowering of the temperature (Bruinsma, 1967; Moody *et al.* 1970; Vostral *et al.* 1970; Isenee *et al.* 1971). The effect of temperature on the initial phase of dalapon uptake is less than on the second phase. The Q_{10} values for the temperature range 17·5° to 27·5°C were 1·42 during the first 30 min of uptake and 2·38 for the 30 min to 2 h time period (Prasad and Blackman, 1965*b*). Temperature was not a major factor in egress of dalapon from *Lemna minor,* but the loss of DNOC from winter rye roots (Bruinsma, 1967) and 2,4-D from *L. minor* roots (Blackman *et al.* 1959) was greater at high (20° to 25°C) than at low (3° to 10°C) temperature.

Hydrogen ion concentration affects root uptake of some herbicides in a manner similar to that for shoot uptake. Greater quantities of DNOC, 2,4-D, dalapon, and picloram were recovered in plants when these compounds were presented to roots at low than at high pH. Absorption of the non-dissociated ion was favoured over that of the dissociated ion.

Within limits a linear relationship exists between root uptake and concentration of the herbicide available to the roots (Prasad and Blackman, 1965*a*; Moody *et al.* 1970). Linearity is lost at high concentrations where the herbicide may exert a toxicological effect on the plant.

3. *Transpiration and Uptake*

Transpiration and herbicide uptake are considered to be closely linked (Sheets, 1961; Bruinsma, 1967). However, Moody *et al.* (1970) concluded that herbicide absorption and transpiration were unrelated. The data of Sikka and Davis (1968) show a marked decrease with time in uptake of prometryne per unit of water absorbed. For soybeans approximately 0·3 µg of prometryne was absorbed per ml of water during the first 12 h of the absorption period, while between 36 and 48 h only 0·02 µg was absorbed per ml of water. Similar data were obtained with cotton. Thus with time there was a divergence between absorption of water and the absorption of herbicide.

Other convincing evidence that transpiration may not be directly involved in root uptake of herbicides comes from absorption data using isolated roots. Uptake curves were similar for isolated roots and intact plants (Bruinsma, 1967; Moody *et al.* 1970), and the response of both to temperature was similar. Another interesting experiment by Bruinsma (1967) provides evidence that DNOC uptake was not tightly coupled to transpiration. Winter rye plants were permitted to absorb DNOC from solutions of varying pH levels, producing varying degrees of DNOC dissociation, and at varying transpiration conditions. At all times of observation over a 4-h period more DNOC was taken up than could be accounted for by passive movement with the transpiration stream. Further, uptake was favoured at pH levels favouring the undissociated species. Obviously, the undissociated DNOC molecules must have gained entry other than via the transpiration stream.

Transpiration apparently plays an important role in root uptake of water-soluble herbicides by facilitating the movement of the absorbed chemical to the aerial plant parts, thus maintaining a concentration gradient from the external medium to the apoplast.

D. MECHANISM OF UPTAKE

Sufficient data are not available to understand fully the mechanism(s) of herbicide uptake. Based on current evidence and on our knowledge of water and nutrient ion uptake, the following is an approximation of the absorption process.

During the initial phase, water-soluble herbicides diffuse into the "free space" of the root. Uptake into the free space is concentration dependent, has a low Q_{10}, and the herbicide ions can be readily lost on transfer to distilled water. These data all suggest that physical factors play a commanding role in the initial phase of uptake. In addition to the "free space" there may be a rapid uptake of cations which are held against distilled water, but can be exchanged for cations in a medium. These cations are electrically bound to anions contained in the cell structure. This exchangeable fraction corresponds to the "Donnan free space" (Brouwer, 1965).

The second phase of uptake, i.e. movement across the plasmalemma and accumulation in cells, is an energy-requiring process. This phase has a higher Q_{10} than the initial phase and is sensitive to metabolic inhibitors. Recently Smith (1972) concluded that 2,4-D uptake by potato (*Solanum tuberosum*) tuber parenchyma tissue was independent of the available energy, but rather was associated with the level of lecithin in the tissue. He suggested that 2,4-D may have been bound to the lecithin molecule.

Non-dissociated and lipid-soluble molecules may partition directly into the lipid constituents of the root. Lipid-rich membranes might well be a preferred site of solubilization. This process could be temperature sensitive and may result in a concentration of herbicide in the root greater than that in the ambient solution.

Although numerous studies have been reported on root uptake of herbicides, our knowledge of the mechanism involved is woefully lacking. The above is a tentative generalization. Before any credible concept can be proposed, extensive critical studies are essential.

V. EXUDATION THROUGH THE ROOT

Loss of endogenous organic substances from roots has been known for some time. Of more recent interest is the root exudation of organic chemicals, including herbicides, applied to the foliage of plants (Mitchell and Linder, 1963). Several important classes of compounds, namely, phenylacetic acids (Linder *et al.* 1964), benzoic acids (Linder *et al.* 1964), phenoxyacetic acids (Neidermyer and Nalewaja, 1969) and picolinic acid (Reid and Hurtt, 1970; Sharma *et al.* 1971) are known to leak out of roots after having been applied to the foliage. Many of these compounds are exuded unaltered (Linder *et al.* 1964; Reid and Hurtt, 1970), and in quantities ranging from less than 1% to more than 10% of the original dose applied to the foliage in a 24-h period (Linder *et al.* 1964; Sharma *et al.* 1971).

The chemical structure of the molecule appears to be a determining factor. With α-methoxyphenylacetic acid (MOPA), the number, position and nature of substituents in both the aromatic and aliphatic portions are critical (Mitchell *et al.* 1959). For example, in *Phaseolus* the *m*-chloro derivative was exuded while the *o*-chloro derivative was not; the *p*-fluoro derivative was readily lost while the *p*-bromo was not. Twenty per cent of the foliar-applied MOPA leaked out in a three-day period compared to only 3% of the 2,4-D.

Using [^{14}C]-MOPA as a model compound, Mitchell *et al.* (1961) attempted to elucidate the mechanism involved. Root exudation increased gradually for the first 5 to 15 h after administration to the foliage, then increased rapidly for the subsequent 20 to 30 h before levelling off. This rapid increase corresponded with saturation of the root, i.e. the level of MOPA in the root did not increase further with time. The rate of root exudation during this time period was proportional to the dose applied. Removal of a portion of the root system resulted in a corresponding reduction in exudation, and only a trace of ^{14}C was exuded from the cut-end of the hypocotyl after all of the roots were removed. Similarly, if some roots were killed by steam, MOPA exudation was limited to living roots. These data suggest that root exudation of herbicides may be an active process.

The extent of exudation appears to be further related to ease of transport and metabolism of the compound. In general, compounds that are not readily transported from the foliage to the roots and are not readily metabolized are less likely to be exuded (Linder *et al.* 1964). The absence of exudation of such compounds may merely be related to the fact that some critical threshold is never attained in the roots rather than that the compound may have some unique properties preventing exudation.

Rovira (1973) demonstrated that the most active region of exudation of ^{14}C-labelled photosynthates from plant roots was the zone of elongation—the region most active in root absorption.

REFERENCES

Adam, N. K. (1948). *Discuss. Faraday Soc.* **3,** 5–11.
Adam, N. K. (1958). *Endeavour,* **17,** 37–41.
Amelunxen, F., Morgenroth, K. and Picksak, T. (1967). *Z. PflPhysiol.* **57,** 79–95.
Anderson, D. B. (1928). *Jb. wiss. Bot.* **69,** 501–515.
Artz, T. (1933). *Ber. dt. bot. Ges.* **51,** 470–500.
Baker, E. A. (1970). *New Phytol.* **69,** 1053–1058.

Baker, E. A. (1971). *In* "Ecology of Leaf Surface Microorganisms" (T. F. Peerce and C. H. Dickerson, eds), pp. 55–65. Academic Press, London and New York.

Baker, E. A. (1972). *The effect of environmental factors on the development of the leaf wax of Brassica oleracea* var. *gemmifera*, M.S. Thesis, Univ. of Bristol, Bristol.

Baker, E. A. and Martin, J. T. (1963). *Nature, Lond.* **199**, 1268–1270.

Baker, E. A. and Holloway, P. J. (1970). *Phytochemistry*, **9**, 1557–1562.

Baker, E. A. and Bukovac, M. J. (1971). *Ann. appl. Biol.* **67**, 243–253.

Barrentine, J. L. and Warren, G. F. (1970). *Weed Sci.* **18**, 373–377.

Becher, P. and Becher, D. (1969). *In* "Pesticide Formulations Research: Physical and Colloidal Chemical Aspects", *Adv. Chem. Ser* **No. 86** (R. F. Gould, ed.), pp. 15–23. American Chemical Society, Washington.

Blackman, G. E., Bruce, R. S. and Holly, K. (1958). *J. exp. Bot.* **9**, 175–205.

Blackman, G. E. and Sargent, J. A. (1959). *J. exp. Bot.* **10**, 480–503.

Blackman, G. E., Sen, G., Birch, W. R. and Powell, R. G. (1959). *J. exp. Bot.* **10**, 33–54.

Brian, R. C. (1967). *Ann. appl. Biol.* **59**, 91–99.

Brian, R. C. (1972). *Pestic. Sci.* **3**, 121–132.

Brongniart, A. (1834). *Annls. Sci. nat. (Bot.)* **1**, 65–71.

Brouwer, R. (1965). *A. Rev. Pl. Physiol.* **16**, 241–266.

Bruinsma, J. (1967). *Acta bot. neerl.* **16**, 73–85.

Brunskill, R. T. (1956). *Proc. 3rd Br. Weed Control Conf.* **2**, 593–603.

Bukovac, M. J. (1965). *Proc. Am. Soc. hort. Sci.* **87**, 131–138.

Bukovac, M. J. (1970). *Proc. 18th int. hort. Congr.* **4**, 21–42.

Bukovac, M. J. and Norris, R. F. (1966). *In* "Trasporto delle molecole organiche nelle piante", pp. 296–309. VI *Simp. int. Agrochim. Varenna, Italy, 5–10 Sept.* 1966.

Bukovac, M. J., Sargent, J. A., Powell, R. G. and Blackman, G. E. (1971). *J. exp. Bot.* **22**, 598–612.

Buta, J. G. and Steffens, G. L. (1971). *Physiologia Pl.* **24**, 431–435.

Butterfass, T. (1956). *Protoplasma*, **47**, 415–428.

Challen, S. B. (1960). *J. Pharm. Pharmac.* **12**, 307–311.

Challen, S. B. (1962). *J. Pharm. Pharmac.* **14**, 707–714.

Crafts, A. S. (1956). *Am. J. Bot.* **43**, 548–556.

Crafts, A. S. (1961). "The Chemistry and Mode of Action of Herbicides", Interscience, New York.

Crafts, A. S. (1964). *In* "The Physiology and Biochemistry of Herbicides" (L. J. Audus, ed.), pp. 75–110. Academic Press, London and New York.

Crafts, A. S. and Yamaguchi, S. (1960). *Am. J. Bot.* **47**, 248–255.

Crisp, C. E. (1965). *The biopolymer cutin*, Ph.D. Thesis, Univ. of Calif., Davis.

Currier, H. B. and Dybing, C. D. (1959). *Weeds*, **7**, 195–213.

Currier, H. B., Pickering, E. R. and Foy, C. L. (1964). *Weeds*, **12**, 301–303.

Darlington, W. A. and Cirulis, N. (1963). *Pl. Physiol., Lancaster*, **38**, 462–467.

Darlington, W. A. and Barry, J. B. (1965). *J. agric. Fd Chem.* **13**, 76–78.

Davies, P. J., Drennan, D. S. H., Fryer, J. D. and Holly, K. (1967). *Weed Res.* **7**, 220–233.

Donoho, C. W., Jr., Mitchell, A. E. and Bukovac, M. J. (1961). *Proc. Am. Soc. hort. Sci.* **78**, 96–103.

Dybing, C. D. and Currier, H. B. (1961). *Pl. Physiol., Lancaster*, **36**, 169–174.

Ebeling, W. (1939). *Hilgardia*, **12**, 665–698.

Eddings, J. L. and Brown, A. L. (1967). *Pl. Physiol., Lancaster*, **42**, 15–19.

Edgerton, L. J. and Haesler, C. W. (1959). *Proc. Am. Soc. hort. Sci.* **74**, 54–60.

Eglinton, G. and Hunneman, D. H. (1968). *Phytochemistry*, **7**, 313–322.

Ennis, W. B., Jr., Williamson, R. E. and Dorschner, K. P. (1952). *Weeds*, **1**, 274–286.

Esau, K. (1962). "Plant Anatomy", John Wiley, New York and London.

Eshel, Y. and Waisel, Y. (1972). *Pl. Physiol., Baltimore*, **49**, 585–589.

Fogg, G. E. (1948). *Discuss. Faraday Soc.* **3**, 162–166.

Fox, H. W. and Zisman, W. A. (1950). *J. Colloid Sci.* **5**, 514–531.

Foy, C. L. (1964). *J. agric. Fd Chem.* **12**, 473–476.

Foy, C. L. and Smith, L. W. (1965). *Weeds*, **13**, 15–19.

Foy, C. L. and Smith, L. W. (1969). *In* "Pesticide Formulations Research: Physical and Colloidal Chemical Aspects", *Adv. Chem. Ser.* No. 86 (R. F. Gould, ed.), pp. 55–69. American Chemical Society, Washington.

Franke, W. (1961). *Am. J. Bot.* 48, 683–691.

Franke, W. (1962). *Planta,* 59, 222–238.

Franke, W. (1964a). *Nature, Lond.* 202, 1236–1237.

Franke, W. (1964b). *Planta,* 63, 279–300.

Franke, W. (1967). *A. Rev. Pl. Physiol.* 18, 281–300.

Franke, W. (1970). *Pestic. Sci.* 1, 164–167.

Franke, W. (1971a). *Residue Rev.* 38, 81–115.

Franke, W. (1971b). *Ber. dt. bot. Ges.* 84, 533–537.

Freed, V. H. and Montgomery, M. (1958). *Weeds,* 6, 386–389.

Frey-Wyssling, A. and Häusermann, E. (1941). *Ber. schweiz. bot. Ges.* 51, 430–431.

Furmidge, C. G. L. (1959). *J. Sci. Fd Agric.* 10, 274–282.

Furmidge, C. G. L. (1962). *J. Sci. Fd Agric.* 13, 127–140.

Greene, D. W. and Bukovac, M. J. (1971). *J. Am. Soc. hort. Sci.* 96, 240–246.

Greene, D. W. and Bukovac, M. J. (1972). *Pl. Cell Physiol., Tokyo,* 13, 321–330.

Greene, D. W. and Bukovac, M. J. (1974). *Am. J. Bot.* 61, 100–106.

Hall, D. M. and Jones, R. L. (1961). *Nature, Lond.* 191, 95–96.

Hallier, E. (1868). "Phytopathologie, Die Krankheiten der Culturgewächse". Wilhelm Engelmann, Leipzig.

Hammerton, J. L. (1967). *Weeds,* 15, 330–336.

Harley, C. P., Regeimbal, L. O. and Moon, H. H. (1956). *Proc. Am. Soc. hort. Sci.* 67, 47–57.

Hartley, G. S. and Brunskill, R. T. (1958). *In* "Surface Phenomena in Chemistry and Biology" (J. F. Danielli, K. G. A. Pankhurst and A. C. Reddiford, eds), pp. 214–223. Pergamon Press, London.

Heinen, W. and Brand, I. V. D. (1963). *Z. Naturf.* 18b, 67–79.

Holloway, P. J. (1969a). *J. Sci. Fd Agric.* 20, 124–128.

Holloway, P. J. (1969b). *Ann. appl. Biol.* 63, 145–153.

Holloway, P. J. (1970). *Pestic. Sci.* 1, 156–163.

Holloway, P. J. and Baker, E. A. (1968). *Pl. Physiol., Lancaster,* 43, 1878–1879.

Holloway, P. J. and Dees, A. H. B. (1971). *Phytochemistry,* 10, 2781–2785.

Hopp, H. and Linder, P. J. (1946). *Am. J. Bot.* 33, 598–600.

Hull, H. M. (1957). *Pl. Physiol., Lancaster (Suppl.)* 32, 43.

Hull, H. M. (1958). *Weeds,* 6, 133–142.

Hull, H. M. (1970). *Residue Rev.* 31, 1–155.

Isenee, A. R., Jones, G. E. and Turner, B. C. (1971). *Weed Sci.* 19, 727–731.

Juniper, B. E. (1959). *Endeavour,* 18, 20–25.

Juniper, B. E. (1960). *J. Linn. Soc. (Bot.)* 56, 413–419.

Jyung, W. H., Wittwer, S. H. and Bukovac, M. J. (1965). *Proc. Am. Soc. hort. Sci.* 86, 361–367.

Kariyine, T. and Hashimoto, Y. (1959). *Experientia,* 9, 136.

Kolattukudy, P. E. (1970). *Lipids,* 5, 259–275.

Kolattukudy, P. E. and Walton, T. J. (1973). *In* "Progress in the Chemistry of Fats and other Lipids" (R. T. Holman, ed.), pp. 121–175. Pergamon Press, Oxford.

Kramer, P. J. (1946). *Pl. Physiol., Lancaster,* 21, 37–41.

Kreger, D. R. (1948). *Recl. Trav. bot. néerl.* 41, 603–736.

Kuiper, P. J. C. (1967). *Meded. LandbHogesch. Wageningen,* 67, 1–23.

Lemin, A. J., Klomparens, W. and Moss, V. D. (1960). *Forest Sci.* 6, 306–314.

Leonard, O. A. (1957). *Weeds,* 5, 291–303.

Linder, P. J., Mitchell, J. W. and Freeman, G. D. (1964). *J. agric. Fd Chem.* 12, 437–438.

Linskens, H. F., Heinen, W. and Stoffers, A. L. (1965). *Residue Rev.* 8, 136–178.

Luckwill, L. C. and Lloyd-Jones, C. P. (1962). *J. hort. Sci.* 37, 190–206.

Maercker, U. (1965). *Naturwissenschaften,* 52, 15–16.

Martin, J. T. and Juniper, B. E. (1970). "The Cuticles of Plants", Edward Arnold, London.

Matic, M. (1956). *Biochem. J.* 63, 168–176.

Mazliak, P. (1968). *In* "Progress in Phytochemistry" (L. Reinhold and Y. Liwschitz, eds), 1, 49–111. Interscience, New York.

Meyer, M. (1938). *Protoplasma,* **29,** 552–586.
Mitchell, J. W. and Linder, P. J. (1963). *Residue Rev.* **2,** 51–76.
Mitchell, J. W., Linder, P. J. and Robinson, M. B. (1961). *Bot. Gaz.* **123,** 134–137.
Mitchell, J. W., Smale, B. C. and Preston, W. H., Jr. (1959). *J. agric. Fd Chem.* **7,** 841–843.
Moody, K., Kust, C. A. and Buchholtz, K. P. (1970). *Weed Sci.* **18,** 642–647.
Neidermyer, R. W. and Nalewaja, J. D. (1969). *Weed Sci.* **17,** 528–532.
Neumann, S. and Jacob, F. (1968). *Naturwissenschaften,* **55,** 89–90.
Nishimoto, R. K. and Warren, G. F. (1971). *Weed Sci.* **19,** 156–161.
Norris, R. F. and Bukovac, M. J. (1968). *Am. J. Bot.* **55,** 975–983.
Norris, R. F. and Bukovac, M. J. (1969). *Physiologia Pl.* **22,** 701–712.
Norris, R. F. and Bukovac, M. J. (1971). *Pl. Physiol., Baltimore,* **49,** 615–618.
Norris, R. F. and Bukovac, M. J. (1972). *Pestic. Sci.* **3,** 705–708.
O'Brien, L. P. and Prendeville, G. N. (1972). *Weed Res.* **12,** 248–253.
Oliver, L. R., Prendeville, G. N. and Schreiber, M. M. (1968). *Weed Sci.* **16,** 534–537.
Orgell, W. H. (1955). *Pl. Physiol., Lancaster,* **30,** 78–80.
Orgell, W. H. (1957). *Proc. Iowa Acad. Sci.* **64,** 189–199.
Orgell, W. H. and Weintraub, R. L. (1957). *Bot. Gaz.* **119,** 88–93.
Ormrod, D. J. and Renney, A. J. (1968). *Can. J. Pl. Sci.* **48,** 197–209.
Osipow, L. I. (1964). "Surface Chemistry". Rheinhold Publishing Corp., New York.
Overbeek, J. van. (1956a). *A. Rev. Pl. Physiol.* **7,** 355–372.
Overbeek, J. van. (1956b). *In* "The Chemistry and Mode of Action of Plant Growth Substances" (R. L. Wain and F. Wightman, eds), pp. 205–210. Butterworths Scientific Publ., London
Overbeek, J. van, Blondeau, R. and Horne, V. (1955). *Am. J. Bot.* **42,** 205–212.
Pallas, J. E., Jr. (1960). *Pl. Physiol., Lancaster,* **35,** 575–580.
Pallas, J. E., Jr. and Williams, G. G. (1962). *Bot. Gaz.* **123,** 175–180.
Parker, C. (1966). *Weeds,* **14,** 117–121.
Parr, J. F. and Norman, A. G. (1965). *Bot. Gaz.* **126,** 86–96.
Patten, B. C. (1959). *Bot. Gaz.* **120,** 137–144.
Phillips, R. E., Egli, D. B. and Thompson, L., Jr. (1972). *Weed Sci.* **20,** 506–510.
Phillips, R. L. and Bukovac, M. J. (1967). *Proc. Am. Soc. hort. Sci.* **90,** 555–560.
Prasad, R. and Blackman, G. E. (1965a). *J. exp. Bot.* **16,** 86–106.
Prasad, R. and Blackman, G. E. (1965b). *J. exp. Bot.* **16,** 545–568.
Prasad, R., Foy, C. L. and Crafts, A. S. (1967). *Weeds,* **15,** 149–156.
Prendeville, G. N. (1968). *Weed Res.* **8,** 106–114.
Priestley, J. H. (1943). *Bot. Rev.* **9,** 593–616.
Reid, C. P. P. and Hurtt, W. (1970). *Nature, Lond.* **225,** 291.
Roberts, E. A., Southwick, M. D. and Palmiter, D. H. (1948). *Pl. Physiol., Lancaster,* **23,** 557–559.
Robertson, M. M., Parham, P. H. and Bukovac, M. J. (1971). *J. agric. Fd Chem.* **19,** 754–757.
Roelofsen, P. A. (1952). *Acta bot. neerl.* **1,** 99–114.
Rovira, A. D. (1973). *Pestic. Sci.* **4,** 361–366.
Sands, R. and Bachelard, E. P. (1973a). *New Phytol.* **72,** 69–86.
Sands, R. and Bachelard, E. P. (1973b). *New Phytol.* **72,** 87–99.
Sargent, J. A. (1964). *Meded. LandbHogesch. OpzoekStns. Gent.* **29,** 656–663.
Sargent, J. A. (1965). *A. Rev. Pl. Physiol.* **16,** 1–12.
Sargent, J. A. (1966). *Proc. 8th Br. Weed Control Conf.* **3,** 804–811.
Sargent, J. A. and Blackman, G. E. (1962). *J. exp. Bot.* **13,** 348–368.
Sargent, J. A. and Blackman, G. E. (1965). *J. exp. Bot.* **16,** 24–47.
Sargent, J. A. and Blackman, G. E. (1969). *J. exp. Bot.* **20,** 542–555.
Sargent, J. A. and Blackman, G. E. (1972). *J. exp. Bot.* **23,** 830–841.
Sargent, J. A., Powell, R. G. and Blackman, G. E. (1969). *J. exp. Bot.* **63,** 426–450.
Saunders, P. F., Jenner, C. F. and Blackman, G. E. (1965). *J. exp. Bot.* **16,** 683–696.
Schnepf, E. (1959). *Planta,* **52,** 644–708.
Schönherr, J. (1969). *Foliar penetration and translocation of succinic acid, 2,2-dimethyl-hydrazide (SADH),* M.S. Thesis, Michigan State Univ., East Lansing.
Schönherr, J. and Bukovac, M. J. (1970a). *Planta,* **92,** 189–201.

Schönherr, J. and Bukovac, M. J. (1970*b*). *Planta*, **92**, 202–207.
Schönherr, J. and Bukovac, M. J. (1972*a*). *Pl. Physiol., Baltimore*, **49**, 813–819.
Schönherr, J. and Bukovac, M. J. (1972*b*). *J. Am. Soc. hort. Sci.* **97**, 384–386.
Schönherr, J. and Bukovac, M. J. (1973). *Planta*, **109**, 73–93.
Schumacher, W. (1942). *Jb. wiss. bot.* **90**, 530–545.
Schumacher, W. and Lambertz, P. (1956). *Planta*, **47**, 47–52.
Scott, F. M. (1950). *Bot. Gaz.* **3**, 378–394.
Scott, F. M., Bystrom, B. G. and Bowler, E. (1963). *Science, N.Y.* **140**, 63–64.
Scott, H. D. and Phillips, R. E. (1973). *Weed Sci.* **21**, 71–76.
Sharma, M. P., Chang, F. Y. and Vanden Born, W. H. (1971). *Weed Sci.* **19**, 349–355.
Sheets, T. J. (1961). *Weeds*, **9**, 1–13.
Shipman, R. D. (1973). *Down to Earth*, **28**, 19–25.
Sievers, A. (1959). *Flora Jena*. **147**, 263–316.
Sifton, H. B. (1963). *Can. J. Bot.* **41**, 199–207.
Sikka, H. C. and Davis, D. E. (1968). *Weed Sci.* **16**, 474–477.
Silva Fernandes, A. M. S. (1965). *Ann. appl. Biol.* **56**, 297–304.
Simon, E. W. and Beevers, H. (1951). *New Phytol.* **51**, 163–190.
Singh, J. N., Basler, E. and Santelmann, P. W. (1972). *Pestic. Biochem. Physiol.* **2**, 143–152.
Sitte, P. and Rennier, R. (1963). *Planta*, **60**, 19–40.
Skoss, J. D. (1955). *Bot. Gaz.* **117**, 55–72.
Smith, A. E. (1972). *Physiologia Pl.* **27**, 338–341.
Smith, A. E., Zukel, J. W., Stone, G. M. and Riddell, J. A. (1959). *J. agric. Fd Chem.* **7**, 341–344.
Smith, J. W. and Sheets, T. J. (1967). *J. agric. Fd Chem.* **15**, 577–581.
Smith, L. W. and Foy, C. L. (1967). *Weeds*, **15**, 67.
Stålfelt, M. G. (1916). *Svensk. bot. Tidskr.* **10**, 37–46.
Strugger, S. (1939). *Biol. Zbl.* **59**, 409–442.
Sutcliffe, J. F. (1962). "Mineral Salts Absorption in Plants". Pergamon Press, New York.
Sutton, D. L. and Foy, C. L. (1971). *Bot. Gaz.* **132**, 299–304.
Tanton, T. W. and Crowdy, S. H. (1972). *J. exp. Bot.* **23**, 600–618.
Tukey, H. B., Ticknor, R. L., Hinsvark, O. N. and Wittwer, S. H. (1954). *Science, N.Y.* **116**, 167–168.
Turner, D. J. (1972). *Pestic. Sci.* **3**, 323–331.
Turner, D. J. (1973). *Weed Res.* **13**, 91–100.
Turner, D. J. and Loader, M. P. C. (1972). *Proc. 11th Br. Weed Control Conf.* **2**, 654–660.
Turrell, F. M. (1947). *Bot. Gaz.* **108**, 476–483.
Vostral, H. J., Buchholtz, K. P. and Kust, C. A. (1970). *Weed Sci.* **18**, 115–117.
Weaver, R. J. and DeRose, H. R. (1946). *Bot. Gaz.* **107**, 509–521.
Weber, F. (1930). *Protoplasma*, **10**, 608–612.
Wedding, R. T. and Erickson, L. C. (1957). *Pl. Physiol., Lancaster*, **32**, 503–512.
Westwood, M. N. and Batjer, L. P. (1958). *Proc. Am. Soc. hort. Sci.* **72**, 35–44.
Westwood, M. N., Batjer, L. P. and Billingsley, H. D. (1960). *Proc. Am. Soc. hort. Sci.* **76**, 30–40.
Whitecross, M. I. and Armstrong, D. J. (1972). *Aust. J. Bot.* **20**, 87–95.
Yamada, Y., Wittwer, S. H. and Bukovac, M. J. (1965). *Pl. Physiol., Lancaster*, **40**, 170–175.
Yamada, Y., Rasmussen, H. P., Bukovac, M. J. and Wittwer, S. H. (1966). *Am. J. Bot.* **53**, 170–172.

CHAPTER 12

HERBICIDE TRANSPORT IN PLANTS

J. R. HAY

Research Station, Agriculture Canada, Regina, Saskatchewan, Canada

I. INTRODUCTION

There are several reviews on the translocation of herbicides (Crafts, 1964; Crafts, 1966; Robertson and Kirkwood, 1970; Crafts and Crisp, 1971; Ashton and Crafts, 1973). This chapter is concerned primarily with work reported since 1962 and not considered in the earlier edition of this book. In addition, the translocation of several new herbicides that have come into use in the last decade are reviewed. Although uptake of herbicides by leaves and by roots is an integral part of movement through plants, it is covered in more detail in Chapter 11.

There are several reasons why it is important to study herbicide translocation. Herbicides are used to kill weeds. Plants that cannot regenerate from below ground can be killed by contact herbicides if coverage is complete. However, many weeds regrow from buds on crowns and stems, or from roots or rhizomes below the ground. To kill these, sufficient herbicide must be moved to the areas where the buds develop. From the practical standpoint then, it is important to study the downward movement of herbicides applied at toxic levels which may differ from that at sub-lethal dosages. Some of the most serious weeds have extensive root systems and can survive chemical and cultivation treatments that kill only the top growth and the roots and underground stems in the top few inches of soil. Control of such weeds would be improved if translocation could be increased. Since it is not easy to contact all of the foliage with spray applications made to brush species, translocation of herbicides in these plants would enhance control. Some movement following basal-bark and cut-surface applications to tree species is also required for effective control. Translocation of soil-applied herbicides would improve effectiveness. These compounds may be taken up by roots or underground parts of stems but usually move to meristematic areas for effective control.

If the translocation of herbicides in plants could be increased, the rates of application and therefore the amount of herbicide introduced into the environment could be reduced. This would lower the hazard of possible harmful effects on the environment.

II. Current Knowledge of Translocation

A. SYMPLAST-APOPLAST CONCEPT

Much of the research on translocation of herbicides tests the mass-flow hypothesis according to which compounds move passively with an assimilate stream from source to sink (Crafts and Crisp, 1971). The other main school of thought is that translocation is an active process (cf. Robertson and Kirkwood, 1970).

In discussions on translocation it is convenient to adopt the symplast-apoplast concept. The symplast has been defined by Crafts and Crisp (1971): "... the total mass of living cells of a plant constitute a continuum, the individual protoplasts being intimately connected throughout the plant by the plasmodesmata ... it follows that the mesoplasm is continuous from cell to cell." Ions and molecules can therefore move from cell to cell without crossing permeability barriers (i.e. membranes). The sieve tubes are considered to be highly specialized components of the symplast. Long distance translocation takes place in sieve tubes at rapid rates (Trip and Gorham, 1968) but lateral or radial movement in stems occurs at rates 1/60 to 1/100 of the longitudinal value (Zimmerman, 1961; Webb and Gorham, 1964, 1965). Movement is largely limited by orthostichy and vascular connections (Wardlaw, 1968). Eschrich (1970) recently reviewed the structure and biochemistry of the phloem.

The apoplast, in contrast "constitutes the total non-living cell wall continuum that surrounds the symplast" (Crafts and Crisp, 1971). It includes the cell walls, intercellular spaces and the xylem vessels and forms a continuous permeable system through which water and solutes may freely move. Ions and molecules enter roots via the apoplast. Substances applied to leaves or stems move through the cuticle and enter cell walls (apoplast). From there they may either move to the leaf margin in the transpiration stream, or leave the apoplast, cross the plasmalemma and enter the cytoplasm (symplast).

B. DOWNWARD MOVEMENT

Assimilates move an average of 2·5 cell diameters before reaching the phloem cells in a minor vein (Wardlaw, 1968). Some or all of this movement may occur in the apoplast. Since assimilates will move into veins against a concentration gradient it is assumed that metabolic energy is required (Geiger and Cataldo, 1969). The organelle-rich companion cells and parenchyma cells in phloem appear to be functional in vein loading. In recent years attention has been focused on transfer cells which are prominent in veins, in xylem parenchyma, at leaf traces and nodes, and have a dense cytoplasm (Pate and Gunning, 1972). They are characterized by having many ingrowths in the cell wall and by having the plasma membrane follow the contours of the wall ingrowths. Thus, these cells increase the interface surface between the apoplast and the symplast. Little work has been done on the physiology of transfer cells; however, their shape, appearance and location strongly suggest that they could play a major role in translocation, i.e. in phloem loading, and transfer of materials between the phloem and xylem particularly at nodes where they are abundant. The active step in phloem loading by phloem parenchyma or transfer cells could occur at the plasma membrane. Advances in the chemistry of transfer of ions and molecules across membranes will undoubtedly lead to a better understanding of uptake and translocation (Epstein, 1973). No references were found on the role of transfer cells in the translocation of herbicides.

Assimilates and possibly exogenous compounds can move from cell to cell via the cytoplasmic connections. It is significant that the mesophyll cells, transfer cells and companion cells are interconnected by these strands but only the companion cells connect directly with the sieve tubes (Shih and Currier, 1969). Thus, movement into sieve tubes via the symplast must be channelled through the cytoplasm of the companion cells.

The phloem-loading component of translocation must be an active step. However, the mechanism of movement within the sieve tubes still remains a mystery. The mass-flow theory states that the high sugar concentration in the phloem causes water to move in from surrounding tissues by osmosis; the high turgor pressure then forces the contents of the sieve tubes to flow *en masse*. Zimmerman (1961) claimed that the observed decrease in concentration of the phloem sap along the stem away from the source, supports the mass flow hypothesis. The flow of sap from excised aphid stylets that have penetrated sieve tubes may go on for hours or days. This suggests that the phloem contents are moving in a stream (Zimmerman, 1961). However, other workers find it difficult to visualize a stream of solutes moving through the pores of the sieve plates at the rates that have been measured, especially when these pores appear to be filled with protein. Evidence for bi-directional movement wherein ^{32}P and ^{14}C arrive simultaneously in the same sieve tube from opposite directions also casts doubt on mass flow (Trip and Gorham, 1968). However, this evidence is not decisive and the debate continues (Canny, 1971). Although there is a lack of convincing evidence for an active translocation mechanism in sieve tubes, energy must be supplied to maintain viability in the living cytoplasm of the sieve tubes and companion cells. This is significant since herbicides may interfere with their own translocation by disrupting the living processes associated with the sieve tubes and with phloem loading (Hay and Thimann, 1956; Crafts, 1964; Vostral *et al.* 1970).

The mass-flow hypothesis of translocation involves a stream of solutes moving from a source (photosynthetic stems or leaves) to a sink (meristematic areas, developing flowers, fruits and storage tissues). It has been assumed that herbicides and other exogenous compounds would be carried along in this stream. Mature leaves are the principal source of assimilates. Young leaves do not export assimilates, nor is vein loading apparent in them (Leonard, 1939). Once they are a third to half expanded, they export and no longer import assimilates (Wardlaw, 1968). Thus, a leaf starts as a sink but at maturity it is a source. Mature leaves near the base of the plant export assimilates to the roots while those near the top export to the apex. Leaves in the middle may support roots and tops. Presumably carbohydrates are not exported from developing leaves because the energy requirements for growth exceed the supply. This suggests that carbohydrates are not available for active phloem loading which could affect export of herbicides and other compounds as well as assimilates.

Sieve tubes develop downward from leaves into the first or second internode of the stem below the axil of the leaf where they may connect with sieve tubes descending from the stem and leaves above (Crafts, 1967). Thus for compounds to move upward via the phloem they must first move down some distance in the stem. Lateral movement from sieve tubes is considered to be slow in comparison to longitudal movement. Compounds can also be carried upwards in the transpiration stream if they "leak" from the sieve tubes into the apoplast. Many herbicides move upward via this route.

Sucrose moved from detached sugar cane and sugar beet leaves against concentration gradients or without the attractive force of a sink (Eschrich, 1970). In a sense, then, assimilates are pushed out of leaves by an active process. In his review of the biochemistry of phloem transport Eschrich (1970) found no indication that sinks were actively involved in the process of long distance transport. The role of assimilate sinks in herbicide transport is challenged below.

C. UPWARD MOVEMENT

Much of the work on upward movement of solutes in plants has been conducted by feeding substances to the roots via the soil or nutrient solutions. Reviews on root uptake, which is the first stage in upward translocation, have appeared recently (Läuchli, 1972; Epstein, 1973). Ions and molecules can diffuse into the apoplast of the cell walls of the cortex. However, movement to the interior of the root via this path is blocked by the Casparian strip which is made up of wax-like substances impregnating the cell walls of the endodermis. It reputedly impedes the inward movement of water and water-soluble substances in the apoplast (Epstein, 1973).

Solutes can by-pass the Casparian strip by crossing the plasma membrane and entering the symplast. This presumably is an active process and can account for high concentration gradients between external solutions and the roots (Geissbühler *et al.* 1963; Verloop and Nimmo, 1969). Solutes can move from cell to cell to the interior of the root and substances that are lost from xylem parenchyma cells, by leakage or by active outward movement across plasma membranes, may be swept upward in the transpiration stream.

If the movement of solutes across cell membranes and through the symplast is effected by active processes, there may be some vital control over upward movement. Movement can be altered by the tendency for parenchyma cells in the xylem (and possibly transfer cells) to withdraw substances from the xylem. Factors affecting transpiration will affect upward movement, and binding of herbicides with various cell constituents may restrict movement. As a result of these factors ions and molecules, such as herbicides, show characteristic patterns of distribution after root uptake.

III. CRITIQUE OF METHODS OF STUDYING TRANSLOCATION OF HERBICIDES

Early studies on the translocation of the phenoxyacetic herbicides used the appearance of epinasty as an indication of movement, and the angle of bending of bean epicotyls as a quantitative measure of the amounts translocated (Rohrbaugh and Rice, 1949). With the sophisticated methods now available these methods are only used for an indication of whether or not a compound is systemic. Curvature measurements are suitable only for the hormone type herbicides.

Translocation has been studied by extracting phenoxy herbicides from the plant tissue and using bioassays to measure the amounts recovered from various parts (Hay and Thimann, 1956; Leonard *et al.* 1962; Herrett and Bagley, 1964; Eliasson, 1965). With appropriate controls reasonably positive identification is obtained although quantitatively there is a hazard that other compounds will inhibit or enhance the growth process and thus give false readings. This method is, of course, suitable only for compounds for which there are good bioassays.

Most of the work on translocation of herbicides has been done with radioactive labelled compounds. They are usually labelled with ^{14}C; however, tritium, ^{36}Cl (Crafts and Foy, 1959) and ^{35}S (Fang and Theisen, 1960; Yamaguchi, 1961) have also been used. Radio-autographs have been widely used in this work. Great care must be exercised in the preparation of radio-autographs to ensure that the label does not move during the drying process (Pallas and Crafts, 1956). Artifacts were present in the early work (Pallas and Crafts, 1956; Levi, 1962) (see Fig. 1). Plants can be cut into sections before drying although this may lead to an accumulation of the label at the cut surfaces. An extensive study by Yamaguchi and Crafts (1958) led to the

Fig. 1. Auto-radiographs of primary leaves of bean plants grown for three days in 750 ml of aerated complete Hoagland solution to which 1.0 μc of P[32] has been added; left, oven dried at 80°C; right, freeze dried (Levi, 1962). (Copyright 1962 by the American Association for the Advancement of Science.) The distribution on the left is an artifact of oven drying.

freeze-drying method of preparing plants for radio-autography. In this method the plants are frozen with dry ice and dried under vacuum before being pressed and exposed to X-ray film. The distribution patterns obtained were similar to those from plants cut into small sections before exposure to the X-ray film. This method has been used by nearly all workers in the last decade.

Care must be taken in interpreting results from radio-autographs. The appearance of an image does indicate the presence of the isotope but it does not necessarily mean that the intact molecule is there. If degradation has occurred, the label may be in the degradation products or assimilated into new products. Thus, parallel studies on degradation must be done. If no breakdown takes place within the time frame of the tests a qualitative picture of translocation can be obtained. Tests done with paraquat labelled in three positions (Slade and Bell, 1966) produced similar patterns with all three. Hence, it was assumed that no degradation had occurred and that the radio-autographs gave a true picture of paraquat movement. The lack of image on the film can be used as an indication of the parts of the plant in which no herbicide or breakdown products have accumulated, or, at least, parts of the plant through which no translocation was occurring when the radio-autograph was prepared.

A reliable quantitative measure of translocation cannot be made from radio-autographs. The intensity of the image will depend on the length of exposure and on the density of the tissues through which the radiation must pass. Radioactivity emitted from xylem tissue will give a weaker image than that from phloem tissue or from the surface. With a thick stem, radiation from vascular tissues may be completely absorbed by overlying tissue. Nevertheless, if all of these limitations are taken into account a general picture of the pattern of translocation of a herbicide can be obtained from radio-autographs.

A quantitative measure of translocation can be obtained by counting the radioactivity in different parts of the plant at fixed times after application. Here again, however, steps must be taken to ensure that the label being counted is actually in the parent compound. Thus, the materials must be extracted, purified and separated by fractionation or chromatography before making the counts. It is preferable that the compounds be separated using more than one solvent system when chromatography is used.

Some of the label may be lost by volatilization of herbicides with high vapour pressures e.g. dichlobenil (Verloop and Nimmo, 1969), or loss of the breakdown products (i.e. $^{14}CO_2$). Again the isotope may be lost from the roots to the nutrient solution or soil (Mitchell and Linder, 1963). These losses must be considered if all of the material is to be accounted for in translocation studies.

Although gas chromatography provides an extremely useful tool in herbicide translocation research, it has not been widely used. It necessitates extraction and purification of the extracts but an accurate measure of very small amounts can be made (Davis et al. 1968b; Morton et al. 1968; Brady, 1969). If run on two or more columns, reasonably positive identifications can be made. Thin layer chromatography has also been used (Smith and Sheets, 1967).

Many studies have been made with herbicides at sub-lethal doses applied in one or more spots on leaves. It is now clear that the amounts translocated can be affected by the dose applied (Hay and Thimann, 1956; Crafts and Crisp, 1971). Thus, since the ultimate object of herbicide research is to control weeds more efficiently, a range of rates including lethal doses should be used (Eliasson, 1965). Translocation can also vary if the whole leaf is treated rather than just a spot on the leaf (Smith and Nalewaja, 1972).

Caution must be used when interpreting results of tests wherein the materials are applied to the roots via the nutrient solution (Duble et al. 1969). Results from such tests may be quite different from those obtained when applications are made to roots via the soil (Geisbühler et al. 1963; Pallas and Williams, 1962). Van der Zweep (1961) reported greater movement from leaves to the roots if the nutrient solution was aerated.

The movement of toxic amounts of herbicides to roots under field conditions can be determined by direct observation. Roots can be excavated at different times after treatment and the amount of root kill observed. This is laborious and time-consuming although it can be done with a back hoe (Hunter, J. H., personal communication). Root kill can also be observed by treating plants

Fig. 2. Roots of Canada thistle (*Cirsium arvense*) growing against glass face of root boxes. Left, no treatment. Centre, partial root kill four months after an application of 1 mg of dicamba to the foliage. Right, complete root kill after 2 mg were applied. (Reproduced through permission of W. J. Saidak, Agriculture Canada, Ottawa.)

grown in root boxes which have a glass face (see Fig. 2). Root kill can be measured at various times by tracing the amount of living and dead roots that appear against the glass. In addition, the roots can be recovered and weighed after the soil is washed away with a fine spray of water. Although this method has not been widely used it is an inexpensive method of obtaining good data.

IV. Translocation of Herbicides

A. INTRODUCTION

In this section, the transport of various herbicides is reviewed. Space does not permit reference to work on all herbicides. Instead, the state of knowledge for the compounds upon which the most work has been done is outlined and related to the knowledge about transport mechanisms.

B. AMITROLE

In the last decade several studies using gross auto-radiographs have reconfirmed earlier conclusions that ^{14}C becomes widely distributed throughout plants after the application of ^{14}C-amitrole to the leaves or roots (Hill *et al.* 1963; Crafts, 1964; Donnalley and Ries, 1964; Smith and Davies, 1965; Forde, 1966; Crafts, 1967; Lund-Høie and Bayer, 1968; Rogerson and Bingham, 1971). As far back as 1958, Racusen (1958) reported that amitrole was not stable in plants: only 7% of ^{14}C in pinto beans could be accounted for in amitrole after five days. It was not clear whether the transformation took place before or after translocation.

Carter and Naylor (1960) were critical of earlier workers, who studied translocation, for not recognizing that amitrole was unstable in plants. In their work with beans, amitrole moved out of the leaves and into the stem in 2 h but amounts in the stem began to decline after 8 h. Since intact amitrole was found in the stem but not at the tips it was deduced that the transformation took place at the tips. Several unknown radioactive compounds were separated out chromatographically with different products predominant in different species. Massini (1963) found that most of the ^{14}C in tomatoes and beans at 24 h was in β-(3-amino-1,2,4-triazol-lyl)-α-alanine (ATX). ATX was only slightly less mobile than amitrole but since it was non-toxic he concluded that the transformation took place after translocation. Herrett and Bagley (1964) isolated three metabolites in Canada thistle (*Cirsium arvense*). Their "Unknown III" was mobile, phytotoxic and degraded to amitrole. They proposed that amitrole was converted to this form prior to translocation since light was required for the transformation and there was a lag period before it was translocated.

Movement of ^{14}C from labelled 3-α-hydroxy-β,β,β-trichloroethylamino-1,2,4-triazole to the tips of pinto beans was more rapid than from ^{14}C-amitrole at 2, 8 and 24 h after application (Gomaa *et al.* 1969).

The herbicidal activity of amitrole is enhanced by the addition of equimolar amounts of ammonium thiocyanate (NH_4SCN). Whether applied to quackgrass (*Agropyron repens*) before, after or in combination with ^{14}C-amitrole, NH_4SCN did not alter the amount of amitrole absorbed (Donnalley and Ries, 1964). However, if applied with or one day before the ^{14}C-amitrole it did increase the amount of ^{14}C moved out of the leaf (Fig. 3). [14]C from $NH_4S^{14}CN$ moved within the leaf of quackgrass but not out of it (Forde, 1966). Thus it was suggested that the enhancing effect of NH_4SCN must occur within the leaf itself.

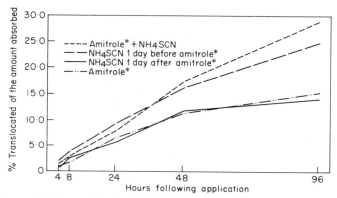

Fig. 3. Percent of ^{14}C from ^{14}C-amitrole absorbed by leaves of quackgrass (*Agropyron repens*) that was translocated after application alone, with, before and after NH_4SCN (Donnalley and Ries, 1964. Copyright 1964 by AAAS).

To summarize amitrole is unstable in plants so that auto-radiographs do not give a true picture of its transport. The label from ^{14}C-amitrole becomes widespread throughout plants suggesting free movement in the symplast and apoplast. However, it is not clear whether amitrole is degraded before or after movement or if it moves in complexes. NH_4SCN appears to enhance movement of amitrole out of treated leaves.

C. DALAPON

Early auto-radiographs showed that the isotope from ^{14}C-dalapon becomes widely distributed in plants when applied to leaves or to roots (Crafts and Foy, 1959). The ^{14}C appeared to move out of the phloem and be carried upward in the xylem. This was confirmed in several species, cotton and sorghum (Foy, 1961), corn (Foy, 1962a), *Tradescentia fluminensis* (Foy, 1962b), sugar beets and *Setaria glauca* (Anderson *et al.* 1962). Dalapon was reported to be relatively stable in plants (Foy, 1961; Anderson *et al.* 1962) since the radioactivity in aqueous and alcoholic extracts had the same R_f as ^{14}C-dalapon. Although small amounts of ^{14}C remained in the residue after extraction it has been assumed by most workers that the ^{14}C in extracts and represented on auto-radiographs is intact dalapon.

Absorption and export of ^{14}C from foliar applications of ^{14}C-dalapon continued to increase over a two-week period (Foy, 1961). Although auto-radiographs show its distribution throughout the plants, there tends to be an accumulation in active meristematic regions (Crafts and Foy, 1959), developing fruits (Foy, 1961), growing tips, and young leaves (Anderson *et al.* 1962) and in developing rhizomes (Hull, 1969).

More ^{14}C from foliar application of ^{14}C-dalapon moved into rhizomes of quackgrass in the dark than in the light (Sagar, 1960). This was confirmed by McIntyre (1962) but there was no evidence that the observed reduction in transpiration during the dark period was the main factor involved. Lund-Høie and Bylterud (1969) found only limited movement into quackgrass rhizomes from foliar applications. However, distribution between the roots and rhizomes was more uniform when applications were made to basal leaves rather than to the youngest fully expanded leaves.

In *Paspalum distichum,* ^{14}C-dalapon moved to the apices of stolons after leaf application (Smith and Davies, 1965). Hull (1969) reported that ^{14}C-dalapon applied to foliage of Johnson grass (*Sorghum halepense*) by-passed internodes and buds on rhizomes and accumulated in the growing buds at the tips. This was in contrast to ^{14}C from ^{14}CO$_2$ which tended to accumulate along the entire length of the rhizome. Movement of ^{14}C-assimilate into rhizomes increased progressively as the growth stage advanced but the ^{14}C from ^{14}C-dalapon did not (Table I). Thus, ^{14}C-dalapon and ^{14}C-assimilates apparently had different "sinks".

Surfactants such as X-77 greatly increased absorption and movement of

TABLE I

DISTRIBUTION OF ^{14}C-ASSIMILATES WHEN ONE CULM OF MULTICULM PLANTS, GROWN FROM RHIZOME CUTTINGS, WAS EXPOSED TO ^{14}CO$_2$, AND OF ^{14}C-DALAPON APPLIED TO THREE LEAVES ON A SINGLE CULM AFTER SEVEN DAYS[1]

Growth stage	Plant part	^{14}C-assimilates			^{14}C-dalapon		
		Total ct/min × 10⁵	Spec. act. ct/min/mg	Percent distri-bution	Total ct/min × 10⁵	Spec. act. ct/min/mg	Percent distri-bution
Preboot	Treated culm	64·40	11,670·0	84·4	1·30	150·0	75·6
	Adjacent culms	4·66	27·2	6·2	0·26	1·0	13·8
	Rhizomes	7·15	29·7	9·4	0·14	0·6	10·6
Boot	Treated culm	74·50	6,370·0	84·8	4·09	361·0	90·0
	Adjacent culms	1·98	5·6	2·2	0·32	1·1	7·9
	Rhizomes	11·00	52·8	13·0	0·09	0·2	2·2
Flower[2]	Treated culm	45·30	5,270·0	68·3	5·73	593·0	87·1
	Adjacent culms	1·49	4·7	3·4	0·31	1·0	5·3
	Rhizomes	20·00	69·3	28·3	0·41	1·9	7·6

[1] Each value is the average of 12 plants (from Hull, 1969).
[2] Dosage at flowering stage was four times that at the pre-boot stage.

[14]C-dalapon in corn leaves (Foy, 1962a) and in bean leaves (Prasad et al. 1967). It increased absorption in *Tradescentia fluminensis* but not translocation (Foy, 1962b).

Movement of ^{14}C from bean and barley leaves treated with ^{14}C-dalapon was greater at 88 than at 28% relative humidity (Table II) (Prasad et al. 1967).

TABLE II

COUNTS/MIN/MG DRY WEIGHT OF ^{14}C IN DIFFERENT PARTS OF BEAN PLANTS AT DIFFERENT RELATIVE HUMIDITIES APPROXIMATELY 8 AND 18 h AFTER APPLICATION OF ^{14}C-DALAPON TO A SINGLE LEAF AT 26±1°C AND 800 f.c.[1]

Treatment time (h)	Relative humidity %	ct/min/mg dry wt[2]			
		Treated leaf	Stem	Terminal bud	Root
6	(Low) 28 ± 3	67	16	54	19
	(Med.) 60 ± 5	101	54	88	21
	(High) 88 ± 3	117	57	81	21
18	(Low) 28 ± 3	172	69	79	67
	(Med.) 60 ± 5	792	251	340	81
	(High) 88 ± 3	864	301	740	85

[1] Dalapon (1,500 mg/l) applied as a 20 μl drop on one leaflet of the first trifoliolate leaf. All plants grown at 26 ± 1°C, 60 ± 5% r.h., and 800 f.c. prior to treatment and transfer.

[2] Counts are total net radioactivity in dalapon translocated out of a 20 mm diameter section of leaflet centred over the treated spot. Counts/min of a standard planchet (equivalent to the applied dose) was 29,477 (from Prasad et al. 1967).

This difference was more pronounced at 18 h than at 6 h. Exposing the plants to high humidity for 18 h before treatment or rewetting leaves after treatment also enhanced movement. It was suggested that hydration of the leaf surface facilitated penetration and movement of dalapon. Movement was greater at 43°C than at 26°C. Thus dalapon appears to be relatively stable in plants so that auto-radiographs and counting data can be assumed to give a reasonably accurate picture of distribution. Transport occurs in both apoplast and symplast and may go on for two weeks after application. Main accumulation is in growing tips, developing fruits and other meristematic regions. Small amounts get to the roots and rhizomes. Here, it by-passes internodes and nodes where carbohydrates are being stored and accumulates in growing tips.

D. DICAMBA

Dicamba was more mobile in bracken (*Pteridium aquilinum*) than other herbicides tested by Hodgson (1964). Leonard et al. (1966) reported that dicamba moved readily in grape cuttings when applied to roots or to foliage. When applied to detached bean leaves ^{14}C from ^{14}C-dicamba moved out of the leaf blades and appeared to concentrate at the base of the petioles (Fig. 4)

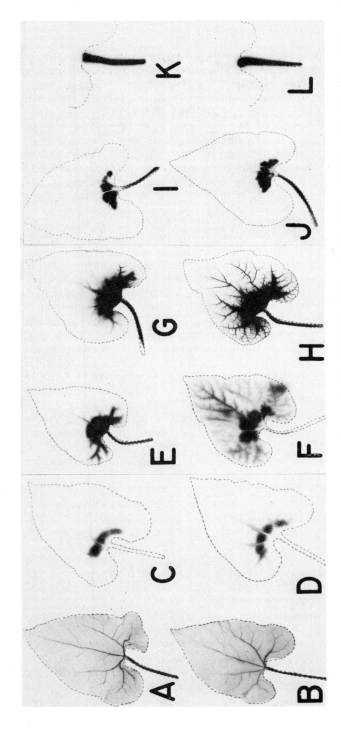

Fig. 4. Auto-radiographs of detached bean leaves showing translocation of labelled assimilates A and B; ^{14}C from ^{14}C-diuron, C and D; ^{14}C-maleic hydrazide, E and F; ^{14}C-2,4-D, G and H; ^{14}C-dicamba, I and J; and of [32]PO$_4$, K and L. In K and L only the petiole was auto-radiographed due to the intense radiation from the treated blade. Leaves were suspended in a moist chamber; in the upper series the petioles were in air and in the lower series the petioles were in 1 cm of water (Leonard and Glenn, 1968).

(Leonard and Glenn, 1968). Dicamba gave extensive root kill when applied to the foliage of Canada thistle (*Cirsium arvense*) plants growing in root boxes (Fig. 2) (personal communication, W. J. Saidak). In aspen (*Populus tremula*) plants grown from cuttings, dicamba moved to the roots more rapidly than 2,4-D, 2,4,5-T or picloram (Eliasson, 1972).

Dicamba applied to corn foliage was translocated through corn in sufficient quantity to affect witchweed (*Striga lutea*) parasitizing corn (Sand *et al.* 1971). Pre-treating wild garlic (*Allium vineale*) with 2-chloroethylphosphonic acid (CEPA) increased basipetal translocation of ^{14}C in ^{14}C-dicamba treated leaves (Binning *et al.* 1971). CEPA breaks bud dormancy, presumably because it degrades to ethylene in plants. It was not clear whether CEPA actually broke dormancy in these plants.

The isotope from ^{14}C-dicamba moved slowly in purple nutsedge (*Cyperus rotundus*) for 10 days (Magalhaes *et al.* 1968). Although the ^{14}C moved through the rhizomes to other shoots there was very little accumulation in the tubers or rhizomes along the way. Using thin-layer chromatography, they found no radioactive metabolites in purple nutsedge even after 10 days. Some degradation of ^{14}C-dicamba was found in Canada thistle, Tartary buckwheat (*Fagopyrum tataricum*), wild mustard (*Brassica kaber*) wheat and barley (Chang and Vanden Born, 1968, 1971a, 1971b). Thus, some caution must be exercised in interpreting auto-radiographs and in counting ^{14}C-dicamba.

Dicamba-treated Canada thistle leaves had to be left on the plants for only 6 h for symptoms to become apparent at the shoot tips (Chang and Vanden Born, 1968). Maximum damage was obtained when leaves were not removed until 24 h. The response at 48 h was similar to that at 24 h. Auto-radiographs indicated that most of the ^{14}C from foliar- or root-applied ^{14}C-dicamba accumulated in the growing tips. Although little ^{14}C appeared in the roots, enough was translocated to the roots and released to the soil to give typical symptoms on safflower (*Carthamus tinctorius*) plants growing in the same pots. It was estimated that 0·5% of the foliar applied material was lost from the roots.

When 12 mg of ^{14}C-dicamba were applied to a single leaf, 17·3% was moved out compared to 3·8% when 120 mg were applied as determined by counting ^{14}C in the extracts. However, the actual amount moved out at the high dosage was greater than at the low dosage (Chang and Vanden Born, 1968).

^{14}C moved to the tips of Canada thistle plants grown in nutrient solution containing ^{14}C-dicamba within 30 min (Chang and Vanden Born, 1968). It was not retained in the roots and although, it appeared in mature leaves at the start, the main accumulation was in the youngest leaves.

^{14}C from ^{14}C-dicamba moved out of Tartary buckwheat (*Fagopyrum tataricum*) leaves in 1 h and accumulated in young leaves at the tips within 4 h (Chang and Vanden Born, 1971a). By 24 h entire leaves were radioactive but by four days the ^{14}C had moved out of the older leaves except around the edges where the tissue was chlorotic and later died due to the treatment (Fig. 5). Time-course studies also revealed that ^{14}C was redistributed from

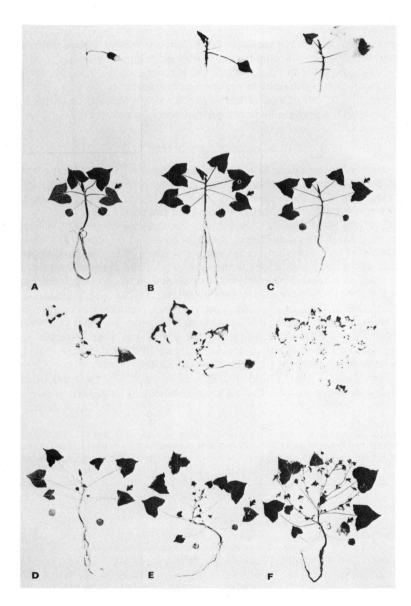

Fig. 5. Auto-radiographs (upper) and plant mounts (lower) of Tartary buckwheat plants at A, 1 h; B, 4 h; C, 1 day; D, 4 days; E, 10 days; F, 20 days; after application of 0·1 μc of ^{14}C-dicamba in a 10 μl droplet on a single leaf as indicated by the arrow (Chang and Vanden Born, 1971).

mature leaves to younger leaves when ^{14}C-dicamba was applied to roots via the nutrient solution.

It seems that dicamba is very mobile in plants. It moves in the symplast and apoplast with the main accumulation taking place at the growing tips. Small amounts may be lost from the roots to the surrounding medium. When fed to roots, it tends not to accumulate in the roots but moves to the tops. It appears to be redistributed out of mature leaves and to concentrate in the young leaves at the shoot tip from where there is a general lack of movement.

E. PHENOXY HERBICIDES

In the last decade considerable work has been done on the translocation of 2,4-D and related compounds without explaining why downward movement is not more extensive. The theoretical aspects of translocation of these compounds has been reviewed recently (Robertson and Kirkwood, 1970).

Much of the translocation research has been done with seedlings of annual plants under controlled conditions. 2,4-D is absorbed by leaves and moves in the apoplast and symplast. The general picture from auto-radiographs, counting and bioassay analyses is that the bulk of the material remains in or on the treated leaves. There is an accumulation in the young leaves at the growing tips but relatively little moves to the roots (Hay and Thimann, 1956; Slife *et al.* 1962; Fites *et al.* 1964; Crafts, 1967; Neidermyer and Nalewaja, 1969; Hallmén and Eliasson, 1972). Although only low amounts were found in the roots from foliage applications, some 2,4-D has been detected in the nutrient solutions in which the plants were grown (Table III) (Fites *et al.* 1964; Neidermyer and Nalewaja, 1969; Coble *et al.* 1970).

Van der Zweep (1961) found that movement of ^{14}C to the roots of barley from foliar applications of ^{14}C-2,4-D varied with leaf stage. His results were

TABLE III

ABSORPTION AND DISTRIBUTION OF ^{14}C AT ONE, FOUR AND EIGHT DAYS AFTER THE APPLICATION OF ^{14}C-2,4-D TO THE LEAVES OF HONEYVINE MILKWEED (*Ampelamus albidus*) WITH AND WITHOUT 1% SURFACTANT (TWEEN 80) (COBLE *ET AL.* 1970)

	% of total activity applied					
	Without surfactant			*With surfactant*		
Sample	*1 day*	*4 days*	*8 days*	*1 day*	*4 days*	*8 days*
Leaf wash	87·0	81·0	74·0	41·0	24·0	13·0
Total absorption	7·2	9·3	10·9	55·8	71·3	78·7
Treated leaf	2·3	3·0	4·0	45·0	61·0	65·0
Other aerial parts	1·4	1·5	1·6	2·1	2·1	2·0
Roots	1·8	1·1	1·1	3·5	2·2	0·7
Growth medium	1·7	3·7	4·2	5·2	6·0	11·0

not compatible with the theory that movement is with photosynthates and he suggested that there was a "block" that prevented movement of 2,4-D to other parts of the plant.

High humidity and high temperature both enhance movement of ^{14}C out of bean leaves treated with ^{14}C-2,4-D (Pallas, 1960). This was considered to be due to increased absorption aided by open stomata under these conditions. Bagging bean plants or just the treated leaf to raise the humidity increased the movement of ^{14}C from cotton leaves treated with ^{14}C-2,4-D (Clor *et al.* 1962). Ringing experiments showed that acropetal movement in the stem was in the xylem (Hay and Thimann, 1956; Clor *et al.* 1962).

Soil moisture levels did not affect absorption of ^{14}C-2,4-D by bean leaves (Basler *et al.* 1961; Pallas and Williams, 1962). However, translocation from primary leaves to the epicotyls of bean plants was reduced as the soils became drier. Approximately twice as much ^{14}C was translocated at a soil moisture tension of 1/3 atmospheres as at 4 atmospheres which was near the wilting point. Translocation was reduced when the leaf turgidity dropped below 80%. The ability of leaves to translocate was not regained until several hours after full turgidity was restored (Fig. 6). Thus, possibly the capacity to load the phloem was reduced when moisture was limiting.

2,4-D does not give good control of deep-rooted perennial plants which have the ability to regenerate from below ground. Therefore, there is considerable

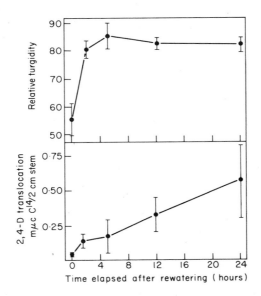

Fig. 6. The relative turgidity of bean leaf tissue at the time of treatment (upper curve) and the translocation of [14]C (from foliar applied ^{14}C-2,4-D) to epicotyl tissue (lower curve) at various intervals after watering plants that had been subjected to a three-day drought. The vertical bars represent standard deviations calculated from 10 determinations (Basler *et al.* 1961).

interest in translocation of this and related compounds in both herbaceous and woody perennial weeds. In most species downward movement is slight.

Young and Fisher (1950) and Blair and Fuller (1952) showed that downward movement of 2,4-D from distal foliage of mesquite (*Prosopis juliflora*) was negligible. In marabú (*Dichrostachys nutans*) less than 1% of foliar-applied 2,4-D moved out of the treated leaves (Hay, 1956). No transport occurred after 24 h. The small amount of downward movement in the bark was sufficient to delay bud development and produce abnormal regrowth if the treated tops were removed at one day after application. When only lower leaves were treated, and a ring of bark was removed above the treated leaves, the 2,4-D moved up through the wood.

In ironweed (*Vernonia baldwini*), ^{14}C from ^{14}C-2,4-D tended to move to young foliage with limited amounts going to roots and rhizomes and with more going at the full bloom stage than at an earlier vegetative stage (Linscott and McCarty, 1962). In neither case were the roots killed. There was extensive degradation of the ^{14}C-2,4-D in this species.

Norris and Freed (1966a) studied ^{14}C movement after the application of acid, amine and ester formulations of ^{14}C-2,4-D and ^{14}C-2,4,5-T in bigleaf maple (*Acer macrophyllum*). They followed only the ^{14}C but there were no differences in the amounts moved regardless of the formulation used although absorption of the ester formulation was greater. Bigleaf maple is more susceptible to 2,4,5-T than to 2,4-D. However, the movement of ^{14}C from these herbicides was virtually equal so that translocation was not considered to be responsible for the difference in efficacy.

Burns et al. (1969) applied ^{14}C-2,4-D to leaves of wolftail (*Carex cherokeensis*) grown in nutrient solution. Dimethyl sulfoxide (DSMO) did not affect the movement of ^{14}C although it enhanced control in the field. Enclosing the plants in polyethylene bags, which raised the humidity and temperature, increased translocation of the ^{14}C out of the treated leaves. At 40% r.h. movement to the rhizomes was negligible.

In honeyvine milkweed (*Ampelamus albidus*), the addition of a surfactant increased the absorption of ^{14}C from ^{14}C-2,4-D but it did not change the amount translocated to the roots (Table III) (Coble et al. 1970). Although degradation took place in the tops the ^{14}C in the roots had the same R_f value as 2,4-D and no other peaks were found by paper chromatography. It is noteworthy that the percentage of ^{14}C found in the nutrient solution was higher than that in the roots. They recovered more ^{14}C in the roots of plants which had the most new root growth and concluded that new root growth created a larger "sink" to draw carbohydrates. However, it is possible that the new root growth retained more of the ^{14}C instead of losing it to the nutrient solution.

When ^{14}C-2,4-D was applied to the leaves of yellow nutsedge (*Cyperus esculentus*) more than 88% of the radioactivity remained in the treated area of the leaf with 2% going to the roots (Bhan et al. 1970). When the leaves were pretreated with unlabelled 2,4-D at 0·56, 1·12, and 2·24 kg/ha immediately prior to the addition of a fixed amount of ^{14}C-2,4-D, the percentage of ^{14}C moved tended to decrease with an increase in the rate of pretreatment. In earlier work, there was no downward translocation in marabú (*Dichrostachys nutans*) of 2,4-D applied 24 h after a previous application (Hay, 1956).

Eliasson (1963) applied 2,4-D and other herbicides to the leaves of aspen (*Populus tremula*) and measured the effect on roots growing on filter paper in plastic bags. Only small changes in the rate of root growth were recorded for 2,4-D. In later experiments, 2,4-D was applied to lower leaves on aspen plants under various experimental conditions (Fig. 7) (Eliasson, 1965). There was only slight downward movement even though these leaves would be expected to be exporting assimilates to the roots (Wardlaw, 1968). As other workers found, the main movement was upward via the xylem since ringing above the treated leaves did not stop upward movement. When the stem above the treated leaves was removed, downward movement in the stem and up into an adjoining untreated shoot was increased. Downward movement in the stem was stopped by ringing the stem immediately below the treated leaves. However, ringing of the adjoining untreated shoot did not stop upward movement in this shoot. Although 2,4-D needs a supply of carbohydrate in the leaf for export to take place (Hay and Thimann, 1956 and many others) once 2,4-D is in the stem Eliasson's data show that it is more likely to be carried in the transpiration stream than in the assimilate flow. Field and Peel (1971) found that 2,4-D moved radially from the xylem to the phloem and vice versa in willow (*Salix viminalis*); 2,4-D moved less readily than MCPA but more than 2,4,5-T.

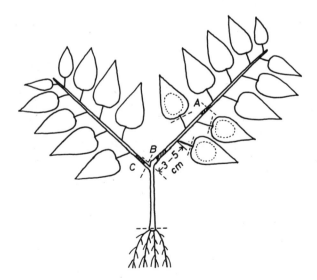

Fig. 7. An outline sketch of aspen (*Populus tremula*) plants used for translocation studies. Dotted lines on lower leaves indicate where the 2,4-D was applied. Immediately before treatment some of the plants were excised at A or steam ringed at A, B or C. The straight broken lines show how the plant was divided before extraction of 2,4-D (Eliasson, 1965).

Eliasson (1965) and Eliasson and Hallmén (1973) proposed that once 2,4-D is moved from the phloem into the xylem it may be carried back into the treated leaf or into other transpiring leaves further up the stem. It may be re-exported from these leaves and move upward again until it is trapped in the immature leaves that are unable to export it. This hypothesis not only accounts for lack of downward movement but also explains why this herbicide always accumulates in the apical tissues. Sagar (1960) proposed a similar hypothesis for the effect of transpiration rate on basipetal movement of dalapon in quackgrass (*Agropyron repens*). As noted above, several workers reported that increasing the relative humidity tends to enhance basipetal translocation of 2,4-D and other herbicides.

Basler *et al.* (1970) injected ^{14}C-2,4,5-T into stems of bean plants grown at $19 \pm 1°C$ and $97 \pm 1\%$ r.h. At the high humidity the amounts of ^{14}C going to the leaves was reduced and that lost to the nutrient solution was increased. Bagging the young shoots of bean plants grown at low humidity reduced the amount of ^{14}C going to these parts. Injecting the ^{14}C-2,4,5-T into the stem was done to rule out the possibility of increased absorption at the high humidity. These results agree with the hypothesis of Eliasson (1965) that high humidity decreases upward movement of herbicides in the transpiration stream and thereby makes it possible for greater downward movement to take place.

Morton *et al.* (1968) compared radioisotope and gas chromatograph methods of analysis in a study on translocation. Comparable results were obtained with both methods (Table IV). The amount of 2,4,5-T absorbed by the leaves increased as the amount in the leaf wash decreased during the 24 h period immediately after application. The amount of 2,4,5-T translocated into the stems supporting the treated leaves was very low and was not related to the amounts absorbed.

Several workers have tried to determine what conditions are required for maximum translocation of 2,4,5-T in woody plants (Dalrymple and Basler, 1963; Morton, 1966; Badiei *et al.* 1966; Brady, 1969). However, no clear picture of the role of environmental factors emerges from this work other than the importance of soil moisture. Absorption in winged elm (*Ulmus alata*) and in mesquite (*Prosopis glandulosa*) was not changed by moisture stress but translocation was reduced (Davis *et al.* 1968b). Translocation of ^{14}C from ^{14}C-2,4,5-T to lower stems and roots of winged elm was reduced by withholding soil moisture prior to foliar application (Wills and Basler, 1971). Absorption by leaves was not affected by soil moisture conditions. Schmutz (1971) claimed that the greatest kills and maximum translocation were obtained if 2,4,5-T was applied to creosotebush (*Larrea tridentata*) 30 days after the start of the summer rains.

^{14}C-2,4-DB was translocated more than some other phenoxy herbicides in bigleaf maple (*Acer macrophyllum*) as indicated by counts of ^{14}C in roots following foliar application (Norris and Freed, 1966b). Briquet *et al.* (1968a, b) found by auto-radiography and counting that ^{14}C from ^{14}C-2,4-DB moved upward to the tips with small amounts going to the roots in beans. Movement of ^{14}C was more extensive in cocklebur (*Xanthium* sp.) than in soybeans when

TABLE IV

MICROGRAMS OF 2,4,5-T RECOVERED AT 0, 1, 6 AND 24 h AFTER APPLICATION OF 250 µg OF AMMONIUM SALT AND BUTOXYETHYL ESTER OF 2,4,5-T TO LEAVES OF HONEY MESQUITE *(Prosopis juliflora)* AS DETERMINED BY RADIOISOTOPE AND GAS CHROMATOGRAPH ANALYSES (MORTON *ET AL.* 1968)

Material analyzed	Radioisotopic analysis				Gas chromatographic analysis			
	0 h	*1 h*	*6 h*	*24 h*	*0 h*	*1 h*	*6 h*	*24 h*
Butoxyethylesters								
Hexane leaf wash	201·8	119·3	44·3	12·5	202·4	162·3	81·1	34·6
Ammonium hydroxide leaf wash	29·7	47·8	19·5	31·7	21·8	37·9	17·7	13·8
Extract of treated leaves	18·5	55·3	124·9	150·2	25·8	49·6	115·2	112·7
Stem supporting treated leaves	0·0	0·1	0·6	0·2	0·0	0·2	0·9	1·4
Total recovered	250·0	222·5	189·3	194·6	250·0	250·0	214·9	162·5
Ammonium salts								
Ammonium hydroxide leaf wash	249·0	185·9	93·1	21·0	245·8	194·3	107·6	27·0
Extract of treated leaves	1·0	16·0	43·2	75·2	4·6	13·2	65·2	61·5
Stem supporting treated leaves	0·0	0·2	0·1	0·2	0·0	0·4	1·1	2·0
Total recovered	250·0	202·1	136·4	96·4	250·4	207·9	173·9	90·5

ring-labelled 2,4-DB was applied to the leaves (Walthana *et al.* 1972). The pattern of movement was similar to that of other phenoxy compounds with the maximum accumulation taking place in the young tips. In neither species was movement extensive but it should be noted that leaves near the apex were treated and export from such leaves may not be large in any case.

To summarize, small amounts of the phenoxy herbicides are moved from leaves to stems. Movement to the roots is negligible regardless of the "flow" of assimilates. Apparently, once in the stems they move from the phloem to the apoplast and are carried back into the treated leaves or up the stem into the developing leaves from which transport does not occur. Thus they move to a transpiration "sink" rather than an assimilate "sink". Removing the upper leaves or raising the humidity may reduce the pull of the transpiration stream and allow greater downward movement. When soil moisture is low enough to reduce growth, translocation is reduced. Perhaps under these conditions the carbohydrate supply for phloem loading is not available. The amount of herbicide moved to the roots does not appear to be related to the amount actually absorbed by the leaves.

F. PICLORAM

Picloram is readily taken up from nutrient solutions and moved to the tops of plants. Stem bending was noted on beans 2 h after the roots were immersed in nutrient solutions containing picloram (Reid and Hurtt, 1969). In oats and soybeans, uptake and translocation to the tops was increased by raising the temperature, increasing the concentration or lowering the pH (Isensee *et al.* 1971). The concentration of picloram in the tops was higher than in the roots. When the plants were transferred to nutrient solution without picloram the concentration in the roots decreased while that in the tops increased. Some of the picloram in the roots may have been lost to the nutrient solution but the increase in the tops must have been at the expense of that in the roots. Concentration of picloram in tops of huisache (*Acacia farnesiana*) and honey mesquite (*Prosopis juliflora*) seedlings was also higher than in the roots (Baur and Bovey, 1969). At 24 h honey mesquite had 2·4 μg/g fresh weight in the tops and 0·4 in the roots, huisache had 1·5 μg/g in the tops and 0·4 in the roots. However, caution should be exercised when interpreting results of translocation studies made on plants grown in nutrient solutions (Bovey *et al.* 1967). When huisache was treated via the soil, higher concentrations of picloram were found in the tops. Control of this species was more effective from soil applications than when equal amounts were applied to the foliage.

Distribution of picloram in bean plants at 3, 6 and 11 h after root uptake from nutrient solutions was determined (Reid and Hurtt, 1969). Much higher concentrations were recorded in the first trifoliate leaf and the growing tips than in the primary leaf, i.e. at 12 h there were 2·6 ng/mg fresh weight in the tips and 0·06 in the primary leaf. It would appear that any picloram that moved into the primary leaf via the transpiration stream was rapidly exported whereas any that arrived at the tips was retained there. Alternatively transfer cells at the node could have removed picloram from the transpiration stream entering the primary leaves.

As happens with many other herbicides, picloram, applied to the foliage remained mostly in or on the treated leaves. In huisache (*Acacia farnesiana*) 545 µg of picloram were washed from the leaves and only 18·5 µg were in the leaves 24 h after application (Bovey *et al.* 1967). At 30 days, concentrations were 1,166 µg/g in leaves, 17·24 µg/g in upper stems, 0·3 µg/g in lower stems with none being detected in the roots. Picloram sprayed on yaupon (*Ilex vomitoria*) leaves at 4 lb/acre (4·5 kg/ha) gave 1,505 µg in the leaf and 40,985 µg in the leaf rinse after 72 h (Davis *et al.* 1968a). In 1·5-month-old mesquite seedlings (*Prosopis glandulosa*), 2·1% left the treated leaves by 24 h (Davis *et al.* 1968b). Of the amounts exported 61% was in the tips, 21% in stems and leaves above the treated leaves, 9% in lower stem, 8% in root-stem transition zone and 1% in the roots. This distribution was not greatly altered at 90 h.

In Canada thistle, stem bending was noted 10 h after application of picloram to leaves (Sharma *et al.* 1971). About 20% of the picloram that entered the leaf was translocated out. By removing leaves at different times after treatment it was deduced that no significant export occurred before 4 h; by 6 h toxic amounts had moved to the stems and by 24 h enough had moved to the stems to cause serious injury to the plant. Transport continued for 48 h.

Treated leaves were removed from 10-day-old bean plants at various times after treatment (Hurtt, 1970). Enough picloram was translocated in 30 min to give suppression of growth in height. Ten hours were required for the development of epinastic effects and 12 h for death of the terminal bud. Girdling petioles of bean leaves stopped the movement of picloram out of treated leaves (Wells *et al.* 1970). Girdling the stem 2 cm above or below the treated leaf did not stop upward movement. Thus acropetal movement in the stems was in the apoplast.

Hamill *et al.* (1972) applied 50 µl of 100 parts/10^6 picloram to three spots on bean leaves. Stem bending was observed at 6 h but at 24 h 99% of the ^{14}C was still on or in the treated leaf. Transport continued for seven days with the bulk of the picloram appearing in the growing tips or newly developing leaves with none in the unifoliate leaf opposite the treated leaf and only small amounts in the roots.

In wheat, picloram was translocated into stems and untreated leaves to a greater extent than was 2,4-D (Hallmén and Eliasson, 1972). However, only a small fraction of either herbicide was recovered from the roots or nutrient solution.

Although low amounts of picloram were recovered in the roots after foliage applications, significant amounts have been detected in the nutrient solutions. Of the picloram absorbed by leaves, 6·2% was exuded into the nutrient solution from roots of red maple (*Acer rubrum*) and 1·6% from the roots of green ash (*Fraxinus pennsylvanica*) (Reid and Hurtt, 1970). The quantities of picloram exuded from roots of red maple, green ash, and white ash (*F.*

americana) were not related to tolerance or resistance of these species (Reid *et al.* 1971).

Moisture stress was imposed on bean seedlings treated with picloram by the addition of mannitol to the nutrient solution (Merkle and Davis, 1967). This did not affect uptake of foliar-applied picloram but it reduced the amount going to the stem and apex. Under conditions that caused visible wilting, the stems contained only 57% of that of plants under no stress. In mesquite (*Prosopis glandulosa*), uptake was reduced by moisture stress but not in winged elm (*Ulmus alata*) (Davis *et al.* 1968*b*). In both species translocation was reduced by moisture stress.

Picloram gave better root kill of aspen (*Populus tremula*) than did 2,4-D, 2,4,5-T or dicamba (Eliasson, 1972). However, this difference could not be accounted for by greater translocation of picloram (Eliasson, 1972; Eliasson and Hallmén, 1973). As with the phenoxy herbicides most of the translocated picloram was in the untreated leaves and growing shoot tips with only minute amounts in the roots (Table V). Picloram was about 10 times more active than 2,4-D or dicamba in inhibiting root growth when the substances were added to the nutrient solution. It was concluded that the inherent high toxicity of picloram was the major reason for its good kill of aspen.

We see then that the transport of picloram from leaves does not appear to be radically different from that of the phenoxy herbicides. The bulk of the

TABLE V

UPTAKE AND TRANSLOCATION OF ^{14}C FROM LEAVES OF ASPEN (*Populus tremula*) AFTER APPLICATION OF 10 µg ^{14}C-PICLORAM OR ^{14}C-2,4-D IN 50 µl OF WATER SOLUTION CONTAINING 10% ETHANOL AND 0·3% TWEEN 20. VALUES ARE % OF TOTAL ACTIVITY APPLIED (ELIASSON, 1973)

Plant part	Average fresh weight during the experiment g	Days after treatment		
		1	*3*	*9*
Picloram				
Leaf wash	—	49·1	21·2	22·4
Treated leaf	0·1	25·3	44·4	38·7
Shoot above treated leaf	0·4	10·8	14·8	23·9
Stem below treated leaf	0·3	7·1	8·4	6·4
Roots	0·6	0·0	0·0	0·0
Recovered activity	—	92·3	88·8	91·4
2,4-D				
Leaf wash	—	8·6	4·8	0·7
Treated leaf	0·2	67·3	57·4	56·4
Shoot above treated leaf	0·4	3·4	15·3	16·6
Stem below treated leaf	0·3	3·7	5·6	5·8
Roots	0·7	0·0	0·3	1·0
Recovered activity	—	83·0	83·4	80·5

translocated picloram accumulates in the treated leaves or growing tips with only minute amounts going to the roots. The high toxicity of picloram could account for the good control of deep-rooted perennials. When taken up by the roots it moves to the tops where it concentrates in the shoot tips and young leaves.

G. 2,3,6-TBA

Linder et al. (1958) reported that 2,3,6-TBA was moved from the foliage of several species to the soil solution since it affected other plants growing in the same pot. 2,3,6-TBA was taken up by the indicator plants and moved to the tips where formative effects were noted. In another study (Linder et al. 1964), leaves of bean plants received 100 µg of 2,3,6-TBA. Although 3 µg of 2,3,6-TBA were lost from the roots during the two days after treatment 1 µg was found in the roots four days after treatment.

2,3,6-TBA moved acropetally in corn plants to injure witchweed (*Striga lutea*) a parasite of corn (Sand et al. 1971).

Thus 2,3,6-TBA appears to move readily in both the apoplast and symplast.

H. UREA HERBICIDES

Monuron is a classical example of a herbicide that is translocated only in the apoplast (Ashton and Crafts, 1973). It is readily taken up by roots and moved to the leaves in the transpiration stream (Muzik et al. 1954; Minshall, 1954). Factors that reduced transpiration reduced movement of monuron to aerial parts of treated plants. Although there is a slight uptake by leaves there is no downward movement. Symptoms and radioactivity from labelled monuron (Crafts, 1967) always appear distal to the point of application. Although movement is largely apoplastic monuron must at some point enter the symplast to exert its toxic effects. It is not known how monuron crosses the endodermis.

The translocation of ^{14}C from ^{14}C-diuron was similar to that from ^{14}C-monuron (Bayer and Yamaguchi, 1965; Leonard et al. 1966; Leonard and Glenn, 1968). [14]C moved freely in the apoplast of barley, soybeans and red kidney beans when ^{14}C-diuron was applied to leaves, stems or roots. There was negligible movement downward and no accumulation in growing tips. [14]C from both ^{14}C-monuron and ^{14}C-diuron moved to the tops of oats, soybeans, corn and cotton from nutrient solutions (Smith and Sheets, 1967). Susceptibility (oats > sorghum > corn > cotton) was not related to the amount of ^{14}C moved to the tops and both compounds were metabolized by all four species. At 120 h, none of the ^{14}C in the leaves of cotton was in diuron and only 9% of the ^{14}C in the tops of soybeans was in diuron as determined by thin layer chromatography. Corn was less able to degrade diuron. It was not clear whether the breakdown took place before or after translocation and with such extensive breakdown it is amazing that consistent patterns are obtained in auto-radiographs of the urea herbicides.

Strang and Rogers (1971) used a micro-radio-autographic technique to
locate concentrations of ^{14}C at the cellular level in cotton plants treated with
^{14}C-diuron via the nutrient solution. It appeared from their observations that
the ^{14}C-diuron moved through the cell walls of the cortical tissue without any
concentration in the vicinity of the endodermis. Radioactivity was noted in the
xylem vessels, particularly in their cell walls, with no accumulation in the
phloem. It also appeared that lateral movement to epidermal cells of stems,
petioles and mature leaves took place via the cell walls. There were striking
accumulations of radioactivity in the lysigneous (oil rich) glands of the cotton
leaves. This accumulation is probably not related to translocation but could
serve as a detoxification process.

^{14}C moved to the shoots of small-flower galinsoga (*Galinsoga parviflora*)
and wild buckwheat (*Polygonum convolvulus*) from nutrient solutions
containing ^{14}C-chloroxuron (Geissbühler *et al.* 1963). Concentrations of ^{14}C
were highest in roots, intermediate in stems and lowest in the leaves. This was
in contrast to monuron which generally had higher concentrations in the leaves
than in the roots. More ^{14}C was in the tops of wild buckwheat than of small-
flower galinsoga although the latter is more susceptible under field conditions.
When ^{14}C-chloroxuron was administered to the soil there was much less ^{14}C in
the tops of these species than when equal amounts were given via the nutrient
solution. Amounts in the aerial parts of the two species were equal when
treated via the soil. As with other urea herbicides there was very little
movement acropetally or basipetally in leaves when ^{14}C-chloroxuron was
applied to the leaf surface.

Metobromuron was taken up from nutrient solutions by *Sinapis arvensis*
and *Veronica persica* (Majumdar and Müller, 1969). It was rapidly moved to
the tops and accumulated at the leaf margins. Susceptibility of these species
was not related to uptake and translocation.

There is therefore little or no basipetal movement of the urea herbicides in
the symplast when applied to leaves. They are taken up readily by roots, move
to the tops via the apoplast and accumulate in young and mature leaves. They
tend to concentrate at the leaf margins where injury symptoms are first noted.
Susceptibility of species to the urea herbicides does not appear to be related to
the amounts moved to the tops.

I. TRIAZINES

Gross autoradiography shows that ^{14}C from labelled simazine (Davis *et al.*
1959; Sheets, 1961; Montgomery and Freed, 1961; Freeman *et al.* 1964;
Lund-Høie, 1969), prometryne (Sikka and Davis, 1968) and atrazine
(Montgomery and Freed, 1961; Wax and Behrens, 1965) is readily taken up
from nutrient solutions and distributed throughout various plants. The ^{14}C
may appear in the tops within 30 min after application (Davis *et al.* 1959).
When so administered the ^{14}C moved to all leaves. It showed up first along the
veins but later accumulated at the tips of grass leaves and around margins of
the leaves of dicotyledonous plants (Davis *et al.* 1959; Sheets, 1961; Lund-

Høie, 1969). Toxic symptoms also tended to show up first around the leaf margins of triazine-treated plants. ^{14}C accumulated in lysigneous glands of cotton (Davis *et al*. 1959; Sheets, 1961; Foy, 1964).

Thus ^{14}C from labelled triazines apparently moves into the apoplast of the roots and is carried upward in the transpiration stream. Factors that increased transpiration increased upward movement of triazines. There was a greater accumulation of simazine (Sheets, 1961) and atrazine (Wax and Behrens, 1965; Vostral *et al*. 1970) in the tops as the temperature was raised or as the humidity was lowered. On the other hand, transpiration may be reduced by the triazines through closure of stomata due to reduced photosynthesis (Smith and Buchholtz, 1962, 1964; Wills *et al*. 1963; Vostral *et al*. 1970). The inhibition of transpiration by triazines may account for the reduced uptake of these compounds with time (Shimabukuro and Linck, 1967).

Translocation of triazines in aquatic plants varies with the species and with the compounds. In plants with shoots above the water level, simazine, ametryne and prometryne moved from roots to tops of alligator weed (*Alternanthera philoxeroides*) (Funderburk and Lawrence, 1963) as did simazine in parrot feather (*Myriophyllum brasiliense*) (Sutton and Bingham, 1969). In watergrass (*Heteranthera dubia*), which grows completely below the surface, simazine moved from roots to shoots but prometryne and ametryne did not (Funderburk and Lawrence, 1963).

Davis *et al*. (1959) concluded that the lack of translocation from foliar applications of simazine in corn, cotton and cucumbers was due to a lack of absorption since there was more movement within the leaves if the surface was "sanded" before application. Both absorption and acropetal translocation within leaves of yellow foxtail (*Setaria glauca*) were increased when atrazine was applied in an oil-water emulsion instead of in water alone (Smith and Nalewaja, 1972). These observations are consistent with the suggestion that the waxy leaf surface restricts uptake and therefore translocation of triazines.

To summarize, triazine herbicides are taken up by the roots and rapidly moved to the tops in the apoplast. They appear to concentrate first at the veins then in the interveinal areas and finally around the margins of the leaves. There is virtually no symplastic movement out of leaves and the waxy leaf surfaces are a barrier to uptake.

J. BIPYRIDYLIUM HERBICIDES

Diquat and paraquat characteristically cause localized injury. However, Calderbank *et al*. (1961) found residues of diquat in potato tubers when it was used to desiccate the tops before harvest. Thus some translocation must have occurred.

^{14}C-diquat and ^{14}C-paraquat did not move from treated spots on tomato leaves to any extent if the plants were kept in the dark. However, if the plants were subsequently placed in the light the label became uniformly distributed throughout the aerial parts of the plant, and the plants died (Baldwin, 1963). This movement took place in the xylem since steam ringing of the petiole did not stop translocation to the rest of the plant. There was no marked accumulation of the label in the meristematic regions. This phenomenon was confirmed by Slade and Bell (1966) for paraquat and by Smith and Sagar (1966) for diquat. Their results supported Baldwin's explanation that these compounds were absorbed into the apoplast of the leaves but were not translocated during the dark period. When the plants were exposed to light the chemicals were activated and the tissues killed. The water was then drawn from the leaves by the other transpiring surfaces. The diquat or paraquat already absorbed into the apoplast during the dark period was carried along with the water. ^{32}P and ^{14}C-urea were also moved from leaves if time was allowed for their uptake in the dark prior to killing the leaf with diquat (Smith and Sagar, 1966).

Putnam and Ries (1968) found slight movement of ^{14}C-paraquat in quackgrass (*Agropyron repens*) leaves in the dark but this was enhanced by a light period after application. In the light paraquat continued to move out of leaves between 8 and 16 h after application even though toxic symptoms were apparent at 8 h. This observation further supports the evidence that movement occurs in the xylem and not in the phloem.

Smith and Davies (1965) reported that paraquat moved acropetally in the xylem of *Paspalum distichum*. Thrower et al. (1965) found that diquat moved in the light and dark and that upward movement was greater at 50% than at 90% r.h. On the other hand, Brian (1966) reported greater translocation of paraquat at high humidity. This was attributed to enhanced uptake in the more humid conditions.

Although diquat was taken up rapidly by the aquatic plant, *Elodea canadensis*, there was little movement within the plant (Davies and Seaman, 1968). This might be expected of a compound that moves in the xylem of a plant which is surrounded by water.

Thus the movement of the bipyridylium herbicides is unlike that of any of the other herbicides described. Although they do not translocate in the symplast they do move out of leaves in the apoplast. Diquat and paraquat penetrate the leaf surface and enter the apoplast. They do not appear to move much until the leaf tissue dies. Then as the leaves dry up the water is drawn into the plant taking the herbicides along.

V. Summary

Most herbicides are taken up by the roots and moved to the tops via the transpiration stream. Urea and triazine herbicides move into all leaves without any marked concentration at the tips. This probably indicates that these compounds cannot move out of leaves in the symplast. Other herbicides such as dalapon, amitrole, picloram and dicamba also move upward in this way but they concentrate in developing tissue at the tips. The lower concentrations in the mature leaves suggest that these herbicides are redistributed out of older leaves via the symplast, and make their way to the new growth which is unable to export materials. Movement of herbicides from upper mature leaves to the growing tips may be wholly in the symplast or partly in the apoplast. However, upward redistribution in stems from lower leaves is almost certainly in the apoplast. Degradation of herbicides and their ability to "complex" with plant constituents may alter distribution patterns.

When applied to the leaves urea and triazine herbicides are not exported. Penetration into leaves is poor and symplastic movement is negligible. Other herbicides are exported from leaves in the symplast but in all cases the bulk of the material remains in the treated leaves with only small amounts being moved to the stems. In the stems movement is predominantly to the apex with very little going to the roots. The herbicides presumably move from the phloem to the xylem in the stems and are carried back up into the treated leaf or into transpiring leaves further up the stem. They could be re-exported from mature leaves and eventually be trapped in the immature tissues at the apex.

A supply of carbohydrate is necessary for the movement of herbicides from leaves. This is needed to set up an assimilate flow or to supply energy for phloem loading. This "push" from the leaves is effective in getting the herbicide out of the leaf and into the stem. Once there the main "pull" is that of the transpiration stream. The tendency to be carried upward will depend on the ease of movement from the phloem to the xylem. Herbicides may "leak" from the phloem more readily than other compounds because of their toxic properties. The movement of herbicides from lower leaves, which normally supply carbohydrates to the roots is mainly upward. Thus, while herbicides move from "sources" when carbohydrate supplies are plentiful there is ample reason to doubt that they are drawn to assimilate "sinks".

Downward translocation has been increased when the relative humidity was raised. This could be due to better penetration but increasing absorption with surfactants has not always led to greater translocation. Thus, the high humidity probably allows greater downward movement to occur by reducing the transpiration pull in the stems.

Dry soil conditions did not affect absorption of herbicides but it did reduce translocation in a number of tests. The lack of soil moisture could have reduced the supply of carbohydrates for phloem loading.

REFERENCES

Anderson, R. N., Linck, A. J. and Behrens, R. (1962). *Weed Sci.* **10**, 1–3.
Ashton, F. M. and Crafts, A. S. (1973). "Mode of Action of Herbicides", John Wiley and Sons, New York.
Badiei, A., Basler, E. and Santelmann, P. W. (1960). *Weed Sci.* **14**, 302–305.
Baldwin, B. C. (1963). *Nature, Lond.* **198**, 872.
Basler, E., Slife, F. W. and Long, J. W. (1970). *Weed Sci.* **18**, 396–398.
Basler, E., Todd, G. W. and Meyer, R. E. (1961). *Pl. Physiol., Lancaster*, **36**, 573–576.
Baur, J. R. and Bovey, R. W. (1969). *Weed Sci.* **17**, 524–528.
Bayer, D. E. and Yamaguchi, S. (1965). *Weeds*, **13**, 232–235.
Bhan, V. M., Stoller, E. W. and Slife, F. W. (1970). *Weed Sci.* **18**, 733–737.
Binning, L. K., Penner, D. and Meggitt, W. F. (1971). *Weed Sci.* **19**, 73–75.
Blair, B. O. and Fuller, W. H. (1952). *Bot. Gaz.* **113**, 368–372.
Bovey, R. W., Davis, F. S. and Merkle, M. G. (1967). *Weeds*, **15**, 245–249.

Brady, H. A. (1969). *Weed Sci.* **17**, 320–322.
Brian, R. C. (1966). *Weed Res.* **6**, 292–303.
Briquet, M. V., Scheller, G., Crevecoeur, E. and Wiaux, A. L. (1968a). *Weed Res.* **8**, 61–63.
Briquet, M. V., Scheller, G., Crevecoeur, E. and Wiaux, A. L. (1968b). *Weed Res.* **8**, 64–67.
Burns, E. R., Buchanan, G. A. and Hiltbold, A. E. (1969). *Weed Sci.* **17**, 401–404.
Calderbank, A., Morgan, C. B. and Yuen, S. H. (1961). *Analyst, Lond.* **86**, 569.
Canny, M. J. (1971). *A. Rev. Pl. Physiol.* **22**, 237–260.
Carter, M. C. and Naylor, A. W. (1960). *Bot. Gaz.* **122**, 138–143.
Chang, F. Y. and Vanden Born, W. H. (1968). *Weed Sci.* **16**, 176–181.
Chang, F. Y. and Vanden Born, W. H. (1971a). *Weed Sci.* **19**, 107–112.
Chang, F. Y. and Vanden Born, W. H. (1971b). *Weed Sci.* **19**, 113–117.
Clor, M. A., Crafts, A. S. and Yamaguchi, S. (1962). *Pl. Physiol., Lancaster,* **37**, 609–617.
Coble, H. D., Slife, F. W. and Butler, H. S. (1970). *Weed Sci.* **18**, 653–656.
Crafts, A. S. (1964). *In* "The Physiology and Biochemistry of Herbicides" (L. J. Audus, ed.), pp. 75–110. Academic Press, New York and London.
Crafts, A. S. (1966). *In* "Isotopes in Weed Research", pp. 3–7. Int. Atomic Energy Agency, Vienna.
Crafts, A. S. (1967). *Hilgardia,* **37**, 625–638.
Crafts, A. S. and Crisp, C. E. (1971). *In* "Phloem Transport in Plants", W. H. Freeman and Co., San Francisco.
Crafts, A. S. and Foy, C. L. (1959). *Down to Earth,* **15**, 1–5.
Dalrymple, A. V. and Basler, E. (1963). *Weed Sci.* **11**, 41–44.
Davies, P. J. and Seaman, D. E. (1968). *Weed Sci.* **16**, 293–295.
Davis, D. E., Funderburk, H. H., Jr. and Sansing, N. G. (1959). *Weeds,* **7**, 300–309.
Davis, F. S., Bovey, R. W. and Merkle, M. G. (1968a). *Weed Sci.* **16**, 336–339.
Davis, F. S., Merkle, M. G. and Bovey, R. W. (1968b). *Bot. Gaz.* **129**, 183–189.
Donnalley, W. F. and Ries, S. K. (1964). *Science, N.Y.* **145**, 497–498.
Duble, R. L., Holt, E. C. and McBee, G. G. (1969). *J. agric. Fd Chem.* **17**, 1247–1250.
Eliasson, L. (1963). *Physiologia Pl.* **16**, 201–214.
Eliasson, L. (1965). *Physiologia Pl.* **18**, 506–515.
Eliasson, L. (1972). *Physiologia Pl.* **27**, 101–104.
Eliasson, L. and Hallmén, U. (1973). *Physiologia Pl.* **28**, 182–187.
Epstein, E. (1973). *Scient. Am.* **228**, 48–58.
Eschrich, W. (1970). *A. Rev. Pl. Physiol.* **21**, 193–214.
Fang, S. C. and Theisen, P. (1960). *J. agric. Fd Chem.* **8**, 295–298.
Field, R. J. and Peel, A. J. (1971). *New Phytol.* **70**, 743–749.
Fites, R. C., Slife, F. W. and Hanson, J. B. (1964). *Weed Sci.* **12**, 180–183.
Forde, B. J. (1966). *Weed Sci.* **14**, 178–179.
Foy, C. L. (1961). *Pl. Physiol., Lancaster,* **36**, 688–697.
Foy, C. L. (1962a). *Weed Sci.* **10**, 35–39.
Foy, C. L. (1962b). *Weed Sci.* **10**, 97–99.
Foy, C. L. (1964). *Weeds,* **12**, 103–108.
Freeman, F. W., White, D. P. and Bukovac, M. J. (1964). *Forest Sci.* **10**, 330–334.
Funderburk, H. H. and Lawrence, J. M. (1963). *Weed Res.* **3**, 304–311.
Geiger, D. R. and Cataldo, D. A. (1969). *Pl. Physiol., Lancaster,* **44**, 45–54.
Geissbühler, H., Haselbach, C., Aebi, H. and Ebner, L. (1963). *Weed Res.* **3**, 181–194.
Goma, E. A. A., Matolcsy, G. and Tanács, B. (1969). *Weed Res.* **9**, 150–153.
Hallmén, U. and Eliasson, L. (1972). *Physiologia Pl.* **27**, 143–147.
Hamill, A. S., Smith, L. W. and Switzer, C. M. (1972). *Weed Sci.* **20**, 226–229.
Hay, J. R. (1956). *Weeds,* **4**, 349–356.
Hay, J. R. and Thimann, K. V. (1956). *Pl. Physiol., Lancaster,* **31**, 446–451.
Herrett, R. A. and Bagley, W. P. (1964). *J. agric. Fd Chem.* **12**, 17–20.
Hill, E. R., Lachman, W. H. and Maynard, D. N. (1963). *Weed Sci.* **11**, 165–166.
Hodgson, G. L. (1964). *Weed Res.* **4**, 167–168.
Hull, R. J. (1969). *Weed Sci.* **17**, 314–320.
Hurtt, W. (1970). *In* "U.S. Clearinghouse Fed. Sci. Tech. Inform., Ad. Issue No. 715704".

Isensee, A. R., Jones, G. E. and Turner, B. C. (1971). *Weed Sci.* **19,** 727–731.
Läuchli, A. (1972). *A. Rev. Pl. Physiol.* **23,** 197–218.
Leonard, O. A. (1939). *Pl. Physiol., Lancaster,* **14,** 55–74.
Leonard, O. A. and Glenn, R. K. (1968). *Weed Sci.* **16,** 352–356.
Leonard, O. A., Lider, L. A. and Glenn, R. K. (1966). *Weed Res.* **6,** 37–49.
Leonard, O. A., Weaver, R. J. and Kay, B. L. (1962). *Weeds,* **10,** 20–22.
Levi, E. (1962). *Science, N.Y.,* **137,** 343–344.
Linder, P. J., Craig, J. C., Cooper, F. E. and Mitchell, J. W. (1958). *J. agric. Fd Chem.* **6,** 356–357.
Linder, P. J., Mitchell, J. W. and Freeman, G. D. (1964). *J. agric. Fd Chem.* **12,** 437–438.
Linscott, D. L. and McCarty, M. K. (1962). *Weed Sci.* **10,** 65–68.
Lund-Høie, K. (1969). *Weed Res.* **9,** 142–147.
Lund-Høie, K. and Bayer, D. E. (1968). *Physiologia Pl.* **21,** 196–212.
Lund-Høie, K. and Bylterud, A. (1969). *Weed Res.* **9,** 205–210.
Magalhaes, A. C., Ashton, F. M. and Foy, C. L. (1968). *Weed Sci.* **16,** 240–245.
Majumdar, J. C. and Müller, F. (1969). *Weed Res.* **9,** 322–332.
Massini, P. (1963). *Acta. bot. neerl.* **12,** 64–72.
McIntyre, G. I. (1962). *Weed Res.* **2,** 165–176.
Merkle, M. G. and Davis, F. S. (1967). *Weed Sci.* **15,** 10–12.
Minshall, W. H. (1954). *Can. J. Bot.* **32,** 795–798.
Mitchell, J. W. and Linder, P. J. (1963). *Residue Rev.* **2,** 51–76.
Montgomery, M. and Freed, V. H. (1961). *Weeds,* **9,** 231–237.
Morton, Howard L. (1966). *Weed Sci.* **14,** 136–141.
Morton, H. L., Davis, F. S. and Merkle, M. G. (1968). *Weed Sci.* **16,** 88–91.
Muzik, T. J., Cruzado, H. J. and Loustalot, A. J. (1954). *Bot. Gaz.* **116,** 65–73.
Neidermyer, R. W. and Nalewaja, J. D. (1969). *Weed Sci.* **17,** 528–532.
Norris, L. A. and Freed, V. H. (1966a). *Weed Res.* **6,** 203–211.
Norris, L. A. and Freed, V. H. (1966b). *Weed Res.* **6,** 283–291.
Pallas, J. E. (1960). *Pl. Physiol., Lancaster,* **35,** 575–580.
Pallas, J. E. and Crafts, A. S. (1956). *Science, N.Y.* **125,** 192–193.
Pallas, J. E. and Williams, G. G. (1962). *Bot. Gaz.* **123,** 175–180.
Pate, J. S. and Gunning, B. E. S. (1972). *A. Rev. Pl. Physiol.* **23,** 173–196.
Prasad, R., Foy, C. L. and Crafts, A. S. (1967). *Weed Sci.* **15,** 149–156.
Putnam, A. R. and Ries, S. K. (1968). *Weed Sci.* **16,** 80–83.
Racusen, D. (1958). *Archs Biochem. Biophys.* **74,** 106–113.
Reid, C. P. P. and Hurtt, W. (1969). *Pl. Physiol., Lancaster,* **44,** 1393–1396.
Reid, C. P. P. and Hurtt, W. (1970). *Nature, Lond.* **255,** 291.
Reid, C. P. P., Hurtt, W. and Wells, W. A. (1971). *In* "U.S. Clearinghouse Fed. Sci. Techn. Inform., Ad. Issue No. 720572".
Robertson, M. M. and Kirkwood, R. C. (1970). *Weed Res.* **10,** 94–120.
Rogerson, A. B. and Bingham, S. W. (1971). *Weed Sci.* **19,** 325–328.
Rohrbaugh, L. M. and Rice, E. L. (1949). *Bot. Gaz.* **111,** 85–89.
Sagar, G. R. (1960). *Proc. Br. Weed Cont. Conf.* pp. 271–278.
Sand, P. F., Egley, G. H., Gould, W. L. and Kust, C. A. (1971). *Weed Sci.* **19,** 240–244.
Schmutz, E. M. (1971). *Weed Sci.* **19,** 510–516.
Sharma, M. P., Chang, F. Y. and Vanden Born, W. H. (1971). *Weed Sci.* **19,** 349–355.
Sheets, T. J. (1961). *Weeds,* **9,** 1–13.
Shih, C. Y. and Currier, H. B. (1969). *Am. J. Bot.* **56,** 464–472.
Shimabukuro, R. H. and Linck, A. J. (1967). *Weed Sci.* **15,** 175–178.
Sikka, H. C. and Davis, D. E. (1968). *Weed Sci.* **16,** 474–477.
Slade, P. and Bell, E. G. (1966). *Weed Res.* **6,** 267–274.
Slife, F. W., Key, J. L., Yamaguchi, S. and Crafts, A. S. (1962). *Weeds,* **10,** 29–35.
Smith, C. N. and Nalewaja, J. D. (1972). *Weed Sci.* **20,** 36–40.
Smith, D. and Buchholtz, K. P. (1962). *Science, N.Y.* **136,** 263–264.
Smith, D. and Buchholtz, K. P. (1964). *Pl. Physiol., Lancaster,* **39,** 572–578.
Smith, J. M. and Sagar, G. R. (1966). *Weed Res.* **6,** 314–321.

Smith, J. W. and Sheets, T. J. (1967). *J. agric. Fd Chem.* **15,** 577–581.
Smith, L. W. and Davies, P. J. (1965). *Weed Res.* **5,** 343–347.
Strang, R. H. and Rogers, R. L. (1971). *Weed Sci.* **19,** 355–362.
Sutton, D. L. and Bingham, S. W. (1969). *Weed Sci.* **17,** 431–435.
Thrower, S. L., Hallam, N. D. and Thrower, L. B. (1965). *Ann. appl. Biol.* **55,** 253–260.
Trip, P. and Gorham, P. R. (1968). *Pl. Physiol., Lancaster,* **43,** 877–882.
Verloop, A. and Nimmo, W. B. (1969). *Weed Res.* **19,** 357–370.
Vostral, H. J., Buchholtz, K. P. and Kust, C. A. (1970). *Weed Sci.* **18,** 115–117.
Wardlaw, I. F. (1968). *Bot. Rev.* **34,** 79–105.
Walthana, S., Corbin, F. T. and Waldrep, T. W. (1972). *Weed Sci.* **20,** 120–123.
Wax, L. M. and Behrens, R. (1965). *Weed Sci.* **13,** 107–109.
Webb, K. L. and Gorham, P. R. (1964). *Pl. Physiol., Lancaster,* **39,** 663–672.
Webb, K. L. and Gorham, P. R. (1965). *Can. J. Bot.* **43,** 97–103.
Wells, W. A., Hurtt, W. and Reid, C. P. P. (1970). *Abstr. Meet. Weed Sci. Soc. Am.* **140,** 72.
Wills, G. D. and Basler, E. (1971). *Weed Sci.* **19,** 431–434.
Wills, G. D., Davis, D. E. and Funderburk, H. H. (1963). *Weeds,* **11,** 253–255.
Yamaguchi, S. (1961). *Weeds,* **9,** 374–380.
Yamaguchi, S. and Crafts, A. S. (1958). *Hilgardia,* **28,** 161–191.
Young, D. W. and Fisher, C. E. (1950). *Proc. N. Cent. Weed Control Conf.* p. 50.
Zimmerman, M. (1961). *Science, N.Y.* **133,** 73–79.
Zweep, van der, W. (1961). *Weed Res.* **1,** 258–266.

CHAPTER 13

HERBICIDE METABOLISM IN PLANTS

AUBREY W. NAYLOR

James B. Duke Professor of Botany,
Duke University, Durham, North Carolina 27706, U.S.A.

I. INTRODUCTION

The selective toxicity of a herbicide to different species may be attributable to numerous factors including absorption, translocation and metabolism. If the herbicide is retained unchanged it is important that this should be known. Increasingly, government agencies are setting tolerance limits on pesticides that may be present in plants or their parts at the time of marketing. Such imposed requirements for information should aid considerably the cause of basic science. It is certainly of more than passing importance that the fates of herbicides in plants and animals be investigated and that the nature and amounts of residues and metabolites be fully known. Although herbicides have been in use for over 30 years the amount of attention paid to their metabolism has not been great until relatively recently.

Among the reports and reviews on the metabolism of pesticides in plants are those by Andreae, 1963; Brian, 1964; Swanson, 1966; Kearney and Kaufman, 1969; Casida and Lykken, 1969; Menzie, 1969; Frear, Hodgson, Shimabukuro and Still, 1972a; Frear, Swanson and Tanaka, 1972b; and Ashton and Crafts, 1973.

Since a herbicide can be inactivated in several different ways, such as by direct binding to one or more plant constituents, by hydrolysis and binding of the decomposition products to plant constituents or by complete degradation, numerous techniques have been required to elucidate herbicidal behaviour in the plant. Most of the metabolic studies thus far made have employed radioactive labelled herbicides. These studies have revealed that oxidation, reduction, hydrolytic and conjugation reactions all occur in higher plants as different classes of herbicides are metabolized.

For convenience, each class of herbicide treated will be discussed separately. Particular attention is directed to the detoxification reactions.

II. PHENOXYALKANOIC ACIDS

During World War II the differential herbicidal properties of the chlorine-substituted phenoxyacetic acids 2,4-D, 2,4,5-T and MCPA were recognized (Slade *et al.* 1945; Kraus and Mitchell, 1947). Quick acceptance by agriculturists occurred following publication of the wartime research. Many formulations containing salts, amine salts, and esters of the phenoxyalkanoic acids were introduced because the derivatives are usually taken up by plants more readily than the free acids. It is likely, however, that the esters are active only after they have been hydrolyzed by the plant to the free acid. The ability of plants to degrade phenoxyacetic acid herbicides has been known since the early 1950s (Holley *et al.* 1950; Holley, 1952; Weintraub *et al.* 1950, 1952). Since then a fairly large literature has developed and reviews of it have appeared regularly, among them being those by Audus, 1961; Hilton *et al.* 1963; Brian, 1964; Loos, 1969; Ashton and Crafts, 1973. In the earliest metabolic work (Holley *et al.* 1950; Holley, 1952; Weintraub *et al.* 1950, 1952) it was found that beans (*Phaseolus vulgaris* L.) degraded the side chain of ^{14}C-labelled-2,4-D since ^{14}CO$_2$ was liberated to a limited degree when either the carboxyl or the methylene group was labelled. Most of the label, however, accumulated in a water-soluble compound presumed by Holley (1952) to be a hydroxy-2,4-dichlorophenoxyacetic acid. Conjugates were also observed. These early studies led to the suggestion that there are three basic mechanisms involved in the metabolism of phenoxyacetic acids by plants: (1) degradation of the acetic acid side chain, (2) hydroxylation of the aromatic ring, and (3) conjugation with a plant constituent. Subsequent work by others has confirmed these conclusions and extended the observations to many other species.

While the ability to degrade the side chain of phenoxyacetic acid herbicides appears to be widespread in plants, in only a few species does it appear to be sufficiently rapid and extensive to be of major importance in accounting for differential sensitivity.

Among the plants that show high rates of $^{14}CO_2$ release from 2,4-D labelled with ^{14}C in the side chain are red currant (Luckwill and Lloyd-Jones, 1960a), Cox's Orange Pippin, and McIntosh apple (Luckwill and Lloyd-Jones, 1960b; Edgerton and Hoffman, 1961), strawberry and garden lilac (Luckwill and Lloyd-Jones, 1960b). Isolated leaves of these plants are capable of releasing up to 33% of the ^{14}C label from 2,4-D-1-^{14}C in 24 h. After three days 50% of the carboxyl carbon and 20% of the methylene carbon was released as $^{14}CO_2$ by red currant and strawberry leaves (Luckwill and Lloyd-Jones, 1960a,b). Apparently, the decarboxylase attacking 2,4-D in red currant, strawberry, and Cox's Orange Pippin apple is relatively non-specific for 2,4-D since $^{14}CO_2$ was released from 2,4,5-T-1-^{14}C, 4-CPA-1-^{14}C, and MCPA-1-^{14}C at about equal speeds. Yet little or no decarboxylation of 2-CPA-1-^{14}C takes place (Luckwill and Lloyd-Jones, 1960a, b). Following decarboxylation there is extensive recycling of the $^{14}CO_2$. As might be anticipated, much of the label is incorporated into plant acids, sugar, dextrins, starch, protein, pectin and cell wall substances (Weintraub et al. 1956; Leafe, 1962).

Luckwill and Lloyd-Jones (1960b) observed low rates of $^{14}CO_2$ liberation in 16 species tested. Some of these were susceptible and others resistant to 2,4-D. The inevitable conclusion reached was that while oxidative degradation of 2,4-D in tolerant plants may partially account for selectivity, some other mechanisms are also involved. This conclusion has been shown to be fully warranted. Among the species with low rates of decarboxylating ability are several varieties of apple (Luckwill and Lloyd-Jones, 1960b; Edgerton and Hoffman, 1961), black-jack oak (*Quercus marylandica* Muench), persimmon (*Diospyros virginiana* L.), green ash (*Fraxinus pennsylvanica* Marsh.), sweet gum (*Liquidambar styracifolia* L.), winged elm (*Ulmus alata* Michx.), (Basler, 1964), big leaf maple (Norris and Freed, 1966), black currant (*Ribes nigrum* L.) (Luckwill and Lloyd-Jones, 1960b), bean (*Phaseolus vulgaris* L.) (Weintraub et al. 1952, 1956), tick bean (*Vicia faba* L. var. *minor*) (Canny and Markus, 1960), cocklebur (Williams et al. 1960), cultivated cucumber (Slife et al. 1962), cotton (Morgan and Hall, 1963), corn (Weintraub et al. 1956) and sorghum (Morgan and Hall, 1963). Since the grasses, in general, are highly resistant, it is clear their resistance is not dependent on 2,4-D-decarboxylating ability.

The fact that carbon dioxide is produced from the carboxyl carbon of the 2,4-D side chain over twice as fast as from the methylene carbon has been considered good evidence for the accumulation of an intermediate (2,4-dichloroanisole) containing the methylene but not the carboxyl carbon of the 2,4-D molecule. Luckwill and Lloyd-Jones (1960a) found a probable intermediate bound to the leaf residue but attempts to extract it were not successful.

A second but not mutually exclusive route of metabolism of 2,4-D is ring hydroxylation. Holley's (1952) investigation of 2,4-D transformation in bean led to his speculation that hydroxy-2,4-dichlorophenoxyacetic acid was a metabolic product. This viewpoint was strengthened and extended by the work of Fawcett et al. (1959) who showed that wheat and pea tissue hydroxylated unsubstituted phenoxyalkanoic acids. Furthermore, they presented evidence

that phenoxyaliphatic acids were hydroxylated in the para position. Positive
identification of the metabolite as a 4-hydroxyphenoxyacetic acid came from
comparative work by Wilcox *et al.* (1963) who found that roots of oats,
barley, and corn could hydroxylate but peanuts, soybeans and alfalfa could
not. Hydroxylation of phenoxyacetic acid in the 4 position by oats was
reported almost simultaneously and independently by Thomas *et al.* (1963). A
further extension of these observations has been made by Fleeker and Steen
(1971) who found that seven weed species hydroxylated 2,4-D in the para
position of the ring. The lack of correlation between the amount of 2,4-D
hydroxylated and herbicide tolerated by the several species led to the
conclusion that hydroxylation *per se* does not account for variation in
susceptibility to 2,4-D by wild buckwheat (*Polygonum convolvulus* L.), wild
oats (*Avena fatua* L.), leafy spurge (*Euphorbia esula*), yellow foxtail (*Setaria
glauca* (L.) Beauv.), wild mustard (*Brassica kaber* (D.C.) L.C. Wheeler),
perennial sowthistle (*Sonchus arvensis,* L.) and kochia (*Kochia scoparia* (L.)
Roth.).

In an extension of their work on the hydroxylation of phenoxyacetic acid by
mesocotyl tissue of *Avena sativa* Thomas *et al.* (1964) demonstrated a broad
relationship, at least along one pathway of metabolism, between the fate of
phenoxyacetic acids and their pattern of ring substitution. Phenoxyacetic acids
with an unsubstituted 4 position were hydroxylated at that position, and the
resulting phenolic acid was accumulated as the 4-O-β-D-glucoside. Thus 2-
chlorophenoxyacetic acid was converted into a more highly polar compound.
β-glucosidase could convert it to glucose and a phenolic acid, namely 2-chloro-
4-hydroxyphenoxyacetic acid. Similarly 2,6-dichlorophenoxyacetic acid was
converted into a glucoside that yielded glucose and 2,6-dichloro-4-
hydroxyphenoxyacetic acid upon hydrolysis. In contrast, phenoxyacetic acids
with a chlorine at position 4 were not hydroxylated to any appreciable extent,
but neutral products were obtained that proved to be phenoxyacetylglucoses.
Compounds of this type were obtained from 4-chlorophenoxyacetic acid and
2,4-dichlorophenoxyacetic acid. But the metabolic fate of 2,4,6-
trichlorophenoxyacetic acid proved to be exceptional. The glucoside of 2,4,6-T
upon hydrolysis yielded glucose and 3-hydroxy-2,4,6-trichlorophenoxyacetic
acid. Thus hydroxylation occurs at position 3.

Esterification of glucose with 2,4-dichlorophenoxyacetic acid was first
reported in wheat coleoptile cylinders (Klämbt, 1961). Shortly afterward,
Thomas *et al.* (1964) found that oat mesocotyls convert 2,4-D, 4-CPA, and
2,4-D to their β-D-glucose esters.

Conjugates with amino acids also seem to be readily formed. The first
conjugate detected was with aspartic acid. This amino acid was found to
combine with 2,4-D forming 2,4-dichlorophenoxyacetylaspartic acid in peas
(Andreae and Good, 1957), red and blackcurrants (Luckwill and Lloyd-Jones,
1960*b*), wheat (Klämbt, 1961) and probably also in wild and cultivated
cucumbers (Slife *et al.* 1962). In cultured soybean callus 2,4-D is rapidly
conjugated with a number of amino acids by means of an amide bond. These
conjugates act like auxins (Feung *et al.* 1973, 1974). Among the amino acids

found to conjugate with 2,4-D in the callus were glutamic acid, aspartic acid, alanine, valine, leucine, phenylalanine and tryptophan.

III. s-TRIAZINES

Some plants, notably grasses including corn, sorghum and sugarcane (*Saccharum officinarum* L.) are essentially unaffected by atrazine and simazine. Lack of sensitivity probably results from the ability to neutralize or metabolize these compounds. The fate of the triazines in plants has been studied intensively. Treatments of the mechanisms of action of s-triazines have been provided by Hilton *et al.* 1963; Moreland, 1967; Knüsli *et al.* 1969; Frear and Shimabukuro, 1971; and Frear *et al.* 1972b).

The s-triazine herbicides are readily absorbed and translocated throughout the plants. Since they are not eliminated by secretion and may disappear essentially completely they may be completely oxidized to CO_2 or incorporated into one or more residues. Degradation may take place by hydrolysis, N-dealkylation and peptide conjugation.

The earliest recognized mechanism of potential detoxication of atrazine and simazine was a non-enzymatic dechlorination reaction accompanied by substitution of a hydroxyl group in the 2-position (Roth, 1957; Gysin and Knüsli, 1960; Castelfranco *et al.* 1961; Hamilton and Moreland, 1962; Hamilton *et al.* 1962). Both roots and shoots of *Zea* yield extracts capable of producing hydroxylation products of atrazine and simazine (Castelfranco *et al.* 1961; Roth and Knüsli, 1961; Hofman and Hofmanova, 1969). The active catalyst was soon isolated and identified as 2,4-dihydroxy-7-methoxy-1,4-benzoxazine-3-one (benzoxazinone) (Wahlroos and Virtanen, 1959; Hamilton and Moreland, 1962; Hamilton, 1964). Corn seems to have benzoxazinone well distributed throughout its axis (Klun and Robinson, 1969). However, hydroxylation occurs predominantly in the roots (Shimabukuro *et al.* 1970; Thompson *et al.* 1970). Since *in vivo* experiments indicate a requirement of a relatively high molar ratio of benzoxazinone to 2-chloro-s-triazines if the hydroxy analogues are to be produced, Tipton *et al.* (1971) have suggested that the catalytically active form of benzoxazinone is an aggregate.

The ability to hydroxylate 2-chloro-s-triazines does not insure resistance to these herbicides. While corn and *Coix lacryma-jobi* L. are resistant, wheat and rye (*Secale cereale* L.) are not (Hamilton, 1964; Shimabukuro, 1967). Furthermore, there are resistant species that do not contain benzoxazinone and consequently cannot carry out the hydroxylation reaction. Therefore, insensitivity of these species to the triazines must be sought in some other direction.

The second means of detoxication is through the production of peptide conjugates (Shimabukuro *et al.* 1971a; Shimabukuro and Swanson, 1969; Shimabukuro *et al.* 1970; Thompson *et al.* 1971). This means of detoxication was first observed in cultivated sorghum (*Sorghum bicolor* (L.) Moench) by Lamoureaux *et al.* (1970) (Fig. 1). Simazine has also been found to form peptide conjugates in shoot tissues of corn and cultivated sorghum. The

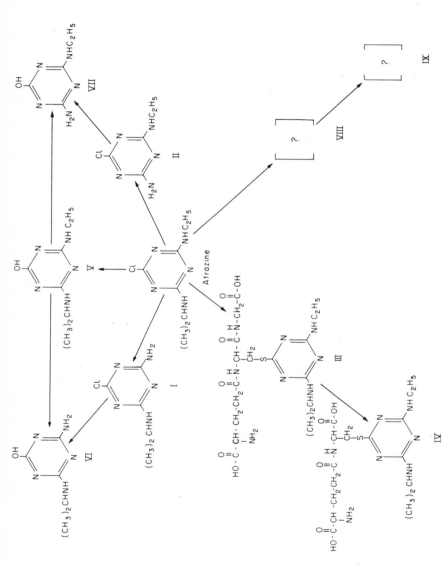

Fig. 1. Proposed routes of metabolism of atrazine in higher plants. The formation of I and II was demonstrated in several species of higher plants treated with atrazine (Shimabukuro, 1967); and the formation of hydroxyatrazine (V), 2-hydroxy-4-amino-6-isopropyl amino-s-triazine (VI), and 2-hydroxy-4-amino-6-ethylamino-s-triazine (VII) was demonstrated in corn (Shimabukuro, 1968). The presence of III and IV was shown in sorghum. Compounds III and IV also appear to be present in corn (Shimabukuro, 1968). Two unidentified water-soluble metabolites, VIII and IX are formed in sorghum treated for extended periods (Shimabukuro, 1967). (From Lamoureaux et al. 1970.)

peptide conjugates of atrazine and simazine migrate similarly in several chromatographic solvent systems. Peptide conjugates were the only major metabolites formed by six *Setaria* and three *Panicum* species; however, ability to produce hydroxy derivatives of these two triazines did bear a relationship to tolerance to atrazine and simazine (Thompson, Jr., 1972*b*).

Thus far, only one enzyme involved in conjugate production has been purified and characterized. This is glutathione-*S*-transferase (Frear and Swanson, 1970). Presumably, other enzymes involved in similar reactions will be found. Lamoureux *et al.* (1970) identified the two metabolites as *S*-(4-ethylamino-6-isopropylamino-*S*-triazinyl-2)-glutathione (GS-atrazine) and *S*-γ-L-glutamyl-(4-ethylamino-6-isopropylamino-*s*-triazinyl-2)-L-cystine. This was the first discovered example of glutathione being involved in the metabolism of a pesticide in plants. These observations have since been extended to wild cane (*Sorghum bicolor* (L.) Moench) (Thompson, Jr., 1972*a*). Glutathione *S*-transferase has been found in the leaves of corn, sorghum, sugar cane, Sudangrass (*Sorghum sudanense* (Piper) Stapf.) and Johnson grass (*Sorghum halepense* (L.) Pers.). The corn-leaf enzyme has been purified over 73-fold and has a molecular weight of 40,000 (Frear and Swanson, 1970; Frear *et al.* 1972*b*).

In an attempt to determine if differential response of wild cane (*Sorghum bicolor* (L.) Moench) to atrazine and simazine can be attributed to differential rates of metabolism, Thompson Jr. (1972*a*) showed that, although both herbicides are absorbed equally by the roots, there are differences in speed with which they are metabolized. Seventy percent of the atrazine while only 30% of the simazine absorbed and translocated to the shoot was transformed in the first 24 h. The major metabolites formed were hydroxy derivatives with lesser amounts of peptide conjugates. Hydroxylation of the two herbicides appeared to occur at approximately the same rate, while wild cane formed peptide conjugates more rapidly with atrazine than with simazine. Thompson concluded that the differential response of wild cane to atrazine and simazine is a reflection of the differential rates of conjugation of these two chloro-*s*-triazines to peptides.

The major pathway of atrazine metabolism in intact sorghum appears to involve the following steps (Lamoureux *et al.* 1973): atrazine → *S*-(4-ethylamino-6-isopropylmino-*s*-triazinyl-2)glutathione → γ-glutamyl-*S*-(4-ethylamino-6-isopropylamino-*s*-triazinyl-2)cystein → *S*-(4-ethylamino-6-isopropyl-*s*-triazinyl-2)cysteine → *N*-(4-ethylamino-6-isopropylamino-*s*-triazinyl-2)cysteine → *N*-4-ethylamino-6-isopropylamino-*s*-triazinyl-2) lanthionine. Lamoureaux *et al.* (1973) estimated that from 40 to 87% of the atrazine entering sorghum through the roots is metabolized via this pathway. Furthermore, these investigators presented evidence indicating that atrazine can be metabolized via this route following *N*-dealkylation. Complete *N*-dealkylation yielded 2-chloro-4,6-diamino-*s*-triazine, which no longer inhibited the Hill reaction and cyclic and noncyclic photophosphorylation (Shimabukuro, Lamoureaux *et al.* 1971, Shimabukuro *et al.* 1973).

A third means of metabolizing the triazines by plants was indicated when the alkyl side chain was labelled with ^{14}C and $^{14}CO_2$ was liberated by treated corn, cotton and soybean plants (Funderburk and Davis, 1963). Subsequently Müeller and Payot (1966) studied the rate of $^{14}CO_2$ liberation by treated corn and found that at least 70% of the labelled simazine was liberated in six days. Obviously, no correlation was found between the ability to degrade substituted alkylamine groups of simazine and atrazine and susceptibility to these herbicides.

Shimabukuro et al. (1966) followed the rate of N-dealkylation of atrazine-treated pea plants. Within 48 h following treatment the concentration of 2-chloro-4-amino-6-isopropylamino-s-triazine was double that of atrazine. In contrast, atrazine-treated sorghum, soybean, wheat and corn plants accumulated 2-chloro-4-amino-6-ethylamino-s-triazine (Shimabukuro, 1967) but in smaller quantities. Subsequent dealkylation in sorghum of 2-chloro-4-amino-6-ethylamino-s-triazine and 2-chloro-4-amino-6-isopropyl-amino-s-triazine yielded the single compound 2-chloro-4,6-diamino-s-triazine (Frear et al. 1971a). Canada thistle (Cirsium arvense L.), an atrazine-sensitive plant, produces the same intermediates but at a very slow rate (Burt, 1974).

N-dealkylation occurs with the hydroxyl derivatives of atrazine, as well as with the 2-chloro-s-triazines. Some evidence supports the contention that complete dealkylation of 2-hydroxy-s-triazines occurs yielding 2-hydroxy-4,6-diamino-s-triazine (ammeline). This occurs slowly according to Shimabukuro, Lamoureaux et al. (1971). Thus far small amounts only of ammeline have been reported in Coix lacryma-jobi L. and corn (Hurter, 1966; Mongomery et al. 1969).

IV. BENZOIC ACIDS

Benzoic acid and its substituted analogues have proven to be highly useful herbicides. The metabolism of benzoic acid in plants, however, has not been followed in detail. The formation of products resulting from ring hydroxylation and complexing with glucose, amino acids and other naturally occurring plant products is known to occur (see review by Swanson, 1969) but ring opening has not been conclusively demonstrated.

Among the chlorinated benzoic acids the most widely used is 2,3,6-trichlorobenzoic acid because of its ability to control deep-rooted, noxious perennial weeds such as field bindweed, leafy spurge and Canada thistle as well as several vines, some conifers and a number of hardwoods. For the most part it seems extremely stable in plants and is not subject to rapid enzymatic decomposition. Nevertheless, dechlorination has been reported but little more information is available (Menzie, 1969; Swanson, 1969). Spitznagle et al. (1969) reported that 2,3,5 tri-iodobenzoic acid was dehalogenated by soybean (Glycine max Morr). Two derivatives 2,5- and 3,5- di-iodobenzoic acids were recovered and described as possible lipid conjugates.

Numerous plant species have the ability to form N-(3-carboxy-2,5-dichlorophenyl)glucosylamine from chloramben (3-amino-2,5-dichloro-benzoic acid). The conjugation reaction apparently proceeds rapidly and the derivative is stable for months in resistant plants (Swanson *et al.* 1966*a, b*). Since glucosylation occurs in both susceptible and resistant species (Ashton, 1966; Colby, 1965; Colby, Warren and Baker, 1964; Frear, Swanson and Kadunce, 1967; Swanson *et al.* 1966*b*) species selectivity of this herbicide cannot be traced to glucosylation capacity alone as had been proposed by Colby (1965). Frear (1968*a*) partially purified UDP-glucose: arylamine *N*-glucosyltransferase and studied its properties. It was found to be non-specific for substituted anilines and there was a correlation between the amount of the enzyme in the tissue and ability to glucosylate. Stoller and Wax (1968) have reported the formation of an additional derivative described as "amiben-x". Subsequently, Stoller (1968) showed that "amiben-x" is not herbicidal and when this derivative was supplied to plant tissue it was converted to chloramben and glucosylchloramben (Stoller, 1968, 1969).

Frear *et al.* (1972*a, b*) and Swanson (1969) have reviewed the evidence that dicamba is metabolized by at least a few plants. In wheat (*Triticum aestivum* L.) and bluegrass (*Poa pratensis* L.) conjugation takes place with no decarboxylation (Broadhurst *et al.* 1966). When the conjugates were accumulated in sufficient quantity for analytical studies it was shown that hydrolysis yields about 90% 5-hydroxy-2,6-dichloro-*o*-anisic acid (5-OH-dicamba), 5% 3,6-dichlorosalicylic acid, and 5% dicamba. Dicamba was present as a glucose ester and 5-hydroxy-dicamba as a glucose ether. The same products have been found in barley (*Hordeum vulgare* L.), Tartary buckwheat (*Fagopyrum tataricum* (L.) Gaertn.) and wild mustard (*Sinapis arvensis* L.) (Chang and Vanden Born, 1971). The additional observation has been made that dicamba has an affinity for chromosomal material and specifically that in buckwheat (*Polygonum convolvulus* L.) where a DNA-histone-dicamba complex is produced (Arnold and Nalewaja, 1971).

Dichlobenil, although a relative newcomer among herbicides, has been the subject of several metabolic studies. The compound is relatively volatile and since it co-distills with water about 90% of that transported in the transpiration stream is lost through transpiration. The remaining 10% may be metabolized. The latter fraction is converted predominantly to 2,6-dichlorobenzoic acid by such widely separated species as bean (*Phaseolus vulgaris* L.) alligator weed (*Alternanthera philoxeroides* (Mart) Griseb.) and fungi (Pate and Funderburk, 1966; Verloop and Nimmo, 1970). Five days after treatment, the principal metabolites recovered were 3-hydroxy-2,6-dichlorobenzonitrile and 4-hydroxy-2,6-dichlorobenzonitrile in a ratio of approximately 4 to 1. Both of these derivatives have proven to be toxic to beans. 2,6-dichlorobenzamide and 2,6-dichlorobenzoic acid, hydrolytic products, were both found, but only in very small amounts. Another portion of the 10% metabolized was found as conjugates of dichlobenil (Verloop and Nimmo, 1969). Most of the conjugates were extractable and hydrolizable but some proved insoluble. With time the conjugates tended to accumulate in the plant.

Chlorthiamid appears to be readily convertible in non-biological systems into dichlobenil (Milborrow, 1963, 1965). Conjugates are formed in a manner similar to those of dichlobenil. Hydrolysis of conjugates from apple

(*Malus* spp.) and wheat yielded 3-OH and 4-OH analogues (Beynon and Wright, 1968) but an insoluble residue remained. Phenolic derivatives were also sometimes produced but this depended upon the species.

Pronamide, one of the few substituted benzamides to be studied with respect to metabolism within the plant, has been supplied as the carbonyl-^{14}C labelled form. Alfalfa metabolizes pronamid slowly. Chromatography of methanolic extracts (Yih and Swithenbank, 1971a, b) led to the recovery of nine derivatives. After 112 days the predominant metabolite, β-(3,5-dichlorobenzamido)-β-methylbutyric acid, accumulated in alfalfa to a level equal to that of pronamide. One unidentified metabolite appeared to be a derivative of 3,5-dichlorobenzoic acid. With time more and more of the label was found in non-extractable forms.

Ioxynil is transformed to 4-hydroxy-3,5-di-iodobenzamide and 4-hydroxy-3,5-di-iodobenzoic acid by plants (Schafer and Chilcote, 1970b). The amide derivative only has been detected in pea and white mustard, while a resistant strain of barley accumulated both the amide and the benzoic acid derivative together with some unknown compounds (Davies *et al.* 1968).

Bromoxynil has been labelled with ^{14}C and fed to wheat (*Triticum aestivum* L.) and coast fiddleneck (*Amsinckia intermedia* Fisch. and Mey.). Schafer and Chilcote (1970a) found that both species evolved ^{14}CO$_2$ but wheat evolved more than fiddleneck. Evidently, bromoxynil was converted to the benzoic acid derivative by decarboxylation though this remains to be confirmed. Considerable quantities of the liberated ^{14}CO$_2$ were fixed by photosynthesis.

V. Substituted Ureas

Studies on the metabolism of substituted ureas such as monuron, diuron, chlorbromuron and linuron have frequently been made with impure radio-tracer-labelled preparations and this has led to spurious reports of the production of small amounts of nitrobenzene. Both the substituted dimethyl- and methoxymethylphenylurea herbicides can undergo stepwise N-dealkylation to mono-dealkylated compounds in seed plants (Geissbühler, 1969; Menzie, 1969). An alternative route of degradation of the substituted methoxymethyl phenylurea herbicides is by way of N-demethylation and N-demethoxylation (Geissbühler, 1969; Kuratle *et al.* 1969). Indeed, N-demethylation is probably the first step in the metabolism of the linuron and chlorbromuron type substituted phenylurea herbicide (Nashed and Ilnicki, 1970; Nashed *et al.* 1970). The end product of N-demethylation of monuron and monomethyl-monuron are N-hydroxymethyl intermediates that become stabilized as conjugated glucosides (Frear *et al.* 1972b; Frear and Swanson, 1972) (Fig. 2). The pyrrolidine urea, *cis*-2,5-dimethyl-1-pyrrolidine-carboxanilide, developed as a wide spectrum pre-emergence herbicide, has been found (Holm and Stallard, 1974) to be metabolized at a slow rate by young corn tissue but the rate of transformation increased with plant age. The metabolite(s) are highly polar and water soluble. Holm and Stallard (1974)

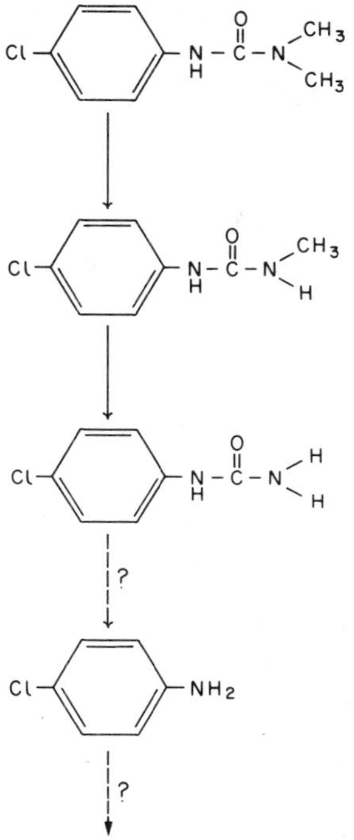

Fig. 2. Proposed pathway for the metabolism of monuron in tolerant plants (Frear, Swanson and Tanaka, 1972*b*).

infer that they are *O*-glucosides probably somewhat like the derivatives Frear and Swanson (1972) reported for monuron.

Only crude enzymic studies have been made of the metabolism of substituted phenylurea herbicides. Nevertheless, these show that an oxidase bound to the microsomal fraction from cotton, plantain and other herbicide-tolerant species is capable of oxidatively *N*-demethylating several *N*-methylphenylurea herbicides (Frear *et al.* 1969, 1972*b*). Some specificity for the enzyme has been established. The reaction is inhibited by end products of the reaction, thiourea, semicarbazide and methyoxymethyl analogues. A similar *N*-demethylase from carrot is less specific and catalyzes the oxidation of methoxymethyl- and methyl-substituted phenylureas (Kuratle *et al.* 1969).

Enzymological work is yet to be reported for the *N*-demethoxylation or *O*-demethylation of the substituted methoxymethyl phenylurea herbicides. Furthermore, nothing is yet available on the enzymology of glucosylation of the *N*-hydroxymethyl intermediates in the *N*-demethylation of the dimethyl- and methyl-substituted phenylureas. It is important that this information be obtained because it seems likely that in the most highly tolerant plants rapid and extensive metabolism of the herbicide takes place. Such information should be of value in designing herbicides that cannot be caused to form glucosidic conjugates and insoluble or "bound" residues.

VI. CARBAMATES

A. METHYL- AND PHENYLCARBAMATES

Although some of the carbamates have been used as herbicides since World War II, very little information is yet available concerning their metabolism within plants. As Herrett (1969) has pointed out, the methyl and phenylcarbamate herbicides are becoming increasingly important. This is so because of their low mammalian toxicity, relatively short residual life in the soil and swift degradation by non-target organisms.

Propham and chlorpropham have been widely used as pre-emergence herbicides in the cultivation of a number of tolerant broad leaf crops. Prendeville's laboratory began reporting metabolic studies in 1968 and since then others have joined the search for information. Prendeville *et al.* (1968) and James and Prendeville (1969) found that, following chlorpropham treatment, water-soluble β-glucosides of a modified chlorpropham molecule could be detected in the four broad leaved plants they used in their tests. There was no evidence of cleavage of the carbamate bond but there was an apparent modification of the 2-propyl ester portion of the molecule by the two weeds and two crop plants.

In a series of feeding experiments with ^{14}C-chlorpropham Still and Mansager (1971, 1972) demonstrated that soybean roots rapidly produced polar derivatives and insoluble residues. Only the polar metabolites seemed to be produced by the shoot tissues. Translocation of label occurred but it was only in the unreacted molecule. Neither the root nor the shoot transported the metabolic products. From the time-course experiments it was concluded that the precursor product sequence was chlorpropham, polar products, followed by insoluble residues. Data from feeding experiments with the ring- and side-chain ^{14}C-labelled chlorpropham showed clearly that cleavage of the carbanilate bond did not occur at any stage of metabolism. The dominant polar metabolite in soybean roots proved to be the O-glucoside of isopropyl-5-chloro-2-hydroxycarbanilate. This, however, was rapidly converted, to insoluble residues. Quite a different metabolic picture was provided by the soybean stems. In the shoot tissue isopropyl-5-chloro-2-hydroxycarbanilate was not the major polar metabolite and it was suspected that it might be the precursor of other, as yet unidentified polar compounds.

Still and Mansager (1973) extended their investigations on the metabolism of ^{14}C-labelled isopropyl-3-chlorocarbanilate to cucumber plants. Polar products and solid residues were found in the roots, stems, and leaves after a three-day treatment period. Time-course experiments involving short exposures to ^{14}C-labelled herbicide made it possible to demonstrate a precursor-product relationship between chlorpropham soluble polar products, and solid residue materials. The polar metabolites were not translocated once they were formed in either the roots or shoots. The radiocarbon distribution patterns of phenyl-^{14}C- and isopropyl-^{14}C-labelled chlorpropham were similar in all comparative experiments. Cucumber tissues rapidly convert chlorpropham to isopropyl-4-hydroxy-3-carbanilate, which is conjugated with unknown plant components. Both the root and shoot tissues have the capacity to produce these conjugates.

It seems reasonable to speculate that the ability to produce isopropyl-5-chloro-2-hydroxycarbanilate rapidly protects the soybean against the herbicidal action of chlorpropham, but solid proof is needed. Many questions are left unanswered. The mechanism of binding to insoluble cell constituents is among these.

Phenmedipham and swep are carbanilate herbicides that have found specialized use in the production of beets and rice. Phenmedipham is a pre-emergence herbicide while swep is used primarily on young seedlings. Investigations on the metabolism of them is still in the initial stages but enough is known to make it clear there are clear cut differences.

Initial studies on the metabolism of phenmedipham by several susceptible and resistant species have been made by Bischof et al. (1970) and Kassebeer (1971). Apparently, phenmedipham was "handled" in different ways by the several species that were tested. It was assumed that *Beta vulgaris* was not damaged because it metabolized the herbicide very rapidly. Similar reasoning suggested that *Amaranthus retroflexus* was more sensitive than beet because that species metabolized phenmedipham at a slower rate. *Galium aparine, Matricaria chamomilla* and *Stellaria media* survived because of very slow uptake of the herbicide. Sensitivity was correlated with rate of uptake and ability to metabolize phenmedipham. Phenmedipham was transferred by beet and *Amaranthus retroflexus* L. to a compound which was not identified positively but was probably hydroxylated phenmedipham. This derivative is able to form complex compounds with plant constituents. What these are is not known.

Swep, in contrast to phenmedipham, appears to form complexes with lignin with little or no benefit of hydrolysis (Chin et al. 1964). This proved true for rice, carrots, oats, wheat, corn and barnyard grass. Free 3,4-dichloroaniline was only barely detectable and could have been an artifact of analysis. The mechanism of binding of the carbanilate to lignin is still to be determined. Upon extraction with HCl-dioxane, a portion of the complex is hydrolyzed to swep, 3,4-dichloroaniline and soluble products of lignin.

Barban is a carbamate possessing a particularly high selectivity. Its chief commercial use is the control of wild oats (*Avena fatua*) in spring wheat, durum wheat, barley, flax, peas, sugar beets, safflower, lentil, mustard and soybeans. All crop plants except oats, buckwheat, and some varieties of rye have shown a high degree of tolerance. Among the first to study the metabolism of barban were Riden and Hopkins (1961, 1962) who found that ^{14}C-barban was rapidly degraded in both resistant wheat and wild oats to an unknown water-soluble substance (X) which released 3-chloroaniline on hydrolysis in 10% aqueous caustic alkali. This observation was extended to 13 species including grasses and broad leaf plants. In time-course experiments it was shown that compound X was metabolized. Apparently, the 3-chloroaniline moiety was complexed into several water-soluble derivatives. In subsequent studies Jacobsohn and Anderson (1972) found that compound X was not phytotoxic even though it, or at least 3-chloroaniline, did enter the roots. They hypothesized that the build-up of compound X may reduce the rate of metabolism of barban. This reasoning was used to account for the greater amounts of free barban found in the leaves of susceptible varieties of oats 12 to 24 h after treatment. Meanwhile, Lamoureaux *et al.* (1971) reported the finding of a barban/glutathione conjugate. It remained, however, for Still and Mansager (1972) to give a satisfactory answer to how the conjugate may have been formed.

Time-course experiments involving short exposures of soy beans to ^{14}C-barban showed that barban was metabolized, first to soluble products and then to insoluble residues (Still and Mansagar, 1972). Water-soluble products, and insoluble residues were rapidly formed in the roots but three days after treatment only water-soluble residues were found in the shoots. Once formed they were not readily translocated. Studies with phenyl-^{14}C-, carbonyl-^{14}C-and butynyl-[1-^{14}C]-barban all resulted in the label being found in the same products. Apparently, soybean cannot cleave the barban molecule. The aromatic nucleus remains intact. Nevertheless, barban was altered in such fashion that its polar metabolites were highly reactive. The hydrolytic products were isolated and separated by gas-liquid chromatography and characterized by mass spectral analysis. The radio-label in all the polar metabolites was in the 3-chloroaniline moiety. It was concluded that the 4-chloro-2-butynyl alcohol moiety was the reactive functional group. In fact, it is highly reactive and the polar metabolites found in the root and shoot were products of 4-chloro-2-butynyl alcohol metabolism and these probably served as the precursors of the insoluble residual materials found in root and stem tissues.

B. N-ALKYL CARBAMATES

Early studies with the N-methyl carbamates showed that they had insecticidal properties apparently because of their anticholinergic actions. Today the N-alkyl carbamates are generally recognized as insecticides. Some of the methyl carbamates, however, are also good herbicides. Among these are tandex (*tert*-butylcarbamic acid with N-(3-hydroxyphenyl),N',N'-dimethylurea), terbutol and dichlormate. Little is known of the metabolism of either terbutol or tandex but Herrett's laboratory has worked fairly intensively on dichlormate (3,4-dichlorobenzylmethylcarbamate) a herbicide that shows promise for use with a number of crop plants.

Dichlormate's herbicidal activity was first correlated with inhibition of chlorophyll formation (Herrett and Berthold, 1965) but the compound also affects other chloroplast components. Bartels and Peglow (1967, 1968) reported that seedlings from dichlormate-treated wheat seed had leaf cells with chloroplasts lacking grana fret membranes and ribosomes while the cytoplasmic reticulum was abundantly present. Furthermore, all organelles other than chloroplasts were morphologically normal. Sensitivity to the herbicidal action of dichlormate does not appear to be related to the ability to metabolize the compound. Corn (moderately sensitive) and bean (resistant) degrade dichlormate with equal efficiency following application to the shoot. Herrett (1970) in a brief report has identified two of the four derivatives in the non-polar fraction; 3,4-dichlorobenzyl alcohol and 3,4-dichlorobenzoic acid. The polar fraction contained glucosides of the N-hydroxymethyl derivative as well as the demethylated dichlormate. In addition, a 2-hydroxydichlormate glucoside is suspected.

C. THIOLCARBAMATES

Some thiolcarbamate molecules are toxic to plants and others to insects. Thiolcarbamate herbicides are readily absorbed by both sensitive and resistant species. They are metabolized at a rate that varies with both the plant species and herbicide molecule. Resistance is closely correlated with the ability to degrade the particular thiolcarbamate being used. Studies with labelled EPTC show that it is rapidly taken up by alfalfa (*Medicago sativa* L.) but extensively metabolized in two to five days (Nalewaja *et al.* 1964). In similar studies with vernolate (*S*-propyl dipropylthiocarbamate) and soybean seedlings Bourke and Fang (1965) showed that extensive metabolism had occurred within 48 h after treatment. ^{14}C-Pebulate (*S*-propyl-[1-^{14}C]-*N*,*N*-ethyl-*n*-butylthiocarbamate) was shown by Fang and Fallin (1965) to be taken up and translocated in young tomato plants initially faster than it was broken down. As the supply of pebulate declined in the soil a reduction of pebulate in the plant was observed, reaching zero in four to five weeks. Earlier, Fang and George (1962) had reported a rapid disappearance of pebulate in resistant mung beans (*Phaseolus aureus* Roxb.), 8 h being sufficient for the process. There is also rapid degradation of ^{14}C-pebulate by burley tobacco (*Nicotiana tabacum* L. "Kentucky 14") seedlings (Long *et al.* 1974a, b). The proposed (Fang, 1969) pathway of pebulate metabolism involves hydrolysis and transthiolation. Subsequent metabolism would involve the propyl moiety being incorporated into amino acids and citric-acid-cycle derivatives. Work by Long *et al.* (1974b) on ^{14}C-pebulate metabolism by burley tobacco seedlings indicates that pebulate is quickly metabolized in the roots and that radioactivity is transported to the shoots as metabolic derivatives. The pattern of labelling in the organic acids, carbohydrates, sterol esters, and oligosaccharides is indicative that ^{14}C is quickly incorporated into the acetyl CoA pools.

VII. AMIDES

Chemically, the amide herbicides comprise a diverse group of compounds. In addition, their biological effects vary widely. While the majority of them are used as pre-emergence herbicides, some are applied to the foliage of weeds that are to be controlled. Although about 13 amide herbicides are in commercial use, the metabolism of only four has been studied in any detail. Jaworski

(1969), Casida and Lykken (1969), Matsunaka (1969*b*) and Frear *et al.* (1972*a*) have reviewed the fate of the amides in plants.

A. CDAA (ALLIDOCHLOR)

Jaworski (1964) made extensive metabolic studies with ^{14}C-CDAA labelled first in the carbonyl group and then in the number 2 carbon of the allyl moiety. Both corn and soybeans (tolerant species) rapidly took up ^{14}C-CDAA from the soil. Radioactivity in the tissue reached a peak after 4–5 days and declined rapidly thereafter. CDAA was shown to be totally degraded within four days by corn and soybeans. After 4–5 days the primary products were glycollic acid and an amine. The glycollic acid appeared to be in equilibrium with glyoxylic acid. Once these were formed, degradation to CO_2 could take place or the label could be incorporated into a number of metabolites. ^{14}C-labelling in the 2-carbon positions of the allylic radicals showed that these components of the herbicide's molecule were also degraded rapidly. No specific intermediates were found. Apparently, no unusual metabolite was formed. Rather, the results of fractionation could best be interpreted to mean the carbon atoms of the allylic moieties were randomized and incorporated through normal metabolic channels into numerous natural products. Studies with other crop plants led to similar conclusions. Jaworski (1969) speculated that the diallylamine portion of the CDAA molecule is oxidized in the presence of an amino oxidase to two acrolein molecules. These highly unstable molecules would be oxidized very rapidly.

B. PROPACHLOR

The uptake and metabolism of uniformly ring labelled (^{3}H) propachlor in maize and soybeans has been studied (Jaworski and Porter, 1965; Porter and Jaworski, 1965; Jaworski, 1969). Both species rapidly took up label from soil treated with the acetanilide. In a time-course experiment it was found that degradation was complete within five days. The acidic polar product was isolated and subjected to base hydrolysis. This was followed by vapour phase chromatography and thin layer chromatography. The results led to the conclusion the *N*-isopropylacetanilide portion of the molecule remained intact in the plant. Acid hydrolysis of the polar metabolite yielded 2-hydroxy-*N*-isopropylacetanilide. This was regarded as strong evidence for the existence of a conjugated hydroxypropachlor intermediate. Jaworski (1969) suggested that the water soluble acidic metabolite contains essentially the entire structure of the original herbicide, except for the chloro group; he thought that group was probably displaced by some nucleophilic plant constituent. This suggestion was followed up by Frear and Swanson (1970) who found that 2-chloro-*N*-isopropylacetanilide will react with glutathione non-enzymatically *in vitro* yielding the propachlor-glutathione conjugate. This same conjugate as well as the γ-glutamylcysteine conjugate of propachlor have since been reported by Lamoureaux *et al.* (1971) to be transitory metabolites of propachlor in the leaves of corn, sorghum, sugar cane and barley. The character of the final product(s) of propachlor metabolism remain(s) to be determined.

C. DIPHENAMID

Degradation of the pre-emergence herbicide diphenamid has been investigated only in tomatoes (Lemin, 1966) and strawberries (Golab *et al.* 1966). Both of these species are resistant to the chemical. In both studies the carbonyl carbon atom of diphenamid was labelled and application was made through the roots. Degradation products were found in tomato seedlings within 12 h after

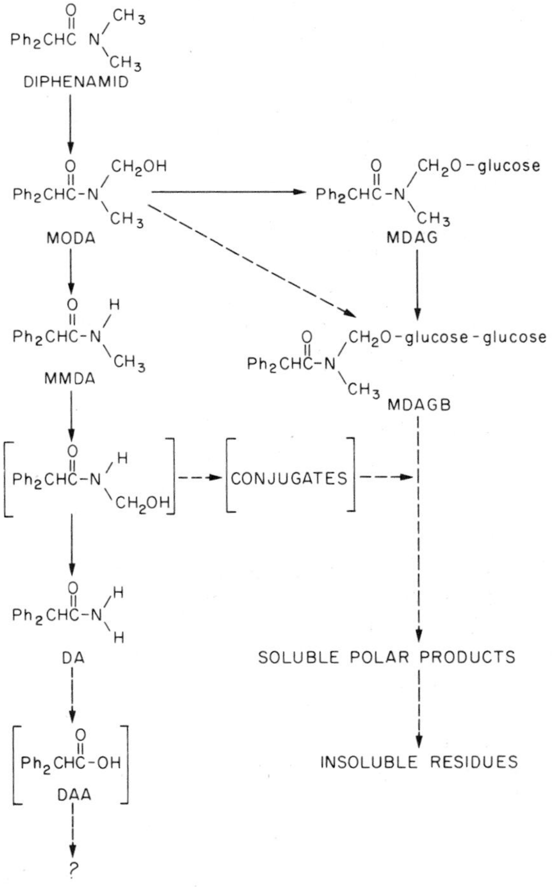

Fig. 3. Proposed scheme for the metabolism of diphenamid in tomato. Solid arrows and unbracketed components have been experimentally determined. Dashed arrows and bracketed components are postulated (Hodgson *et al.* 1974).

administering the herbicide. After a week, 59% of the radioactivity was in diphenamid, 36% in diphenylacetamide and some in diphenylacetic acid while the remainder was not identified. Two weeks later, degradation had proceeded to the point where the major radioactive compound was N-methyl-2,2-diphenylacetamide. Some diphenylacetamide was also present. Enough information was thus available for Lemin (1966) to propose that degradation involves direct hydroxylation of the N-alkyl group with the production of formaldehyde. In the strawberry experiments Golab et al. (1966) assayed the berries approximately one month and six weeks following treatment. Leaves were also taken in the final sample. As in the three week tomato plant samples, the major radioactive metabolite was N-methyl-2,2-diphenylacetamide. Labelled diphenylacetamide, diphenylacetic acid and p- and α-hydroxy-2,2-diphenylacetic acids were also identified. Hodgson et al. (1973, 1974) have also made a time course study of diphenamid metabolism in tomato. Some of these were simultaneously treated with 30 parts/10^6 ozone. After two days, the concentrations of N-methyl-2,2-diphenylacetamide were equivalent in both fumigated and control plants. But less glucoside conjugate with N-hydroxymethyl-N-methyl-2,2-diphenylacetamide was present in fumigated plants two to four days after treatment than in the control plants. Stepwise synthesis of the gentiobioside conjugate was apparently accelerated by ozone fumigation because the gentiobioside conjugate was found in the fumigated plants after 0·5 day but not in the control plants until after two days. Tertiary metabolites occurred initially at trace levels and increased nearly linearly for 0·5 to 8 days. Unextracted residues from plants treated for 24 days increased similarly. Diphenamid and its unconjugated metabolites are mobile in plant tissue, but mobility of the carbohydrate conjugates has not been demonstrated. On the basis of their time-course studies, Hodgson et al. (1974) have proposed a scheme for the metabolism of diphenamid in tomato (Fig. 3).

D. PROPANIL

Propanil has found widespread use in the post-emergence control of grasses in crops such as tomatoes and rice. This herbicide has the desirable characteristics of being completely metabolized by resistant plants and a short half life in soils. A large amount of research has been done on the metabolism of propanil. Most of it has been reviewed by Casida and Lykken (1969), Matsunaka (1969b), and Frear et al. (1972a).

Several reports have appeared indicating that rice and some other species can hydrolyze propanil to 3,4-dichloroaniline (McRae et al. 1964; Adachi et al. 1966a, b; Ishizuka and Mitsui, 1966; Still and Kuzirian, 1967; Yih et al. 1968a; Still, 1968a, b). While many of these investigators have suggested that propionic acid is a degradation product derived by direct hydrolysis, Still (1968a, b) using both a sensitive and a resistant species has provided the greatest array of evidence in support of the hypothesis. In contrast, Yih et al. (1968a) find that at least one other route of degradation may operate. They presented evidence from studies with barnyard grass (susceptible) that metabolism of propanil to 3,4-dichloroaniline was not a direct hydrolysis but an oxidative reaction yielding 3,4-dichloroacetanilide, which subsequently was hydrolyzed to 3,4-dichloroaniline and lactic acid (Fig. 4). 3,4-dichloro-acetanilide was a transient intermediate in rice but accumulated in barnyard grass. Yih *et al.* (1968*b*) also showed with *Oryza sativa* L. cv. Bluebonnet 50 that the 3,5-dichloroaniline moiety conjugates with carbohydrates. But the soluble aniline-carbohydrate complexes account for only a small fraction of the hydrolyzed 3′,4′-dichloropropionilide. The major portion of the 3,4-dichloroaniline moiety was found complexed with polymeric cell constituents, mainly lignin and hemicellulose. Still (1968*b*) found three carbohydrate conjugates whereas Yih *et al.* (1968*a*) recovered four. The one in greatest abundance was *N*-(3,4-dichlorophenyl)glucosamine. The one second in abundance was a 3,4-dichloroaniline saccharide conjugate containing glucose, xylose and fructose. The remaining two aniline-carbohydrate complexes were present in only small amounts. One of these was not stable and readily decomposed to *N*-(3,4-dichlorophenyl)glucosamine while the remaining derivative was not well characterized.

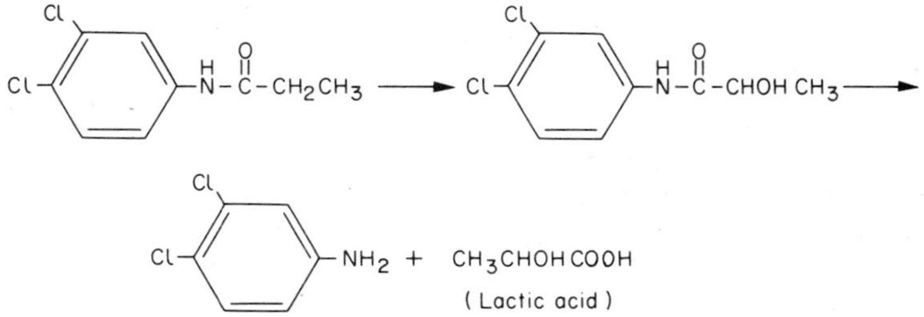

Fig. 4. Proposed scheme of oxidative and hydrolytic metabolism of propanil (Yih *et al.* 1968*b*).

As with other herbicides, the effectiveness of propanil is dependent to a marked degree on environmental factors prevailing at the time of application. Hodgson (1971) has shown that the metabolism of propanil in rice is quantitatively modified by temperature and day length. Uptake and metabolism, as measured in recoverable 3,4-dichloroaniline and *N*-(3,4-dichlorophenyl)glucosamine, were most rapid under high (32°C) and long (16 h) day conditions.

Crude enzyme preparations from barnyard grass and rice leaves capable of hydrolyzing propanil to 3,4-dichloroaniline have been described by Adachi *et al.* (1966*a, b*) and Still and Kuzieran (1967). The preparations from rice were 10 to 20 times as active as those from barnyard grass. Fractionation of a rice enzyme preparation (Frear and Still, 1968) led to the characterization of an aryl acylamidase (aryl-acylamine amidohydrolase, EC 3.5.1a). While the enzyme is non-specific, the form derived from rice leaves is apparently efficient in the degradation of 3,4-dichloropropionanilide. Frear and Still (1968) thought their data supported the hypothesis that resistance vs. susceptibility to propanil was a reflection of the specific activity of the aryl acylamidase in a plant's tissues. An alternative explanation of relative susceptibility has been provided by Ishizuka and Mitsui (1966) who have taken the isozymic approach and propose that selectivity results from differences in substrate specificity of aryl acylamidases derived from different species. While their explanation seems reasonable, the issue is unresolved.

One of the fruits of the enzymatic work with aryl acylamidase has been the development of a rational explanation of why propanil, when used in combination with certain commonly used insecticides, is toxic to rice (Bowling and Hudgins, 1966; Hisada, 1967). Frear and Still's (1968) investigations with the partially purified aryl acrylamidase from rice shows that the enzyme is strongly inhibited by insecticidal carbamates and to some extent by the organophosphates. Detoxication cannot occur and the rice plant suffers.

The enzymatic approach to an understanding of the interaction of herbicides, fungicides and insecticides is deserving of increasing exploitation.

VIII. DINITROANILINES

Trifluralin appears to be extensively metabolized (Probst *et al.* 1967) by soybean (*Glycine max* L.), and cotton (*Gossypium hirsutum* L.), yet only small amounts appear as $^{14}CO_2$ from ^{14}C-trifluralin labelled in the *n*-propyl or the trifluromethyl group. The radioactivity from both types of labelled trifluralin was distributed in lipids, glycosides, hydrolysis products, proteins and cellular fractions. With propyl-labelled material, approximately 25–30% of the original radioactivity was found in the cellular fraction of soybeans and cotton. The radioactivity from ^{14}C-trifluromethyltrifluralin resided principally in the glycoside fraction. Hydrolysis of this glycoside fraction and thin-layer chromatography of the hydrolytic products revealed no major degradation products in soybean plants or cotton seeds. Carrots (*Daucus carota* L.) appear to be much less capable than soybean or cotton of metabolizing trifluralin labelled with ^{14}C in the trifluoromethyl group (Golab *et al.* 1967). Ninety-seven percent of the radioactivity in the carrots was extractable with 93% remaining in hexane after water partitioning. Investigation by thin-layer chromatography, gas chromatography, and radioautography showed that 84% of the radioactivity in hexane was trifluralin. The major conversion product was α,α,α-trifluro-2,6-dinitro-*N*-(*n*-propyl)-*p*-toluidine, evidently derived through dealkylation. Small amounts of α,α,α-trifluro-5-nitro-*N*⁴-(*n*-propyl)-toluene-3,4-diamine and 4-(di-*n*-propyl-amino)-3,5-dinitrobenzoic acid were also produced. The first of these may have been derived by a reduction of one nitro group together with the dealkylation of one group. The second is an oxidized derivative. Golab *et al.* (1967, 1970) have hypothesized two routes of degradation which may operate simultaneously (Fig. 5). Similar feeding experiments have been carried out with peanut (*Arachis hypogea* L.) and sweet potato (*Ipomoea batatas* L.) by Biswas and Hamilton (1969). In both species the dealkylated and reduced derivatives made their appearance. There was also evidence that phenolic and benzoic-acid derivatives were produced. Crude

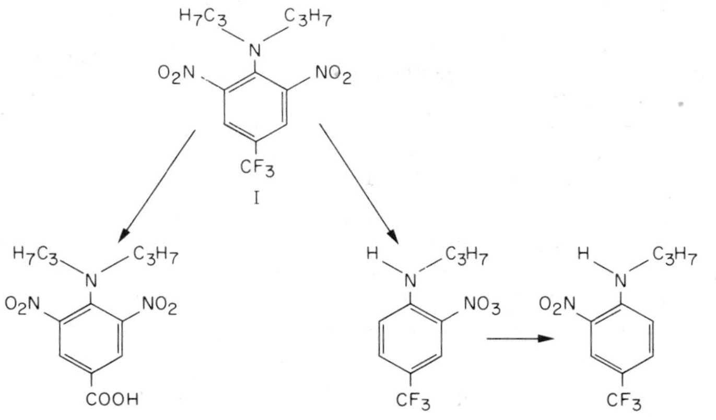

Fig. 5. Proposed route of trifluralin degradation in carrot root: I α,α,α-trifluoro-2,6-dinitro-N,N-dipropyl-p-toluidine; II α,α,α-trifluoro-2,6-dinitro-N-(n-propyl)-p-toluidine; III α,α,α-trifluoro-5-nitro-N-propyltoluene-3,4-diamine; IV 4-(dipropylamino)-3,4-dinitrobenzoic acid (Probst and Tepe, 1969).

extracts of peanut and sweet potato leaves were capable of degrading trifluralin (Biswas and Hamilton, 1969).

[14]C-Benefin (benfluralin), (N-butyl-N-ethyl-α,α,α-trifluoro-2,6-dinitro-p-toluidine) has been supplied by way of the soil to peanuts and alfalfa (Golab et al. 1970). Perhaps it is significant that the derivative compounds found in the plants are the same as those detected in the soil. After four months only small amounts of radioactivity could be recovered from peanut plants. Most of the extractable tracer was in the stem, roots and hulls and this was chiefly in the polar materials α,α,α,-trifluoro-5-nitrotoluene-3,4-diamine,2,6-dinito-α,α,α-trifluoro-p-cresol and α,α,α-trifluorotoluene-3,4,5-triamine. A major portion of the radioactivity, however, was unextractable. In contrast, after 7·5 months 65% of the radioactivity was extractable, almost all of it being in polar derivatives. Small amounts of benefin were also recovered. The question still remains as to whether seed plants have a significant capacity to degrade benefin. Critical experiments are yet to be run.

IX. HETEROCYCLICS

Picloram, 4-amino-3,5,6-trichloropicolinic acid, has been widely employed for commercial and military purposes. Because of its effectiveness in killing a great variety of tree species, picloram has proven useful in keeping open rights-of-way for powerlines, railways, and highways. The substance is easily applied and is readily translocated throughout the plant. None of the seed plants' organs appears to be able rapidly to metabolize picloram. Cotton (Meikle et al. 1966) and bean plants (Sargent and Blackman, 1970) when supplied with carbonyl-^{14}C-picloram appear to decarboxylate the herbicide at a slow rate but not fast enough for the reaction to be an important one in detoxification. The decarboxylation reaction still awaits rigorous testing.

Perhaps the most extensive studies of the metabolism of picloram have been carried out by Redemann *et al.* (1968). These investigators along with Maroder and Prego (1971) have detected small amounts of neutral and acidic metabolites of picloram. Gas chromatography has been used to detect 4-amino-3,5-dichloro-6-hydroxypicolinic acid (Redemann *et al.* 1968) and a derivatization technique picked up oxalic acid. After 84 days wheat supplied with ring-labelled ^{14}C-picloram had 83% of the total label in picloram, 8% in oxalic acid, 5% in 4-amino-3,5-dichloro-6-hydroxy picolinic acid, and 4% in 4-amino-2,3,5-trichloro-pyridine. Thus, after being in the plant for nearly three months only 17% of the picloram had been metabolized.

Some presumed lipid conjugates with picloram have been reported (Redemann *et al.* 1968). Identification of the lipid component was indirect and based principally on hydrolysis of the conjugate by pancreatin. Still another conjugate appears to be produced by slimleaf wallrocket [*Diplotaxis tennifolia* (L.) DC.]. This conjugate upon hydrolysis yielded only picloram (Maroder and Prego, 1971). The size of the conjugate remains undetermined.

Pyrazon, an excellent inhibitor of the Hill reaction (Eshel, 1969; Frank and Switzer, 1969; Hilton *et al.* 1969) has not been studied intensively with respect to its metabolism in plants. Nevertheless, two of the primary metabolites have been identified. In both resistant and non-resistant species radioactive pyrazon is rapidly transformed. Very little unchanged pyrazon persists in the plant.

The major metabolite appears to be a *N*-glucoside of pyrazon (*N*-glucosylpyrazon) (Ries *et al.* 1968). Four weeks after application of ^{14}C-labelled pyrazon to sugar beet, a resistant plant, some 97% of the radioactivity was found in *N*-glucosylpyrazon (Stephenson and Ries, 1969). The remaining 3% was found in a cleavage product of pyrazon, 5-amino-5-chloro-3(2H)-pyridazinone (ACP) and an undefined conjugate of ACP.

An attempt has been made by Stephenson and Ries (1967, 1969) to correlate metabolism with herbicidal resistance. These authors have shown that tomato, a susceptible plant, seems incapable of producing *N*-glucosylpyrazon while sugar beets that are herbicide resistant formed *N*-glucosylpyrazon in the shoots but not in the roots. In a time-course study to show when and where ^{14}C-labelled pyrazon was translocated and transformed only pyrazon was detected in the roots after three days whereas 50–60% of the translocated pyrazon was transformed to the *N*-glucoside in the shoots. Four months after the ^{14}C-pyrazon was introduced into the sugar beets about 12% of the total radioactivity was found in the insoluble fraction. The intermediate steps are unknown (Stephenson and Ries, 1969).

It is likely that *N*-glucosylation of pyrazon and chloramben occur in a parallel manner (Swanson *et al.* 1966a; Frear *et al.* 1967), but no data are available to support the view. It is true, however, that the arylamine *N*-glucosyltransferase involved in the *N*-glucosylation of chloramben is non-specific and furthermore it has been found present in a number of species

(Frear *et al.* 1967; Frear, 1968*a*). It, therefore, does not seem unreasonable to hypothesize the existence of an analogous enzyme system capable of the glucosylation of pyrazon.

X. DIPHENYLETHERS

Among the diphenylethers now in use as herbicides only one, fluorodifen has been seriously studied at the metabolic level. No enzymatic work has yet been reported. Among the earliest studies on the metabolism of the diphenylethers were those by Williams (1959). He concluded that as a class they are relatively stable in the plant. There is, however, a rapid cleavage of the ether bond in fluorodifen (Geissbühler *et al.* 1969; Rogers, 1971; Eastin, 1971*a, b, c*). A glucoside is formed by conjugation of glucose with the 4-nitrophenol moiety following cleavage of the ether. What happens to the 2-nitro-4-trifluoromethyl-phenyl moiety is not known. When fluorodifen-1-^{14}C is supplied to plants, a very substantial portion of the label is ultimately found in the insoluble fraction (Rogers, 1971; Eastin, 1971*b, c*; Geissbühler, 1969). The same authors have also found a feeble pathway that utilizes one or both of the nitrogen groups.

The metabolism of nitrofen (2,4-dichlorophenyl-*p*-dinitrophenyl ether) is beginning to be studied. Apparently one or more photochemical steps may be involved in its degradation. Basically cleavage occurs at the ether linkage. This is followed by the production of several unidentified substances (Hawton and Stobbe, 1971). Matsunaka (1969*a*) has proposed two mechanisms for the light effect in ortho-substituted diphenylether uptake and herbicidal activity. Possibly light activates ortho-substituted diphenylethers enabling them to be transported across cell membranes and then penetrate chloroplast membranes where they interfere with photosynthesis. Other hypotheses, of course, should not be excluded.

While a beginning has been made in the study of the metabolism of the two groups of diphenylether herbicides, there is an urgent need for research on the nature of the light reaction. Too little is known about the mechanism of ether cleavage, and much is yet to be learned about the metabolic pathways affected.

XI. TRIAZOLES

Amitrole has proven to be useful in controlling perennial broadleaf weeds and grasses in non-cropped areas. Most crop plants appear to be sensitive. Treatment with amino-triazole at phytotoxic levels is followed by the development of albino leaves and shoots. If the dosage is relatively low, however, non-pigmented leaves may be followed by leaves that develop normally.

After the introduction of 3-amino-1,2,4-triazole, a number of 1,2,4-triazole derivatives were synthesized and evaluated. A brief review of the work done with these compounds has been provided by Ashton and Crafts (1973). None of the derivatives tested was significantly superior to the parent compound.

The metabolism of amitrole has been followed in a number of plants (see reviews by Naylor, 1964; Carter, 1969; Ashton and Crafts, 1973). Carter and Naylor (1961) separated, by means of paper chromatography, 13 radioactive compounds derived from ^{14}C-5-amitrole. None of these appeared to be metabolites to be expected from CO_2 fixation. It was suggested that most of them were probably amitrole conjugates. Ring cleavage followed ultimately by evolution of $^{14}CO_2$ from ^{14}C-5-amitrole has been reported to proceed by non-biological reactions in the soil (Kaufman et al. 1968). No evidence has been brought forward that a similar reaction occurs in green plants.

Apparently amitrole conjugates readily with a variety of compounds. Some of these can only be assumed to be artifacts producing during extraction and analytical procedures. Aside from difficulties in interpretation introduced by the ease of artifact production, different investigators have unknowingly assigned various symbols to the same unidentified metabolically produced derivatives.

The first metabolic studies with amitrole were reported by Rogers (1957a, b). As a result of his investigations he reported that amitrole was metabolized by several plants to an amine glucoside derivative (N-s-triazol-3-yl-glucosylamine). Fredericks and Gentile (1960) reported that this adduct is formed readily upon allowing a mixture of amitrole, glucose and water to stand at room temperature. Even though Fredericks and Gentile (1960, 1961, 1962, 1966) have used in vitro systems almost exclusively, they have consistently maintained that the glucose derivative is the primary product of amitrole metabolism. Other workers do not agree (Racusen, 1958; Massini, 1959; Miller and Hall, 1961; Carter and Naylor, 1961; Herrett and Linck, 1961; Smith et al. 1967). The evidence against the glucose adduct being a common product produced in vivo is overwhelming.

Another apparent artifact of isolation procedures is unknown III reported by Herrett and Bagley (1964). This compound is of special interest because it is five to eight times as effective as amitrole in suppressing the growth of tomato and lettuce roots. However, Smith et al. (1967) and Carter (1969) have attempted without success to reisolate the compound. In Herrett and Bagley's (1964) original study, they were able to show that compound III is not reactive with ninhydrin and would not form azo dyes. This probably means there was substitution of the 3-amino group. But the compound was not the glucoside.

Racusen (1958) quickly confirmed Rogers (1957a, b) finding that 3-amino-1,2,4-triazole is rapidly metabolized in the plant and proceeded to isolate two major metabolites which he called "X" and "Y". The most abundant of these was "X"—a compound that exhibited positive reactions with azo dyes and ninhydrin reagents and behaved as a zwitterion during electrophoresis at different pHs. "Y" exhibited only acidic properties above pH 4·5, was not ninhydrin positive, but did form azo dyes. Both compounds were stable to 6N HCl for 5 h at 100°. Thus neither compound was a simple amide nor an amino-glycoside.

Other reports quickly followed on the metabolism of amino-triazole-5-^{14}C (Massini, 1959, 1963; Carter and Naylor, 1959, 1961; Miller and Hall, 1961; Herrett and Linck, 1961). The principal metabolites isolated by these workers exhibited azo-dye reactions, ninhydrin sensitivity and zwitterion behaviour. Almost certainly "X" (Racusen, 1958), ATX (Massini, 1959), "1" (Carter and Naylor, 1961) and compounds with similar characteristics reported by Miller and Hall (1961), Herrett and Linck (1961) and Smith *et al.* (1967) is the compound described by Massini (1959) as an alanilyl derivative of amitrole and found by Braun (1963) (see Massini, 1963) to be 3-(3-amino-1,2,4-triazol-lyl)-2-aminopropionic acid (3-ATAL).

Carter and Naylor (1961) and Carter (1965) showed that ^{14}C from serine or glycine readily enters 3-ATAL, but labelling from glucose, succinate, alanine, glyoxylate, and formate is relatively slow. Massini (1963) and Carter (1965) have suggested that 3-ATAL is formed by the condensation of amitrole and serine (Fig. 6) in the manner Dunnill and Fowden (1963) described for the formation of β-pyrazol-1-yl-α alanine from pyrazole and serine. The enzymatic reaction involved is apparently sensitive to ammonium thiocyanate. At least in the presence of this divalent sulphur compound 3-ATAL is not readily formed.

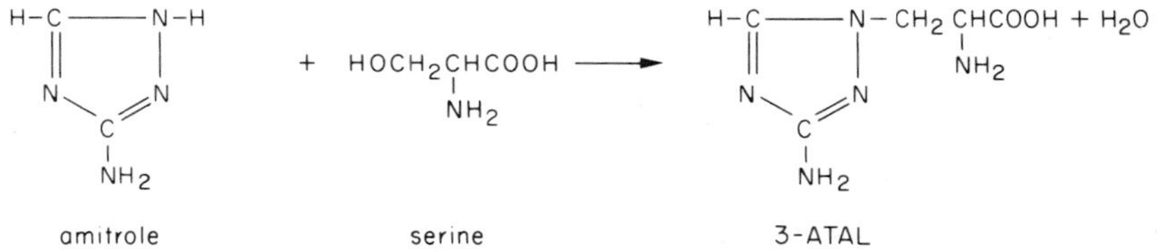

Fig. 6. Proposed route of metabolism of amitrole (Massini, 1963; Carter, 1965).

A second common derivative of ^{14}C-amitrole is a compound designated "Y" by Racusen (1958) and Miller and Hall (1961) and compound "2" by Carter and Naylor (1960). As yet, this derivative is not characterized other than that it is not ninhydrin sensitive, it gives a positive azo dye reaction and is stable to 6N HCl for 5 h at 100°C. Except for cotton leaves where "Y" is reported (Miller and Hall, 1961) to be the most abundant metabolite of amitrole, 3-ATAL accumulates in the greatest amount. This is true for cotton seed, bean (*Phaseolus vulgaris* L.) leaves, bindweed (*Convolvulus arvensis* L.) silver maple (*Acer saccharinum* L.), honeysuckle (*Lonicera japonica,* Thunb.) and alfalfa (Miller and Hall, 1961; Racusen, 1958; Massini, 1963; Carter and Naylor, 1960; Smith *et al.* 1967). The structure of "Y" has not been determined and little is known beyond chromatographic behaviour for the several other derivatives reported by Carter and Naylor (1960).

XII. CONCLUSIONS

While an impressive beginning has been made in learning about the metabolism of a number of herbicides, except for a few of them our information is still in a rudimentary state. Thus the taxonomic phase of the biochemistry of herbicides is still developing. Large gaps in our knowledge persist and the metabolic story for most of the herbicides now in common use is at best fragmentary and in need of being augmented. Fortunately, the number of laboratories engaged in such investigations has grown steadily. Since the broad scale procedures for systematic investigations have been established, progress should be more rapid in the future.

Many of the primary metabolites have been characterized chemically, but in instance after instance the polar metabolites are only broadly described and remain unidentified. The difficult job of identifying the insoluble residuals must be undertaken. The importance of performing such work is indicated by the fact the carbon skeleton of many herbicides persists in the insoluble fraction. The ultimate fate of these derivatives needs to be determined. A more thorough knowledge of the reaction products of enzymatic breakdown—whether they are innocuous or metabolic poisons—is essential. The long term significance of the herbicide derivatives in the food chain of man is not known. Studies relative to this point are overdue.

The enzymology of degradation of herbicides is in its infancy. Not only is a study of this aspect of metabolism likely to be valuable in arriving at an understanding of herbicidal selectivity but the knowledge gained will be helpful in understanding the limits of specificity of important enzymes. Work on the nature of the limitation of certain herbicides may well provide an insight into isozyme production in organs of specific plants and between species. Furthermore, differences involved in the induction of certain degradative enzymes may be revealed. With an understanding of the chemistry and physics of target enzymes involved in herbicide metabolism should come a more rational approach to the design of new herbicides.

Knowledge thus far gained about the biochemical nature of the synergistic effects of certain insecticides and herbicides has already proven valuable in widening the range of useful applications. These examples will doubtless serve as models for future exploitation.

REFERENCES

Adachi, M., Tonegawa, K. and Ueshima, T. (1966a). *Pest. Technol.* **14**, 19–22.
Adachi, M., Tonegawa, K. and Ueshima, T. (1966b). *Pest. Technol.* **15**, 11–14.
Andreae, W. A. (1963). *In* "Metabolic Inhibitors" (R. M. Hochster and J. H. Quastel, eds), Vol. II, 243–61. Academic Press, New York and London.
Andreae, W. A. and Good, N. E. (1957). *Pl. Physiol., Lancaster,* **32**, 566–572.
Arnold, W. E. and Nalewaja, J. D. (1971). *Weed Sci.* **19**, 301–305.
Ashton, F. M. (1966). *Weeds,* **14**, 55–57.
Ashton, F. M. and Crafts, A. S. (1973). "Mode of Action of Herbicides", John Wiley & Sons, New York.
Audus, L. T. (1961). *In* "Encyclopedia of Plant Physiology" (W. Ruhland, ed.), Vol. 14, 1061–1067. Springer, Berlin.
Bartels, P. G. and Pegelow, E. J., Jr. (1967). *Pl. Physiol., Lancaster,* **42** (Supplement), xxviii.
Bartels, P. G. and Pegelow, E. J., Jr. (1968). *J. Cell Biol.* **37**, 1–6.
Basler, E. (1964). *Weeds,* **12**, 14–18.
Beynon, K. I. and Wright, A. N. (1968). *J. Sci. Fd Agric.* **19**, 727–732.

Bischof, Von F., Koch, W., Jajumdar, J. C. and Schwerdtle, F. (1970). *Z. PflKrankh. PflPath. PflSchutz. Sonderh,* 95–102.
Biswas, P. K. and Hamilton, W., Jr. (1969). *Weed Sci.* **17,** 206–211.
Bourke, J. B. and Fang, S. C. (1965). *J. agric. Fd Chem.* **13,** 340–343.
Bowling, C. C. and Hudgins, H. R. (1966). *Weeds,* **14,** 94–95.
Braun, P. B. (1963) (referred to by Massini, 1963).
Brian, R. C. (1964). *Weed Res.* **4,** 105–117.
Broadhurst, N. A., Montgomery, M. L. and Freed, V. H. (1966). *J. agric. Fd Chem.* **14,** 585–588.
Burt, G. W. (1974). *Weed Sci.* **22,** 116–119.
Canny, M. J. and Markus, K. (1960). *Aust. J. biol. Sci.* **13,** 486–500.
Carter, M. C. (1965). *Physiologia. Pl.* **18,** 1054–1058.
Carter, M. C. (1969). "Degradation of Herbicides" (P. C. Kearney and D. D. Kaufman, eds), pp. 187–206. Marcel Dekker, Inc., New York.
Carter, M. C. and Naylor, A. W. (1959). *Pl. Physiol., Lancaster,* **34** (Supplement), vi.
Carter, M. C. and Naylor, A. W. (1960). *Bot. Gaz.* **122**(2), 138–143.
Carter, M. C. and Naylor, A. W. (1961). *Physiologia Pl.* **14,** 20–27.
Casida, J. E. and Lykken, L. (1969). *A. Rev. Pl. Physiol.* **20,** 607–636.
Castelfranco, P., Foy, C. L. and Deutsch, D. B. (1961). *Weeds,* **9,** 580–591.
Chang, F. Y. and Vanden Born, W. H. (1971). *Weed Sci.* **19,** 580–591.
Chin, W. T., Stanovick, R. P., Cullen, T. E. and Holsing, G. C. (1964). *Weeds,* **12,** 201–205.
Colby, S. R. (1965). *Science, N.Y.* **150,** 619–620.
Colby, S. R., Warren, G. F. and Baker, R. S. (1964). *J. agric. Fd Chem.* **12,** 320–321.
Davies, P. J., Drennan, D. S. H., Fryer, J. D. and Holly, K. (1968). *Weed Res.* **8,** 241–252.
Dunnill, P. M. and Fowden, L. (1963). *J. exp. Bot.* **14,** 237–248.
Eastin, E. F. (1971a). *Weed Res.* **11,** 63–68.
Eastin, E. F. (1971b). *Weed Res.* **11,** 120–123.
Eastin, E. F. (1971c). *Weed Sci.* **19,** 261–265.
Edgerton, L. T. and Hoffman, M. B. (1961). *Science, N.Y.* **134,** 341–342.
Eshel, Y. (1969). *Weed Sci.* **17,** 492–496.
Fang, S. C. (1969). *In* "Degradation of Herbicides" (P. C. Kearney and D. D. Kaufman, eds), pp. 147–164. Dekker, New York.
Fang, S. C. and Fallin, E. (1965). *Weeds,* **13,** 153–155.
Fang, S. C. and George, M. (1962). *Pl. Physiol., Lancaster,* **37** (Supplement), xxvi.
Fawcett, C. H., Pascal, R. M., Pybus, M. B., Taylor, H. F., Wain, R. L. and Wightman, F. (1959). *Proc. R. Soc.* **B150,** 95–119.
Feung, C.-S., Hamilton, R. H. and Mumna, R. O. (1973). *J. agric. Fd Chem.* **21,** 637–640.
Feung, C.-S., Mumna, R. O. and Hamilton, R. H. (1974). *J. agric. Fd Chem.* **22,** 307–309.
Fleeker, J. and Steen, R. (1971). *Weed Sci.* **19,** 507–510.
Frank, R. and Switzer, C. M. (1969). *Weed Sci.* **17,** 344–348.
Frear, D. S. (1968a). *Phytochemistry,* **7,** 381–390.
Frear, D. S. (1968b). *Science, N.Y.* **162,** 674–675.
Frear, D. S., Hodgson, R. H., Shimabukuro, R. H. and Still, G. G. (1972a). *Adv. Agron.* **24,** 327–378.
Frear, D. S. and Shimabukuro, R. H. (1971). *Tech. Pap. FAO int. Conf. Weed Contr., Weed Sci. Soc. Am.,* pp. 560–578.
Frear, D. S. and Still, G. G. (1968). *Phytochemistry,* **7,** 913–920.
Frear, D. S. and Swanson, H. R. (1970). *Phytochemistry,* **9,** 2123–2132.
Frear, D. S. and Swanson, H. R. (1972). *Phytochemistry,* **11,** 1919–1929.
Frear, D. S., Swanson, C. R. and Kadunce, R. E. (1967). *Weed Sci.* **15,** 101–104.
Frear, D. S., Swanson, H. R. and Tanaka, F. S. (1969). *Phytochemistry,* **8,** 2157–2169.
Frear, D. S., Swanson, H. R. and Tanaka, F. S. (1972b). *Recent Adv. Phytochem.* **5,** 225–246.
Fredericks, J. F. and Gentile, A. C. (1960). *Physiologia Pl.* **13,** 761–765.
Fredericks, J. F. and Gentile, A. C. (1961). *Archs Biochem. Biophys.* **92,** 356–359.
Fredericks, J. F. and Gentile, A. C. (1962). *Physiologia Pl.* **15,** 186–193.
Fredericks, J. F. and Gentile, A. C. (1966). *Phytochemistry,* **4,** 851–856.

Funderburk, H. H., Jr. and Davis, D. E. (1963). *Weeds*, **11**, 101–104.

Geissbühler, H. (1969). *In* "Degradation of Herbicides" (P. C. Kearney and D. D. Kaufman, eds), pp. 79–111. Dekker, New York.

Geissbühler, H., Baunok, I. and Gross, D. (1969). *IUPAC Symp. Johannesburg, South Africa.*

Golab, T., Herberg, R. J., Gramlich, J. V., Rann, A. P. and Probst, G. W. (1970). *J. agric. Fd Chem.* **18**, 838–844.

Golab, T., Herberg, R. J., Parka, S. J. and Tepe, J. B. (1966). *J. agric. Fd Chem.* **14**, 592–596.

Golab, T., Herberg, R. J., Parka, S. J. and Tepe, J. B. (1967). *J. agric. Fd Chem.* **15**, 638–641.

Gysin, H. and Knüsli, E. (1960). *Adv. Pest Contr. Res.* **3**, 289–358.

Hamilton, R. H. (1964). *J. agric. Fd Chem.* **12**, 14–17.

Hamilton, R. H., Bandurski, R. S. and Reusch, W. H. (1962). *Cereal Chem.* **39**, 107–113.

Hamilton, R. H. and Moreland, D. S. (1962). *Science, N.Y.* **135**, 373–374.

Hawton, D. and Stobbe, E. H. (1971). *Weed Sci.* **19**, 555–558.

Herrett, R. A. (1969). *In* "Degradation of Herbicides" (P. C. Kearney and D. D. Kaufman, eds), pp. 113–145. Marcel Dekker, Inc., New York.

Herrett, R. A. (1970). *Abstr. 160th Meet. Am. chem. Soc. Chicago*, No. 77.

Herrett, R. A. and Bagley, W. P. (1964). *J. agric. Fd Chem.* **12**, 17–20.

Herrett, R. A. and Berthold, R. V. (1965). *Science, N.Y.* **149**, 191–193.

Herrett, R. A. and Linck, A. J. (1961). *Physiologia Pl.* **14**, 767–776.

Hilton, J. L., Jansen, L. L. and Hull, H. M. (1963). *A. Rev. Pl. Physiol.* **14**, 353–384.

Hilton, J. L., Scharen, A. L., St. John, J. B., Moreland, D. E. and Norris, K. H. (1969). *Weed Sci.* **17**, 541–547.

Hisada, T. (1967). *Proc. 1st Asian Pacific Weed Confr. Interchange*, p. 107.

Hodgson, R. H. (1971). *Weed Sci.* **19**, 501–507.

Hodgson, R. H., Dusababek, K. E. and Hoffer, B. L. (1974). *Weed Sci.* **22**, 205–210.

Hodgson, R. H., Frear, D. S., Swanson, H. R. and Regan, L. A. (1973). *Weed Sci.* **21**, 543–549.

Hofman, J. and Hofmanova, D. (1969). *Eur. J. Biochem.* **8**, 109–112.

Holley, R. W., Boyle, F. D. and Hand, D. B. (1950). *Archs Biochem. Biophys.* **27**, 143–151.

Holley, R. W. (1952). *Archs Biochem. Biophys.* **35**, 171–175.

Holm, R. E. and Stallard, D. E. (1974). *Weed Sci.* **22**, 10–14.

Hurter, J. (1966). *Experientia*, **22**, 741–742.

Ishizuka, K. and Mitsui, S. (1966). *Abstr. a. Meet. agric. chem. Soc. Japan*, p. 62.

Jaworski, E. (1964). *J. agric. Fd Chem.* **12**, 33–37.

Jaworski, E. G. (1969). *In* "Degradation of Herbicides" (P. C. Kearney and D. D. Kaufman, eds), pp. 165–185. Marcel Dekker, Inc., New York.

Jaworski, E. and Porter, C. A. (1965). *Abstr. 149th Meet. Am. chem. Soc. Detroit*, **21A**.

Jacobsohn, R. and Anderson, R. N. (1972). *Weed Sci.* **20**, 74–80.

James, C. S. and Prendeville, G. N. (1969). *J. agric. Fd Chem.* **17**, 1257–1260.

Kassebeer, Von H. (1971). *Z. PflKrankh. PflPath. PflSchutz.* **78**, 158–174.

Kaufman, D. D., Plimmer, J. R., Kearney, P. C., Blake, J. and Guardia, F. S. (1968). *Weed Sci.* **16**, 266–272.

Klämbt, H. D. (1961). *Planta*, **57**, 339–353.

Kearney, P. C. and Kaufman, D. D. (eds) (1969). "Degradation of Herbicides." Marcel Dekker, Inc., N.Y.

Klun, J. A. and Robinson, J. F. (1969). *J. econ. Ent.* **62**, 214–220.

Knüsli, E., Berrer, D., Dupius, G. and Esser, H. (1969). *In* "Degradation of Herbicides" (P. C. Kearney and D. D. Kaufman, eds), pp. 51–78. Marcel Dekker, Inc., N.Y.

Kraus, E. J. and Mitchell, J. W. (1947). *Bot. Gaz.* **108**, 301–350.

Kuratle, H., Rahn, E. M. and Woodmansee, C. W. (1969). *Weed Sci.* **17**, 216–219.

Lamoureaux, G. L., Shimabukuro, R. H., Swanson, H. R. and Frear, D. S. (1970). *J. agric. Fd Chem.* **18**, 81–86.

Lamoureaux, G. L., Stafford, L. E., Shimabukuro, R. H. and Zaylshic, R. G. (1973). *J. agric. Fd Chem.* **21**, 1020–1030.

Lamoureaux, G. L., Stafford, L. E. and Tanaka, F. S. (1971). *J. agric. Fd Chem.* **19**, 346–350.

Leafe, E. L. (1962). *Nature, Lond.* **193**, 485–486.

Lemin, A. J. (1966). *J. agric. Fd Chem.* **14**, 109–111.

Long, J. W., Thompson, L., Jr. and Rieck, C. E. (1974a). *Weed Sci.* **22**, 42–47.
Long, J. W., Thompson, L., Jr. and Rieck, G. E. (1974b). *Weed Sci.* **22**, 91–94.
Loos, M. A. (1969). *In* "Degradation of Herbicides" (P. C. Kearney and D. D. Kaufman, eds), pp. 1–47. Marcel Dekker, Inc., N.Y.
Luckwill, L. C. and Lloyd-Jones, C. P. (1960a). *Ann. appl. Biol.* **48**, 613–625.
Luckwill, L. C. and Lloyd-Jones, C. P. (1960b). *Ann. appl. Biol.* **48**, 626–636.
McRae, D. H., Yih, R. Y. and Wilson, H. F. (1964). *Abstr. Weed Sci. Soc. Am.* p. 87.
Maroder, H. L. and Prego, I. A. (1971). *Weed Res.* **11**, 193–195.
Massini, P. (1959). *Biochim. Biophys. Acta*, **36**, 548–549.
Massini, P. (1963). *Acta bot. neerl.* **12**, 64–72.
Matsunaka, S. (1969a). *J. agric. Fd Chem.* **17**, 171–175.
Matsunaka, S. (1969b). *Residue Rev.* **25**, 45–58.
Meikle, R. W., Williams, E. A. and Redemann, C. T. (1966). *J. agric. Fd Chem.* **14**, 384–387.
Menzie, C. N. (1969). *U.S. Fish Wildlife Service, Special Scientific Report, Wildlife*, **127**, 1–487.
Milborrow, B. V. (1963). *Biochem. J.* **87**, 255–258.
Milborrow, B. V. (1965). *Weed Res.* **5**, 332–342.
Miller, C. S. and Hall, W. C. (1961). *J. agric. Fd Chem.* **9**, 210–212.
Montgomery, M. L., Botsford, D. L. and Freed, V. H. (1969). *J. agric. Fd Chem.* **17**, 1241–1243.
Moreland, D. E. (1967). *A. Rev. Pl. Physiol.* **18**, 365–386.
Morgan, P. W. and Hall, W. C. (1963). *Weeds*, **11**, 130–135.
Müeller, P. W. and Payot, P. H. (1966). *Proc. IAEA. Symp. Isotopes in Weed Res., Vienna*, 1965, pp. 61–70.
Nalewuja, J. D., Behrens, R. and Schmid, A. R. (1964). *Weeds*, **12**, 269–272.
Nashed, R. B. and Ilnicki, R. D. (1970). *Weed Sci.* **18**, 25–28.
Nashed, R. B., Katz, S. E. and Ilnicki, R. D. (1970). *Weed Sci.* **18**, 122–128.
Naylor, A. W. (1964). *J. agric. Fd Chem.* **12**, 21–25.
Norris, L. A. and Freed, V. H. (1966). *Weed Res.* **6**, 212–220.
Pate, D. A. and Funderburk, H. H., Jr. (1966). *Proc. IAEA Isotopes, Weed Res., Vienna*, 1965, pp. 17–25.
Porter, C. A. and Jaworski, E. (1965). *Pl. Physiol., Lancaster*, **40** (Supplement), xiv–xv.
Prendeville, G. N., Eshel, Y., Jones, C. S., Warren, G. F. and Schreiber, M. M. (1968). *Weed Sci.* **16**, 432–435.
Probst, G. W., Golab, T., Herberg, R. J., Holzer, F. J., Parka, S. J., Van der Schaus, C. and Tepe, J. B. (1967). *J. agric. Fd Chem.* **15**, 592–599.
Probst, G. W. and Tepe, J. B. (1969). *In* "Degradation of Herbicides" (P. C. Kearney and D. D. Kaufman, eds), pp. 255–282. Marcel Dekker, Inc., New York.
Racusen, D. (1958). *Archs. Biochem. Biophys.* **74**, 106–113.
Redemann, C. T., Meikle, R. W., Hamilton, P., Banks, V. S. and Youngson, C. R. (1968). *Bull. Environ. Contam. Toxicol.* **3**, 80–96.
Riden, J. R. and Hopkins, T. R. (1961). *J. agric. Fd Chem.* **9**, 47–49.
Riden, J. R. and Hopkins, T. R. (1962). *J. agric. Fd Chem.* **10**, 455–458.
Ries, S. K., Zabik, M. J., Stephenson, G. R. and Chen, T. M. (1968). *Weed Sci.* **16**, 40–41.
Rogers, B. J. (1957a). *Hormolog*, **1**, 10.
Rogers, B. J. (1957b). *Weeds*, **5**, 5–11.
Rogers, R. L. (1971). *J. agric. Fd Chem.* **19**, 32–35.
Roth, W. (1957). *C.r. hebd. Séanc. Acad. Sci. Paris*, **245**, 942–944.
Roth, W. and Knüsli, E. (1961). Experentia, **17**, 312–313.
Sargent, J. A. and Blackman, G. E. (1970). *J. exp. Bot.* **21**, 219–227.
Schafer, D. E. and Chilcote, D. O. (1970a). *Weed Sci.* **18**, 725–729.
Schafer, D. E. and Chilcote, D. O. (1970b). *Weed Sci.* **18**, 729–732.
Shimabukuro, R. H. (1967). *Pl. Physiol., Lancaster*, **42**, 1269–1276.
Shimabukuro, R. H., Frear, D. S., Swanson, H. R. and Walsh, W. C. (1971a). *Pl. Physiol., Baltimore*, **47**, 10–14.
Shimabukuro, R. H., Kadunce, R. E. and Frear, D. S. (1966). *J. agric. Fd Chem.* **14**, 392–395.

Shimabukuro, R. H., Lamoureaux, G. L., Frear, D. S. and Bakke, J. E. (1971*b*). *IUPAC Symp. Tel Aviv, Israel,* 323–342.

Shimabukuro, R. H. and Swanson, H. R. (1969). *J. agric. Fd Chem.* **17,** 199–205.

Shimabukuro, R. H., Swanson, H. R. and Walsh, W. C. (1970). *Pl. Physiol., Baltimore,* **46,** 103–107.

Shimabukuro, R. H., Walsh, W. C., Lamoureaux, G. L. and Stafford, L. E. (1973). *J. agric. Fd Chem.* **21,** 1031–1036.

Slade, R. E., Templeman, W. G. and Sexton, W. A. (1945). *Nature, Lond.* **155,** 497–498.

Slife, F. W., Key, J. L., Yamaguchi, S. and Crafts, A. S. (1962). *Weeds,* **10,** 29–35.

Smith, L. W., Bayer, D. E. and Foy, C. L. (1967). *Abstr. Weed Sci. Soc. Am.* pp. 61–62.

Spitznagle, L. A., Christian, J. E. and Ohlrogge, A. J. (1969). *J. pharm. Sci.* **58,** 1234–1237.

Stephenson, G. R. and Ries, S. K. (1967). *Weed Res.* **7,** 51–60.

Stephenson, G. R. and Ries, S. K. (1969). *Weed Sci.* **17,** 327–331.

Still, G. G. (1968*a*). *Pl. Physiol., Lancaster,* **43,** 543–546.

Still, G. G. (1968*b*). *Science, N.Y.* **159,** 992–993.

Still, G. G. and Kuzerian, O. (1967). *Nature, Lond.* **216,** 799–800.

Still, G. G. and Mansager, E. R. (1971). *J. agric. Fd Chem.* **19,** 879–884.

Still, G. G. and Mansager, E. R. (1972). *Phytochemistry,* **11,** 515–520.

Still, G. G. and Mansager, E. R. (1973). *J. agric. Fd Chem.* **21,** 787–791.

Stoller, E. W. (1968). *Weed Sci.* **16,** 384–386.

Stoller, E. W. (1969). *Pl. Physiol., Lancaster,* **44,** 854–860.

Stoller, E. W. and Wax, L. M. (1968). *Weed Sci.* **16,** 283–288.

Swanson, C. R. (1966). *Proc. IAEA Symp. Isotopes in Weed Research, Vienna,* 1965, pp. 135–146.

Swanson, C. R. (1969). *In* "Degradation of Herbicides" (P. C. Kearney and D. D. Kaufman, eds), pp. 299–320. Marcel Dekker, Inc., New York.

Swanson, C. R., Kadunce, R. E., Hodgson, R. H. and Frear, D. S. (1966*a*). *Weeds,* **14,** 319–323.

Swanson, C. R., Hodgson, R. H., Kadunce, R. E. and Swanson, H. R. (1966*b*). *Weeds,* **14,** 323–327.

Thomas, E. W., Loughman, B. C. and Powell, R. G. (1963). *Nature, Lond.* **199,** 73–74.

Thomas, E. W., Loughman, B. C. and Powell, R. G. (1964). *Nature, Lond.* **204,** 286.

Thompson, L., Jr. (1972*a*). *Weed Sci.* **20,** 153–155.

Thompson, L., Jr. (1972*b*). *Weed Sci.* **20,** 584–587.

Thompson, L. Jr., Houghton, J. M., Slife, F. W. and Butler, H. S. (1971). *Weed Sci.* **19,** 409–421.

Thompson, L., Jr., Slife, F. W. and Butler, H. S. (1970). *Weed Sci.* **18,** 509–514.

Tipton, C. L., Usted, R. R. and Tsao, F. H. C. (1971). *J. agric. Fd Chem.* **19,** 484–486.

Verloop, A. and Nimmo, W. B. (1969). *Weed Res.* **9,** 357–370.

Verloop, A. and Nimmo, W. B. (1970). *Weed Res.* **10,** 59–64.

Wahlroos, O. and Virtanen, A. I. (1959). *Acta chem. scand.* **13,** 1906–1908.

Weintraub, R. L., Brown, J. W., Fields, M. and Rohan, J. (1950). *Am. J. Bot.* **37,** 682.

Weintraub, R. L., Brown, J. W., Fields, M. and Rohan, J. (1952). *Pl. Physiol., Lancaster,* **27,** 293–301.

Weintraub, R. L., Reinhart, J. H. and Sherff, R. A. (1956). *In* "A Conference on Radioactive Isotopes in Agriculture, A.E.C. Rept. TID-7152", pp. 203–208.

Wilcox, M., Moreland, D. E. and Klingman, G. C. (1963). *Physiologia Pl.* **16,** 565–571.

Williams, R. T. (1959). "Detoxication Mechanisms", John Wiley & Sons, New York.

Williams, M. C., Slife, F. W. and Hanson, J. B. (1960). *Weeds,* **8,** 244–255.

Yih, R. Y., McRae, D. H. and Wilson, H. F. (1968*a*). *Science, N.Y.* **161,** 376–378.

Yih, R. Y., McRae, D. H. and Wilson, H. F. (1968*b*). *Pl. Physiol., Lancaster,* **43,** 1291–1296.

Yih, R. Y. and Swithenbank, C. (1971*a*). *J. agric. Fd Chem.* **19,** 314–319.

Yih, R. Y. and Swithenbank, C. (1971*b*). *J. agric. Fd Chem.* **19,** 320–324.

CHAPTER 14

DISLOCATION OF DEVELOPMENTAL PROCESSES

N. P. KEFFORD

Department of Botany
University of Hawaii, Honolulu, Hawaii 96822, USA

I. INTRODUCTION

Within broad terms of reference, any chemical could be responsible for or contribute to a herbicidal effect by inhibiting or changing a plant's development, or by inducing any such change which places the plant at an ecological disadvantage with regard to competing species. In this chapter, however, the principal concern is with the death of a plant and the means by which disruptions to its development might result in death.

A. THE PARADOX OF PLANT DEATH

The induction of rapid death in an entire plant as a consequence of treatment with a growth-regulator herbicide, appears contrary to the nature of plant development and its control by regulators. Plants can regenerate following injury and this capacity involves endogenous growth regulators to which many herbicides are chemically related. Studies on the regeneration of plant parts, such as cuttings, suggest that growth regulators function in the regeneration of those organs necessary for a fully functioning plant. Furthermore, for some tissues, death follows normal development but such happenings are localized and are not part of a general degeneration of the plant. The formation of mature xylem vessels requires that the elements die, and the progress of such development is, at least in part, under the control of growth regulators (Digby and Wareing, 1966). The same is true for senescencing organs (see references in Woolhouse, 1967). One must wonder how an applied regulator can so disorganize these conservative systems that plant death occurs.

Another element of the paradox of rapid plant death is that plants have means of isolating or buffering against lethal agents through the death or abscission of the cells or tissues affected.

B. PRIMARY ACTIONS LEADING TO PLANT DEATH

The enigma of degenerative injury that terminates in plant death raises the question of the nature of the primary actions that make the death of plant cells and tissues inevitable. In death resulting from regulator herbicides, the nature of this critical action is not known, but this ignorance is not unique to death from herbicides. For example Dugger and Ting (1970), writing about plant damage by air pollutants, state that the mechanism of the damage of plants by oxidants is still an outstanding problem. Even for death resulting from water-stress, the process or structure that is primarily affected has not been defined.

For plant injury and death induced by chilling however, recent findings have permitted the ordering of a mass of information into a hierarchy of effects, which lead sequentially to death (Lyons, 1973). According to this new model, the *primary* effect of decreasing temperature upon susceptible tissues is a physical phase-transition in cell membranes from a flexible, liquid-crystalline structure to a solid-gel structure. The effects of chilling proposed to be *secondary* and to be the direct resultants of the phase transition are: (i) cessation of protoplasmic streaming, (ii) increased activation energy of membrane-bound enzymes, and (iii) increased membrane permeability. These secondary effects of chilling may produce effects at other levels of the hierarchy. Cessation of protoplasmic streaming would cause a general disturbance of cellular metabolism including a reduction in ATP supply. Increased activation energy of membrane-bound enzymes would result in an imbalance in metabolism leading to a worsening ATP deficit and the accumulation of toxic metabolites such as ethanol and acetaldehyde. Increased permeability would also contribute to the imbalance of metabolism and would permit solute leakage and a disruption of ion balance.

An ordering of information, just as has occurred in the field of chilling injury, is needed for those herbicidal effects that appear to work through a dislocation of development. Faced with a plethora of descriptions of phenomena, we await the keys to a sorting of the phenomena into hierarchies and thence into sequences of processes which lead inexorably to death.

C. APPROACHES TO MECHANISMS OF TOXICITY

A discussion of the mechanisms of toxicity resulting from effects upon development might ideally reveal a unifying theory for the regulation of normal plant development and then pinpoint, within this theory, loci for the disruptions induced by each class of herbicide. No unifying theory for normal development exists; nor has a unifying theory for the mechanisms of development-disrupting herbicides been achieved through any other approach. In such circumstances, an apt strategy is to organize current information with the hope of encouraging the development of hypotheses. Chapters 15, 16, and 17 approach mechanisms of toxicity, some of which relate to development, from the subcellular and molecular levels. The present chapter takes a complementary approach; to view the plant as a regulated whole and to enquire into the means by which the processes, which organize the whole, might be disrupted to produce death.

In pursuing this approach, the present chapter first attempts to formulate a unifying hypothesis for the regulation of the developmental processes which produce organization; this hypothesis is based upon the field concept, which Wardlaw (1965) has advocated as a worthy basis for the explanation of plant organization. Thereafter, points at which herbicides might interfere in the hypothetical scheme are proposed.

II. PROPERTIES OF REGULATORS APPLICABLE TO A FIELD CONCEPT OF DEVELOPMENT

The proposed unifying hypothesis for normal plant development will be based upon those properties of plant growth regulators, which indicate a capability of these substances to form chemical fields upon which the organization of a plant could be established and maintained.

A. CLASSES OF REGULATORS OF DEVELOPMENT AND THEIR MOVEMENT

Five classes of natural plant growth regulators are generally recognized at present. Compounds with the biological activity of four of the classes are solids at physiological temperatures and exhibit evidence of directed movement in plant tissues; these classes, which are physiologically and chemically distinct, have been given the trivial names: auxins, gibberellins, cytokinins and abscisins. The fifth class has no trivial name and its principal member is ethylene. The fact that ethylene is a gas at physiological temperatures and moves by diffusion, which is random and undirected, appears to set it apart from the other four classes of regulator.

Members of the five classes of regulators occur naturally, act at low concentration, and are synthesized in one part of a plant and move to another part where they induce specific effects. These are the properties of a plant hormone; however Abeles (1973) proposes that, as ethylene does not have directed movement which the other classes may have, it may be useful conceptually to put ethylene into a distinct class and not call it a plant hormone. Further properties of a hormone are that it is not consumed in the processes which it regulates, but that it can be inactivated by reactions which thereby establish concentration levels for the hormone in a tissue.

The fact that regulators are synthesized in one part of a plant, are active in another part, and may be inactivated in yet another part, implies that there are currents of regulators in the plant. Such currents could be one source of the fields of regulators proposed to function as determinants in plant development. These currents may involve directed movement which would provide an additional vector to the field.

Substantial evidence for the directed movement or transport of auxin exists and directed movement of gibberellin, cytokinin and abscisin may also occur (Goldsmith, 1969). To varying degrees, depending on the tissue and its age, the auxin transport system shows polarity and thus it has the potential for maintaining directionality in a tissue or organ. In some organs longitudinal and lateral auxin transport occur and both may exhibit polarity but appear in-dependently controlled; hence the possibility exists for a plant hormone to be involved in directionality in more than one dimension of an organ. Directionality may be conceived as a desirable property of a substance setting up a chemical field.

B. INTERDEPENDENCE AND QUANTITATIVE INTERACTIONS OF CLASSES OF REGULATORS

Another property of regulators that appears relevant to a field concept of development, is their interdependence in producing some of their effects. For instance, both auxin and gibberellin can promote organ expansion. Possible explanations for this dual regulation of organ expansion are: (i) gibberellin increases the concentration of auxin in the tissue or *vice versa*; (ii) auxin and gibberellin can each induce expansion by acting independently through separate pathways; or (iii) for any organ expansion to occur, both auxin and gibberellin must be present because their actions are interdependent. Experimental evidence favours the explanation that auxin and gibberellin are both required and are interdependent in the control of expansion in some organs (Brian, 1959). Auxin and cytokinin also exhibit interdependence, for instance in the induction of cell division (Patau *et al.* 1957). Interactions between classes of regulator, each of which may have its own path of movement, could mean that fields which determine events in plant development are constructed from currents of more than one regulator.

Another facet of the interaction of regulators is that their concentration ratios can be determinants of developmental events. The nature of the organs that regenerate on excised portions of tobacco-stem pith is determined by the ratio of cytokinin to auxin in the medium on which the pith grows (Skoog and

Miller, 1956). If the ratio of cytokinin to auxin is high, buds form; if the ratio is low, roots form and at intermediate ratios callus rather than organ development occurs.

Quantitative interactions between regulators in determining the nature of regenerating organs suggest the hypothesis that in a tissue in which gradients of regulators exist specific developmental events can be initiated where particular combinations of concentrations of the regulators occur.

Regulators not only interact in producing developmental effects, they also interact in each other's movement and synthesis. For instance, cytokinin and gibberellin treatments can enhance auxin transport while ethylene may inhibit it (Goldsmith, 1969).

C. REGULATION OF CHANGES IN DEVELOPMENTAL PATTERNS

Other evidence that the patterns of development in plants are based upon chemical fields formed by regulators is the fact that regulator treatments can induce the switch of an organ from one established developmental pattern to another.

An example is the effect which gibberellin treatment can have upon shoot apices of ivy, *Hedera helix* (Robbins, 1960). In the adult stage, ivy has heart-shaped leaves and has the ability to flower. If adult ivy is sprayed with gibberellin, the form of development changes: the leaves become lobed, the internodes are longer, and the ability to flower is lost. The new form of development is that characteristic of juvenile plants and, most significantly, the new type of development persists. Cuttings of the gibberellin-induced, juvenile form grow in the juvenile fashion. The change in plant form arises from a change in developmental pattern, induced in the shoot apex by gibberellin.

In another case, a switch between two phases in the development of a bud required treatment with both gibberellin and auxin in the potato species *Solanum andigena* (Booth, 1959; 1963). On plants in which the apical bud was replaced by a mixture of auxin plus gibberellin lateral buds developed as stolons, while lateral buds, on similar plants untreated or treated with gibberellin alone, developed into leafy shoots. Hence a treatment of auxin plus gibberellin was required for the switch from normal leafy shoots into stolons.

In addition to triggering switches between the developmental patterns of an organ, there is evidence that regulators form the fields that determine each developmental pattern and that the regulator field must change if the developmental pattern is to change. It is not yet possible to map the regulator fields in a plant, but some gross changes in regulators have been observed as organs switch from one developmental pattern to another.

Such a change for example precedes bulb development of onions. Lengthening photoperiods initiate this switch in developmental pattern and Clark and Heath (1962) found that the placing of onions in long days is followed by an upsurge of auxin in the tissues that will form the bulb. The increase occurs during the first 3–5 long days, then the auxin level decreases to below that of the tissues held in short days. Clark and Heath argue that the high auxin level may characterize the transition phase and the low auxin level leads to bulb development. Similar surges in concentration are observed for gibberellin as barley changes from vegetative to prefloral development, and for cytokinin in some fruits at the resumption of development following pollination (Nicholls and May, 1964; Letham, 1963). Thus regulators appear to be involved both in triggering transitions between developmental phases and in determining the nature of the transitions.

D. ETHYLENE—REGULATOR WITHOUT DIRECTED MOVEMENT

The properties of ethylene (see Abeles, 1973 for a review) set it apart from the other classes of regulator; it is a gas, it moves through tissues by diffusion which is undirected, and it dissipates into the external atmosphere. Mechanisms for the detoxification of ethylene are not known in plants, so its levels in tissues must be chiefly dependent upon the rates of synthesis. However, ethylene and the other classes of regulators do mutually interact in their synthesis, movement, and action.

The movement of ethylene by diffusion rather than by a regulated, directed transport may be a necessary attribute for one of its roles in development. Ethylene has a role in the regulation of senescence and death in tissues and organs which involves the disorganization of cellular systems. As cellular and tissue systems degenerate, the regulator system that is maintaining them may be assumed also to degenerate and so lose control. Ethylene diffuses through tissues, alive or dead, and is not dependent upon transport systems for movement; therefore the partial disorganization of a tissue is not necessarily a barrier to its effectiveness as a messenger and regulator. Developmental processes in which ethylene may be acting in this way are the ripening of fruit and the senescence of leaves.

E. INTERACTIONS OF REGULATORS WITH THE PROTOPLASM

Growth regulators are being advanced as the agents which, in the protoplasm of plants, form chemical fields that determine plant development. The growth regulators and the protoplasm in which they function, interact in a variety of ways. The protoplasm synthesizes the regulators, directs their movement, and inactivates them. Also protoplasm must respond to a regulator field if the effects of the field on development are to be materialized through the organization of cells and tissues. As the ultimate aim of the present discussion is to explore potential sites for the disruption of development, all the sites, that a field concept of the regulation of development suggests, are of interest. Four possible sites of interaction between regulators and plant cytoplasm can be suggested and each of these sites may make particular contributions to processes involved in regulation:

(i) A *synthetic site* could exert quantitative control on development by regulating the rate of synthesis of the regulator. Its location could introduce directional control.

(ii) *Directed movement sites* or tracks may occur for each class of regulator except ethylene, and these sites, through their directionality, could orient a regulator field and impose quantitative control by varying movement rates.

(iii) Inactivation mechanisms for each class of regulator except ethylene are known and if *inactivation sites* exist they could contribute quantitative and directional control to regulators and hence to development.

(iv) The cytoplasm must perceive the presence of a regulator at a *recognitive site,* or a primary site of regulator action, the response of which would be related to regulator concentration. The response to a particular concentration of one regulator can be dependent upon the other classes of regulator present and their concentrations. Thus a recognitive site may be required to assess the relative concentrations of all the regulators in its environment. This assessment may initiate an interaction between the recognitive site and the genome of the cell.

The above sites postulated for interactions between regulators and the cytoplasm do not represent the genome although an alternative of a direct regulator-genome interaction has been proposed (see Hall, 1973; Key, 1969). In any case, the response that a cell makes to a regulator is probably partly determined by the status of its genome, i.e. whether particular genes are active or not (Fig. 1). Possible evidence of this is the differing effects of a single regulator upon different organs, which may be presumed to have different portions of their genomes active or inactive.

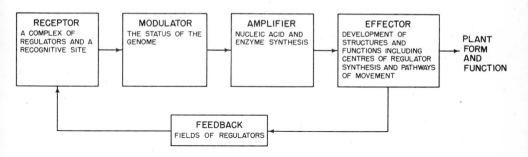

Fig. 1. A diagrammatic scheme, for the maintenance of a developmental pattern in a plant, drawn in terms of simple control-system concepts and terminology.

Little is known about the nature, location, or the control mechanisms for the hypothetical sites and reactions associated with growth-regulator function. Proteins are the molecular species capable of recognizing small molecules, but the nature of the proteins that are most uniquely involved in growth-regulator function, namely those at the transport and recognitive sites, is quite unknown. The clues given by the chemical structures of the growth regulators have not solved the problems so far.

III. HERBICIDAL POTENTIALS FOR DISRUPTIONS
OF REGULATOR–FIELD PATTERNS

The model of plant organization, developed here from a field concept, emphasizes particular properties of regulators and particular hypothetical loci for the interactions between regulators and the protoplasm. These loci and properties will now be explored for potential disruptions leading to lethal effects. In addition the properties of classes of synthetic herbicides and theories of their mechanisms of action will be compared with aspects of the proposed scheme of the regulation of development.

A. REGULATOR CONCENTRATION EFFECTS

Responses to regulators are commonly proportional to the logarithm of regulator concentration up to an optimum, above which the response diminishes. In the comparatively simple system in which exogenous auxin maintains the expansion of cells in isolated sections of a grass coleoptile, supraoptimal auxin concentrations induce expansion rates which initially exceed those at the optimal concentration, but these highest rates rapidly diminish (Thimann, 1969). The diminishing growth rates at supraoptimal auxin concentrations are sometimes associated with a loss of turgor by the sections, suggesting a loss of structural integrity of the membranes. Thus, even in a system such as an elongating coleoptile section with the potential for development apparently limited to cell expansion and maturation, exogenous auxin above a certain concentration has effects that are potentially herbicidal. Whether these are the result of the very high growth rates initially induced or of other effects is not known.

That the breakdown of a tissue in high auxin concentrations may derive from processes controlled by growth regulators, rather than from unspecific toxicity, is suggested by the fact that another class of regulator can prevent tissue collapse. Milborrow (1966) found that oat mesocotyl sections treated only with a supraoptimal concentration of indol-3yl-acetic acid (IAA) did not grow and became flaccid and pink coloured, both indications of cellular disruption. If however abscisic acid were added along with IAA, the sections grew slightly and retained their turgidity and normal appearance.

In earlier chapters it was shown that as the concentration of an auxin herbicide applied to plants is increased, an array of effects ranging from the promotion of normal growth (in auxin deficient tissues), through abnormal growth, to injury and death are observed. This could represent a continuum of effects arising from a single interaction between the herbicide and the plant, as may occur in the auxin effects on oat seedlings just described. Alternatively, auxin may initiate a series of different reactions as its concentration increases and this occurs in pea stem tissues. When auxin treatments are applied to sections excised from elongating pea stems, low concentrations cause section elongation to resume but, as the concentration is increased, auxin-induced ethylene synthesis begins and the ethylene opposes auxin-induced increase in length causing a decrease in elongation rate (Chadwick and Burg, 1967).

For some of the synthetic auxins, their relative activities at low concentrations, as measured by the stimulation of normal developmental processes, are similar to their relative hebicidal activities at higher concentrations. This correlation may indicate a connection between the acts promoting normal growth an 1 those initiating injury and death.

B. INTERFERENCE WITH REGULATOR SYNTHESIS AND INACTIVATION

Gibberellin synthesis is inhibited by compounds such as chloroethyltrimethylammonium chloride (chlormequat chloride) and one of their effects upon growth is to inhibit stem elongation. Such inhibition has herbicidal potential, for example, by preventing a bud, regenerating on a buried root fragment, from reaching the light before the nutrient reserves are exhausted (Kefford and Caso, 1972). The compound N-dimethylaminosuccinic acid, B-995, which may inhibit auxin synthesis (Reed *et al.* 1965), also inhibits the elongation of buds regenerating on roots (Caso and Kefford, 1973).

The enzyme system, which inactivates the native auxin IAA by causing it to conjugate with amino acids, is inducible by 2,4-dichlorophenoxyacetic acid (2,4-D) but is relatively ineffective in promoting 2,4-D conjugation (Zenk, 1962). Thus a plant treated with 2,4-D cannot inactivate this exogenous auxin while its ability to inactivate the native auxin is increased. Such a condition has the potential of disrupting development, both by preventing the native auxin from playing its role in development and by permitting exogenous auxin to persist in amounts and locations that are unregulated.

In pea tissue (Kang *et al.* 1971) the production of ethylene closely follows the levels of free auxin. When IAA (the native auxin) is added to the tissue, ethylene production increases but, as conjugation of the free IAA proceeds, the rate of ethylene production falls away. When the added auxin is 2,4-D however, inactivation of the auxin does not occur and the high rate of ethylene production is maintained. For ethylene, the rate of production determines the tissue concentration as there is no inactivation mechanism, so a 2,4-D-treated tissue may be subject to a sustained high concentration of ethylene which could induce abnormal effects.

C. DISPLACEMENT OF REGULATOR MOVEMENT

Movement of regulators would be a critical factor in the creation of regulator fields; thus any regulator of development that does not conform to the rules of movement could occur within a field in the wrong place or at the wrong concentration. Of course, such circumstances would be expected to result particularly from a general, external application of a regulator to a plant. In addition, other chemicals act rather specifically on transport processes.

A class of synthetic compounds, with broad effects upon plant development, which may act through an effect upon auxin transport, are certain fluorene derivatives which have been named *morphactins* (Schneider,

1970). The morphactins have only a chemical definition; their biological effects are, so far, too diverse to propound a useful biological definition. However, they do not appear to belong to any of the recognized classes of regulators. The morphactins have two, possibly related properties that may account for their effects on development and these are the inhibition of auxin transport and the disorganization of meristematic zones.

Morphactins, if their effects were indeed due to an inhibition of auxin transport, would illustrate the potential for the disruption of development by an agent that interferes with regulator movement and hence presumably with regulator fields. In meristematic zones, morphactins produce a trend toward randomness of orientation of cell divisions; cell division may increase but,the new cells fail to organize properly.

Naptalam inhibits the transport of auxin and in coleoptile sections, the rate of transport declines within a minute of application. Progress has been made in locating the site at which naptalam attaches to cells and affects auxin transport (Lembi et al. 1971; Thomson et al. 1973). Fractionation of homogenates of coleoptiles that had been treated with labelled naptalam, revealed that the naptalam was chiefly bound to a fraction containing plasma-membrane vesicles. The binding of naptalam was strong but not covalent and the presence of other auxins, a gibberellin, an abscisin, or TIBA (2,3,5-triiodobenzoic acid) did not interfere with the binding. Thus, although naptalam inhibits auxin transport, it does so in a manner other than by competing with auxin for the transport site. However, an auxin-transport site also located in the plasma membrane seems likely from the following evidence. Goldsmith and Ray (1973) and Cande et al. (1973), working with coleoptiles and the chemical cytochalasin B, which inhibits protoplasmic streaming, found that the rate of polar auxin transport was only slightly slower in cells with no streaming. Further, if non-streaming cells were centrifuged so that the cytoplasm was concentrated at either their apical or the basal ends, transport was inhibited in sections with cytoplasm at the apical ends but not in those with cytoplasm at the basal ends. A reasonable conclusion from such experiments is that the metabolic component of auxin transport is a secretion process located in the basal, plasma membranes of transporting cells. In this case, naptalam may be acting in the plasma membrane to inhibit this process, but not at the auxin site. Naptalam itself is not transported.

On the other hand TIBA, which also inhibits auxin transport, is itself transported. It binds to the membrane fraction of cells but can be displaced by auxins. TIBA therefore appears to bind at the auxin transport site and in so doing inhibits transport. Neither naptalam nor TIBA have effects on growth comparable with their effects on transport, so Thomson et al. (1973) propose that the transport site is different from the primary site of action for auxin in the induction of coleoptile growth.

D. IMBALANCE IN REGULATOR INTERACTIONS

Some interactions between classes of regulators produce effects on plants that involve disruptions in development which may suggest a mode of herbicidal action. Such a case is the formation of roots on stem cuttings, which is promoted by auxin but is inhibited by both gibberellin and cytokinin treatments (Brian *et al.* 1960; Humphries, 1960). However the inhibiting effects occur through differing mechanisms: the gibberellin treatment appears to inhibit cell division and divert developmental activity toward cell expansion, while cytokinin activity changes the cytokinin-auxin ratio from that appropriate for the differentiation of root primordia to that inducing callus formation. Thus change in the concentrations of the classes of regulators that can influence a single developmental process may cause development to be diverted from the pattern necessary for survival.

Plants have the potential for regulating processes that lead toward the death of tissues, as in senescing and abscising organs and the differentiation of xylem. In most young growing tissues such processes must be under constraint, but there may be combinations of regulators resulting from herbicide treatments that can release these constraints and cause the existing developmental pattern to be replaced by one directed toward tissue death.

E. COMPETITION FOR THE RECOGNITIVE SITE

Some regulator antagonists have chemical structures that relate sufficiently closely to the presumed active portions of regulator molecules to suggest that the antagonism arises from competition at the primary recognitive site. Antiauxins, such as 2,4-dichlorophenoxyisobutyric acid (McRae and Bonner, 1953), are in this category as are some benzylurea derivatives (Kefford *et al.* 1968) and purine derivatives (Hecht *et al.* 1971) that antagonize cytokinin action. Herbicidal properties have not been demonstrated for regulator antagonists, but their abilities to counteract regulator effects provide such a potential. In addition, antiauxins, in the presence of high auxin concentrations, increase the auxin inhibition rather than counteracting it, and in the case of *trans*-cinnamic acid (van Overbeek *et al.* 1951), the antiauxin was found to become toxic in the presence of high auxin concentrations. This phenomenon is as predicted by a two-point attachment theory of the reaction of auxins with the recognitive site and the competition between auxins and antiauxins at the site (Foster *et al.* 1955). A further antagonistic effect with herbicidal potential is demonstrated by synthetic auxins antagonizing cell expansion induced by the native auxin IAA (McRae *et al.* 1953).

F. REGULATOR FIELDS AND THE GENOME—POSSIBILITIES FOR ABNORMALITIES

The holistic approach to plant development adopted in this chapter should generate broader possibilities for interactions between the cytoplasm and regulators than a molecular or cellular approach. Indeed some insights into a comprehensive control system might be expected, and it is possible to express the interrelationships described earlier for regulators, the cytoplasm, and the genome, in terms of a closed-loop control system for development. This has been done in simple terms in Fig. 1 for a control system that might maintain a particular pattern of development.

In terms of the disruption of development, interest is in circumstances under which feedback ceases to control or induces conditions which tend to disrupt rather than regulate the whole system. For instance, under conditions of positive feedback, increasing the output of a feedback system accelerates its rate. Some phenomena regulated by ethylene act in this way; for example, in an orchid flower following pollination (Burg and Dijkman, 1967), ethylene produced in one locality diffuses into adjacent cells which are induced to senesce and produce further ethylene, and so on until the senescence and fading of the flower is complete. Thus ethylene, initially produced as a consequence of pollination, may be said to feed back positively so as to accelerate the rate of ethylene production and of flower death.

Control systems may go awry when they are overloaded, as may be the case following treatment with a large dose of a regulator or one which is immune from concentration-control systems in the cytoplasm. Yet another potential for miscontrol inherent in control systems is the condition of ever-increasing deviations from the equilibrium state. If an exogenous regulator is imposed upon the control system illustrated in Fig. 1, it could initiate abnormal development. This abnormal development would produce unnatural regulator fields which, when fed back to the receptor portion of the control system, could further corrupt the developmental pattern. In this way development could become progressively more divergent from normal.

It may be possible for the regulator fields occurring in a tissue to become so divergent and unnatural that they cannot be translated through the genome into patterned development. Callus formation might be described as growth with cell division but without patterned development. Callus formation can be induced or its rate of growth can be increased by the treatment of tissues with exogenous auxin, cytokinin, gibberellin, or ethylene. Also the concentration relations of combinations of these classes of regulators and of abscisin can regulate callus growth. So regulators may be able to create conditions for unpatterned development and for other abnormal demands on the genome.

IV. Herbicide Effects on Development Other than through Regulators

The herbicide maleic hydrazide (MH) and those of the carbamate class affect development, but appear to do so at steps not involving regulators.

A. MALEIC HYDRAZIDE—PROBING FOR A MECHANISM OF TOXICITY

Traditionally MH is classified as a herbicide which acts by inhibiting development, but investigations into its mechanisms of action are still probing for definitive leads. Some leads that have been explored are as follows.

MH was proposed as an antiauxin, i.e. a competitive inhibitor of auxin-induced growth, because MH-inhibition of auxin-induced coleoptile growth was counteracted by increasing the auxin concentration, as is required for competitive inhibition (Leopold and Klein, 1952). McRae and Bonner (1953) however, concluded that MH was disproportionately inhibitory of coleoptile growth at low auxin concentrations to be a competitive inhibitor. Another finding contrary to the antiauxin hypothesis was that MH did not inhibit auxin-induced cell expansion in all tissues. Also, in tissues such as corn roots, it inhibited growth by decreasing the number of cells which may grow, rather than inhibiting cell expansion directly (Noodén, 1969). Thus the hypotheses that MH acts generally as an inhibitor of cell expansion or as an antiauxin may be dismissed.

The possibility that MH and gibberellin (GA) affect growth through the same essential process has also been investigated. Brian and Hemming (1957) were impressed by the opposite effects of MH (short internodes and dark-green leaves) and GA (elongated internodes and light-green leaves) on plants and investigated their interrelationship. From an analysis of the effects of MH and GA on dwarf and tall peas, they concluded that GA could not restore MH-induced inhibition of main-axis growth; indeed, for a tall variety, GA did not increase the growth of MH-inhibited plants at all. Brian and Hemming proposed that GA and MH do not act competitively, but that MH appears to inhibit growth at a step prior to the GA-regulated step and thus prevents GA-sensitive plants from responding to GA. In another line of investigation, Sachs and Lang (1961) compared the effects of GA and MH on cell division in shoot apices. GA specifically stimulated cell division in the subapical zones of stem apices while MH was inhibitory through all zones including leaf primordia, and GA did not reverse MH-induced inhibition. This is a further reason for dissociating the effects of MH and GA on development.

Inhibition of cell division is a relatively general effect of MH and has been analyzed at a variety of levels. Evans and Scott (1964), in a study on *Vicia faba* L. roots, found that MH retarded division in cells in which DNA synthesis was occurring at the time of treatment; so they favored an influence of MH on chromosome duplication. Cytological studies of MH-treated plants revealed that chromatid aberrations accompanied cell division inhibition, and the fact that the aberrations were of the chromatid type supported an effect on DNA replication (for a review of the effects on dividing cells see Kihlmann, 1966). MH treatments do not always inhibit cell division. In the regeneration of buds on root sections, some MH concentrations increased the numbers of buds formed on each root section and the effect was additive to that of cytokinin (Kefford and Caso, 1972).

A biochemical explanation of the inhibition of cell division by MH has been offered by Hughes and Spagg (1958). In peas, they found that glutathione accumulated in tissues where cell division was inhibited by MH. They postulated that maleic hydrazide reacted with protein SH groups and inhibited the reduction of protein —S—S— groups by GSH during mitosis.

At the molecular level, speculations on the mechanism of action have centered upon MH being a structural isomer of uracil (Rakitin *et al.* [1971], summarize one school of thought propounding this theory). Butenko and

Baskakov (1960) based an hypothesis, that MH replaces uracil in nucleic acids, on the structural similarities of the molecules, the similarities of developmental effects of MH and thiouracil, and the lack of activity of MH derivatives. These authors found 5×10^{-3} g/l of MH inhibited the growth of carrot-root callus by 50% and the addition of 10^{-2} g/l uracil restored growth to that of the control. Noodén (1969) however was unable to reverse MH inhibition of cell division in roots with uracil or other nucleobases or nucleosides; furthermore inhibition by fluorouracil was reversed by thymidine while MH inhibition was not. In an attempt to gain direct evidence of the incorporation of MH into RNA, Coupland and Peel (1971) fed labelled MH to willow roots. A relatively small amount of the total label extracted from the roots was found in an RNA fraction and the experiment requires repetition under more critical conditions.

None of the avenues of probing for the mechanism of action of MH appear close to being definitive. There is no evidence that MH interacts directly with one of the known classes of plant growth regulators. Based on available evidence, an effect on a process within cell division appears to have most support. In some of the systems studied, the effect on cell division appears specific and this was demonstrated in corn seedlings by Noodén (1969): seedlings from normal seed grew with cell division while seedlings from γ-irradiated seed grew without cell division, and MH inhibited growth only in normal seedlings. Just where in the cell division process MH acts awaits elucidation.

B. CARBAMATES—POSSIBLE INHIBITORS OF MITOSIS

The effects on plants of the carbamate class of herbicides and the progress of research on their mechanism of action, resemble those of maleic hydrazide.

Phenomena demonstrating relations between carbamates and auxin or GA have been observed: chlorpropham inhibits both GA-induced amylase activity in barley half-grains (Devlin and Cunningham, 1970) and auxin-induced growth of soybean hypocotyl sections (James et al. 1970). However there is no evidence of a close relationship between the carbamates and these classes of regulators.

A rather basic effect on metabolism was proposed by Mann et al. (1967). They suggested that carbamates inhibited changes in the metabolic state or, colloquially, they inhibited "the turning on and turning off of genes". Keitt (1967) tested this hypothesis by comparing the effects of carbamates upon systems in which growth was either just being maintained or being induced afresh. Propham up to 10^{-3} M did not inhibit the extension of coleoptile sections, in which growth may be considered to have been suspended by the act of explantation, but to have resumed in the presence of auxin. This result supported the hypothesis that carbamates are active only when the metabolic state is changing. However, for tobacco-stem pith, in which auxin and cytokinin induce cell division and growth afresh, propham inhibits cell division but not cell expansion. To provide support for the hypothesis, both cell division and expansion should have been inhibited because each may involve a metabolic change in the original pith cells. The hypothesis of Mann *et al.* (1967) is not strongly founded.

A conclusion to be derived from Keitt's (1967) experiments is that carbamates are inhibitors of cell division and he cites situations in which this occurs: inhibition of cell division in roots, inhibition of sprouting in potatoes and carrots, and inhibition of wound periderm formation. Indeed, inhibition of mitosis is probably the most general effect of carbamates and it is the effect in which the most definitive lead to a mechansim of action has been obtained. In addition to an overall inhibition of cell division, cells treated with carbamates show blocked metaphases, multinucleate cells, giant nuclei, and an increase in the number of partly contracted nuclei. A cytological study of blood-lily cells treated with propham (Jackson, 1969), showed disrupted mitoses, cessation of chromosome movement, and a loss of birefringence of the spindle, indicating a disordering of the microtubules. Observations on the ultrastructure of these cells (Hepler and Jackson, 1969) focused attention on the arrangement of the microtubules of the spindle. The microtubules appeared individually unchanged by propham treatment and the same was the case for other cytoplasmic components such as mitochondria and plastids. However, the alignment of the microtubules was changed from parallel to radial arrays. In treated cells, chromosomes failed to align at metaphase and instead appeared to form micronuclei around the foci of the radial arrays of microtubules. Therefore carbamates may inhibit growth by disordering mitosis.

V. Conclusion

Foregoing arguments propose various means by which herbicides might disrupt development and the foregoing evidence shows that the classes of herbicides probably differ in the means by which they initiate such disruptions. Nothing, however, can be offered to cover the gap between a disruption of development and the common outcome of herbicidal action, degenerative injury and plant death.

References

Abeles, F. B. (1973). "Ethylene in Plant Biology", Academic Press, New York and London.
Booth, A. (1959). *J. Linn. Soc. (Bot.)* **56,** 166–169.
Booth, A. (1963). *Proc. Easter School agric. Sci. Univ. Nottingham,* **10,** 99–113.
Brian, P. W. (1959). *Biol. Rev.* **34,** 37–84.
Brian, P. W. and Hemming, H. G. (1957). *Ann. appl. Biol.* **45,** 489–497.
Brian, P. W., Hemming, H. G. and Lowe, D. (1960). *Ann. Bot.* **24,** 408–419.
Burg, S. P. and Dijkman, M. J. (1967). *Plant Physiol., Lancaster,* **42,** 1648–1650.
Butenko, R. G. and Baskakov, Y. A. (1960). *Soviet Pl. Physiol.* **7,** 323–329.
Cande, W. Z., Goldsmith, M. H. M. and Ray, P. M. (1973). *Planta,* **111,** 279–296.
Caso, O. H. and Kefford, N. P. (1973). *Weed Res.* **13,** 24–33.
Chadwick, A. V. and Burg, S. P. (1967). *Pl. Physiol., Lancaster,* **42,** 415–420.
Clark, J. E. and Heath, O. V. S. (1962). *J. exp. Bot.* **13,** 227–249.
Coupland, D. and Peel, A. J. (1971). *Physiologia Pl.* **25,** 141–144.

Devlin, R. M. and Cunningham, R. P. (1970). *Weed Res.* **10,** 316–320.
Digby, J. and Wareing, P. F. (1966). *Ann. Bot.* **30,** 607–622.
Dugger, W. M. and Ting, I. P. (1970). *A. Rev. Pl. Physiol.* **21,** 215–234.
Evans, H. J. and Scott, D. (1964). *Genetics,* **49,** 17–38.
Foster, R. J., McRae, D. H. and Bonner, J. (1955). *Pl. Physiol., Lancaster,* **30,** 323–327.
Goldsmith, M. H. M. (1969). *In* "Physiology of Plant Growth and Development" (M. B. Wilkins, ed.), pp. 127–162 McGraw-Hill, London.
Goldsmith, M. H. M. and Ray, P. M. (1973). *Planta,* **111,** 297–314.
Hall, R. H. (1973). *A. Rev. Pl. Physiol.* **24,** 415–444.
Hecht, S. M., Bock, R. M., Schmitz, R. Y., Skoog, F. and Leonard, N. J. (1971). *Proc. natn. Acad. Sci., USA,* **68,** 2608–2610.
Hepler, P. K. and Jackson, W. T. (1969). *J. Cell Sci.* **5,** 727–743.
Hughes, C. and Spagg, S. P. (1958). *Biochem. J.* **7,** 205–212.
Humphries, E. C. (1960). *Physiologia Pl.* **13,** 659–663.
Jackson, W. T. (1969). *J. Cell Sci.* **5,** 745–755.
James, C. S., Prendeville, G. N., Warren, G. F. and Schreiber, M. M. (1970). *Weed Sci.* **18,** 137–139.
Kang, B. G., Newcomb, W. and Burg, S. P. (1971). *Pl. Physiol., Lancaster,* **47,** 504–509.
Kefford, N. P. and Caso, O. H. (1972). *Aust. J. biol. Sci.* **25,** 691–706.
Kefford, N. P., Zwar, J. A. and Bruce, M. I. (1968). *In* "Biochemistry and Physiology of Plant Growth Substances" (F. Wightman and G. Setterfield, eds.), pp. 61–69. The Runge Press, Ltd., Ottawa.
Keitt, G. W., Jr. (1967). *Physiologia Pl.* **20,** 1076–1082.
Key, J. L. (1969). *A. Rev. Pl. Physiol.* **20,** 449–474.
Kihlman, B. A. (1966). "Actions of Chemicals on Dividing Cells", Prentice Hall, Inc., New York.
Lembi, C. A., Morré, D. J., Thomson, K. S. and Hertel, R. (1971). *Planta,* **99,** 37–45.
Leopold, A. C. and Klein, W. H. (1952). *Physiologia Pl.* **5,** 91–99.
Letham, D. S. (1963). *N.Z. J. Bot.* **1,** 336–350.
Lyons, J. M. (1973). *A. Rev. Pl. Physiol.* **24,** 445–466.
Mann, J. C., Cota-Robles, E., Yung, K. H., Pu, M. and Haid, H. (1967). *Biochim. biophys. Acta,* **138,** 133–139.
McRae, D. H. and Bonner, J. (1953). *Physiologia Pl.* **6,** 485–510.
McRae, D. H., Foster, R. J. and Bonner, J. (1953). *Pl. Physiol., Lancaster,* **28,** 343–355.
Milborrow, B. V. (1966). *Planta,* **70,** 155–171.
Nicholls, P. B. and May, L. G. (1964). *Aust. J. biol. Sci.* **17,** 619–630.
Noodén, L. D. (1969). *Physiologia Pl.* **22,** 260–270.
Patau, K., Das, N. K. and Skoog, F. (1957). *Physiologia Pl.* **10,** 949–966.
Rakitin, Y. V., Povolotskaya, K. L., Geiden, T. M., Garaeva, K. G., Khovanskaya, I. V. and Kalibernaya, Z. V. (1971). *Soviet Pl. Physiol.* **18,** 514–518.
Reed, D. J., Moore, T. C. and Anderson, J. D. (1965). *Science, N.Y.* **148,** 1469–1471.
Robbins, W. J. (1960). *Am. J. Bot.* **47,** 485–491.
Sachs, R. M. and Lang, A. (1961). *In* "Plant Growth Regulation", Proc. 4th Internat. Conf. on Plant Growth Regulation, pp. 567–578. Iowa State Univ. Press, Ames, Iowa.
Schneider, G. (1970). *A. Rev. Pl. Physiol.* **21,** 499–536.
Skoog, F. and Miller, C. O. (1956). *Symp. Soc. exp. Biol.* **11,** 118–131.
Thimann, K. V. (1969). *In* "Physiology of Plant Growth and Development" (M. B. Wilkins, ed.), pp. 3–45. McGraw-Hill, London.
Thomson, K. S., Hertel, R., Müller, S. and Tavares, J. E. (1973). *Planta,* **109,** 337–352.
van Overbeek, J., Blondeau, R. and Horne, V. (1951). *Am. J. Bot.* **38,** 589–595.
Wardlaw C. W. (1965). "Organization and Evolution in Plants", Longmans, Green and Co., Ltd., London.
Woolhouse, H. W. (ed.) (1967). *Symp. Soc. exp. Biol.* **21,** 179–213.
Zenk, M. H. (1962). *Planta,* **58,** 75–94.

CHAPTER 15

ACTION ON RESPIRATION AND INTERMEDIARY METABOLISM

R. C. KIRKWOOD

*Department of Biology, University of Strathclyde,
Glasgow, Scotland*

Sabotage of the respiratory metabolism is an obvious method by which to kill or injure unwanted plants, since effects on the energy-releasing processes must inevitably affect a variety of energy-dependent functions such as protein and lipid synthesis, mineral absorption, short-distance and long-distance transport of assimilates etc. It is intended in this chapter to review the literature concerning the effect of the major herbicide groups on respiratory processes including glycolysis, the pentose phosphate pathway (PPP), the tricarboxylic acid cycle (TCA cycle), oxidative phosphorylation, the electron transport chain, terminal oxidases etc. The effects of herbicide action on enzyme systems concerned with intermediary metabolism, including lipid metabolism and general protein poisons will also be included.

I. THE RESPIRATORY PROCESS

The steps involved in respiratory metabolism are well known. Respiration of starch or sugars can be divided into three stages. First the sugars are broken down into the three carbon acid, pyruvic acid, by a series of stages known as the Embden, Meyerhoff, Parnas pathway (EMP pathway) (Fig. 1); this process (glycolysis) is believed to occur in the cytoplasm, as opposed to the mitochondria, and can take place in the absence of oxygen. The pyruvic acid formed is subsequently broken down into CO_2 in a number of steps involving a series of organic acids (Fig. 2); oxygen is required for this phase which takes place on the surface of the inner membrane (cristae) of mitochondria. These steps make up what is alternatively called the Krebs cycle, citric acid cycle or tricarboxylic acid (TCA) cycle.

Closely allied with this second stage is a third in which electrons and hydrogen atoms are removed from the organic acids and after a series of oxidation-reduction steps, the H atoms unite with oxygen to form water (Fig. 3). The transference of electrons takes place from compounds with low reduction potentials to those with higher reduction potentials; this pathway is termed the electron transport system, the cytochrome system or the respiratory chain. It is in this final phase that energy released in respiration is trapped and stored as adenosine triphosphate (ATP) (oxidative phosphorylation), a chemical "energy store". All oxidative phosphorylation reactions can be described by the reversible reaction:

$$AH_2 + B + ADP + Pi \rightleftharpoons A + BH_2 + ATP \qquad \text{Eq. 1}$$

by which inorganic phosphate (Pi) combines with adenosine diphosphate (ADP) to form ATP.

Two types of oxidative phosphorylation exist (a) substrate-linked where AH_2 is one of the substrates, phosphoglyceraldehyde, pyruvate or α-ketoglutarate and (b) respiratory chain where AH_2 and B are both members of the respiratory chain (Slater, 1963). In the latter case, the energy required to form ATP comes from the drop in electrochemical potential as electrons are transferred from one compound to another in the electron transport system. The mechanism by which ATP formation is coupled to electron transfer has

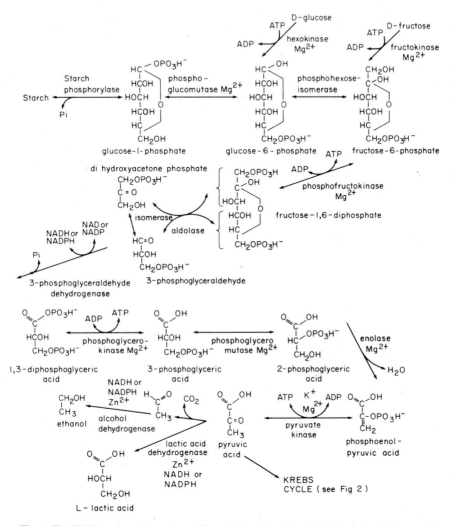

Fig. 1. The EMP or glycolytic pathway (from "Plant Physiology" by Frank B. Salisbury and Cleon Ross © 1969 by Wadsworth Publishing Company, Inc., Belmont, California 94002. Reprinted by permission of the publisher).

still to be clarified. It has been suggested that high-energy phosphorylated proteins ("activated intermediates") are formed upon electron transport and these transfer phosphate to ADP (the so-called chemical coupling theory of Chance and Williams, 1956). An alternative hypothesis, which is also speculative but receiving increasing support has been suggested (Mitchell, 1961, 1966a, b; Mitchell and Moyle, 1967) and assumes that a reversible

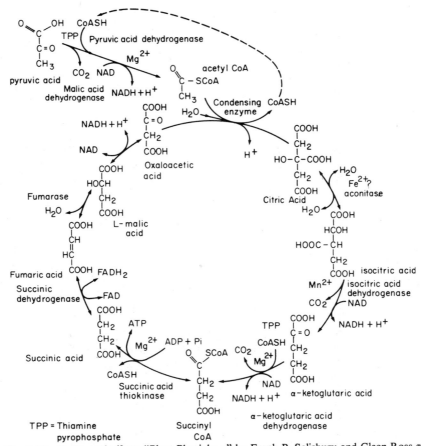

Fig. 2. The Krebs cycle (from "Plant Physiology" by Frank B. Salisbury and Cleon Ross © 1969 by Wadsworth Publishing Company, Inc., Belmont, California 94002. Reprinted by permission of the publisher).

ATPase is present in the mitochondrial membrane such that protons and hydroxyl ions generated by ATP formation are discharged. If the protons are directed toward that side on which O_2 reduction occurs and the hydroxyl ions are directed towards the focus of proton formation by dehydrogenation of substrates, reformation of water would constitute a driving force for ATP formation.

Irrespective of the mechanisms involved, it is known that for each electron pair which traverse the electron transport chain via NAD or NADH, three ATP molecules will be produced; if succinate is the source, only two molecules of ATP are synthesized per atom of oxygen absorbed. This indicates that ATP synthesis occurs between NADH and FAD (Fig. 3), possibly between cytochrome *b* and *c* or CoQ and cytochrome *b*, and also between cytochrome *c* and cytochrome oxidase. Much of the knowledge about the electron transport sequence has been learned from studies of isolated mitochondria and the action on them of uncouplers and inhibitors of oxidative phosphorylation (see reviews by Lieberman and Baker, 1965; Ikuma, 1972).

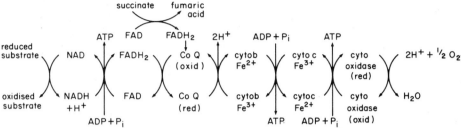

Fig. 3. The electron transport system (from "Plant Physiology" by Frank B. Salisbury and Cleon Ross © 1969 by Wadsworth Publishing Company, Inc., Belmont, California 94002. Reprinted by permission of the publisher).

Much of the work on the action of herbicides on respiratory metabolism has been carried out with an oscillating platinum electrode to measure O_2 uptake by isolated mitochondria. Several procedures have been described for the isolation from higher plants of mitochondria which exhibit "tight coupling" of oxidation and phosphorylation (e.g. Palmer, 1967; Ikuma and Bonner, 1967; Sarkissian and Srivastava, 1968; Ku *et al.* 1968; Haard and Hultin, 1968; Romani *et al.* 1969). The functional integrity of isolated mitochondria can be determined from their respiratory control (RC) and phosphorylation/oxidation (P/O) ratios. The RC is the ratio of the rate of respiration in the presence of ADP to the rate in the absence of ADP (Chance and Williams, 1955, 1956). Intact and functionally active mitochondria usually have high RC and P/O ratios and are termed "highly-" or "tightly-coupled", while mitochondria with low phosphorylation capacities and RC ratios are termed "loosely-coupled".

A modification of the isolation procedures of Sarkissian and Srivastava (1968) and Palmer (1967), produced highly active mitochondria from root and epicotyl tissue of *Vicia faba* (Matlib *et al.* 1971*b*) and was used to study the effect of certain phenoxy-acid herbicides on oxidative phosphorylation (Matlib *et al.* 1972). Examples of typical oxygraph traces (Fig. 4) recorded by Matlib *et al.* (1971*b*) shows the pattern of O_2 uptake by broad bean mitochondria in the presence of a variety of substrates. These traces indicate that endogenous respiration (state 1) was increased on the addition of substrates (state 4). An immediate acceleration in respiration was observed when ADP was added (state 3) and this high rate continued until the ADP supply was exhausted (state 4). This state-4/state-3/state-4 cycle, which is characteristic of tightly-coupled mitochondria, was repeated several times with successive 100–150 μM additions of ADP until all the dissolved oxygen in the reaction medium was exhausted. The calculation of true RC and P/O values requires a sharp transition from state 3 to state 4 at the point of ATP exhaustion. Lack of a sharp transition probably indicates that the ATP synthesized is being recycled

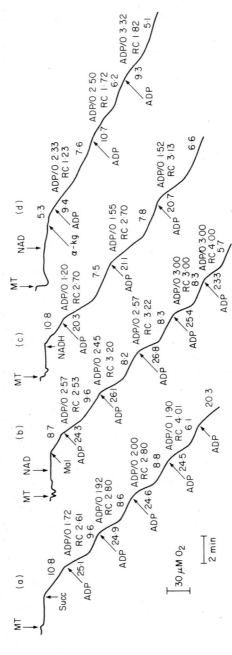

Fig. 4. Oxygraph traces showing the uptake of O_2 by bean (*Vicia faba*) mitochondria using a variety of substrates (after Matlib *et al.* 1971*b*. Reproduced by permission).

to ADP and Pi as a result of ATPase activity as does the decrease in RC values on successive additions of ADP (Wiskitch, 1966).

The use of bovine serum albumin (BSA) for the isolation of tightly-coupled mitochondria from a variety of sources, especially those which contain high concentrations of fatty acids and phenolic compounds, is well documented (Lance *et al.* 1965; Verleur, 1965; Ku *et al.* 1968; Ikuma and Bonner, 1967; Sarkissian and Srivastava, 1968; Palmer, 1967; Watson and Smith, 1967; Stinson and Spencer, 1968). Weinbach and Garbus (1966) have demonstrated that BSA removes endogenous fatty acids and phenolic compounds by binding. However, its incorporation into isolated mitochondrial systems used to test phenolic herbicides may bring problems due to binding of a proportion of the applied herbicide dose. This aspect will be discussed later.

II. UNCOUPLERS AND INHIBITORS OF OXIDATIVE PHOSPHORYLATION

Chemicals which interfere with the mitochondrial system can be classified as uncoupling agents, energy-transfer inhibitors or electron transport inhibitors. There may be considerable overlap in the action of these compounds. They may exhibit an uncoupling action at low concentrations and affect electron transport at relatively high concentrations (e.g. Hemker, 1964; Dam, 1967; Bottrill and Hanson, 1969). Further, their action may be dependent on the pH of the medium, the source of the mitochondria and other factors.

The properties of uncouplers have been described in reviews by Slater (1963) and Matlib (1970). Characteristically they completely abolish the synthesis of ATP coupled with mitochondrial respiration; at low concentrations they stimulate the hydrolysis of ATP added to mitochondria and inhibit it at higher concentrations (Slater, 1962; Bottrill and Hanson, 1969). Uncouplers also inhibit the Pi-ATP exchange reactions (Eq. 1) and the exchange of an oxygen atom between Pi and water and also ATP and water, catalysed by mitochondria in the absence of added substrate. They release the inhibition of respiration exerted by oligomycin (see below) and induce the swelling of mitochondria (Bottrill and Hanson, 1969). Classical uncouplers include the herbicide groups—the nitrophenols and halophenols, dicoumarol, antibiotics such as gramicidin and arsenate; all of these groups act at the three sites of phosphorylation (Fig. 5) except arsenate which additionally affects substrate level oxidative phosphorylation linked to α-ketoglutarate (Sanadi *et al.* 1954). A recent paper by Bertagnolli and Hanson (1973) reports on the functioning of the adenosine nucleotide transporter in the arsenate uncoupling of corn mitochondria.

In addition to the action of uncouplers, oxidative phosphorylation may be inhibited due to the effect of compounds which either inhibit the transfer of energy to intermediates in the formation of ATP or inhibit the flow of electrons along the electron-transport chain. The features of both types of inhibitor have been reviewed by Matlib (1970). Energy-transfer inhibitors inhibit respiration where oxidation is coupled to phosphorylation; this inhibition is released by uncouplers such as DNP, but not by arsenate. They completely inhibit DNP-stimulated mitochondrial ATPase activity and Pi⇌ATP exchange reactions. Classical inhibitors of energy-transfer include oligomycin, alkylguanidine and phenethylbiguanide (PBI) and synthalin (Slater, 1966); their sites of action are shown in Fig. 5.

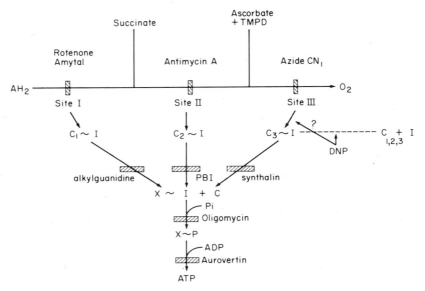

Fig. 5. A schematic representation showing the sites of action of classical uncouplers of oxidative phosphorylation (reproduced with permission from Matlib, 1970).

Contrary to the situation with energy-transfer inhibitors, electron transport inhibitors suppress respiration even when phosphorylation is not linked to oxidation; the inhibition is not released by classical uncouplers, though electron donors or acceptors may create a by-pass of the site of inhibition, releasing the inhibition. They may interfere with phosphorylation only at very high concentrations and under certain conditions. Classical electron-transport inhibitors include cyanide, azide, antimycin A, malonate, rotenone and amytal and their sites of action are demonstrated in Fig. 5.

Using these inhibitors it is possible to elucidate the pathways of substrate oxidation by isolated mitochondria (e.g. Day and Wiskitch, 1974). In any investigation of the action of a herbicide compound on the respiratory process, it would seem desirable to include selected examples of these classical uncouplers and inhibitors in order to check the efficiency of mitochondrial coupling in the system being used and also as comparative checks to help elucidate the action of the herbicide.

III. Herbicide Uncouplers

A. SUBSTITUTED PHENOLS

Many herbicides are active as uncouplers of oxidative phosphorylation, the most notable groups including the substituted phenols, hydroxybenzonitriles and substituted alkanoic acids. Most work has been carried out on the substituted phenols; indeed DNP is the classical uncoupler which has been extensively used in mitochondrial respiration studies. Further, the substituted phenols were the first organic chemicals to be used as herbicides (Hartley and West, 1969; Schroeder and Warren, 1971). There is now considerable evidence that phenolic herbicides act as uncoupling agents. The stimulation of respiration by nitrophenols has been reported by several earlier workers (Ehrenfest and Ronzoni, 1933; Dodds and Grenville, 1934; Ronzoni and Ehrenfest, 1936). Later studies led to the generalization that nitrophenols stimulate intercellular respiration, resulting in the inhibition of energy-requiring functions, e.g. cell division in sea urchins eggs (Clowes and Krahl, 1936), growth of yeast (Martin and Field, 1934), assimilation in micro-organisms (Pickett and Clifton, 1943) and phosphate uptake by yeast (Hotchkiss, 1944). Bonner (1949) reported that growth of *Avena* coleoptiles was inhibited by 5 and 10 mg/l DNP whereas respiration was increased (19%) or reduced (15%) respectively by these concentrations. He attributes this effect to the uncoupling of respiratory oxidation from phosphorylation.

It is a characteristic of many uncoupling compounds that they uncouple phosphorylation at low concentrations and inhibit at higher levels. In the case of DNP, Foy and Penner (1965) reported concentration-dependent responses of O_2 uptake by plant mitochondria with succinate as substrate. Bottrill and Hanson (1969), using corn root mitochondria, studied the action of a wide range of concentrations of DNP on the partial reactions of oxidative phosphorylation and found a similar uncoupling/inhibition relationship with concentration. At a concentration of 1×10^{-4} M, DNP caused classic uncoupling of oxidative phosphorylation; higher concentrations inhibited phosphorylation and oxidation, the latter being due to an inhibition of malate dehydrogenase. Wilson and Merz (1967) also reported that the inhibition of respiration by nitrophenols was by an inhibition at the substrate level. Bottrill and Hanson (1969) found a second peak of stimulation of ATPase activity at 1×10^{-3}M which coincided with a stimulation of mitochondrial swelling. A similar peak was also reported for DNP (and 2,4-DB, 2,4,5-T & 2,4,5-TB) by Matlib *et al.* 1972 (see later). Increased swelling at 5×10^{-3} M DNP resulted in disrupted mitochondrial structure and concomitant inhibition of enzyme functions; ATPase activity was inhibited and the mitochondria were unable to contract due to their inability to utilize an energy source. Bottrill and Hanson (1969) concluded that the action of DNP on mitochondria is not a simple enzyme-substrate-inhibitor interaction such as was suggested by Slater (1962) or Hemker (1964) but involves a number of sites in the overall mitochondrial function.

A similar response has been found for other substituted phenols. For example, Gaur and Beevers (1959) studied the respiratory and other responses of carrot tissues to a series of nitro-, chloro- and bromo-substituted phenols as well as phenol, methylene blue (MB) and sodium azide. They found that as the concentration of each compound was increased over a narrow range, respiration was stimulated up to a maximum after which it declined sharply. The concentrations which produced maximum respiration markedly reduced glucose uptake. Since the test compounds stimulated O_2 uptake by castor bean mitochondria in a manner similar to that of DNP, it was assumed that they were acting as uncoupling agents with varying degrees of efficiency. Gaur and Beevers (1959) found that di-substitution, either nitro or chloro, was more effective than mono-substitution in inducing uncoupling as measured by O_2 or sugar uptakes. The order of effectiveness among the substituted phenols tested was di > para > meta > ortho. Wojtaszek et al. (1966) found that dinoseb both inhibited the incorporation of ^{32}P into ATP by tomato leaf discs and stimulated O_2 utilization. The site and mechanism of action of nitrophenols has received considerable attention; indeed knowledge of the mechanism of action of uncouplers results largely from work with DNP. Slater (1963) has reviewed the mechanisms of action of uncouplers; at least four have been proposed. Ernster and his co-workers (e.g. Ernster and Lee, 1964) favour the suggestion that DNP acts by allowing the oxidation reaction to proceed without concomitant phosphorylation. Most workers favour the view that DNP promotes the hydrolysis of an intermediate non-phosphorylated high-energy compound (Lardy and Wellman, 1952; Loomis and Lipmann, 1948; Borst and Slater, 1961; Stoner et al. 1964). Pinchot (1963, 1967) presented data supporting a concept previously put forward by Racker (1961) who believed that DNP somehow dissociates the phosphorylation process from electron transport. According to this view, DNP does not accelerate the hydrolysis of intermediates nor interfere with intermediate formation; it does, however, prevent the association of the coupling enzyme with coupling particles, thus interfering with further formation of an enzyme-particle complex.

A further mechanism, based on mitochondrial cation and anion exchange, has been proposed by Lynn and Brown (1966). They suggest that uncoupling agents which alter intra-mitochondrial concentration of various divalent anions, as well as protons, can influence the efficiency of phosphorylation by activating exchange between H^+ and external cations and anions. More recently Weinbach and Garbus (1969) proposed a mechanism whereby the uncoupling agents in the unionized state, traverse the lipid envelope or the lipid-protein interface of the mitochondrial membrane. Within the mitochondrion, the reagent ionizes and interacts with the charged group of the protein resulting in the re-organization of the protein structure. These conformational transitions probably result in a structural disorganization and altered function of the enzymes catalysing the coupling of phosphorylation to electron transport.

It is worth considering the classical work of Weinbach and Garbus in more detail. Much of their investigation was carried out with pentachlorophenol (PCP) which is a powerful uncoupler of oxidative phosphorylation (e.g. Weinbach, 1956a, b). There is evidence that PCP interacts (bonds) with intact rat liver mitochondria and mitochondrial preparation free of lipid and water-soluble components (Weinbach and Garbus, 1964, 1965). This type of bonding is apparently not readily broken since washing procedures, freezing, thawing and sonication will not completely release the bound PCP. However, treatment of the protein complex with sodium hydroxide (0·1 N NaOH) will effectively free the PCP. Weinbach and Garbus (1965) found that BSA restored the capacity for oxidative phosphorylation to isolated rat liver mitochondria, uncoupled by various substituted phenols. In addition, BSA, together with ATP, largely restored the morphological integrity of mitochondria, swollen by PCP. They suggested that BSA exerts these beneficial effects by tightly binding the uncoupling phenol thus removing it from the mitochondria. Previously Giglio and Moura Goncalves (1963) had recorded evidence of an interaction between BSA and PCP and proposed an anion-anion type of interaction. Weinbach and Garbus (1964) believe that mitochondrial protein may possess a similar capacity to bind lipophilic reagents. This suggestion was substantiated by the following evidence (1) chromatography of extracts of mitochondria containing bound PCP or DNP on Sephadex G-25 revealed that these compounds migrated with the protein-containing moiety and not with the released lipids; (2) treatment of intact mitochondria with reagents such as 8M urea, which change protein conformation, released PCP from its mitochondrial binding site; (3) the time required for heat coagulation of mitochondria containing PCP was increased compared to untreated mitochondria; and (4) there was evidence that mitochondria, depleted of their lipid content to varying degrees, bound PCP to the same or to a greater degree than intact mitochondria. Subsequently protein, devoid of lipids and water-soluble components, was prepared from isolated rat-liver mitochondria and tested for its capacity to interact with uncoupling phenols. This protein was comparable in this respect to intact mitochondria (Table I).

The results show that the mitochondrial protein has an affinity for

TABLE I

BINDING OF UNCOUPLING PHENOLS TO INTACT MITOCHONDRIA AND TO MITOCHONDRIAL PROTEIN (after Weinbach and Garbus, 1964, 1965)

Phenol	Phenol bound %	
	Intact mitochondria	Mitochondrial protein
Pentachlorophenol	53	70
2,3,5-Trichlorophenol	44	48
2,4-Dichlorophenol	25	38
2,4-Dinitrophenol	7	7

representative halo- and nitro-phenols equal to or greater than that of intact mitochondria. Further experiments revealed that the degree of binding of these phenols was similarly influenced by pH, concentration of phenols, concentrations of mitochondria or mitochondrial protein. In the case of a homologous series of halophenols, the absolute amount bound to isolated mitochondrial protein was related to the efficacy of these compounds as uncouplers of oxidative phosphorylation with intact mitochondria; the amount bound was also related to the ease with which phenols were removed from intact mitochondria by simple washing procedures. There was also evidence from difference spectrophotometry that phenols which uncouple interact with the protein moiety of intact mitochondria. Weinbach and Garbus (1965) conclude that it seems quite possible that the biochemical effect of phenols which uncouple may be a consequence of such protein/phenol interaction. They suggest that these phenols interact with mitochondrial proteins participating in oxidative phosphorylation at sites other than the active centres, inducing configurational changes analogous to the allosteric effects proposed by Monod et al. (1963). These configurational changes resulting in modified enzymes, cause a loss in coupling activity and other latent enzymatic activities may be activated. These changes may or may not be reversible depending upon the affinity of the protein for the uncoupling reagent. Weinbach and Garbus (1965) do not completely discount the importance of lipid solubility, a factor considered to be important by Hemker (1962), pointing out that uncoupling reagents must traverse a lipid barrier or a lipid-protein interface in order to reach the protein sites where they are bound.

The cause of difference in whole-plant susceptibility to substituted phenols may be due to a variety of factors including differentials in retention, cuticle penetration, absorption or fundamental differences in susceptibility of the enzyme systems involved in oxidative phosphorylation. Wojtaszek (1966) studied the relationship between susceptibility of certain plant species to dinoseb and their capacity to accumulate externally-applied ^{32}P to form ATP in leaf discs. Twenty-four weed and crop species were sprayed with doses equivalent to 1/16, 1/8 and 1/4 kg/ha of dinoseb in 100 gal (450 l). A correlation was found between species susceptibility to dinoseb and their capacity for ^{32}P accumulation and ATP synthesis. Considerably greater amounts of ^{32}P were accumulated in leaf discs which were highly resistant to dinoseb than in those which were highly susceptible. In light conditions the correlation was 0·91, and in dark 0·70. Wojtaszek postulated that susceptibility to dinoseb depended upon the level of ATP in tissues; this, in turn, depended either on its formation or storage in the plant. In dark conditions, the amount of ATP decreased either because it was used in normal metabolism or because synthesis is lower in the dark than light, because ATP is produced only by oxidative phosphorylation.

B. BENZIMIDAZOLES

In his review of the mechanisms of action of herbicides, Moreland (1967) points out that some of the 2-trifluoromethylbenzimidazoles were found to be more effective as uncouplers than DNP in studies using rat mitochondria (e.g. Jones and Watson, 1965). These derivatives stimulated ADP-limited respiration, released the inhibition of O_2 uptake imposed by oligomycin and inhibited reversed electron transport; they also stimulated rat mitochondrial ATPase activity. There was evidence that the most phytotoxic derivatives were the most effective uncouplers of oxidative phosphorylation (Burton *et al.* 1965) and that their uncoupling activity and phytotoxicity were correlated with their pKa values (Jones and Watson, 1965).

C. NITRILES

The discovery that ioxynil had outstanding herbicidal activity was first reported by Wain (1963). Bromoxynil was also found to be an effective herbicide but was much less active. Similar findings were later made independently by Carpenter and Heywood (1963) and by Amchem Products (1963). Kerr and Wain (1964) later showed that whilst the *p*-hydroxybenzonitrile itself is not very effective in uncoupling oxidative phosphorylation, substitution of an iodine atom into the 3-position or dihalogenation in the 3,5-positions can confer considerable activity in this respect. The relative activity of these halogenated compounds, in terms of uncoupling oxidative phosphorylation, was di-iodo > dibromo > dichloro analogues, coinciding with the herbicidal activity of these compounds. The results showed that ioxynil was comparable with DNC and better than DNP; the herbicidal activity of ioxynil was superior to both of these compounds. The concentrations at which the P/O ratio became 0 were: ioxynil $(4 \times 10^{-5}$ M), DNC $(4 \times 10^{-5}$ M), DNP $(8 \times 10^{-5}$ M), bromoxynil $(9 \times 10^{-5}$ M), 3,5-dichloro derivative $(13\cdot5 \times 10^{-5}$ M). Parker (1965) working on rat-liver mitochondria also ranked the derivatives in the order I > Br > Cl for stimulation of O_2 uptake and inhibition of phosphate uptake. Moreland and Blackmon (1970) found that bromoxynil stimulated ADP-limited O_2 utilization and inhibited non-ADP-limited O_2 uptake, although slightly higher concentrations were required to produce effects similar to ioxynil. Foy and Penner (1965) studied the effect of several herbicides upon the oxidation of TCA substrates by isolated cucumber mitochondria. They found that ioxynil progressively inhibited O_2 uptake over a concentration range from $0\cdot75 \times 10^{-6}$ M to $0\cdot75 \times 10^{-3}$ M; in contrast, dichlobenil increased O_2 consumption to about 150% of the controls at $1\cdot45 \times 10^{-4}$ and $0\cdot73 \times 10^{-4}$ M concentrations. They concluded that while dichlobenil and ioxynil show similarities in their ability to block the TCA cycle, dichlobenil appears to act principally by disrupting the phosphorylation mechanism; ioxynil on the other hand, apparently inhibits the TCA oxidases leading to a reduction rather than increase in O_2 uptake.

There is evidence, however, that low concentrations of ioxynil uncouple oxidative phosphorylation in plant tissue (Paton and Smith, 1965, 1967; Smith *et al.* 1966). Paton and Smith (1967) studied the effect of ioxynil on the component parts of the respiratory electron transport chain using mitochondria isolated from roots of *V. faba*; O_2 uptake was stimulated at low concentrations $(5 \times 10^{-8}$ to 5×10^{-7} M) and inhibited at high values $(5 \times 10^{-6}$

to 5×10^{-3} M). A concentration of 5×10^{-7} M reduced the P/O ratio by about 50%. Ioxynil inhibited the oxidation of both NADP and succinate equally by 50% but had no effect on terminal cytochrome oxidase, or on succinate-PMS reductase; succinate-methylene blue reductase was inhibited by 50%. These results indicate that the site of action lay prior to the point of entry of ascorbate, suggesting that ioxynil was acting at a site between the flavoproteins and cytochrome b (Fig. 6). One of the components believed to be located

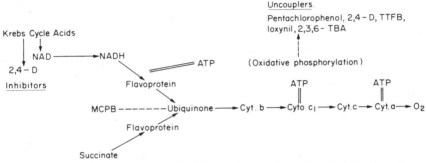

Fig. 6. The pathway of oxidation of cellular products and possible sites of action of some herbicides (from Smith, Paton and Robertson, 1966. Reproduced by permission of the National Research Council of Canada from the *Canadian Journal of Botany*, **45**, 1891–1899 (1967)).

between these two electron carriers is ubiquinone and the behaviour of this compound and its reduced form, ubiquinol, were investigated by Paton and Smith (1965, 1967). They showed that while ubiquinone alone cannot be utilized, in the presence of succinate it was slowly capable of supporting and slightly stimulating the oxidation of succinate. Ubiquinol, however, did act as a substrate and reduced electron flow; a similar reversal of inhibition was found for menadione (vitamin K_3). They concluded that it is likely that the ability to uncouple oxidative phosphorylation at low concentrations and to block electron flow at higher values contributes to the herbicidal activity of ioxynil.

Ferrari and Moreland (1967, 1969) while investigating the respiration of white potato tubers (*Solanum tuberosum* L.) confirmed that ioxynil was acting as an uncoupler since it meets several of the criteria established for uncouplers (Slater, 1963). For example, the respiration of mitochondria suspended in a medium deficient in either phosphate acceptor or phosphate was stimulated and the hydrolysis of added ATP was promoted. Apart from this they also showed that ioxynil circumvents oligomycin-inhibited respiration and slightly stimulates mitochondrial ATPase activity. Like Kerr and Wain (1964) they found that the dibromo and dichloro-substituted analogues of ioxynil stimulated ATP-limited respiration and released oligomycin-inhibited oxygen uptake, the decreasing order of uncoupling activity being I > Br > Cl. The order of stimulation production was reversed (Cl > Br > I); however a difference of only 10% separated ioxynil and its analogues suggesting that halogen substitution might play a minor role.

While investigating the effects of various growth substances on the metabolic uptake of indole-3yl-acetic acid (IAA) by *Zea mays* L. mesocotyl segments, Sabnis and Audus (1967) found that ioxynil and bromoxyil induced marked inhibition which they presumed occurred by preventing oxidative phosphorylation. The degree of inhibition was not dependent on the concentration of IAA.

Dichlobenil is a powerful inhibitor of seed germination and is believed to act on the actively dividing cells of the meristem. There is, however, some diversity of opinion regarding its effect on oxidative phosphorylation. Most authors have concluded that dichlobenil is not an uncoupler of oxidative phosphorylation *per se*; however, uncoupling effects may occur in plants which can degrade the herbicide into its hydroxy derivatives (Verloop, 1972). For example, uncoupling activity of dichlobenil was not observed by Wit and van Genderen (1966) in isolated rat liver mitochondria, in yeast cell suspension, or in spinach mitochondria (Brandon, unpublished data, *ex* Verloop, 1972).

There was evidence that the 3-hydroxy and 4-hydroxy derivatives are uncouplers of oxidative phosphorylation with an activity comparable to DNP. Dichlobenil had no effect on the respiration of tobacco leaves (Koopman and Daams, 1960), the green algae *Chlorella* or *Scenedesmus* (Daams and Barnsley, 1961) or detached dormant purple nut sedge tubers (Hardcastle and Wilkinson, 1968). On the other hand, it has been reported that dichlobenil increased the respiration of red beetroot sections, corn roots and cucumber seedlings (Price, 1969) and six-week-old purple nut sedge (Akobundu *et al.* 1970). Again, Foy and Penner (1965) using isolated cucumber mitochondria, with succinate and α-ketoglutarate as substrates, found that $0 \cdot 73 \times 10^{-4}$ M and $1 \cdot 45 \times 10^{-4}$ M dichlobenil increased O_2 uptake; they suggested that part of the effect of dichlobenil on proteolytic activity in squash cotyledons results from its effect on ATP synthesis. Unfortunately, the rates of hydroxylation in many of these species are not known and it is not possible to compare the effects on respiratory metabolism with hydroxylation (Verloop, 1972).

There is some evidence that these degeneration products of dichlobenil have herbicidal properties (Wain, 1963; Verloop and Nimmo, 1969), but it is thought to be slight in comparison to that caused by dichlobenil (Verloop, 1972).

D. PHENOXY-ACIDS

These compounds have been extensively used as herbicides since 1945 and while their mode of action is incompletely understood, there is considerable evidence that they act as uncouplers and inhibitors of oxidative phosphorylation. It was Brody (1952) using rat liver mitochondria, who first reported that 2,4-D uncoupled phosphorylation associated with pyruvate oxidation. He found that 2,4-D stimulated respiration with a concomitant lowering of P/O values in a phosphate-deficient medium; this did not occur in a medium containing phosphate. Similar results were obtained with mitochondria isolated from lupin (Dow, 1952) and etiolated soybeans (Switzer, 1957), cabbage

(Lotlikar, 1960), cauliflower and red beet (Wedding and Black, 1962) and maize (Bottrill, 1965). Stenlid and Saddik (1962) demonstrated uncoupling in cucumber (*Cucumis sativus* L.) mitochondria with D and L enantiomorphs of dichlorprop. Conversely, Baxter (1967) using mitochondria isolated from dark-grown soybean hypocotyls found that 10^{-3} M 2,4-D had no effect on respiration or phosphorylation.

Freed *et al.* (1961) reported that the activity of glyceraldehyde-3-phosphate dehydrogenase and isocitric dehydrogenase was stimulated at low concentration (5×10^{-4} M) and inhibited at high concentration (5×10^{-3} M). They suggest that physical adsorption of 2,4-D on to the enzyme surface so modified it that its catalytic activity was increased at low concentrations and reduced at higher levels.

Further evidence that 2,4-D interferes with ATP synthesis has been presented by Wedding and Black (1961). Using intact *Chlorella* cells, they found that 2,4-D stimulated oxidation and inhibited the incorporation of ^{32}P into ATP. Using mitochondria isolated from cauliflower and red beetroot, Wedding and Black (1962) found that 2,4-D inhibited oxidation and phosphorylation with malate and citrate as substrates. At concentrations up to $3 \cdot 2 \times 10^{-3}$ M, 2,4-D had no significant effect on O_2 uptake by cauliflower and red-beet mitochondria with succinate or NADH as substrates while a reduction of 40–60% in phosphorylation was recorded. The greater inhibitory effect of 2,4-D on malate and citrate oxidation was attributed to the inhibition of malic and isocitric dehydrogenase. Later studies suggested that 2,4-D formed a non-enzymatic complex with NAD (Wedding and Black, 1963). Whilst these authors failed to report uncoupling activity, Bottrill (1965) reported that 2,4-D uncoupled oxidative phosphorylation in corn mitochondria and Lotlikar *et al.* (1968) found that 2,4-D uncoupled phosphorylation in cabbage mitochondria.

Whilst most work has been carried out with 2,4-D, the effect of other phenoxy compounds on respiration has also been investigated. Smith and Shennan (1966) found that MCPA and MCPB stimulated growth of *Aspergillus niger* at low concentrations (10^{-6} M) but inhibited at higher (10^{-4} M). While 5×10^{-3} M MCPB completely inhibited endogenous respiration, a similar concentration of MCPA resulted in only a 52% inhibition. They also reported that 5×10^{-3} M MCPA and MCPB inhibited mitochondrial succinate oxidase by 10 and 90% respectively; the respective figures for inhibition of NADH oxidase were 10 and 85%. The effect of MCPA, 2,4-D, 2,4,5-T, MCPB, 2,4-DB and 4(2,4,5-TB) on oxidative phosphorylation of mitochondria isolated from young hypocotyls of *Vicia faba* has been investigated by Matlib (1970). Using mitochondria from a single isolation, concentrations ranging from 5×10^{-6} to 4×10^{-3} M of each herbicide were applied in each case on completion of one state-4/state-3/state-4 cycle. Uncoupling equated with the loss of respiratory stimulation on addition of ADP; complete uncoupling was obtained when the ADP/O ratio was nil and the respiratory control (RC) was one. The results are summarized in Table II (Matlib *et al.* 1972). When NADH was used as substrate all the test herbicides

TABLE II

SUMMARY OF THE EFFECT OF SELECTED PHENOXY-ACID HERBICIDES ON THE UNCOUPLING OF OXIDATIVE PHOSPHORYLATION OF MITOCHONDRIA FROM *V. faba* L. (After Matlib *et al.* 1972. Reproduced with permission)

Herbicide	Concentration (M) of herbicides which completely uncoupled mitochondria			Concentration (M) of herbicides which completely inhibited respiration		Concentration (M) of herbicides which gave maximum stimulation during NADH oxidation
	Malate (10 mM)	Succinate (10 mM)	NADH (2.5 mM)	Malate (10 mM)	Succinate (10 mM)	
MCPA	10^{-3}	2×10^{-3}	4×10^{-3}	4×10^{-3}	$>4 \times 10^{-3}$ (60%)	—
MCPB	5×10^{-4}	5×10^{-4}	10^{-3}	4×10^{-3}	$>4 \times 10^{-3}$ (70%)	4×10^{-3}
2,4-D	10^{-3}	2×10^{-3}	4×10^{-3}	4×10^{-3}	$>4 \times 10^{-3}$ (68%)	—
2,4-DB	10^{-4}	5×10^{-4}	5×10^{-4}	10^{-3}	2×10^{-3}	10^{-3}
2,4,5-T	5×10^{-4}	10^{-3}	10^{-3}	4×10^{-3}	$>4 \times 10^{-3}$ (75%)	2×10^{-3}
4(2,4,5-TB)	10^{-4}	10^{-4}	10^{-4}	5×10^{-4}	10^{-3}	5×10^{-4}

TABLE III

THE OXIDATIVE AND PHOSPHORYLATIVE ACTIVITIES OF MITOCHONDRIA ISOLATED FROM STEMS AND ROOTS OF PLANTS TREATED *IN VIVO* WITH 120 μg OF EITHER MCPA, MCPB, 2,4-D OR 2,4-DB PER PLANT (After Matlib *et al.* 1972. Reproduced with permission)

Item	Substrate	Untreated		MCPA		MCPB		2,4-D		2,4-DB	
		Stem	Root	Stem	Root	Stem	Root	Stem	Root	Stem	Root
ADP/O	Malate	2.56 ± 0.19	2.88 ± 0.10	0	0	2.49 ± 0.34	2.70 ± 0.13	0	0	2.38 ± 0.04	2.59 ± 0.06
	Succinate	1.97 ± 0.43	1.95 ± 0.04	0	0.97 ± 0.01	1.86 ± 0.13	1.97 ± 0.05	0	0.98 ± 0	1.58 ± 0.16	1.94 ± 0.05
	NADH	1.83 ± 0.03	1.61 ± 0.26	0	0.08 ± 0.01	1.76 ± 0.08	1.88 ± 0.17	0	0.73 ± 0.14	1.60 ± 0.27	1.85 ± 0.10
	Malate	3.32 ± 0.33	3.00 ± 0.72	1.00	1.00	1.99 ± 0.42	1.60 ± 0.37	1.00	1.00	1.89 ± 0.34	1.71 ± 0.04
RC	Succinate	2.42 ± 0.08	1.66 ± 0.12	1.00	1.51 ± 0.30	1.71 ± 0.05	1.47 ± 0.09	1.00	1.35 ± 0.27	1.51 ± 0.28	1.76 ± 0.55
	NADH	2.24 ± 0.27	2.14 ± 0.52	1.00	1.44 ± 0.17	1.74 ± 0.11	1.68 ± 0.19	1.00	1.51 ± 0.03	1.87 ± 0.06	1.76 ± 0.18

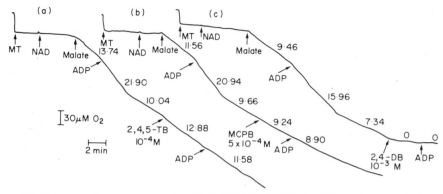

Fig. 7. Oxygraph traces showing the effect of MCPB, 2,4-DB, 2,4,5-TB on O_2 uptake by bean (*Vicia faba*) mitochondria using malate as substrate (after Matlib *et al.* 1972. Reproduced by permission).

were found to stimulate state-4 respiration with the loss of phosphorylation and RC in varying degrees. When malate and succinate were used, separately, as substrates, treatment with 2,4-DB, 2,4,5-T and 4(2,4,5-TB) at low concentrations resulted in a marked stimulation of state-4 respiration (Fig. 7); this effect was not obtained with MCPA, MCPB or 2,4-D. At higher concentrations all herbicides strongly inhibited respiration possibly due to inhibition of malate and succinate dehydrogenases (Wedding and Black, 1962; Wilson and Merz, 1967; Bottrill and Hanson, 1969), to the inhibition of an energy-dependent substrate translocation (Dam, 1967), or to the accumulation of oxaloacetate which can inhibit respiration (Papa *et al.* 1968). In varying degrees these phenoxy-acid compounds released oligomycin inhibition during NADH oxidation, stimulated mitochondrial ATPase activity and induced swelling of isolated mitochondria, supporting the view that they act, to varying extent, as true uncouplers. Phenoxybutyric acid compounds were found to be more toxic as uncouplers *in vitro* than the corresponding acetic analogue, and uncoupling activity was increased by chlorine substitution. There was evidence of a higher degree of binding of butyric compounds to BSA than acetic (Matlib *et al.* 1971a). Interestingly enough, *in vivo*, the position was reversed, phenoxyacetic acids being very active as uncouplers while phenoxybutyric acids had a negligible effect (Table III). It was concluded that, *in vivo*, non activity of foliage-applied phenoxybutyric acids is due to their restricted absorption/translocation to the sites of action (Matlib *et al.* 1972).

The probable sites of action of these and other herbicides acting as uncouplers are indicated in Fig. 8.

Fig. 8. A schematic representation tentatively showing the sites of action of herbicide uncouplers and inhibitors of oxidative phosphorylation; the concentration (M) of the latter required to bring about a 50% inhibition of O_2 uptake is also shown. (Determined from the results of Lotlikar *et al.* 1968 (1), Foy and Penner, 1965 (2), and Switzer, 1957 (3) using mitochondria isolated from cabbage, cucumber and soybean respectively.)

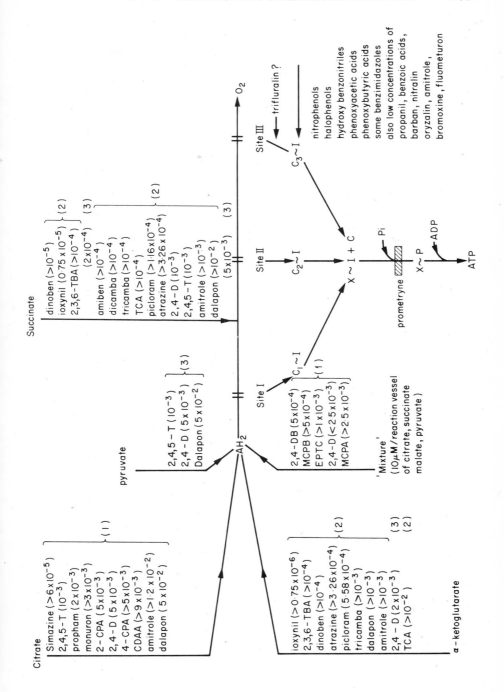

IV. Inhibitors of Oxidative Phosphorylation

The following compounds are generally considered to act as inhibitors of oxidative phosphorylation, though in certain instances, uncoupling activity has been reported. Indeed inclusion in this section is somewhat tentative, since later work may reveal uncoupling rather than or in addition to, inhibitory activity. As Moreland (1967) points out, many experiments with herbicides on plant mitochondria have been less critical than those to determine the specific sites of action of inhibitors of animal mitochondria-mediated reactions. Many studies have involved relatively insensitive manometric procedures in which the use of hexokinase and glucose as a phosphate "trap" eliminates the respiratory control normally exercised by ADP. Such mitochondria are unsuitable for critical uncoupling experiments since the stimulation of O_2 utilization in the presence of uncouplers is not usually observed. This point appears to be exemplified in the case of the phenoxy-acid herbicides by comparison of the findings of Lotlikar et al. (1968) and Matlib et al. (1972).

A. AMIDES

Members of this rather diverse group of chemicals have been recorded as inhibitors of oxidative phosphorylation. For example allidochlor affected oxidation and phosphorylation of cabbage mitochondria, a concentration of 9×10^{-3} M reducing the P/O ratio by just under 50% when using citrate as a substrate (Lotlikar et al. 1968). Previously Jaworski (1956) had reported that 10 parts/10^6 allidochlor strongly inhibited the respiration of ryegrass seeds, though wheat seed respiration was only moderately affected. He believed that allidochlor inhibited certain sulphydryl-containing enzymes involved in respiration. It has also been reported that allidochlor affects the respiration of pine seedlings (Sasaki and Kozlowski, 1966).

Another amide, propanil also appears to inhibit respiration, though the results of different investigations are somewhat variable. Sierra and Vega (1967) found that treatment of barnyard grass with 100 parts/10^6 propanil increased respiration after 1 h; respiration was reduced when the leaves were soaked in 1,000 parts/10^6 propanil. Hofstra and Switzer (1968) found that application of 3.45×10^{-3} M propanil to leaves of tomato and common lamb's quarters (Chenopodium album L.) initially inhibited respiration in both species. After 72 h, respiration of tomato almost returned to the control level, though it continued to decline in the case of lamb's quarters. Hofstra and Switzer also reported that treatment of soybean mitochondria with propanil inhibited O_2 uptake and Pi esterification; there was an uncoupling effect. There is also evidence that propanil inhibited oxidative phosphorylation and reduced the ATP content of yeast (Inoue et al. 1967) and excised soybean hypocotyls (Gruenhagen and Moreland, 1971).

B. BENZOIC ACIDS AND PICLORAM

There is good evidence that 2,3,6-TBA can inhibit oxidative phosphorylation. Foy and Penner (1965) using succinate as substrate, found that O_2 uptake by mitochondria isolated from five-day-old cucumber cotyledons was inhibited over the concentration range 10^{-5}, 10^{-4}, 10^{-3}, 10^{-2} M by 30, 43, 75 and 100% respectively. They obtained greater inhibitions with α-ketoglutarate as substrate, O_2 uptake being completely inhibited at 10^{-3} M. Lotikar et al. (1968), using citrate as substrate and mitochondria isolated from cabbage, reported that 1×10^{-3} M 2,3,6-TBA inhibited O_2 uptake and phosphorylation by 16 and 21% respectively; these inhibitions increased with concentration to 82 and 100% respectively at $1 \cdot 2 \times 10^{-2}$ M 2,3,6-TBA. While it is clear that 2,3,6-TBA can inhibit both O_2 uptake and phosphorylation, the relatively high concentrations involved throws some doubt on the relevance of this mechanism when 2,3,6-TBA is applied in vivo.

The effect of amiben, dinoben, dicamba and tricamba on O_2 uptake has also been investigated by Foy and Penner (1965). Over the concentration range 10^{-5}, 10^{-4} and 10^{-3} M, and with succinate as substrate, O_2 uptake was inhibited by 20, 35 and 70% (amiben), 60, 64 and 84% (dinoben), 5, 51 and 68% (dicamba) and 26, 29 and 39% (tricamba) respectively. The α-ketoglutarate system was much less sensitive to all of these compounds; amiben and dicamba showed essentially no inhibition of this system at a concentration of 10^{-3} M.

Using cucumber mitochondria and succinate as substrate, Foy and Penner (1965) found that picloram in the range $1 \cdot 16 \times 10^{-5}$–$1 \cdot 16 \times 10^{-3}$ M increasingly inhibited O_2 uptake. Inhibitions were less marked with α-ketoglutarate as substrate. Chang and Foy (1971) found that 10^{-5} M picloram caused swelling of mitochondria isolated from barley (Hordeum vulgare L.) and safflower (Carthamus tinctorius L.) and previously contracted with ATP. Further, 10^{-3} M picloram was found to increase ATPase activity in safflower mitochondria by 31% as compared with the control, suggesting that it may act as an uncoupler of oxidative phosphorylation.

It has been reported that fusaric acid, a derivative of picolinic acid, depressed the coupled phosphorylation of isolated mitochondria of tomato (Lycopersicon esculentum L.) and cauliflower (Brassica oleracea, var italica L.) (Sanwal, 1959).

C. CARBAMATES

The effect of propham and chlorpropham on oxidative phosphorylation of mitochondria isolated from cabbage has been investigated by Lotlikar et al. (1968). They reported that 5×10^{-4}, 1×10^{-3}, 3×10^{-3} M and 5×10^{-3} M inhibited O_2 uptake by 13, 35, 72 and 73% respectively and Pi esterification by 53, 93, 100 and 100% respectively. Chlorpropham was even more effective; with a mixture of substrates, 1×10^{-3} and 1×10^{-2} M chlorpropham inhibited O_2 uptake by 85 and 90% respectively and Pi esterification by 91 and 97% respectively. Treatment of excised soybean hypocotyls with $0 \cdot 6$ mM chlorpropham reduced the ATP level by 23% (Gruenhagen and Moreland, 1971).

The effect on oxidative phosphorylation and other metabolic processes of in vivo treatment of two-day-old etiolated seedlings of wild oat with 25, 250 and

1,000 parts/10^6 barban has been investigated (Ladonin, 1966, 1967 and Ladonin and Svittser, 1967). Barban inhibited O_2 uptake and esterification of Pi by mitochondria isolated from these treated plants; *in vitro* treatment with barban of mitochondria isolated from untreated plants produced similar results and also evidence of uncoupling activity. Chromatography revealed that barban treatment produced a marked increase in ATP consumption, possibly due to increased ATPase activity (Ladonin, 1967).

The effect of asulam on respiration of bracken (*Pteridium aquilinum* L.) frond buds has been investigated by Veerasekaran in this laboratory (Veerasekaran, 1975). He found that *in vitro* treatment with 10–1,000 parts/10^6 asulam inhibited respiration by up to 50% after three days. *In vivo* application of 2, 4, 6 and 8 kg/ha to the foliage inhibited respiration of the rhizome buds even 24 h after treatment; respiratory inhibition of up to 100% was obtained 6 and 12 months after such treatment. These results substantiate the findings of Yukinaga *et al.* (1973) who studied the respiratory activity of rhizome buds and tips of bracken, 60 days and 6 months after foliage-treatment with asulam (6 kg/ha). Inhibitions of 30% (60 days) and approximately 85% (6 months) were recorded.

It seems likely that the inhibition of oxidative phosphorylation by carbamate herbicides is not the primary site of action, but rather that inhibition of certain biosynthetic reactions such as RNA and protein syntheses (Gruenhagen and Moreland, 1971) may be in part an indirect effect of interference with energy production.

D. DINITROANILINES

It is doubtful if the effect of these compounds on respiratory metabolism represents an important aspect of their mode of action. For example, the concentrations of trifluralin which inhibit growth do not appear to affect respiration (Feeny, 1966; Talbert, 1967; Negi *et al.* 1967; Hendrix and Muench, 1969). Negi *et al.* (1968) found, however, that 10^{-4} M trifluralin inhibited O_2 uptake and Pi esterification by mitochondria isolated from several species; nitralin treatments produced a similar response. Again, Gruenhagen and Moreland (1971) reported that 0.2 mM trifluralin did not affect the ATP level of hypocotyls excised from soybean.

More recently Wang, Grooms and Frans (1974) have reported on the response of soybean mitochondria to substituted dinitroaniline herbicides. They found that oryzalin and nitralin, uncoupled succinate oxidation at the ADP-limited second state-4 (IV_2). Half maximal increases in IV_2 oxidation were obtained with concentrations of 55 µM oryzalin and 0.9 mM nitralin, maximal stimulations were obtained at 80 µM and 3 mM respectively. Trifluralin did not affect IV_2 but inhibited second state-3 (III_2) oxidation, indicating that like oligomycin, it acts as an energy-transfer inhibitor. The fact that DNP overcame inhibition by trifluralin suggested to these authors that the locus of inhibition of trifluralin is at or near to the site of action of DNP in the energy-transfer system. They also suggested that oryzalin acted at a site on the energy-transfer system closer to the formation of H_2O than does DNP or trifluralin and that nitralin appears to act earlier than trifluralin on the respiratory chain.

E. HALOGENATED ALIPHATICS

The mode of action of these compounds is incompletely understood, but it seems likely that they interfere little with the production of energy, rather with its utilization.

Foy and Penner (1965) observed that of the chloro-aliphatic compounds which they studied, the only noteworthy effect on cucumber mitochondria was that of TCA upon succinate oxidation; inhibitions of 26, 60 and 66% were obtained at concentrations of 10^{-4}, 10^{-3} and 10^{-2} M respectively. Even at concentrations of 10^{-2} and 10^{-3} M, dalapon failed to inhibit succinate oxidation. This latter finding substantiates other reports (Ingle and Rodgers, 1961; Ross, 1966; Jain et al. 1966). Rai and Hamner (1956) and Corbadzijska (1962) found that dalapon increased respiration, while Switzer (1957) obtained inhibitions of 50%, using 5×10^{-2} M dalapon on isolated soybean particles with succinate or pyruvate as substrates. Lotlikar et al. (1968) obtained only small inhibitions (O_2 uptake, 14%; Pi esterification, 15%) following treatment of cabbage mitochondria with a range of concentrations of dalapon up to $1 \cdot 2 \times 10^{-2}$ M.

F. PYRICLOR

The effect of pyriclor on O_2 uptake and oxidative phosphorylation has been investigated by Killion and Frans (1969) using mitochondria isolated from hypocotyls of etiolated soybean seedlings. They reported that pyriclor inhibited both coupled and uncoupled reactions and suggested that this compound may act on a non-phosphorylated intermediate close to the electron carrier chain, and/or with a compound of the electron carrier chain. DNP did not reverse the inhibition caused by pyriclor. Gruenhagen and Moreland (1971) found that $0 \cdot 6$ mM pyriclor inhibited the ATP content of soybean hypocotyls by 69%.

G. THIOCARBAMATES

The literature concerned with the thio and dithiocarbamates indicates that these compounds affect plant metabolism in a number of ways, via photosynthesis, respiration, oxidative phosphorylation, protein synthesis and nucleic acid metabolism (Ashton and Crafts, 1973). At present it is difficult to pinpoint the relative importance of the effect on each process in the overall herbicidal action.

It has been reported that relatively low concentrations of EPTC stimulate respiration of excised embryos of maize and mung bean (Ashton, 1963). This is true only when the data is expressed on a fresh weight basis; when calculated on a per-embryo basis the stimulation was no longer evident. Relatively high concentrations severely inhibited O_2 and Pi uptake, esterification of Pi being more sensitive than O_2 utilization. Ashton questioned the physiological significance of these high-concentration effects. Lotlikar *et al.* (1968), reported that 1×10^{-3} M EPTC inhibited O_2 uptake by 77% and Pi esterification by 100%.

Gruenhagen and Moreland (1971) found that CDEC (0·2 mM) and EPTC (0·6 mM) did not significantly reduce ATP levels in excised soybean hypocotyls but suggested that this could be related to the failure of the herbicides to penetrate the tissues. They also pointed out the inability of the techniques used to measure small but possibly meaningful changes in ATP levels; such changes could be of sufficient magnitude to interfere with RNA and protein synthesis.

H. TRIAZINES

Studies on the effect of triazines on respiration of plant tissues, organs or mitochondria have brought differing results depending upon the concentration, treatment duration, species, organ stage of growth and the specific triazine under study. An immediate increase in respiration following treatment of whole bean plants with 10 parts/10^6 atrazine has been reported by Ashton (1960); Roth (1958) reported likewise. However, Funderburk and Davis (1963) found that though short-term treatment (1–24 h) of a variety of crop plants with 1 kg/ha atrazine increased respiration, the long-term effect (7–11 days) was reversed. These authors found that the decrease in respiration correlated well with a decline in activity of peroxidase and catalase in many of the test species, particularly sensitive species. Inhibition of respiration was similarly reported for atrazine by Olech (1966) and ipazine (Nasyrova *et al.* 1968). Olech (1966) concluded that the inhibition of respiration was indirect and caused by a lowered level of assimilates resulting from an inhibition of photosynthesis.

No effect on respiration of bean embryos was recorded 48 h following an 8 h treatment of seeds with 1 and 10 parts/10^6 atrazine (Ashton and Uribe, 1962). Olech (1967) subsequently found that atrazine did not inhibit the respiration of 5–6-day-old seedlings of bean, sunflower, barley and *Agrostemma githago* treated in the dark with 0·5 to 0·65 and 1·6 10^{-7} mol/dm^2 of leaf surface; similar applications were made to the roots. Krishning (1965) and Davis (1968) also failed to find an effect of atrazine on respiration, while Gruenhagen and Moreland (1971) found that atrazine neither reduced soybean hypocotyl respiration levels nor significantly inhibited RNA and protein syntheses. Nasyrova *et al.* (1968) reported that ipazine and prometryne initially inhibited the respiration of cotton seedlings but they recovered after one month. Ipazine inhibited the respiration of *Amaranthus* sp. and *Chenopodium album* but these species failed to recover.

Shaukat (unpublished review of triazine literature) has concluded that these studies on intact plants tend to suggest that respiration is enhanced immediately (viz. 1–2 h) after treatment, probably because only low concentrations of herbicide have been absorbed; these low levels probably stimulate the activity of respiratory enzymes (Eastin *et al.* 1964; Goren and Monselise, 1965). After a few days, however, when greater amounts of herbicide have been absorbed by susceptible species, a decrease in respiration results. Shaukat believes that respiration of intermediate and resistant species may also be affected by a higher concentration though they may be able to recover.

The literature concerning the effect of triazine on isolated tissues or mitochondria also reveals variations. Using various substrates and excised barley roots, Palmer and Allen (1962) observed a general trend of respiratory stimulation. Allen and Palmer (1963) later reported differential responses of simazine with various substrates and postulated that this phenomenon could be responsible for differential protection of various species to simazine. Using isolated cucumber mitochondria, Foy and Penner (1965) found that at a concentration of $3 \cdot 25 \times 10^{-6}$ M, atrazine inhibited the oxidation of succinate and α-ketoglutarate. However, Davis (1968) could detect no effect of atrazine on O_2 uptake or P/O ratio of corn shoot mitochondria at a concentration of 4×10^{-4} M. Lotlikar *et al.* (1968) using cabbage mitochondria and citrate as substrate, found that 1×10^{-5}, 3×10^{-5} and 6×10^{-5} M simazine inhibited O_2 uptake by 10, 8 and 6% and Pi esterification by 16, 13 and 21% respectively. Voinilo *et al.* (1967) obtained similar results with atrazine and later suggested that the site of action occurred between cytochrome *b* and *c* (Voinilo *et al.* 1968).

Several investigations have been carried out on the effect of prometryne on mitochondrial respiration. Truelove and Davis (1969) found that prometryne inhibited state-3 respiration of maize shoot and rat liver mitochondria; this inhibition was relieved by uncoupling agents. McDaniel and Frans (1969) reported that prometryne treatment caused a decrease in second state-3 oxidation of both malate and succinate in soybean hypocotyl mitochondria; similarly this inhibition was overcome by the uncoupling agent DNP. This suggests that the action of prometryne occurs at the conversion of ADP to ATP, after the point at which DNP uncouples oxidative phosphorylation. Thompson *et al.* (1969) using rat liver and corn shoot mitochondria found that both state-3 and state-4 respiration were inhibited by prometryne, state-3 to a greater degree than state-4. The inhibition of state-3 respiration of the rat-liver mitochondria was partially relieved by uncoupling agents, and they suggested that the block due to prometryne occurs in the sequence of reactions leading to ATP formation, beyond the point where DNP reacts. They subsequently reported that ametryne inhibits the respiration of isolated mitochondria (Thompson *et al.* 1970).

J. TRIAZOLES

It seems doubtful if amitrole has a significant direct influence on respiration. Hall *et al.* (1954) treated whole cotton plants by dipping the leaves in 10^{-3},

10^{-2} and 10^{-1} M concentrations of amitrole. Leaf discs were sampled 3, 24, 48 and 72 h after treatment and their respiration determined. Increases of 20–60% in the level of O_2 uptake were recorded within 3 h of treatment; at longer treatment durations, respiration was inhibited by the two higher concentrations. Miller and Hall (1957) sprayed cotton leaves with $4 \cdot 8 \times 10^{-2}$ M amitrole and found that respiration was initially stimulated (4 h) but later nullified (26 h).

Amitrole treatment of grass seedlings (Russell, 1957), maize (McWhorter, 1959) and wheat or bean plants (Wort and Shrimpton, 1958) stimulated respiration. The latter found that application of amitrole to homogenates of wheat seedlings grown in the dark for six days also increased the oxidation of cytochrome c, indicating that cytochrome oxidase activity was increased by amitrole.

Amitrole, applied *in vivo* to the leaf surface or by vacuum infiltration had little effect on the respiration of detached leaves of maize for periods of up to four days. However, application of $9 \cdot 5 \times 10^{-4}$ M amitrole via the roots in sand culture increased respiration by about 20% after 24 h (McWhorter and Porter, 1960). These authors subsequently reported that the chlorotic tissue had an RQ of 0·82. Foy and Penner (1965) found that amitrole inhibited oxidation (57% by 10^{-3} M) while Lotlikar *et al.* (1968) reported that even $1 \cdot 2 \times 10^{-2}$ M did not significantly affect O_2 uptake, Pi esterification, or P/O ratios of isolated cabbage mitochondria.

Gruenhagen and Moreland (1971) observed a reduction of 17% in ATP level following amitrole treatment of soybean hypocotyls.

K. UREAS

Little work has been done on the effect of substituted ureas on respiration. Foy and Penner (1965) found that diuron had little or no effect on O_2 consumption at the range of concentrations examined $(1 \cdot 46 \times 10^{-6} – 1 \cdot 46 \times 10^{-4}$ M); Gruenhagen and Moreland (1971) reported likewise. St. John (1971) using *Chlorella* reported that diuron inhibited the photochemical but not oxidative production of ATP. There is some evidence, however, that monuron may cause appreciable inhibition of oxidative phosphorylation. Lotlikar *et al.* (1968) found that monuron at 5×10^{-4}, 1×10^{-3} and 3×10^{-3} M, inhibited O_2 uptake of cabbage mitochondria by 20, 15 and 25% respectively and Pi esterified by 57, 48 and 63% respectively.

The known possible sites of action of some of these inhibitors are indicated in Fig. 8.

V. GLYCOLYSIS AND THE PPP

The process of glycolysis by which starch, glucose or fructose are broken down into pyruvic acid has already been described (Fig. 1). In the presence of O_2, the NADH (or NADPH) formed may be oxidized to form H_2O and the pyruvic further degraded via the Krebs cycle (Fig. 2). If O_2 is limiting, the NADH formed during the oxidation of phosphoglyceraldehyde is utilized in the reduction of pyruvic to lactic acid. Alternatively, if pyruvate undergoes decarboxylation, CO_2 and acetaldehyde are formed and the latter is reduced to alcohol. A further possibility concerns the conversion of pyruvate to form alanine.

The pentose phosphate pathway (PPP), by which glucose-6-phosphate can be oxidized via a series of intermediate five-carbon sugar phosphates (Fig. 9),

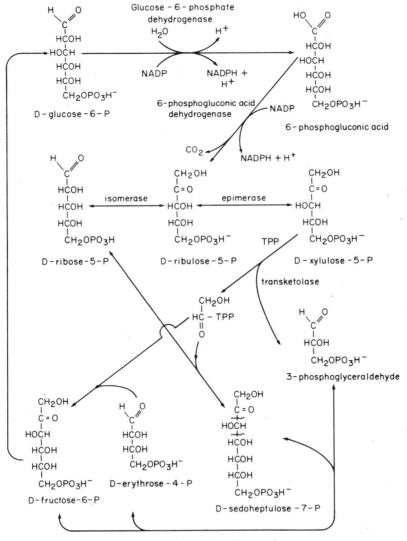

Fig. 9. The pentose phosphate pathway.

apparently takes place in the cytoplasm and not in the mitochondria. While the PPP is entirely dependent on NADP, the glycolytic dehydrogenase enzyme and Krebs cycle enzymes rely mainly on NAD to accept electrons.

The initial step of the PPP involves the oxidation of glucose-6-phosphate to 6-phosphogluconic acid which is then dehydrogenated and decarboxylated to form ribulose-5-phosphate. As a result of these oxidation reactions a molecule of CO_2 is formed and at each step NADP is reduced to NADPH. The subsequent reactions are concerned with the cycling back of ribulose-5-phosphate to glucose-6-phosphate. The first step involves the conversion of ribulose-5-phosphate to xylulose-5-phosphate or ribose-5-phosphate. Subsequently the 2-carbon ketol group of xylulose-5-phosphate is transferred to ribose-5-phosphate with the formation of sedoheptulose-7-phosphate and phosphoglyceraldehyde (3-PGA). The upper three carbons of sedoheptulose are then transferred to 3-PGA with the formation of fructose-6-phosphate and erythrose-4-phosphate; the latter can accept a ketol group from xylulose-5-phosphate yielding fructose-6-phosphate and 3-PGA. The fructose-6-phosphate is finally converted to glucose-6-phosphate. It would appear that some of the 3-PGA and hexose phosphates are broken down by the glycolytic enzymes and in some tissues at least, the two processes may be closely interwoven.

A number of investigators have determined the influence of herbicides on glycolytic and PPP reactions by feeding plant tissue with glucose labelled with ^{14}C in the C-1 (aldehyde) or C-6 positions. The occurrence of ^{14}C in the CO_2 or other metabolic products can indicate the nature of the reaction pathway and possible sites of blockage due to herbicide.

A. HALOGENATED ALIPHATICS

Bourke et al. (1964), investigating the effect of dalapon on glucose metabolism in peas, found that it interfered with glycolysis. However, this was not borne out by Jain et al. (1966) who studied the effect of dalapon on the utilization of glucose-[1-^{14}C], glucose-[6-^{14}C] and glucose-[U-^{14}C] by roots and shoots of one-week-old barley seedlings. They found no difference in the C-6/C-1 ratio or $^{14}CO_2$ release due to dalapon treatment and concluded that it did not favour the PPP to the exclusion of the Krebs cycle. Ross (1966), using bean leaf discs, came to a similar conclusion. Jain et al. (1966) found that in dalapon-treated plants the radioactivity in the ethanol-soluble extract decreased while the radioactivity in the ethanol-insoluble extract increased. Chromatography of the ethanol extract revealed equally high levels of radioactivity in citric acid and cis-aconitic acid, suggesting that glycolysis was not inhibited by dalapon. Dalapon increased the ^{14}C-labelling of sucrose, aspartic acid, glutamic acid, asparagine and glutamine whereas the labelling of α-ketoglutaric acid was reduced. While it was not possible to demonstrate a specific site of blockage either at the initiation of the glycolytic pathway or in the Krebs cycle, Jain et al. (1966), concluded that:

1. The PPP was not involved as a distinctly exclusive route of glucose utilization in barley seedlings.

2. Dalapon interfered with glucose utilization.
3. Partial inhibition may occur at the initiation of the glycolytic pathway in dalapon-treated shoots, or within the Krebs cycle in dalapon-treated roots.

Jain *et al.* (1966) and Ross (1966) also concluded that dalapon does not inhibit Co-enzyme A (CoA) activity contrary to the views of Oyolu and Huffaker (1964) who suggested that dalapon reduced the level of CoA since dalapon treatment of wheat increased fumaric acid and reduced aconitic acid and citric acid levels.

B. PHENOXY-ACIDS

The action of 2,4-D on glycolysis and associated metabolism has been investigated by several groups of workers. It appears to have a variety of specific effects on intermediary metabolism in plant tissues, though the exact points at which the herbicide intervenes are somewhat obscure. These effects are also dependent on the concentration level (Bourke and Fang, 1962).

Humphreys and Dugger (1957*b*) reported that 2,4-D stimulated the respiration rate of etiolated pea (*Pisum sativum* L.) seedlings and increased the metabolism of glucose-[1-^{14}C] via the PPP (Humphreys and Dugger, 1957*a*). They suggested that this increase in respiration via the PPP was due to a blockage of the synthetic metabolic pathways (Humphreys and Dugger, 1958). Subsequently it was reported that 10^{-3} M 2,4-D reduced (by 50%) the uptake of labelled glucose by roots of *Zea mays* seedlings, but doubled $^{14}CO_2$ production. Apparently sufficient glucose entered the root tips to saturate the enzyme responsible for substrate oxidation (Humphreys and Dugger, 1959). They suggested that 2,4-D (and DNP) promoted the oxidation of exogenous substrate by blocking synthetic pathways. Their studies with labelled glucose and gluconate indicated that in corn root tips treated with 2,4-D (10^{-3} M), glucose catabolism is almost totally accommodated via the PPP.

Similar treatment of three-day-old etiolated corn seedlings with 2,4-D (10^{-3} M) produced a general increase in glucose catabolism through an increase in the activity of enzymes associated with the PPP (Black and Humphreys, 1962). They found evidence of increased utilization of ribose-5-phosphate, enhanced formation of heptulose and hexose from ribose-5-phosphate and increased oxidation of glucose-6-phosphate and 6-phosphogluconate, indicating enhanced activity of glucose-6-phosphate dehydrogenase and 6-phosphogluconate dehydrogenase. Black and Humphreys also carried out *in vitro* studies on the effect of 2,4-D on the enzymes of the glycolytic pathway and reported a decline in the activity of 6-phosphofructokinase, aldolase and glyceraldehyde-3-phosphate dehydrogenase; the activities of phosphoglucoisomerase, phosphoglyceric kinase and enolase were unaffected. They were of the opinion that the action of 2,4-D on these enzymes did not satisfactorily explain the shift from the glycolytic to the PPP. They suggested three alternative possibilities in explanation (1) adaptive enzyme formation; (2) inhibition of glycolysis and promotion of the PPP due to a low level of inorganic Pi, or (3) stimulation of the PPP by increased nucleotide synthesis.

The effect of phenoxy-acid compounds on the uptake and metabolism of labelled acetate by plant tissue has been investigated. Humphreys and Dugger (1959) using corn roots found that 2,4-D increased the respiration of $^{14}CO_2$ from ^{14}C-acetate and reduced the conversion of acetate into cell wall constituents; ^{14}C-acetate absorption was increased by 20%. On the other hand, Stevens et al. (1962) found that, irrespective of their biological activity, eight phenoxy acids (10^{-4} M) inhibited acetate absorption by about 20%; other herbicides (including amitrole, dalapon, EPTC, maleic hydrazide (MH) and propham) had little or no effect. Similarly, Bourke et al. (1962) investigating the effect of a range of phenoxy compounds on the absorption of ^{14}C-glucose by pea root tips found that absorption of glucose was inhibited. There was no correlation between inhibition of absorption and herbicidal activity. In general, all of the test compounds reduced ^{14}C-glucose incorporation into the cell wall, only slightly affected its oxidation to $^{14}CO_2$ and increased its incorporation into alcohol-soluble intermediates. No correlation was found between changes in the C-6/C-1 ratios and herbicidal activity. They concluded, that the effect on oxidation was due either to a stimulation of glycolysis by low concentrations of 2,4-D or to a reduction in glycolysis and PPP activity; glycolysis was affected to a relatively greater degree by higher concentrations of the herbicide.

Kim and Bidwell (1967) studied the effect of 30-min treatments by 5×10^{-5} M, 2,4-D (or IAA) on the metabolism of ^{14}C-specifically-labelled pyruvic, acetic, succinic and glutamic acids by pea root tips. While pyruvic acid decarboxylation was unaffected, the carboxylation of pyruvic acid and entry into the Krebs cycle of acetate derived from pyruvate was inhibited by 2,4-D (and IAA). Kim and Bidwell (1967) also found that these compounds stimulated the oxidation of acetate into CO_2 and inhibited the accumulation of ^{14}C in Krebs cycle intermediates or amino acids derived from them. They suggested that these results indicate that 2,4-D (and IAA) have an inhibiting effect on co-carboxylase or more probably on α-lipoic acid metabolism, but not on CoA. Leopold and Guernsey (1953) had earlier suggested that auxin may effect the metabolism of CoA.

Mostafa and Fang (1971) carried out a time-course experiment on the in vitro effect of 2,4-D (10^{-4} M) on the metabolism and incorporation of specific ^{14}C-labelled glucose into pea (Pisum sativum L. var Alaska) and corn (Zea mays L. var Golden Cross). They found a preferential release of C-1 as CO_2 which was affected by 2,4-D; there was also evidence that 2,4-D enhanced C-6 oxidation in pea roots and slightly in corn stems, apparently via the glucuronic acid pathway. The pattern of effect on the PPP was reversed. Mostafa and Fang also found that 2,4-D affected the incorporation of ^{14}C into certain amino acids, increased fructose content and the accumulation of malic acid. The probable sites of attack are shown in Fig. 10. At a stimulatory level of 2,4-D, less glucose was completely oxidized to CO_2 and more was utilized in synthetic reactions, presumably through the normal pathways of amino-acid and protein synthesis. At an inhibitory level of 2,4-D, the synthesis of TCA-cycle amino acids was

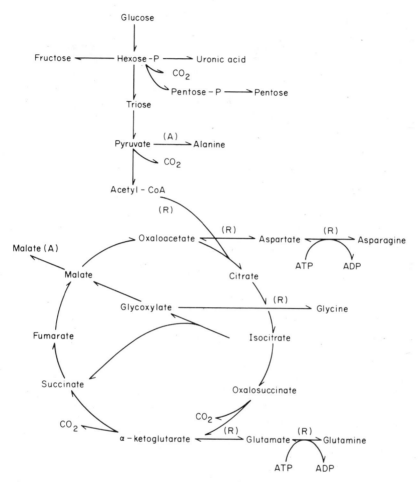

Fig. 10. The respiratory metabolism of glucose and the sites of action of 2,4-D. (A = accumulation, I = inhibition and R = regulation) (reproduced by permission from Mostafa and Fang, *Weed Sci.* **19**, 248–253 (1971).

reduced, and the accumulation of asparagine and glutamine drastically influenced. Earlier, Bourke and Fang (1962) had reported the stimulatory and inhibitory effects of low and high concentrations of 2,4-D on glycolysis in pea tissues.

VI. OTHER OXIDATION SYSTEMS

The cytochrome system generally accepts most of the electrons removed from glycolytic or Krebs cycle intermediates; however there are certain other enzymes which are potentially capable of transferring electrons to oxygen. These include phenol oxidases, glycolic acid oxidases, peroxidase and catalase.

Dicryl inhibited the respiration of maize leaf tissue and altered the action of certain enzymes. For example, catalase was only slightly affected two days after treatment though reductions of 50% were observed after five days; similar results were observed for peroxidase and glycollic acid oxidase (Funderburk and Porter, 1961a). These authors also reported that three to four days after treating maize with dicryl, ascorbate oxidation was markedly increased (Funderburk and Porter, 1961b). Dicryl treatment of cotton in the cotyledon stage reduced ascorbic acid oxidase but respiration was only slightly decreased (Bingham and Porter, 1961a); the normal increase in catalase and peroxidase was prevented but glycolic acid oxidase was unaffected (Bingham and Porter, 1961b).

VII. Intermediary Metabolism

Carbohydrates play a significant role in the plant since they represent important food reserves and constitute an important part of the structural framework of each cell. The metabolic processes concerned with changing photosynthetic products into such derivatives involve a multitude of biochemical processes which may exhibit varying degrees of susceptibility to herbicides (Fig. 11). This section is concerned with those herbicides which are known to affect the carbohydrate level, certain aspects of nitrogen metabolism (this aspect is considered fully elsewhere) or lipid metabolism. Lipids occur in plants as true fats, phospholipids or waxes. The fats and phospholipids are synthesized by esterification of glycerol with fatty acids, while the waxes are esters of fatty acids and long-chain mono-hydroxy alcohols, or occasionally dihydroxy alcohols. Glycerol is derived from dihydroxyacetone phosphate while fatty acids are synthesized from acetyl-CoA formed by glycolytic reactions or derived from pantothenic acid. The steps involved in synthesis of butyric acid from two molecules of acetyl-CoA are shown in Fig. 12; repetition of this procedure can produce fatty acids of increasingly greater chain length.

A. AMIDES

Mann and Pu (1968), using hypocotyls of hemp sesbania (*Sesbania exaltana*) reported that allidochlor (1–20 parts/10^6) inhibited the incorporation of ^{14}C from malonic acid-[2-^{14}C] into lipid by 70–90%.

B. BENZOICS AND PICLORAM

There is evidence that substituted benzoic acids may influence carbohydrate metabolism, directly or indirectly. Stevens et al. (1962) reported that TIBA and 2,3,6-TBA caused a significant disturbance of acetate metabolism as well as inhibition of acetate uptake. TIBA was more inhibitory than 2,3,6-TBA. However, Hilton (1965), using *Escherichia coli,* tentatively concluded that

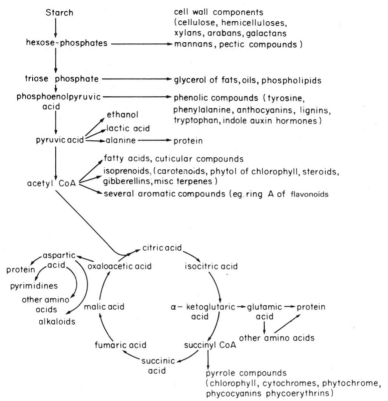

Fig. 11. The relationship of glycolytic and Krebs cycle reactions to the production of compounds involved in growth (from "Plant Physiology" by Frank B. Salisbury and Cleon Ross © 1969 by Wadsworth Publishing Company Inc., Belmont, California 94002. Reprinted by permission of the publisher).

inhibition of pantothenate biosynthesis was of little physiological significance as a site of action responsible for herbicidal activities of benzoic acids (see Fig. 12).

The effect of 2,3,6-TBA, applied annually over a three-year period, on carbohydrate metabolism in the roots of Russian knapweed (*Centaurea repens*) revealed a progressive decline in the inulin and disaccharide content of the shallow roots (Chernyshev, 1968). It has been suggested (Ashton and Crafts, 1973) that the biochemical basis for this reduction in carbohydrate content may be found in the work of Zhirmunskaya (1966). He sampled roots of Canada thistle (*Cirsium arvense*) and perennial sowthistle (*Sonchus arvensis*) over a period of 3 to 24 months after 2,3,6-TBA treatment; a 2–3-fold increase in total nitrogen was accompanied by an intensive synthesis of protein and a build-up of non-protein forms of N at the expense of the amides. Zhirmunskaya (1966) suggested that the decline in carbohydrate content was

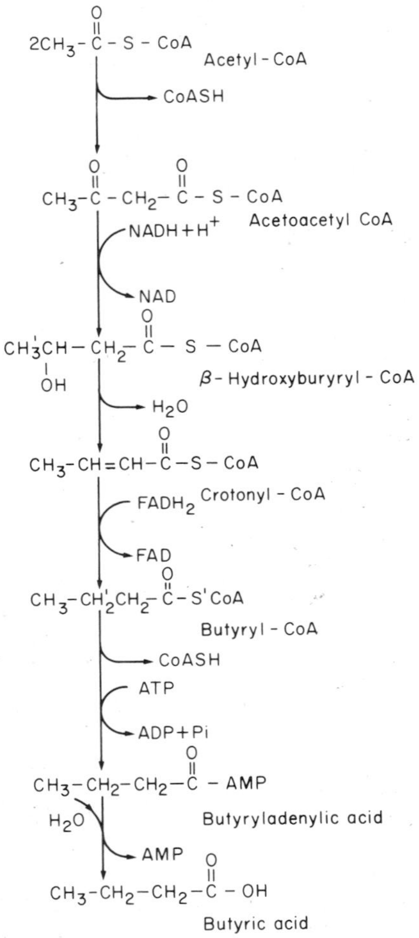

$$2CH_3-\overset{\overset{\displaystyle O}{\|}}{C}-S-CoA \quad \text{Acetyl-CoA}$$

CoASH

$$CH_3-\overset{\overset{\displaystyle O}{\|}}{C}-CH_2-\overset{\overset{\displaystyle O}{\|}}{C}-S-CoA \quad \text{Acetoacetyl CoA}$$

NADH+H$^+$

NAD

$$CH_3CH-CH_2-\overset{\overset{\displaystyle O}{\|}}{C}-S-CoA$$
OH $\qquad$ β-Hydroxyburyryl-CoA

H$_2$O

$$CH_3-CH=CH-\overset{\overset{\displaystyle O}{\|}}{C}-S-CoA$$

FADH$_2$ $\quad$ Crotonyl-CoA

FAD

$$CH_3-CH_2CH_2-\overset{\overset{\displaystyle O}{\|}}{C}-S'CoA$$
Butyryl-CoA

CoASH

ATP

ADP+Pi

$$CH_3-CH_2-CH_2-\overset{\overset{\displaystyle O}{\|}}{C}-AMP$$

H$_2$O $\qquad$ Butyryladenylic acid

AMP

$$CH_3-CH_2-CH_2-\overset{\overset{\displaystyle O}{\|}}{C}-OH$$
Butyric acid

Fig. 12. The biosynthesis of butyric acid from acetyl-CoA.

due to their utilization as substrates under circumstances where nitrogen metabolism is increased.

There is also evidence that substituted benzoic acids and picloram may influence lipid synthesis. For example, Mann and Pu (1968) investigated the effect of 30 herbicides (1–20 parts/10^6) on lipid synthesis by excised hypocotyls of hemp sesbania (*Sesbania exaltata*). At a concentration of 1 part/10^6, amiben stimulated the incorporation of radioactivity from malonic acid-[2-^{14}C] into lipid; on the other hand dinoben had a strong inhibitory effect, even at 1 part/10^6. Picloram (1–20 parts/10^6) stimulated lipogensis by 48–53%.

C. CARBAMATES

Certain carbamates have been shown to inhibit oxidative phosphorylation, RNA synthesis and protein synthesis (Ashton and Crafts, 1973). The effects on RNA and protein synthesis are discussed elsewhere (Chapter 17) but it is worth mentioning that there is evidence that lipid biosynthesis in excised hypocotyls of hemp sesbania was inhibited by CDEC. While Still *et al.* (1970) found that chlorpropham had no significant effect on total amounts of extractable epicuticular lipids nor the wax structure on leaves of peas (*Pisum sativum* L.).

Sumida and Ueda (1973) found that swep inhibited the biosynthesis of complex lipids in *Chlorella* by 61%. They found that the proportion of mono-acid digalactolipids decreased while the proportion of phospholipids increased.

D. DINITROANILINES

Trifluralin is known to induce several biochemical responses in higher plants including changes in carbohydrate, lipid, nitrogen and nucleic acid content (Ashton and Crafts, 1973).

Dukes and Biswas (1967) grew sweet potatoes and peanuts for 72 h in a range of concentrations of trifluralin; the carbohydrate content increased at 5 and 10 parts/10^6 trifluralin and decreased at 20, 50 and 100 parts/10^6. Diem *et al.* (1968) reported that trifluralin increased the nitrogen content of both shoot and root of grain sorghum and maize and Schweizer (1970) reported that 0·75 kg/ha trifluralin increased the N content but reduced the sugar content in the aberrant sugar beet hypocotyledonary neck and root tissues. Johnson and Jellum (1969) failed to observe any effect of 0·75 kg/ha trifluralin on the protein, oil or fatty acid composition of soybean seeds. Likewise, Penner and Meggitt (1970) found that 0·75 and 1·0 kg/ha trifluralin had no effect on the total oil content of soybean seeds though there was a significant effect on oil quality at the higher concentration; the linoleic acid content increased while the stearic acid level fell. The α-amylase activity of barley seeds was virtually unaffected by 5×10^{-4} M trifluralin (commercial formulation).

St. John and Hilton (1973) evaluated the *in vitro* effect of a number of herbicides in glyceride synthesis using spinach (*Spinacia oleracea* L.). They found that MBR 8251 (1,1,1-trifluoro-4[1]-(phenylsulphonyl)-methanesulphono-*o*-toluidide) and dinoseb inhibited glyceride synthesis. Using intact wheat (*Triticum aestivum* L) seedlings they studied the physiological significance of this inhibition and reported a build-up of free fatty acids and a decrease in neutral and polar lipids. They suggested that both compounds alter membrane structure and function through an inhibition of membrane lipid synthesis.

The effect of the dinitroanilines on nucleic acid may well be the key to their mechanism of action; however, this aspect is dealt with elsewhere.

E. HALOGENATED ALIPHATICS

There is conflicting evidence as to the manner in which halogenated aliphatics influence carbohydrate metabolism. For example, it has been reported that soil-application of TCA increased the reducing-sugar content of wheat seedlings, while non-reducing sugar decreased. On the other hand, McWhorter (1961) found that the dalapon treatment which gave the best control of Johnson grass (*Sorghum halapense*) resulted in a general reduction of glucose and a corresponding increase in sucrose content.

There is evidence that chlorinated aliphatic acids may interfere with the biosynthesis of pantothenic acid which is a precursor of CoA (Hilton *et al.* 1959) (Fig. 13). Anderson *et al.* (1962) reported that a dalapon-induced increase in pantothenic acid in sugar beet and yellow foxtail was only temporary in sugar beet, the resistant species. Van Oorschot and Hilton (1963) found that unchlorinated aliphatic acids block the utilization of β-alanine, while certain chloro-substituted derivatives including dalapon and TCA, inhibited pantothenate production by competitive action with pantoate. On the other hand, several workers failed to obtain any significant reversal of dalapon action with pantothenate (Ingle and Rogers, 1961; Leasure, 1964; Prasad and Blackman, 1965; Åberg and Johansson, 1966) and concluded that although

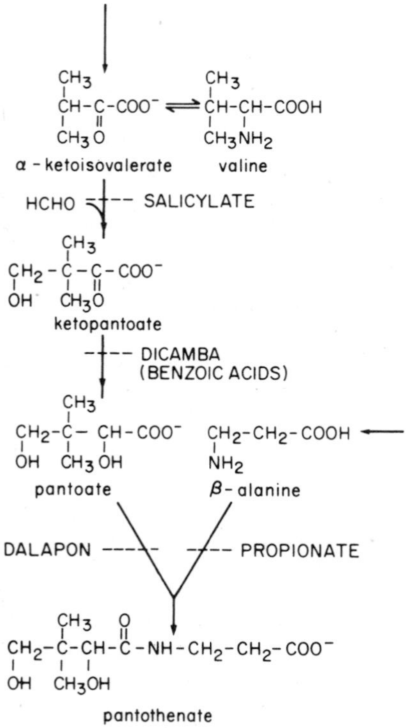

Fig. 13. The pathway for pantothenate biosynthesis from α-ketoisovalerate and β-alanine showing the proposed site of inhibition (broken lines) by salicylate, benzoic acid derivatives, dalapon and propionate (reproduced with permission from Maas and Vogel, 1953).

the inhibitory action of dalapon on growth can be counteracted to some degree with pantothenate, there is little evidence to substantiate the view that this represents an important aspect of the mode of action of dalapon. They could not reverse the growth inhibition of cucumber and maize roots with pantothenic acid, pantoic acid or β-alanine and concluded that dalapon interferes with the utilization of metabolic energy rather than with its production. Rodgers (1963) later reported that 10^{-4} M dalapon produced a long-term inhibition of anthocyanin in several plant species without appreciable inhibition of growth.

Stevens *et al.* (1962) studied the effect of a range of herbicides on the metabolism of ^{14}C-acetate, in three-day-old seedlings of *Pisum sativum* L. They found that dalapon (10^{-4} M) had little or no effect on the uptake or metabolism of ^{14}C-acetate; acetate carbons are incorporated into many cellular compounds and intermediates.

Ross (1966) reported that dalapon treatment of bean leaf discs caused an increase in ATP and a reduction in an unidentified sugar phosphate; however, several other energy-producing steps were not affected. Ross concluded that if dalapon does have a specific site of action it is probably phosphorus metabolism and the build up of ATP occurs because it is not utilized for phosphorylation of some compound needed in synthetic processes. This view is consistent with the effect of TCA and dalapon on lipid synthesis. Both herbicides are known to affect lipid metabolism and cuticle-wax deposition. For example, Dewey *et al.* (1956), Pfeiffer *et al.* (1957), Juniper and Bradley (1958) and Kolattukudy (1968) found that these compounds altered the cuticular wax of peas and maize rendering them more wettable to subsequent sprays. Mashtakov *et al.* (1967*a*) reported that TCA reduced the thickness of the cuticle and lamina in leaves of sensitive varieties of *Lupinus luteus*; they also found that TCA reduced the level of pantothenic acid in forage lupins (Mashtakov *et al.* 1967*b*). Still *et al.* (1970) reported that TCA had little effect on wax deposition in peas (*Pisum sativum* L.) compared to certain carbamates. Again, these results can be explained on the basis that dalapon appears to affect the utilization of energy rather than its production.

The aliphatic acid herbicides are also known to affect nitrogen metabolism. Anderson *et al.* (1962) studied the fate of dalapon in sugar beet (resistant species) and yellow foxtail (sensitive species) and reported that dalapon treatment resulted in an increase in the degradation of protein to amino acids and an increase in the level of amides. The free amino acids were further broken down to ammonia, while the amides (glutamine in sugar beet, asparagine in yellow foxtail) apparently acted in detoxication of the released ammonia by serving as storage sites. In the case of the resistant species, the amide and amino acids returned to normal levels after a period of time; this did not occur in the sensitive species. It has been reported that TCA increased the levels of asparagine and glutamine in a resistant variety of *Lupinus luteus* (Mashtakov and Moshchuk, 1967); the levels of protein and β-alanine were slightly increased, though free ammonia was not affected. In a sensitive variety, however, TCA treatment reduced the amount of amides and β-alanine, whereas free ammonia and protein increased. Dalapon produced similar results (Mashtakov *et al.* 1967*a*). After a 48-h treatment period, free ammonia contents for the sensitive variety were 63 and 30% (TCA and dalapon respectively); the corresponding figures for the resistant variety were 3 and 21% respectively. Mashtakov *et al.* concluded that TCA and dalapon inhibited the enzymes involved in the conversion of ammonia to amides with the consequent accumulation of free ammonia to toxic levels. The steps involved in the synthesis of glutamine are shown in Fig. 14.

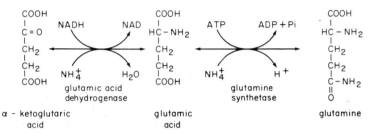

Fig. 14. The conversion of ammonia into glutamine.

Mashtakov and Moshchuk (1967) found that TCA inhibited proteinase activity, particularly in the resistant variety of *Lupinus luteus,* while Ashton *et al.* (1968) reported only slight inhibition of proteolytic activity in squash (*Cucurbita maxima* Duchesne) seeds grown in 10^{-3} M dalapon. The α-amylase activity of barley seeds was reduced 42% by dalapon (Jones and Foy, 1971).

Ashton and Crafts (1973) have pointed out that an aspect of aliphatic acid mode of action which justifies investigation concerns the effect of these compounds on changes in protein conformation. They believe that if such changes do indeed occur, then they could explain many of the results which have been reported. Further, these compounds are known to be highly effective as protein denaturants; indeed TCA is the classical compound used by biochemists for this purpose, though it is doubtful if the concentrations normally used (around 5%) occur *in vivo* in the plant.

F. NITRILES

The action of several herbicides, including bromoxynil and dichlobenil, on amylase activity in barley and squash seedlings has been investigated by Penner (1968). He concluded that the development of amylase activity controlled by the embryo in the distal halves of intact barley (*Hordeum vulgare* L) seeds during the first two days of germination was prevented by the presence in the culture solution of 10^{-4} M bromoxynil and 10^{-4} M dichlobenil. Bromoxynil also inhibited the development of low levels of amylase activity found in two-day-old squash (*Cucurbita maxima* Duchesne, var Chicago Warted Hubbard) cotyledons, though added benzyladenine overcame the inhibition. Penner (1968) concluded that the tolerance to bromoxynil (and amiben) of squash and other seeds low in carbohydrates during germination, may be related to their independence from energy and carbon sources required for anabolic purposes, which are liable to be inhibited by herbicides.

Using germinating barley grains, Moreland *et al.* (1969) found that gibberellin (GA)-induced α-amylase development was only slightly inhibited by ioxynil (4%) and more strongly by dichlobenil (82%). In a separate *in vitro* study, they found that none of the 22 test herbicides inhibited α-amylase activity and attributed the *in vivo* effects to interference with the formation of the enzyme. Jones and Foy (1971) found that the activity of α-amylase in barley half-seeds, induced by exogenous GA, was almost completely inhibited by 5×10^{-4} M bromoxynil (commercial formulation).

Mann and Pu (1968) reported that 10 and 20 parts/10^6 ioxynil and dichlobenil inhibited the incorporation of ^{14}C from malonate-[2-^{14}C] into lipid by 60–65% and 30–40% respectively. Since the addition of dichlobenil to CDEC did not enhance the inhibition produced by CDEC alone, these authors concluded that these two herbicides affect the same process in lipogenesis. Sumida and Ueda (1973) found that dichlobenil (5 mg/l) stimulated biosynthesis of complex lipids in *Chlorella*.

G. PHENOXY-ACIDS

There is ample evidence that phenoxy-acid herbicides alter the carbohydrate content of plants (Wort, 1964). For example, treatment of Red Kidney bean with 1,000 parts/10^6 2,4-D resulted in the depletion, in the stems, of reducing and non-reducing sugars, starch, crude fibre and acid-hydrolyzable polysaccharides; in roots and leaves only non-reducing sugars were depleted (Sell *et al.* 1949; Weller *et al.* 1950). Similarly, eight days after treating buckwheat plants (*Fagopyrum esculentum*) with 50–1,000 parts/10^6 2,4-D, Wort (1951) recorded a decline in the total sugar content and starch dextrin fraction. On the other hand, Wolf *et al.* (1950) found that the starch content of soybeans grown in nutrient solution containing 20 parts/10^6 2,4-D increased, while the reducing sugar and total sugar actually declined. Rhodes (1952) reported that depletion of starch and sugar in tomato plants sprayed with 200 parts/10^6 MCPA (sodium salt) was due to a 50% decline in the rate of synthesis rather than a depletion of the reserves.

Chacravarti *et al.* (1955) reported that the amount of sugar obtained from sugar cane increased by 5% when treated with 50 parts/10^6 2,4-D (sodium) applied at 60 gal/ac (670 l/ha); previously Lugo-Lopez *et al.* (1953) failed to obtain any such increase in sugar content following field treatment of sugar cane with a range of concentration of 2,4-D. Treatment of detached bananas with 1,000 parts/10^6 2,4-D, 2,4,5-T and 4-CPA increased fruit ripening and four days after treatment, the level of reducing sugars was significantly increased compared with controls; this effect was attributed to a herbicide-induced increase in the activity of amylase.

Fang *et al.* (1961) studied the effect of 2,4-D in bean stem sections and found that it increased substrate absorption, increased respiratory $^{14}CO_2$ production, decreased the incorporation of acetate carbons into cell wall constituents and increased lipid synthesis by certain pathways. Stevens et al. (1962) reported that all eight phenoxy acids tested caused approximately 20% reduction of acetate uptake and utilization. Bourke et al. (1962) similarly studied the effect of various herbicides on glucose metabolism in root tissue of Pisum sativum L. At a concentration of 1×10^{-3} M 2,4-D, intermediary metabolites accumulated and there was a decrease in the production of metabolic end products. Lower concentrations had no significant effect. They concluded that the herbicidal activity of 2,4-D and 2,4,5-T was not singularly due to glucose metabolism since both herbicidal and non-herbicidal compounds caused similar effects.

Coble and Slife (1971) sprayed honeyvine milkweed (Ampelamus albidus (Nutt.) Britt.)) with 2,4-D (dimethylamine salt) and recorded from the roots evidence of starch reduction, protein increase, exudation of α-amylase and nucleotides. A 30-fold increase in cellulase activity after three days coincided with a 180% increase in protein during that period, suggesting that part of the new protein may be the enzyme cellulase. Grant and Fuller (1971) provided evidence that 2,4-D (10^{-5} M) induced a marked change in the α-cellulose content of the cell walls of Vicia faba root tips. They suggested that in control root tips, events are very dependent upon the carbohydrate level; 2,4-D appears to interfere with this state and the changed carbohydrate balance leads to changes in starch, DNA and cell wall composition. Hanson and Slife (1969) and Venis (1971) reported that auxin herbicides appeared to stimulate DNA-dependent RNA synthesis.

The effect of phenoxy-acid herbicides, particularly 2,4-D on enzymes has been reviewed by Wort (1964). There is evidence from the literature that 2,4-D may affect the activity of amylase, ascorbic acid oxidase, catalase, cytochrome oxidase, glycolic oxidase, IAA oxidase, invertase, pectin methylesterase, peptidase, peroxidase, phosphatase, phosphorylase, polyphenol oxidase, proteinase and proteolytic enzymes. It is generally thought that the activity of 2,4-D is invariably altered in an indirect way, through effects on the conditions under which the enzyme reacts e.g. pH, hydration etc., on the supply of materials required for the supply of apoenzyme and coenzyme etc. 2,4-D inhibited proteinase activity of squash cotyledons by 55% (Ashton et al. 1968) and dipeptidase activity by over 90% (Tsay and Ashton, 1971). Recently it has been proposed that selective phytotoxicity of auxin-like herbicides is based on a differential alteration of RNA species and interference with protein synthesis (Chen et al. 1973) (see Chapter 17).

Phenoxy-acid herbicides are known to affect lipid synthesis. Mann and Pu (1968) found that 2,4-D and 2,4,5-T stimulated lipogenesis; however, the stimulation declined over the concentration range tested ($1-20$ parts/10^6). Similarly, Sumida and Ueda (1973) reported that 2,4-D stimulated lipid biosynthesis in Chlorella.

There is considerable evidence that the phenoxy-acid herbicides affect respiratory and intermediary metabolism; it appears that the main mechanism of action involves a complex series of reactions initiated by the derepression of the gene regulating synthesis of the enzyme RNAase. This aspect is dealt with elsewhere in the book (Chapter 17).

H. THIOCARBAMATES

The effect of these compounds on protein and nucleic acid metabolism is described elsewhere (Chapter 17). In addition, there is evidence that they do influence lipid metabolism. For example, CDEC inhibited lipid biogenesis by hypocotyls of hemp sesbania by 40–60% over the concentration range 1–20 parts/10^6 (Mann and Pu, 1968). These authors interpreted this partial inhibition as an indication that CDEC (also dichlobenil and endothal) only inhibited a fraction of the total lipids. Sumida and Ueda (1973) apparently failed to observe a significant effect on lipid synthesis in *Chlorella*. Previously Gentner (1966) had described the increased toxicity of a contact herbicide to cabbage after treatment with EPTC. A similar increase in response to propanil had been noted for diallate by Still *et al.* (1970). These authors determined the effect of several thiocarbamates (and TCA) on the biosynthesis of epicuticular lipids from leaves of pea (*Pisum sativum*) and found that diallate, EPTC and CDEC all had a significant effect. Diallate introduced through the roots reduced epicuticular lipids by 50%, while exposure of the leaves to vapour resulted in an 80% reduction. The ratio of wax lipid components in diallate-treated plants remained unchanged, with the exception of the primary alcohols which were reduced. Still *et al.* suggested that diallate may interfere with the biosynthesis of a precursor to the elongation-decarboxylation pathway of lipid synthesis. Chlorpropham had no significant effect on the total amounts of extractable epicuticular lipids, nor did it alter the structure of the wax formation on the leaves; EPTC did alter wax structure but not its composition. Still *et al.* suggested that the carbonyl group may be the functional portion of diallate and EPTC.

J. TRIAZINES

Members of the triazine group are known to have an adverse effect on the carbohydrate content of plants and this is not unexpected from compounds which principally act as photosynthesis inhibitors (see Chapter 16).

Gast (1958) reported that starch accumulation by *Coleus blumei* was inhibited following triazine treatment. He found that starch-free *Coleus* leaves, kept in the dark in a sucrose solution were able to form starch in the presence of simazine indicating that simazine inhibits sucrose formation. This alleviation of triazine toxicity on the addition of sucrose or glucose has been noted to varying degrees by other workers (Moreland *et al.* 1958; Eastin, 1963; Ashton and Bisalputra, 1964). Ashton and Uribe (1962) investigated the effect of atrazine on the metabolism of ^{14}C-sucrose by excised bean leaves. Atrazine caused a marked increase in the radioactivity of aspartic and glutamic acids, while the level of ^{14}C detected in serine, alanine and glyceric acid declined;

malic and citric acids were unaffected. Since sucrose metabolism was unaffected when photosynthesis was blocked by placing untreated plants in dark conditions, these authors concluded that atrazine not only blocks photosynthesis but in addition, interferes with other metabolic processes. Ashton and Uribe (1962) reached the same conclusion following similar studies with ^{14}C-serine. Stevens et al. (1962) found that simazine reduced ^{14}C-acetate uptake by about 10%. The metabolism of the C-1 carbon showed a slight increase in catabolic function while the metabolism of the C-2 carbon was unchanged. The absence of starch in atrazine-treated Chlorella vulgaris L. (Ashton et al. 1966) and in the chloroplasts of higher plants treated with atrazine (Ashton et al. 1963; Hill et al. 1968) suggested that starch was either not formed or not used in metabolic activity.

The extent of the carbohydrate reduction may vary according to the kind of triazine, its concentration, the plant species and its stage of growth; the reduction appears to be greater in susceptible species. Ploszynski et al. (1969) studied the effect of several triazines on the carbohydrate content of Echinochloa crus-galli, Sinapis alba and Lepidium sativum. Pre-emergence application reduced the saccharide level of all the test species, whereas the glucose content per unit dry weight increased in Sinapis alba and Lepidium sativum. Herbicides showed the following order of toxicity: atrazine > simazine > propazine > atratone > prometone > prometryne. This order corresponds approximately to the relative order in which these triazines inhibit photosynthesis (or the Hill reaction).

It has been suggested that the decline in carbohydrate level may be due to their utilization in amino-acid and protein synthesis (Wu et al. 1971a). These authors found that foliar application of 2−5 mg/l solution of s-triazine compounds increased the protein content of pea (Pisum sativum L.), sweet corn (Zea mays L.) (Wu et al. 1971b) and bush beans (Singh et al. 1972).

Application of simazine, 2-methoxy-4-isopropylamino-6-butylamino-s-triazine (igran), or 2-methoxy-4-isopropylamino-6-butylamino-s-triazine (GS. 14254), at 2 mg/l significantly stimulated the activity of starch phosphorylase, pyruvate kinase, cytochrome oxidase and glutamate dehydrogenase in the leaves of pea and sweet corn seedlings, 5, 10 and 15 days after treatments (Wu et al. 1971a). Previously, it had been reported that starch and total sugar levels were decreased following treatment of beans, peas and maize with s-triazines (Wu et al. 1971b) and that s-triazines increased total soluble amino acids, nitrate reduction, transaminase, α-amylase, starch phosphorylase, ATPase and α-aminolevulinic acid dehydratase (Tweedy and Ries, 1967; Singh and Salunkhe, 1970).

Ries et al. (1967) found that treatment of rye plants with sub-herbicidal concentrations of simazine resulted in increased protein accumulation; treated pea plants produced seeds containing 40% more protein than untreated plants. Similarly triazine herbicides increased the protein content of bean plants (Campbell et al. 1971; Singh et al. 1972). Pulver and Ries (1973) found that root application of 10^{-8} M simazine to 10-day-old barley (Hordeum vulgare) reduced soluble carbohydrate content but increased water-soluble protein.

They also found that simazine increased ^{14}C-leucine incorporation into protein, indicating that simazine may have a direct effect on protein synthesis.

The effect of prometryne on the uptake and incorporation of ^{14}C-leucine by cucumber cotyledon tissue was studied by Truelove et al. (1973). Prometryne decreased ^{14}C-leucine uptake both in light and dark, and reduced leucine incorporation into protein only when the discs were also exposed to over 4 h of light. They suggested that the increased percentage of leucine incorporated into protein may be related to the increased availability of absorbed leucine for use in protein synthesis, rather than an effect of prometryne on the rate of protein synthesis..

Mann and Pu (1968) found that atrazine and simazine had little effect on lipid synthesis by excised hypocotyls of hemp sesbania. On the other hand, Sumida and Ueda (1973) reported that simazine (5 mg/l) inhibited the biosynthesis of complex lipids by 80% and there was evidence that the pattern of ^{14}C-incorporation into lipids from [^{14}C-U]-acetate was noticeably altered. For example, the proportion of mono and digalactolipids markedly decreased while that of phospholipids increased.

K. TRIAZOLES

Amitrole has a wide range of biochemical action, including effects on carbohydrate, lipid, nitrogen and other aspects of metabolism. The effect on protein and nucleotide metabolism is treated elsewhere.

The influence of amitrole on carbohydrate composition of cotton plants has been studied by spraying plants in the field with $1 \cdot 3 \times 10^{-2}$, $3 \cdot 0 \times 10^{-2}$ and $6 \cdot 0 \times 10^{-2}$ M solution of amitrole, and analysing the plant 24 and 48 h after treatment (Hall et al. 1954). Total carbohydrates were reduced, approximately 50% of the original reducing sugars and sucrose being lost from the aerial organs within 48 h, while a slight increase in starch was essentially balanced by an equivalent loss in the hemicellulose fraction. Increases in starch accumulation following amitrole treatment have also been reported in the alga Scenedesmus quadricauda (Castelfranco and Bisalputra, 1965) and in duckweed fronds (Dekock and Innes, 1970). Castelfranco and Bisalputra concluded that this accumulation resulted from an arrested cell division possibly due to blocked purine metabolism which allowed carbohydrate accumulation to proceed much faster than utilization.

Amitrole had little effect on ^{14}C-acetate uptake or metabolism when applied to the roots of three-day-old pea seedlings (Stevens et al. 1962), nor did it significantly affect lipogenesis in hypocotyls of hemp sesbania (Mann and Pu, 1968).

L. UREAS

There is evidence that, like the triazines, substituted ureas adversely affect carbohydrate levels in plants. Walsh and Grow (1971), using selected species of marine unicellular algae, investigated the effect of a range ($5-30$ parts/10^6)

of concentrations of several ureas and found inhibition of up to 66% in the most susceptible species. With regard to their effect on lipid synthesis, Sumida and Ueda (1973) reported that diuron (5 mg/l) inhibited the biosynthesis of complex lipids by 78% and the pattern of lipid synthesis was altered as for simazine.

VIII. IMPLICATIONS

There is ample evidence that the disruption of intermediary or respiratory metabolism by a variety of herbicides may have far-reaching implications for a whole range of energy-requiring processes. It seems reasonable to assume, for example, that the non-translocation of several "contact" herbicides may reflect inhibition of energy-requiring steps associated with short-distance or long-distance transport of assimilates. Since foliage-applied herbicides might normally be expected to be translocated along the same pathways and involve the same mechanisms, it would appear that such herbicides may inhibit their own movement; considerations of this nature have been discussed with reference to the phenoxy-acid herbicides (e.g. Robertson and Kirkwood, 1969, 1970; Kirkwood *et al.* 1972). While some controversy surrounds the actual mechanism of long-distance transport in the phloem, it is generally accepted that the loading of assimilates (and herbicides) into the minor veins, involves an energy-requiring process. It is significant that most contact or "non-translocated" herbicides either uncouple or inhibit oxidative phosphorylation or inhibit electron transfer.

The inhibitory effect of a number of herbicides on synthetic processes such as protein or lipid synthesis may be due to direct effects on specific biosynthetic mechanisms or reflect inhibition due to indirect effects on energy metabolism. In this latter connection the findings of Moreland *et al.* (1969) represent a particularly relevant example. These authors reported that 14 out of 22 test compounds inhibited RNA and protein biosynthesis. While they found that the correlation between effects on O_2 uptake or oxidative phosphorylation with protein synthesis was not particularly clear, it is interesting that the compounds which were most inhibitory to RNA and protein biosynthesis included those which strongly uncouple or inhibit oxidative phosphorylation (dinoseb, propham, chlorpropham, ioxynil and pyriclor). They concluded that the effect of some compounds may be quite complex and involve such factors as the concentration of herbicides at the site(s) of action, the possibility of involvement with indirect metabolic factors such as synthesis of precursors, interruption of the energy balance, species selectivity, and effects on specific enzymes. Clearly, while advances in the techniques available for mode of action studies have enabled great advances to be made in the last decade, the field is one of considerable scope and potentially fruitful study.

REFERENCES

Åberg, B. and Johansson, I. (1966). *LantbrHögsk. Annlr.* **32,** 245–254 (Cit Ashton, F. M. and Crafts, A. S. 1973).
Akobundu, J. O., Bayer, D. E. and Leonard, O. A. (1970). *Weed Sci.* **18,** 403–408.

Allen, W. S. and Palmer, R. D. (1963). *Weeds*, **11**, 27–31.
Anderson, R. N., Linck, A. J. and Behrens, R. (1962). *Weeds*, **10**, 1–3.
Amchem Products (1963).
Ashton, F. M. (1960). *Pl. Physiol., Lancaster*, **35** (Suppl.) xxviii.
Ashton, F. M. (1963). *Weeds*, **11**, 295–297.
Ashton, F. M. and Bisalputra, T. (1964). *Pl. Physiol., Lancaster*, **39** (Suppl.) xxxiii.
Ashton, F. M., Bisalputra, T. and Risley, E. B. (1966). *Am. J. Bot.* **53**, 217–219.
Ashton, F. M. and Crafts, A. S. (1973). *In* "Mode of Action of Herbicides", pp. 504, Wiley-Interscience.
Ashton, F. M., Gifford, E. M. and Bisalputra, T. (1963). *Bot. Gaz.* **127**, 336.
Ashton, F. M., Penner, D. and Hoffman, S. (1968). *Weed Sci.* **16**, 169–171.
Ashton, F. M. and Uribe, E. G. (1962). *Weeds*, **10**, 295–297.
Baxter, R. (1967). *The effect of 2,4-D upon metabolism and composition of soybean mitochondria.* Ph.D. thesis, University of Illinois, p. 105. (*Diss. Abst.* **28**, 4887-B).
Bertagnolli, B. L. and Hanson, J. B. (1973). *Pl. Physiol., Lancaster*, **52**, 431–435.
Bingham, S. W. and Porter, W. K. Jr. (1961a). *Weeds*, **9**, 290–298.
Bingham, S. W. and Porter, W. K. Jr. (1961b). *Weeds*, **9**, 299–306.
Black, C. C. and Humphreys, T. E. (1962). *Pl. Physiol., Lancaster*, **37**, 66–73.
Bonner, J. (1949). *Am. J. Bot.* **36**, 323–332.
Borst, P. and Slater, E. C. (1961). *Biochem. biophys. Acta*, **48**, 362–379.
Bottrill, D. E. (1965). *Studies on the uncoupling action of 2,4-dichlorophenoxyacetic acid. Diss. Abstr.* **26**, 57.
Bottrill, D. E. and Hanson, J. B. (1969). *Aust. J. biol. Sci.* **22**, 847–855.
Bourke, J. B. and Fang, S. C. (1962). *Pl. Physiol., Lancaster*, Suppl. **37**, xxiv.
Bourke, J. B., Butts, J. S. and Fang, S. C. (1962). *Pl. Physiol., Lancaster*, **37**, 233–237.
Bourke, J. B., Butts, J. S. and Fang, S. C. (1964). *Weeds*, **12**, 272–279.
Brody, T. M. (1952). *Proc. Soc. exp. Biol. Med.* **80**, 533–536.
Burton, D. E., Lambie, A. J., Ludgate, J. C. L., Newbold, G. T., Percival, A. and Saggers, D. T. (1965). *Nature, Lond.* **208**, 1166–1169.
Campbell, W. F., Singh, B. and Salunkhe, D. K. (1971). *Abstr. Mountain States Soc. Electron Microscopists* (cit. Anderson and Thomson, 1973).
Carpenter, K. and Heywood, B. J. (1963). *Nature, Lond.* **200**, 28–29.
Castelfranco, P. and Bisalputra, T. (1965). *Am. J. Bot.* **52**, 222–227.
Chacravarti, A. S., Srivastava, D. D. and Khanna, K. L. (1955). *Sug. J.* **18**, 23–25.
Chance, B. and Williams, G. R. (1955). *J. biol. Chem.* **217**, 409–427.
Chance, B. and Williams, G. R. (1956). *Adv. Enzymol.* **17**, 65–134.
Chang, I. K. and Foy, C. L. (1971). *Weed Sci.* **19**, 54–58.
Chen, L. G., Ali, A., Fletcher, R. A., Switzer, C. M. and Stephenson, G. R. (1973). *Weed Sci.* **21**, 181–184.
Chernyshev, J. D. (1968). *Biol. Nauki.* **11**, 75–79.
Clowes, G. H. A. and Krahl, M. E. (1936). *J. gen. Physiol.* **20**, 145.
Coble, H. D. and Slife, F. W. (1971). *Weed Sci.* **19**, 1–3.
Corbadzijska, B. (1962). *Nauchni Trud. vissh. selskostop. Inst. Vasil Kolarov (Agron. Fak.)* **11**, 87–98.
Daams, J. and Barnsley, G. E. (1961). *Proc. EWRC—Columa Conf.* 68–74.
Dam, K. V. (1967). *Biochem. biophys. Acta*, **131**, 407–411.
Davis, D. E. (1968). *Proc. 21st Sth. Weed Control Conf.* 346.
Day, D. A. and Wiskitch, J. T. (1974). *Pl. Physiol., Baltimore*, **53**, 104–110.
Dekock, P. C. and Innes, A. M. (1970). *Can. J. Bot.* **48**, 1285–1288.
Dewey, O. R., Gregory, P. and Pfeiffer, R. K. (1956). *Proc. 3rd Br. Weed Control Conf.* 313–326.
Diem, J. R., Funderburk, H. H. and Negi, N. S. (1968). *Proc. 21st Sth Weed Control Conf.* p. 342 (cit. Ashton, F. M. and Crafts, A. S. 1973).
Dodds, E. C. and Greville, G. D. (1934). *Lancet*, **112**, 398.
Dow, W. A. (1952). *The effect of growth substances on the oxidative activity of plant mitochondria.* M.Sc. Thesis, University of Wisconsin.
Dukes, I. E. and Biswas, P. K. (1967). *Abstr. Meet. Weed Soc. Am.* p. 59.

Eastin, E. F. (1963). *Mode of action of 2-chloro-s-triazines in susceptible and resistant lines of Zea mays.* M.Sc. Thesis, State College, Mississippi.

Eastin, E. F., Palmer, R. D. and Grogan, C. O. (1964). *Weeds,* **12**, 64–65.

Ehrenfest, E. and Ronzoni, E. (1933). *Proc. Soc. exp. Biol. Med.* **31**, 318–319.

Ernster, L. and Lee, C. P. (1964). *A. Rev. Biochem.* **33**, 729–788.

Fang, S. C., Teeny, F. and Butts, J. S. (1961). *Pl. Physiol., Lancaster,* **36**, 192–196.

Feeny, R. W. (1966). *Proc. 20th N. East Weed Control Conf.* 595–603.

Ferrari, T. E. and Moreland, D. E. (1967). *Abstr. Meet. Weed Sci. Am.* 57–58.

Ferrari, T. E. and Moreland, D. E. (1969). *Pl. Physiol., Lancaster,* **44**, 429–434.

Foy, C. L. and Penner, D. (1965). *Weeds,* **13**, 226–231.

Freed, V. H., Reithel, F. J. and Remmert, L. F. (1961). *Proc. 4th int. Conf. Pl. Growth Regulation* 289–303.

Funderburk, H. H. and Porter, W. K. Jr. (1961a). *Weeds,* **9**, 538–544.

Funderburk, H. H. and Porter, W. K. Jr. (1961b). *Weeds,* **9**, 545–557.

Funderburk, H. H. and Davis, D. E. (1963). *Weeds,* **11**, 101–104.

Gast, G. (1958). *Experientia,* **14**, 134–136. English summary.

Gaur, B. K. and Beevers, H. (1959). *Pl. Physiol., Lancaster,* **34**, 427–432.

Gentner, W. A. (1966). *Weeds,* **14**, 27–30.

Giglio, J. R. and Moura Goncalves, J. (1963). *Anais Acad. bras. Cienc.* **35**, 293.

Goren, R. and Monselise, S. P. (1965). *Pl. Physiol., Lancaster,* **40** (Suppl.), xv.

Grant, M. E. and Fuller, K. W. (1971). *J. exp. Bot.* **22**, 49–59.

Gruenhagen, R. D. and Moreland, D. E. (1971). *Weed Sci.* **19**, 319–323.

Haard, N. F. and Hultin, H. O. (1968). *Analyt. Biochem.* **24**, 299–304.

Hall, W. C., Johnson, S. P. and Leinweber, C. L. (1954). *Bull. Texas agric. Exp. Stn.* **789**, 1–15.

Hanson, J. B. and Slife, F. W. (1969). *Residue Rev.* **25**, 59–67.

Hardcastle, W. S. and Wilkinson, R. E. (1968). *Weed Sci.* **16**, 339–340.

Hartley, G. S. and West, T. F. (1969). *In* "Chemicals for Pest Control", p. 316, Pergamon Press.

Hemker, H. C. (1962). *Biochim. biophys. Acta,* **63**, 46–54.

Hemker, H. C. (1964). *Biochim. biophys. Acta,* **81**, 1–8.

Hendrix, D. L. and Muench, S. R. (1969). *Pl. Physiol., Lancaster, Abstr.* **44**, S-26.

Hill, E. R., Putala, E. C. and Vengris, J. (1968). *Weed Sci.* **16**, 377–380.

Hilton, J. L. (1965). *Weed Sci.* **13**, 267–271.

Hilton, J. L., Ard, J. S., Jansen, L. L. and Gentner, W. A. (1959). *Weeds,* **7**, 381–396.

Hofstra, G. and Switzer, C. M. (1968). *Weed Sci.* **16**, 23–28.

Hotchkiss, R. D. (1944). *Adv. Enzymol.* **4**, 153–199 (cit. E. C. Slater, 1963, p. 505).

Humphreys, T. E. and Dugger, W. M. (1957a). *Pl. Physiol., Lancaster,* **32**, 136–140.

Humphreys, T. E. and Dugger, W. M. (1957b). *Pl. Physiol., Lancaster,* **32**, 530–536.

Humphreys, T. E. and Dugger, W. M. (1958). *Pl. Physiol., Lancaster,* **33**, 112–116.

Humphreys, T. E. and Dugger, W. M. (1959). *Pl. Physiol., Lancaster,* **34**, 580–582.

Ikuma, H. (1972). *A. Rev. Pl. Physiol.* **23**, 419–436.

Ikuma, H. and Bonner, W. D. (1967). *Pl. Physiol., Lancaster,* **42**, 67–75.

Ingle, M. and Rodgers, B. J. (1961). *Weeds,* **9**, 264–272.

Inoue, Y., Ishizuka, K. and Mitsui, S. (1967). *Agric. biol. Chem.* **31**, 422–427.

Jain, M. L., Kurtz, E. B. and Hamilton, K. C. (1966). *Weeds,* **14**, 259–262.

Jaworski, E. G. (1956). *Science, N.Y.* **123**, 847–848.

Johnson, B. J. and Jellum, M. D. (1969). *Agron. J.* **61**, 185–187.

Jones, D. W. and Foy, C. L. (1971). *Weed Sci.* **19**, 595–597.

Jones, O. T. G. and Watson, W. A. (1965). *Nature, Lond.* **208**, 1169–1170.

Juniper, B. E. and Bradley, D. E. (1958). *J. Utrastruct. Res.* **2**, 16–27.

Kerr, M. W. and Wain, R. L. (1964). *Ann. appl. Biol.* **54**, 441–446.

Killion, D. D. and Frans, R. E. (1969). *Weed Sci.* **17**, 468–470.

Kim, W. K. and Bidwell, R. G. S. (1967). *Can. J. Bot.* **45**, 1751–1760.

Kirkwood, R. C., Dalziel, J., Matlib, M. A. and Somerville, L. (1972). *Pestic. Sci.* **3**, 307–321.

Kolattukudy, P. E. (1968). *Science, N.Y.* **159**, 498–505.

Koopman, H. and Daams, J. (1960). *Nature, Lond.* **186**, 89.

Krishning, J. (1965). *Phytopath. Z.* **53**, 65–84.

Ku, H. S., Pratt, H. K., Spurr, A. R. and Harris, W. M. (1968). *Pl. Physiol., Lancaster,* **43,** 883–887.
Ladonin, V. F. (1966). *Vest. sel'-khoz. Nauki Mosk.* **11,** 137–141.
Ladonin, V. F. (1967). *Agrokhimiya,* **2,** 85–95.
Ladonin, V. F. and Svittser, K. M. (1967). *Soviet Pl. Physiol.* **14,** 853–860.
Lance, C., Hobson, G. E., Young, R. E. and Biale, J. B. (1965). *Pl. Physiol., Lancaster,* **40,** 116–123.
Lardy, H. A. and Wellman, H. (1952). *J. biol. Chem.* **195,** 215–224.
Leasure, J. K. (1964). *J. agric. Fd Chem.* **12,** 40–43.
Leopold, A. C. and Guernsey, F. S. (1953). *Proc. natn. Acad. Sci. U.S.A.* **39,** 1105–1111.
Lieberman, M. and Baker, J. E. (1965). *A. Rev. Pl. Physiol.* **16,** 343–382.
Loomis, W. F. and Lipmann, F. (1948). *J. biol. Chem.* **173,** 807–808.
Lotlikar, P. D. (1960). *The effects of herbicides on oxidative phosphorylation in mitochondria from cabbage,* Brassica oleracea. *Diss. Abstr.* **21,** 446–447.
Lotlikar, P. D., Remmert, L. F. and Freed, V. H. (1968). *Weed Sci.* **16,** 161–165.
Lugo-Lopez, M. A., Samuels, G. and Grant, R. (1953). *J. Agric. Univ. P. Rico,* **37,** 44–51 (cit. Wort, D. J., 1964).
Lynn, W. S. and Brown, R. H. (1966). *Archs Biochem. Biophys.* **114,** 271–281.
Maas, W. K. and Vogel, H. J. (1953). *J. Bact.* **65,** 388–393.
McDaniel, J. L. and Frans, R. E. (1969). *Weed Sci.* **17,** 192–196.
McWhorter, C. G. (1959). *Some effects of 3-amino-1,2,4-triazole on the respiratory activities of* Zea mays L. Ph.D. Thesis, Louisiana State University. (*Diss. Abstr.* **19,** 1929).
McWhorter, C. G. (1961). *Weeds,* **9,** 563–568.
McWhorter, C. G. and Porter, W. K. (1960). *Weeds,* **8,** 29–38.
Mann, J. D. and Pu, M. (1968). *Weed Sci.* **16,** 197–198.
Martin, A. W. and Field, J. (1934). *Proc. Soc. exp. Biol. Med.* **32,** 54–55.
Mashtakov, S. M. and Moshchuk, P. A. (1967). *Agrokhimiya,* **9,** 80–99 (cit. Ashton and Crafts, 1973).
Mashtakov, S. M., Deeva, V. P. and Volynets, A. P. (1967a). *Nauka Tekh. Minsk.* (cit. Ashton and Crafts, 1973).
Mashtakov, S. M., Deeva, V. P. and Volynets, A. P. (1967b). *In* Mashtakov, S. M. "Physiological Effects of Herbicides on Varieties of Crop Plants", pp. 124–146, *Nauka Tekh, Minsk* (cit. Ashton and Crafts, 1973).
Matlib, M. A. (1970). *Effect of selected phenoxy-acid herbicides on the oxidative and phosphorylative activities of mitochondria from* Vicia faba. Ph.D. thesis, University of Strathclyde, Glasgow.
Matlib, M. A., Kirkwood, R. C. and Patterson, J. D. E. (1971a). *Weed Res.* **11,** 190–192.
Matlib, M. A., Kirkwood, R. C. and Smith, J. E. (1971b). *J. exp. Bot.* **22,** 291–303.
Matlib, M. A., Kirkwood, R. C. and Smith, J. E. (1972). *J. exp. Bot.* **23,** 886–898.
Miller, C. S. and Hall, W. C. (1957). *Weeds,* **5,** 218–226.
Mitchell, P. (1961). *Nature, Lond. t3191,* 144–148.
Mitchell, P. (1966a). *Biol. Rev.* 445–502.
Mitchell, P. (1966b). *In* "Regulation of Metabolic Processes in Mitchondria", (J. M. Tager *et al.* ed.) BBA Library, Vol. 7, p. 397. Elsevier, Amsterdam.
Mitchell, P. and Moyle, J. (1967). *In* "Biochemistry of Mitochondria" (E. C. Slater *et al.* eds), p. 53. Academic Press, London and New York.
Monod, J., Changeux, J. P. and Jacob, F. (1963). *J. molec. Biol.* **6,** 306–329.
Moreland, D. E. (1967). *A Rev. Pl. Physiol.* **18,** 365–386.
Moreland, D. E. and Blackmon, W. J. (1970). *Weed Sci.* **18,** 419–426.
Moreland, D. E., Hill, J. L. and Hilton, J. L. (1958). *Abstr. Meet. Weed Soc. Am.* pp. 40–41.
Moreland, D. E., Malhotra, S. S., Gruenhagen, R. D. and Shokraii, E. H. (1969). *Weed Sci.* **17,** 556–563.
Mostafa, I. Y. and Fang, S. C. (1971). *Weed Sci.* **19,** 248–253.
Nasyrova, T., Mirkasimova, Kh. and Pazilova, S. (1968). *Uzbek. biol. Zh.* **12,** 23–26.
Negi, N. S., Funderburk, H. H., Schultz, D. P. and Davis, D. E. (1967). *Abstr. Meet. Weed Soc. Am.* p. 58.
Negi, N. S., Funderburk, H. H., Schultz, D. P. and Davis, D. E. (1968). *Weed Sci.* **16,** 83–85.

Olech, K. (1966). *Annls Univ. Mariae Curie-Skłodowska* (E), **21**, 289–308. (In Polish, English summary.)

Olech, K. (1967). *Annls Univ. Mariae Curie-Skłodowska* (E), **21**, 289–308. (In Polish, English summary.)

Oorschot, J. L. P. van and Hilton, J. L. (1963). *Archs Biochem. Biophys.* **100**, 289–294.

Oyolu, C. and Huffaker, R. C. (1964). *Crop Sci.* **4**, 95–96.

Palmer, J. M. (1967). *Nature, Lond.* **216**, 1208.

Palmer, R. D. and Allen, W. S. (1962). *Proc. 15th Sth Weed Control Conf.* p. 271.

Papa, S., Lofrumento, N. E., Paradies, G. and Quagliariello, E. (1968). *Biochem. biophys. Acta,* **153**, 306–308.

Parker, M. V. (1965). *Biochem. J.* **97**, 658–662.

Paton, D. and Smith, J. E. (1965). *Weed Res.* **5**, 75–77.

Paton, D. and Smith, J. E. (1967). *Can. J. Biochem.* **45**, 1891–1899.

Penner, D. (1968). *Weed Sci.* **16**, 519–522.

Penner, D. and Meggitt, W. F. (1970). *Crop Sci.* **10**, 553–554.

Pfeiffer, R. K., Dewey, O. R. and Brunskill, R. T. (1957). *Proc. int. Congr. Crop Protection, 4th Meeting* (Hamburg, Germany, Sept. 8–15).

Pickett, M. J. and Clifton, C. E. (1943). *J. Cell. comp. Physiol.* **22**, 147–165 (cit. E. C. Slater, 1963, p. 505).

Pinchot, G. B. (1963). *Fedn. Proc. Fedn. Am. Socs. exp. Biol.* **22**, 1076–1079.

Pinchot, G. B. (1967). *J. biol. Chem.* **242**, 4577–4583.

Ploszynski, M., Swietochowski, B. and Zurawski, H. (1969). *Roczn. Nauk roln.* (ser. A), **95**, 401–415. (In Polish, English summary.)

Prasad, R. and Blackman, G. E. (1965). *J. exp. Bot.* **16**, 545–568.

Price, H. C. (1969). *The toxicity, distribution and mode of action of dichlobenil in plants.* Ph.D. dissertation, Mich. State Univ. p. 72.

Pulver, E. L. and Ries, S. K. (1973). *Weed Sci.* **21**, 233–237.

Racker, E. (1961). *Adv. Enzymol.* **23**, 323–399.

Rai, G. S. and Hamner, C. L. (1956). *Bull. Mich. agric. Coll. Exp. Stn.* **38**, 555–558.

Rhodes, A. (1952). *J. exp. Bot.* **2**, 129–154.

Ries, S. K., Chmiel, H., Dilley, D. R. and Filner, P. (1967). *Proc. natn. Acad. Sci. U.S.A.* **58**, 526.

Robertson, M. M. and Kirkwood, R. C. (1969). *Weed Res.* **9**, 224–240.

Robertson, M. M. and Kirkwood, R. C. (1970). *Weed Res.* **10**, 94–120.

Rodgers, B. J. (1963). *Pl. Physiol., Lancaster,* Suppl. **38**, liv.

Romani, R. J., Yu, I. K. and Fisher, L. K. (1969). *Pl. Physiol., Lancaster,* **44**, 311–312.

Ronzoni, E. and Ehrenfest, E. (1936). *J. biol. Chem.* **115**, 749–768.

Ross, M. A. (1966). *Abstr. Meet. Weed Soc. Am.* p. 50.

Roth, W. (1958). *Recherches sur l'action sélective de substances herbicides du groupe des triazines.* Ph.D. dissertation, Université de Strasbourg.

Russell, J. (1957). *Can. J. Bot.* **35**, 409.

Sabnis, D. D. and Audus, L. J. (1967). *Ann. Bot.* **31**, 263–281.

Salisbury, F. B. and Ross, C. (1969). "Plant Physiology", Wadsworth and Co. Publ.

Sanadi, D. R., Gibson, D. M. and Ayengar, P. (1954). *Biochem. biophys. Acta,* **14**, 434–436.

Sanwal, B. D. (1959). In "Recent Advances in Botany" (9th Int. Bot. Congr., Montreal), Vol. 2. University of Toronto Press, Toronto, pp. 1012–1017.

Sarkissian, I. V. and Srivastava, H. K. (1968). *Pl. Physiol., Lancaster,* **42**, 1406–1410.

Sasaki, S. and Kozlowski, T. T. (1966). *Nature, Lond.* **210**, 439–440.

Schweizer, E. E. (1970). *Weed Sci.* **18**, 131–134.

Schroeder, M. and Warren, G. F. (1971). *Weed Sci.* **19**, 671–675.

Sell, H. M., Luecke, R. W., Taylor, B. M. and Hamner, C. L. (1949). *Pl. Physiol., Lancaster,* **24**, 295–299.

Sierra, J. N. and Vega, M. R. (1967). *Philipp. Agric.* **51**, 438–452.

Singh, B. and Salunkhe, D. K. (1970). *Can. J. Bot.* **48**, 2213–2217.

Singh, B., Campbell, W. F. and Salunkhe, D. K. (1972). *Am. J. Bot.* **59**, 568–572.

Slater, E. C. (1962). *Comp. Biochem. Physiol.* **4**, 281–301.

Slater, E. C. (1963). *In* "Metabolic Inhibitors" (R. M. Hochster and J. H. Quastel, ed.), Vol. 2. Academic Press, New York and London.
Slater, E. C. (1966). *In* "Regulation of Metabolic Processes in Mitochondria" (J. M. Tager, S. Papa, E. Quagliariello and E. C. Slater, eds), Elsevier, Amsterdam.
Smith, J. E., Paton, D. and Robertson, M. M. (1966). *Proc. 8th Br. Weed Control Conf.* **1**, 279–282.
Smith, J. E. and Shennan, J. L. (1966). *J. gen. Microbiol.* **42**, 293–300.
St. John, J. B. (1971). *Weed Sci.* **19**, 274–276.
St. John, J. B. and Hilton, J. L. (1973). *Weed Sci.* **21**, 477–480.
Stenlid, G. and Saddik, K. (1962). *Physiologia Pl.* **15**, 369–379.
Stevens, V. L., Butts, J. S. and Fang, S. C. (1962). *Pl. Physiol., Lancaster,* **37**, 215–222.
Still, G. G., Davis, D. G. and Zander, G. L. (1970). *Pl. Physiol., Baltimore,* **46**, 307–314.
Stinson, R. A. and Spencer, M. (1968). *Can. J. Biochem.* **46**, 43–50.
Stoner, C. D., Hodges, T. K. and Hanson, J. B. (1964). *Nature, Lond.* **203**, 258–261.
Sumida, S. and Ueda, M. (1973). *Pl. Cell Physiol. Tokyo,* **14**, 781–785.
Switzer, C. M. (1957). *Pl. Physiol., Lancaster,* **32**, 42–44.
Talbert, R. E. (1967). *Abstr. Meet. Weed Soc. Am.* pp. 50–51.
Thompson, O. C., Truelove, B. and Davis, D. E. (1969). *J. agr. Fd Chem.* **17**, 997–999.
Thompson, O. C., Truelove, B. and Davis, D. E. (1970). *Proc. Weed Sci. Soc. Am.* 318.
Truelove, B. and Davis, D. E. (1969). *Abstr. Meet. Weed Sci. Soc. Am.* 179. (*Weed Abstr.* **21**, 65. 1972.)
Truelove, B., Jones, L. R. and Davis, D. E. (1973). *Weed Sci.* **21**, 24–27.
Tsay, R. C. and Ashton, F. M. (1971). *Weed Sci.* **19**, 682–684.
Tweedy, J. A. and Ries, S. K. (1967). *Pl. Physiol., Lancaster,* **42**, 280–282.
Veerasekaran, P. (1975). *The mode of action of asulam in bracken* (Pteridium aquilinum L. Kuhn). Ph.D. thesis, University of Strathclyde, Glasgow.
Venis, M. A. (1971). *Proc. natn. Acad. Sci. U.S.A.* **68**, 1824–1827.
Verleur, J. D. (1965). *Pl. Physiol., Lancaster,* **40**, 1003–1007.
Verloop, A. (1972). *Residue Rev.* **43**, 55–104.
Verloop, A. and Nimmo, W. B. (1969). *Weed Res.* **9**, 357–370.
Voinilo, V. A., Deeva, V. P. and Mashtakov, S. M. (1967). *Dokl. Akad. Nauk. belorussk. SSR.* **11**, 638–642.
Voinilo, V. A., Deeva, V. P. and Mashtakov, S. M. (1968). *Dokl. Akad. Nauk belorussk SSR.* **12**, 460–462.
Wain, R. L. (1963). *Nature, Lond.* **200**, 28–29.
Walsh, G. E. and Grow, T. E. (1971). *Weed Sci.* **19**, 568–570.
Wang, B., Grooms, S. and Frans, R. E. (1974). *Weed Sci.* **22**, 64–66.
Watson, K. and Smith, J. E. (1967). *Biochem. J.* **104**, 332–339.
Wedding, R. T. and Black, M. K. (1961). *Pl. Soil,* **14**, 242–248.
Wedding, R. T. and Black, M. K. (1962). *Pl. Physiol., Lancaster,* **37**, 364–370.
Wedding, R. T. and Black, M. K. (1963). *Pl. Physiol., Lancaster,* **38**, 157–164.
Weinbach, E. C. (1956a). *Archs Biochem. Biophys.* **64**, 129–143.
Weinbach, E. C. (1956b). *Science, N.Y.* **124**, 940.
Weinbach, E. C. and Garbus, J. (1964). *Science, N.Y.* **145**, 824–826.
Weinbach, E. C. and Garbus, J. (1965). *J. biol. Chem.* **240**, 1811–1819.
Weinbach, E. C. and Garbus, J. (1966). *J. biol. Chem.* **241**, 3708–3713.
Weinbach, E. C. and Garbus, J. (1969). *Nature, Lond.* **221**, 1016–1018.
Weller, L. E., Luecke, R. W., Hamner, C. L. and Sell, H. M. (1950). *Pl. Physiol., Lancaster,* **25**, 289–293.
Wilson, D. F. and Merz, R. D. (1967). *Archs Biochem. Biophys.* **119**, 470–476.
Wiskitch, J. T. (1966). *Nature, Lond.* **212**, 641–642.
Wit, J. G. and van Genderen, H. (1966). *Biochem. J.* **101**, 707–710.
Wojtaszek, T. (1966). *Weeds,* **14**, 125–129.
Wojtaszek, T., Cherry, J. H. and Warren, G. F. (1966). *Pl. Physiol., Lancaster,* **41**, 34–38.
Wolf, D. E., Vermillion, G., Wallace, A. and Ahlgren, G. H. (1950). *Bot. Gaz.* **112**, 188–197.
Wort, D. J. (1951). *Pl. Physiol., Lancaster,* **26**, 50–58.

Wort, D. J. (1964). *In* "The Physiology and Biochemistry of Herbicides" (L. J. Audus, ed.), pp. 291–330. Academic Press, New York and London.

Wort, D. J. and Shrimpton, M. (1958). *Res. Rep. natn. Weed Committee, West Sect. Canad. Dept. Agric.,* p. 124 (cit. Wort, D. J., 1964).

Wu, M. T., Singh, B. and Salunkhe, D. K. (1971a). *Pl. Physiol., Lancaster,* **48,** 517–520.

Wu, M. T., Singh, B. and Salunkhe, D. K. (1971b). *Phytochemistry,* **10,** 2025–2027.

Yukinaga, H., Ide, K. and Ito, K. (1973). *Weed Res. (Japan),* **15,** 34.

Zhirmunskaya, N. M. (1966). *Khimiya sots. Khoz.* **4,** 46–51 (cit. Ashton, and Crafts, 1973).

CHAPTER 16

ACTIONS ON PHOTOSYNTHETIC SYSTEMS

D. E. MORELAND

United States Department of Agriculture
Agricultural Research Service
Southern Region, Crop Science Department
North Carolina State University
Raleigh, North Carolina 27607, U.S.A.

J. L. HILTON

United States Department of Agriculture
Agricultural Research Service
Northeastern Region, Agricultural Research Center
Beltsville, Maryland 20705, U.S.A.

I. Introduction

The process in which carbohydrates are synthesized from carbon dioxide and water in the presence of light (photosynthesis) is unique to green plants.

Consequently, herbicides that specifically inhibit this process can be anticipated to be relatively non-toxic to mammals. The process is extremely complicated, and many of its details are not understood completely. The objectives of this chapter are to consider how herbicides interfere with photosynthesis and to relate the inhibition to the expression of phytotoxicity. Means by which photosynthesis can be inhibited by herbicides include interference with the (a) reproduction, development, structure, and integrity of chloroplasts; (b) many biosynthetic pathways that are involved in the production of output products; and (c) photochemical induction pathways involved in the conversion of radiant energy to chemical energy.

Inhibition of photosynthesis in micro-organisms was reported by Warburg in 1919 and 1920 for ethyl N-phenylcarbamate (phenylurethan). However, the selective action and usefulness of the N-phenylcarbamates as herbicides were not described until 1945 by Templeman and Sexton. Subsequently, in the early 1950's, phenylurea herbicides were introduced. In the mid-1950's, Wessels and van der Veen (1956) and Cooke (1956) related Warburg's work with phenylurethan to the N-phenylcarbamate and the phenylurea herbicides by demonstrating that these herbicides interfered with the photochemically induced electron transport of isolated chloroplasts. Some of the herbicidal phenylureas were 2,500 times more inhibitory than phenylurethan and inhibited photochemically induced electron transport at concentrations less than $0 \cdot 1 \ \mu M$ (Wessels and van der Veen, 1956). Recognition of the potent inhibition of photosynthesis by the phenylurea herbicides is considered to be among the most significant discoveries involving the modes and mechanisms of action of herbicides. The initial findings have been confirmed by a variety of physiological and biochemical approaches. Diuron also became a very useful tool to block oxygen evolution for photobiologists engaged in the elucidation of the basic processes of photosynthesis. The s-triazine herbicides, introduced shortly after the phenylureas, were also recognized as being able to inhibit the photochemical activity of isolated chloroplasts and the photosynthesis of intact plants.

During the late 1950's, through the 1960's, and into the 1970's, considerable attention has been given to the effects of herbicides on one or more of the steps involved in photosynthesis. More than half of the currently used herbicides have been shown to interfere with photosynthesis in some way, and these herbicides account for a major portion of the total herbicide production, sales, and application around the world. Because of the efforts of a large number of investigators, more is known generally about the mechanism of action and behaviour of the photosynthesis inhibitors in plants than any of the other groups of herbicides.

II. Photochemical Reactions of Chloroplasts

The primary source of energy in the biosphere results from the conversion of solar energy to chemical energy by chlorophyll-containing plant cells. This photosynthetic reaction occurs in organelles called chloroplasts. Chloroplasts contain specific DNA and RNA, have the capacity for protein synthesis, and reproduce by division. Chloroplasts, like mitochondria, undergo shape and conformational changes during electron transport, and the chloroplast is capable of ion accumulation. The higher-plant chloroplast is saucer-shaped, and from 4 to 10 µm in diameter and 1 to 3 µm thick. The chlorophyll is concentrated in bodies within the chloroplasts called grana, which are about 0·4 µm in diameter. The lamellae of the grana are surrounded by an embedding matrix referred to as the stroma. Reactions that are catalyzed within the chloroplasts are classified as light reactions, which depend directly on light energy, and dark reactions, which can occur in the absence of light. The two sets of reactions can be represented by the following unbalanced equations, each of which summarizes a complex series of events:

$$H_2O + ADP + P_i + NADP^+ \xrightarrow[\text{energy}]{\text{light}} O_2 + ATP + H^+ + NADPH \tag{1}$$

$$CO_2 + ATP + NADPH + H^+ \longrightarrow (CH_2O) + ADP + P_i + NADP^+ \tag{2}$$

The light reactions and associated electron transport reactions of photosynthesis that comprise Equation (1) take place in the chlorophyll-containing lamellae. The number of ATP molecules produced by the reactions summarized in Equation (1) is not known exactly. Enzymes involved in CO_2 fixation (the dark reactions of photosynthesis) are confined to the stroma, and the reactions catalyzed are represented by Equation (2). In Equation (2), (CH_2O) designates a molecule of CO_2 reduced to the carbohydrate level. Reduction of one mole of CO_2 requires 2 moles of NADPH and 3 moles of ATP, and synthesis of one mole of hexose requires 12 moles of NADPH and 18 moles of ATP, if carbohydrates are formed by the Calvin cycle (C_3-cycle). However, 30 moles of ATP per mole of hexose are required if CO_2 is reduced via the Hatch–Slack pathway (C_4-cycle), and 39 moles per mole of hexose if Crassulacean acid metabolism is involved. Evidence for additional pathways of CO_2 fixation has been obtained, but details, including ATP requirements, have not been established.

The light reaction is very rapid and becomes saturated with light in 10 µs. Up to 100 ms of dark time are required to use up the products of the light reaction. A considerable number of chlorophyll molecules need to cooperate to trap sufficient quanta of light energy to fix carbon dioxide and evolve oxygen. Estimates for the number of chlorophyll molecules that comprise such a "photosynthetic unit" range from 200 to 300 up to 2,500.

Photoinduced electron transport and the coupled phosphorylation, as they are postulated to occur in chloroplast lamellae, are presented schematically in Fig. 1. Not all investigators agree on the details of the scheme as shown here, and even the sequence of the intermediates identified in the scheme is questioned by some. The numbers and locations of the phosphorylation sites also remain to be identified precisely. However, the scheme as shown is a reasonable approximation, based on available information, which will be useful in identifying sites of action of herbicides. In the Figure, reactions that occur in the light are represented by the open arrows, and the solid arrows represent electron transfers that occur in the dark.

Absorption of light energy causes electrons to flow from an electron donor (water) to an electron acceptor ($NADP^+$). Involved in the process are two light reactions identified as photosystem II and photosystem I. The light-trapping unit of photosystem II, which absorbs short wavelength light (< 680 nm), involves chlorophylls a and b and contains accessory pigments. An unidentified chlorophyll a molecule serves as the reaction centre. The primary electron acceptor of photosystem II, symbolized as Q, has not been identified, but has been shown to quench the fluorescence of photosystem II. The oxidized chlorophyll receives an electron from a postulated donor, designated as Y. The oxidized Y extracts electrons from water, ultimately yielding oxygen, and involving, in some manner, a mangano-protein and chloride ions. The symbol Y is also used by some investigators to designate the water-splitting enzyme system.

The next carrier after Q on the electron transport chain is a plastoquinone (PQ). The reduced plastoquinone, in turn, reduces a b-type cytochrome (b_{559}). This cytochrome makes contact with cytochrome f, which is a c-type cytochrome. Cytochrome f is in contact with a copper-protein, plastocyanin (PC), which appears to be immediately adjacent to the reaction centre of photosystem I. Electron passage along the chain generates at least one molecule of ATP.

Photosystem I absorbs light having wavelengths greater than 680 nm, and the reaction centre involves a special chlorophyll a molecule identified as P_{700}. The immediate electron acceptor from P_{700} may be a form of bound ferredoxin, which has been called ferredoxin-reducing substance (FRS). Electrons flow subsequently to $NADP^+$ and involve the participation of ferredoxin (Fd) and a flavoprotein (Fp, ferredoxin-NADP oxidoreductase).

Cyclic electron flow is represented as a shunt or bypass in Fig. 1. It is not certain whether electron return begins with FRS or ferredoxin, but it is postulated to involve a distinct b-type cytochrome (b_{563}) from which electrons flow back to the main chain. A site of ATP generation on the cyclic pathway is shown also in the Figure. In cyclic electron flow, there is no way to measure the flux of electrons around the closed chain. Consequently, exactly where these electrons enter the main electron transport chain is not known. Some investigators consider the re-entry to be close to cytochrome f, but others suggest that the shunted electrons may enter the central chain at some point around plastoquinone. Hence, the phosphorylation site in the central chain may be shared by both cyclic and noncyclic electron flow.

In the reactions that occur in the lamellae, water is oxidized, and the products are ATP and NADPH. Hence, the energy (ATP) and reducing power (NADPH) required for CO_2 fixation are produced by the combined serial action of photosystems I and II. In this situation, electron transport is considered to be coupled to phosphorylation. Artificial electron acceptors can be substituted for $NADP^+$, which give rise to oxygen evolution, but involve only a short segment of the oxidation chain. This observation was first

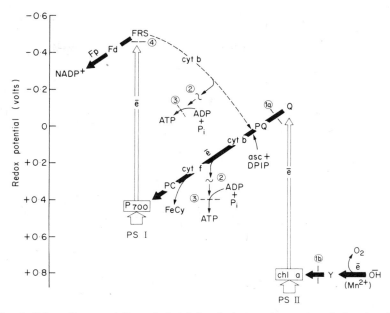

Fig. 1. Schematic presentation of photoinduced electron transport and phosphorylation reactions considered to occur in chloroplast lamellae. Open arrows indicate light reactions, solid arrows indicate dark reactions, and the narrow dashed line represents the cyclic pathway. Abbreviations used: PS I, photosystem I; PS II, photosystem II; Q, unknown primary electron acceptor for photosystem II; PQ, plastoquinones; cyt b, b-type cytochromes; cyt f, cytochrome f; PC, plastocyanin; P_{700}, reaction centre chlorophyll of photosystem I; FRS, ferredoxin-reducing substance; Fd, ferredoxin; Fp, ferredoxin-NADP oxidoreductase; FeCy, ferricyanide; asc, ascorbate; and DPIP, 2,6-dichlorophenolindophenol. The numbers 1a, 1b, 2, 3, and 4 indicate postulated sites of action by herbicides. See text for details.

reported by R. Hill in 1937. He demonstrated that oxygen evolution occurred when a suspension of isolated chloroplasts was illuminated in the presence of an artificial electron acceptor in accordance with the following equation:

$$2H_2O + 2A \xrightarrow{\text{light}} 2AH_2 + O_2 \qquad (3)$$

In the equation, A is the hydrogen (electron) acceptor and AH_2 is its reduced form. This reaction is known universally as the Hill reaction, the acceptor A as a Hill reagent, and an inhibitor of the reaction as a Hill inhibitor. A Hill reagent such as ferricyanide or a reducible dye thus acts as an artificial acceptor for electrons arising from water. With ferricyanide as acceptor, ATP generation can be demonstrated to occur during the Hill reaction. This type of phosphorylation is referred to as noncyclic photophosphorylation.

A number of compounds in their oxidized form, such as phenazine methosulfate (PMS), viologens, and certain flavins, can intercept electrons from photosystem I destined to reduce NADP$^+$. These compounds are considered to act in or around FRS on the scheme (Fig. 1) and catalyze a cyclic, but unmeasurable flow of electrons. Phosphorylation associated with the cyclic flow is termed cyclic photophosphorylation.

Various compounds other than water can donate electrons along the transport pathway between the two photosystems. Among these are reduced indophenols [2,6-dichlorophenolindophenol (DPIP) maintained in a reduced state by an excess of ascorbic acid], reduced TMPD (N,N,N',N'-tetramethyl-p-phenylenediamine), and reduced DAD (2,3,5,6-tetramethyl-p-phenylene-diamine).

[The many background observations that were summarized in this section are documented in contributions authored by Avron (1967, 1971), Avron and Neumann (1968), Bishop (1971), Black (1973), Cheniae (1970), Good and Izawa (1973), Hatch and Slack (1970), Kok (1965), and Park (1965).]

III. Inhibition of the Photochemical Reactions of Isolated Chloroplasts

A full comprehension of the specific sites involved in the inhibitory action of herbicides and the mechanisms through which inhibition is produced will become possible only when the uncertainty is resolved surrounding the sequence and interrelation of components in the electron transport pathway, numbers and locations of phosphorylation sites, and the mechanism of phosphorylation. Inhibitors have played, and will continue to play, an important role in the elucidation of the physiology and biochemistry of photosynthetic processes. Their action has been summarized by various authors including Avron (1967), Avron and Shavit (1965), Good and Izawa (1973), Good et al. (1966), and Izawa and Good (1972), but only limited consideration has been extended to herbicides other than diuron.

A. TYPES OF INHIBITORY ACTION

Herbicides and non-herbicidal inhibitors that affect the photochemically induced reactions of isolated chloroplasts can be divided into the following classes depending on effects inposed: (a) electron transport inhibitors, (b) uncouplers, (c) energy transfer inhibitors, (d) inhibitory uncouplers (multiple types of inhibition), and (e) electron acceptors.

Herbicides that inhibit the photochemical reactions of isolated chloroplasts have been called routinely inhibitors of the Hill reaction, primarily for convenience, and also because the action of the compounds was evaluated for many years under non-phosphorylating conditions, frequently with ferricyanide as the electron acceptor. In the past few years, more sophisticated studies have been conducted with herbicides, and more is known about their differential action. Consequently, the types of action and characteristic features permit their separation into the classes given above. Unfortunately, the actions of only some herbicides have been studied on more than just electron transport under non-phosphorylating conditions.

To provide an appropriate background, some of the terms that will be used to describe the photochemical reactions of isolated chloroplasts will be defined, together with the conditions under which measurements are made routinely. Basal electron transport is the reductive reaction in which water serves as the electron donor and $NADP^+$ or ferricyanide as the electron acceptor. The reaction is conducted in the absence of phosphorylating reagents (ADP, inorganic phosphate, and magnesium). Reduction of the electron acceptor can be followed spectrophotometrically, and oxygen evolution can be measured polarographically. Effects on noncyclic photophosphorylation are usually examined with water serving as the electron donor and with ferricyanide or $NADP^+$ as the electron acceptor in the presence of phosphorylating reagents (ADP, inorganic phosphate, and magnesium). Under these conditions, electron transport is coupled to phosphorylation. If ferricyanide is used as an electron acceptor, effects on oxygen evolution, reduction of ferricyanide, or esterification of inorganic phosphate can be measured. If $NADP^+$ is used as the electron acceptor and water serves as the electron donor, the reduction of $NADP^+$ can be monitored. With $NADP^+$, ferredoxin has to be added back to the reaction mixture, because it is usually eluted from the chloroplasts during extraction and isolation. Reduced DPIP is used sometimes to donate electrons to the transport pathway subsequent to photosystem II. Usually, with reduced DPIP, $NADP^+$ is used as the electron acceptor, and the reduction of $NADP^+$ is coupled to phosphorylation. With reduced DPIP, inhibitory effects imposed on photosystem II are circumvented.

Effects on cyclic photophosphorylation are conducted with PMS as the electron mediator under an argon gas phase. The objective in this system is to force oxidation to occur through the cytochrome chain when PMS is reduced by photosystem I. Reduced PMS is oxidized preferentially by oxygen, hence, in conducting studies under cyclic conditions, all oxygen must be removed from the system.

An uncoupled system is one in which electron transport is not coupled to phosphorylation. Ammonia and simple aliphatic amines such as methylamine will, in the presence of phosphorylating reagents, prevent phosphorylation but will not inhibit the rate of electron transport. In fact, the simple amines actually increase the rate of electron flow by removing the rate-limiting control imposed by phosphorylation on electron transport.

Representative examples of results that were obtained with different types of inhibitors are shown in Fig. 2. The data were obtained with freshly isolated spinach (*Spinacia oleracea*) chloroplasts. Unless indicated otherwise, assays were conducted under phosphorylating conditions at pH 8·0 with ferricyanide as the electron acceptor in the noncyclic studies and with PMS as the electron mediator in an argon atmosphere for the cyclic photophosphorylation measurements (Moreland and Boots, 1971).

1. *Electron Transport Inhibitors*

Electron transport inhibition results from the removal or inactivation of one or more of the intermediate electron transport carriers. The chlorinated

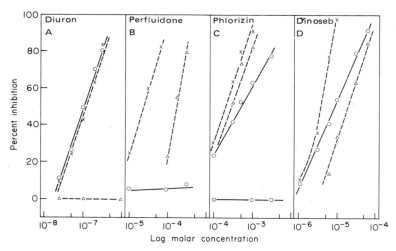

Fig. 2. Inhibition produced by varying concentrations of an electron transport inhibitor, uncoupler, energy transfer inhibitor, and inhibitory uncoupler on electron transport and ATP formation in isolated spinach chloroplasts. Data were obtained with freshly isolated spinach chloroplasts. Assays were conducted at pH 8·0 with ferricyanide as the electron acceptor in the noncyclic studies and with PMS as the electron mediator in an argon atmosphere for the cyclic photophosphorylation measurements. See Moreland and Boots (1971) for details of the procedures. (Legend: O—O, noncyclic electron transport; ×— —×, noncyclic photophosphorylation; △— —△, cyclic photophosphorylation; □—□, methylamine-uncoupled electron transport.)

phenylureas, *bis*-carbamates such as phenmedipham, chlorinated *s*-triazines, and substituted uracils appear to have a single site of action on the photochemical pathway (Moreland, 1967, 1969). These compounds inhibit basal electron transport, methylamine-uncoupled electron transport, and noncyclic electron transport with water as electron donor and ferricyanide or $NADP^+$ as electron acceptor. They also inhibit the coupled phosphorylation to the same extent as they do the reductive reaction, but do not inhibit cyclic photophosphorylation markedly. Photoreduction of $NADP^+$ with electron acceptors that circumvent photosystem II (ascorbate + DPIP, DAD) is not affected. Results obtained with diuron on noncyclic electron transport and photophosphorylation are shown in Fig. 2A. These patterns are obtained typically with electron transport inhibitors. Diuron is one of the most potent inhibitors identified to date, and I_{50} values reported in the literature are for the most part in the 10^{-8} to 10^{-7} M range. Additional herbicides that act in this way include the pyridazinones (Hilton *et al.* 1969), diphenylethers (Moreland *et al.* 1970a), 1,2,4-triazinones (Draber *et al.* 1969), azido-*s*-triazines (Gabbott, 1969), cyclopropanecarboxamides (Hopkins *et al.* 1967), *p*-alkylanilides (Bellotti *et al.* 1968), and *p*-alkylthioanilides (Baruffini *et al.* 1970b). The action of some of the above-listed herbicides has been examined only against the Hill reaction. Consequently, after more detailed study, some might need to be reclassified.

Diuron is used routinely by photobiologists to block the photosystem II mediation of oxygen evolution. Identification of its site of action has received considerable attention. However, the site has not been resolved to the satisfaction of all investigators. In the presence of diuron, the photoreduction of cytochrome f, plastocyanin, and P_{700} is inhibited, but their photooxidation is not. In addition, the fluorescence yield of photosystem II is increased by diuron. This increase in fluorescence is interpreted as evidence that Q becomes almost totally reduced, which in turn suggests that it is no longer capable of passing its electrons to photosystem I. Because of the effect on Q, Duysens and Amesz (1962) considered that diuron acted at the reducing rather than the oxidizing side of photosystem II (Fig. 1, site 1a). However, some investigators consider that diuron may act on the oxidizing side of photosystem II (Fig. 1, site 1b) (Kok and Cheniae, 1966; Rosenberg, et al. 1972) or directly on the chlorophyll a reaction centre (Döring et al. 1969). Renger (1973) proposed that diuron may act on both sides of photosystem II: (a) on the reducing side where it acts as an inhibitor, and (b) on the oxidizing side where it accelerates the deactivation of the water-splitting enzyme system Y.

Of the many inhibitors of photosystem II, only the action of diuron has been studied in considerable detail. Although the action of the many other diversely structured compounds is compared frequently to diuron, their site(s) of action have not been resolved beyond the general area around photosystem II. The mechanism by which inhibition is imposed, even by diuron, is unknown.

2. *Uncouplers*

Uncouplers are compounds that dissociate electron transport from ATP formation; hence, they inhibit phosphorylation but not electron transport. Ideally, an uncoupler should, in a complete noncyclic photophosphorylating system, inhibit the rate of phosphorylation without inhibiting the rate of electron flow. However, in the absence of one or more of the components necessary for photophosphorylation (ADP, inorganic phosphate, or magnesium), an uncoupler should stimulate the rate of electron flow to at least that rate obtained upon the addition of the missing component. Uncouplers are thought to act by promoting the breakdown of a high-energy intermediate or "state". If the rate of breakdown of the high-energy intermediate or state by the uncoupler is faster than it is during ATP formation, an acceleration of electron flow may be seen. Under conditions where the rate of discharge of the high-energy state limits the overall flow of electrons (such as in the absence of ADP, inorganic phosphate, or magnesium), uncouplers do stimulate markedly the rate of electron flow (Good and Izawa, 1973).

Results obtained with a herbicidal uncoupler, perfluidone [1,1,1-trifluoro-4'-(phenylsulphonyl)methanesulphono-o-toluidide], are shown in Fig. 2B. This compound did not inhibit electron transport with ferricyanide as the electron acceptor under either phosphorylating or non-phosphorylating conditions.

However, both noncyclic and cyclic photophosphorylation were inhibited, with cyclic being slightly less sensitive than noncyclic photophosphorylation, as evidenced by the higher I_{50} value.

Perfluidone is the only herbicide that has been shown to function as an ideal uncoupler at pH 8·0. Common non-herbicidal uncouplers of photophosphorylation are ammonium ions, simple amines, chlorpromazine, atebrin, derivatives of carbonyl-cyanide phenylhydrazones (CCCP and FCCP), antibiotics such as valinomycin and gramicidin, and some phenols (Good and Izawa, 1973). Because uncouplers relieve the inhibition of electron transport imposed by energy transfer inhibitors, they are considered to act at a point closer to the electron transport chain than the site of phosphate uptake. The postulated sites of action of uncouplers are at site 2 in Fig. 1.

The processes involved in the uncoupling of electron transport from phosphorylation will be elucidated comprehensively only when the coupled reaction is understood more completely. Unfortunately, biochemists have only a limited understanding of either oxidative phosphorylation or photophosphorylation. Membranes are involved in the energy conservation step, and the movement of ions across the membranes results in the formation of ion gradients. Some or all of the redox energy conserved in the phosphorylation reaction can be used, reversibly, in creating transmembrane electro-chemical potential gradients. Consequently, compounds that cause membranes to become "leaky" to ions may create a wasteful energy sink. Electron transport continues, but without any measurable phosphorylation. At times, it is possible to interpret the action of uncouplers and uncoupling conditions in terms of induced permeability of the membranes to one or more kinds of ion, although a full explanation of uncoupling in these terms has not yet been provided. In Fig. 1, uncouplers are depicted as dissipating some form of conserved energy represented as ∼ on the ATP-generating pathways. This dissipation could represent loss of an essential electro-chemical potential gradient through ion leaks or could represent catalysis of the breakdown of some chemical intermediate (Good and Izawa, 1973).

3. *Energy Transfer Inhibitors*

Energy transfer inhibitors act directly on the phosphorylation process. Like electron transport inhibitors, they inhibit both electron transport and ATP formation in coupled systems. However, the inhibition of electron flow (but not of ATP formation) is released by the addition of an appropriate uncoupler. Energy transfer inhibitors are considered to act by preventing the formation of high-energy intermediates and thus inhibit electron flow by removing the path normally followed for utilization of the high-energy intermediates. Addition of an uncoupler, which provides an alternate path for the conversion of the high-energy intermediates, relieves such inhibition of electron flow. No herbicides have yet been identified that act as energy transfer inhibitors in photophosphorylation. Compounds that behave in this way include the antibiotic Dio-9 and phloretin $2'$-β-glucoside (phlorizin) (Good and Izawa, 1973).

As shown in Fig. 2C, phlorizin inhibited both noncyclic electron transport and the coupled photophosphorylation with the phosphorylation reaction being slightly more sensitive. Cyclic phosphorylation was also inhibited and had a sensitivity close to that of noncyclic phosphorylation. Inhibition of noncyclic electron transport was alleviated by the uncoupler methylamine. In Fig. 1, site 3 designates the postulated site of action of energy transfer inhibitors on the ATP generating pathway.

4. Inhibitory Uncouplers

Compounds that are classified as inhibitory uncouplers appear to have more than a single site of action; the responses inhibited are those associated with both electron transport inhibitors and uncouplers. Inhibitory uncouplers inhibit basal, methylamine-uncoupled, and coupled electron transport with ferricyanide as the electron acceptor and water as the electron donor, much as the electron transport inhibitors do. Coupled noncyclic photophosphorylation is inhibited, and the phosphorylation reaction is slightly more sensitive than the reduction of ferricyanide. Cyclic photophosphorylation is also inhibited. $NADP^+$ reduction, when photosystem II is circumvented with ascorbate plus DPIP, is not inhibited; however, the associated phosphorylation is inhibited. Action expressed against the phosphorylation reactions resembles that associated with uncouplers.

Inhibitions obtained with dinoseb on noncyclic electron transport, noncyclic photophosphorylation, and cyclic photophosphorylation are shown in Fig. 2D. Herbicides that express action of this type include the N-phenylcarbamates, acylanilides, the 3,5-dihalogenated 4-hydroxybenzonitriles, substituted imidazoles, and substituted benzimidazoles (Moreland, 1967, 1969). Other herbicides that inhibit both cyclic and noncyclic photophosphorylation include several π-excessive heterocyclics (Büchel and Draber, 1969), bromofenoxim (Moreland and Blackmon, 1970), substituted 2,6-dinitroanilines (Moreland et al. 1972), pyriclor (Meikle, 1970), and substituted 1,2,4-thiadiazoles (Bracha et al. 1972). Some of the herbicides classified previously as electron transport inhibitors might be reclassified as inhibitory uncouplers after examination of their action on assays other than the Hill reaction alone.

Some compounds, like the dinitrophenols and alkylated dinitrophenols, are frequently identified in the literature as being uncouplers. The uncoupling action is observed at pH values around 6·0 and is attributed to the protonated form of the compound (Good and Izawa, 1973). However, at pH 8·0, the compounds do inhibit electron transport, hence, they are considered to be inhibitory uncouplers in this classification. The action of the inhibitory uncouplers can be explained by considering two sites of action, one associated with photosystem II, where they act much like diuron, and a second site of action associated with the phosphorylation pathways.

Herbicides that act as uncouplers or that are included as inhibitory uncouplers of photophosphorylation also act as uncouplers of oxidative phosphorylation. This similarity in responses elicited in the organelles suggests that the mechanism involved in the generation of ATP and the environment in which it occurs in both mitochondria and chloroplasts have much in common.

5. *Electron Acceptors*

Compounds classified as electron acceptors are able to compete with some component of the electron transport pathway and subsequently undergo reduction. A number of compounds such as ferricyanide, PMS, and FMN, which are used to study partial reactions of the photochemical pathway, operate in this manner. However, they are not phytotoxic.

Bipyridyliums, such as diquat and paraquat, with redox potentials in the range of -300 to -500 mV, are able to accept electrons in competition with FRS (Fig. 1, site 4) and have herbicidal activity. Interception of electron flow from photosystem I essentially shunts the electron transport chain. Evidence for the site at which electrons are accepted by diquat was provided by its ability to (*a*) support both noncyclic and cyclic photophosphorylation; (*b*) undergo photoreduction by illuminated chloroplasts under anaerobic conditions; (*c*) inhibit the photoreduction of NADP$^+$, with the inhibition not being circumvented by the addition of DPIP and ascorbate; and (*d*) be reduced quantitatively by NADPH in the presence of ferredoxin-NADP oxidoreductase under anaerobic conditions in the absence of chloroplasts (Zweig *et al.* 1965).

B. STRUCTURE-ACTIVITY RELATIONS

1. *Electron Transport Inhibitors and Inhibitory Uncouplers*

Structure-activity studies have been conducted with a large number of compounds that inhibit electron transport close to photosystem II. Classes of herbicides that have been examined in this way include the phenylureas, *N*-phenylcarbamates, polycyclic ureas, acylanilides, *s*-triazines, uracils, 3,5-dihalogenated 4-hydroxybenzonitriles, azido-*s*-triazines, 1,2,4-triazinones, imidazoles, and benzimidazoles (Büchel, 1972; Moreland, 1969). For the most part, effects on electron transport were measured under nonphosphorylating conditions. Some of the above classes of herbicides (inhibitory uncouplers) also have a second site of action, however, only results obtained for the inhibition of electron transport (Hill reaction) will be considered in this discussion. The objectives of the structure-activity studies have been to (*a*) determine the substituents required for maximum inhibitory effectiveness, (*b*) relate the chemical and physical properties of the herbicides to the inhibitory action, (*c*) elucidate the environment in which the inhibitors operate, and (*d*) identify interactions between substituents of the inhibitors and the postulated reaction centres in the chloroplasts.

Not all herbicides are equally efficient in inhibiting the Hill reaction, as evidenced by the wide range of I_{50} values reported in the literature. How inhibitors vary in potency and specificity remains to be established. In view of the diverse chemistry represented among the inhibitors of the Hill reaction, it seems unlikely that all chemicals interfere at the same site through a common mechanism. However, there is little direct evidence to suggest that they do not do so.

The I_{50} values, as reported in the literature, are calculated directly from the concentrations of the inhibitors added to the reaction mixtures. They do not provide an estimate of the concentration of the inhibitor actually inside the chloroplasts and available for the inhibitory reaction. Inhibition of the Hill reaction is achieved by the more active herbicides at concentrations much lower than the quantity of chlorophyll present. In addition, inside the chloroplast some of the inhibitor may be associated with sites that are not involved in the photochemical reaction, hence, may not participate in the inhibitory response. By dividing the chlorophyll concentration by the concentration of herbicide required in the reaction mixture to produce 50% inhibition, chlorophyll to herbicide ratios of 100 to 1 were estimated for diuron (Moreland, 1969). After correction for binding of monuron, diuron, and atrazine to sites that could be occupied without affecting the Hill reaction and for partitioning into the chloroplasts, inhibitor-sensitive sites have been estimated at one site for approximately 2,500 chlorophyll molecules. One inhibitor-sensitive site per several hundred chlorophyll molecules has also been estimated for simazine, prometryne, and propanil. Hence, many of the herbicides seem to be very specific inhibitors of the photochemical reaction, as evidenced by an apparent stoichiometry of one molecule of inhibitor per photosynthetic unit.

Inhibitors obviously must possess properties that will enable penetration to the active site and assumption of the precise spatial configuration required to complement the molecular architecture of the active centre, and hence, to block the key reaction. Properties that could conceivably be of importance include partitioning characteristics (hydrophilic/lipophilic balance), steric relations (the ultimate obtainable configuration), resonance, keto-enol tautomerization, *cis* or *trans* relation of the amide hydrogen and the carbonyl oxygen, and the possession of a critical charge by particular substituent groups that participate in intermolecular interactions at the active centres (Moreland, 1969).

Hansch (Hansch, 1969; Hansch and Deutsch, 1966) evaluated, by multiple regression analysis, the behaviour of *N*-phenylcarbamate, acylanilide, and phenylurea Hill inhibitors. Terms for electronic, steric, and polarity factors or constants, in addition to partition coefficients, were included in their equations. This extrathermodynamic approach, which is a mathematical formulation of a linear combination of free-energy-related substituent constants, separates and accounts for the contribution of the different properties of a compound in the expression of inhibition. The Hansch approach has been extended to the azido-*s*-triazines (Gabbott, 1968, 1969), 1,2,4-triazinones (Draber *et al.* 1969, 1972), benzimidazoles and imidazoles (Büchel *et al.* 1966), and π-excessive heterocyclics (Büchel and Draber, 1969). Results obtained in some of these studies have also been summarized by Büchel (1972) and Hansch (1969). The analyses have shown that inhibitory potency expressed against the Hill reaction can be correlated with physico-chemical parameters within a particular group of herbicides. However, no investigator has correlated successfully activities between chemical families.

Hansch (1969) identified several features that he considered to be essential structural elements. He postulated that the site of action in the chloroplasts involved the amide linkage of particular strategically located proteins. Good inhibitors possessed a large lipophilic moiety and a polar function that anchored the inhibitor to the sites of action. They were also characterized by having an N-H group attached to an electron-deficient sp^2 carbon atom. Binding of the inhibitor occurred through a charge-transfer complex, which resulted when the lone-pair electrons on the nitrogen of the herbicide molecule interacted with the electron-deficient carbon of the protein amide carbonyl group. In addition, the electrons of the nitrogen atom of the protein amide linkage could interact with the electron-deficient carbon of the inhibitor carbonyl function. Hansch visualized that binding involved something between a complete charge-transfer complex and a simple dipole-dipole interaction, possibly reinforced by hydrogen bonding.

A characteristic of Hill inhibitors is the reversibility of their action. They can be eluted from the chloroplasts with a restoration of Hill activity. Consequently, this free reversibility suggests that only weak bonds are formed between the inhibitor and the sensitive sites within the chloroplasts. Some investigators have attributed a stronger role to hydrogen bonds than did Hansch (Barth and Michel, 1969; Moreland, 1969). A structural requirement for a free and sterically unhindered amide or imino hydrogen has been demonstrated repeatedly for inhibition of the Hill reaction. Changes in molecular structure that can be anticipated to decrease hydrogen bonding behaviour are correlated with decreased inhibitory activity expressed against the Hill reaction. Substituents of herbicide molecules that could participate in hydrogen bond formation include the amide hydrogens, imino hydrogens, carbonyl oxygens, ester oxygens, and the azomethine nitrogens in the s-triazines, benzimidazoles, and imidazoles. Hydrogen bonds involving these same groups are responsible for maintaining the functional structure of proteins and nucleic acids in biological systems. Hence, this inhibitory action concerns chemical groups and bond relations that are not foreign to the cell. Hydrogen bonds are relatively weak bonds, and energy in the range of 3 to 8 kcal/mol is required for their rupture. However, in biological systems, hydrogen bonds are probably not broken without other changes occurring simultaneously such as the formation of new hydrogen bonds in an exchange type of reaction. The energy required to consummate an exchange reaction may be considerably less than that needed to break a hydrogen bond (Moreland, 1969).

Models that have been developed to account for inhibitory action are undoubtedly only approximations to the actual interactions between the inhibitors and the biochemical sites of action. However, they do provide a basis for the development of hypotheses that can be subjected to investigative study.

The Hill reaction does provide an *in vitro* system that can be used to identify potential phytotoxicity of particular groups of chemicals. Within certain herbicide families, there appears to be a close correlation between potency expressed against the Hill reaction and phytotoxicity; i.e. the most phytotoxic derivatives are the most potent inhibitors of the Hill reaction (Büchel, 1972; Moreland, 1969). Agreement between phytotoxicity and potency against the Hill reaction has been reported also for the azido-s-triazines (Gabbott and Barnsley, 1968), 1,2,4-triazinones (Draber *et al.* 1972), *p*-alkylthioanilides (Baruffini *et al.* 1970a), and *p*-alkylanilides (Bellotti *et al.* 1968). The above correlations were established for both pre- and post-emergence treatments. However, Baruffini *et al.* (1970a) reported that chloroacetanilides and α-chlorophenylacetanilides which were highly active as inhibitors of the Hill reaction were almost inactive in *in vivo* tests. The *in vitro* tests indicate sensitivity that might be expected if the unaltered herbicide molecule should reach the chloroplasts. However, in *in vivo* tests, many factors may prevent the unaltered herbicide molecule from reaching the chloroplasts, including behaviour and availability in the soil, absorption by roots, translocation within the plant, decomposition in the soil environment, and binding or degradation within the plant.

2. Electron Acceptors

Correlations between structure and phytotoxicity of the bipyridyliums have been reviewed comprehensively by Baldwin (1969) and Calderbank (1972). These chemicals are considered to intercept electron flow from photosystem I in the general region of FRS or at some point prior to ferredoxin (Fig. 1). Phytotoxicity is associated with quarternary salts in which the nitrogen atoms are in the 2,2'-, 2,4'-, and 4,4'-positions. Bipyridyls joined at the 2,3'- or 3,3'-positions do not produce herbicidally active quarternary salts. High herbicidal activity is obtained only when the two pyridine rings can assume a planar configuration. Salts that possess herbicidal properties can be reduced to stable, water-soluble, free radicals by the addition of one electron, and the radicals are converted back to the original ions in the presence of oxygen. Ease of reduction to the free radical, as reflected in the redox potential, is correlated with herbicidal activity over the range -300 to -500 mV.

Diquat and paraquat, which are biquarternary salts of 2,2'- and 4,4'-bipyridyl, are the two most important bipyridylium herbicides that have been introduced. Diquat and paraquat form stable radicals because an electron added to any ring position neutralizes the positive charge on one of the nitrogens. This can occur only with 2,2'-, 4,4'-, or 2,4'-bipyridylium salts and results in a complete delocalization of the odd electron. The radicals derived from the simple pyridylium salts are unstable because the added electron can be located only at positions 2, 4, and 6 without producing a structure with charge separation. In the same way, 3,3'-bipyridyl biquarternaries do not form stable radicals on reduction because the electron can be located only at the 2,2'-, 4,4'-, and 6,6'-positions. The electron is not delocalized over the whole structure, and consequently the radical is unstable.

Not all biquarternary salts of 2,2'- and 4,4'-bipyridyl form radicals with equal ease. Ease of reduction (radical formation) is indicated by the redox potential; i.e. the difficulty of reduction increases with the negativity of the redox potential. The property, within a group of bipyridylium quarternaries that affects the redox potential most profoundly, is inter-ring twisting. The degree of inter-ring twisting has been shown to increase with the electronegativity of the redox potential. Substituents at positions on the ring that tend to increase the twist between the two rings also increase the electronegativity, so that the compound is difficult to reduce. Substituents placed at positions around the ring which do not increase the twist, make the compound easier to reduce. After intensive study, it was determined that only those compounds that produced a colour with zinc and, hence, produced a radical with a redox potential less negative than -800 mV, had any chance of becoming herbicides.

Some derivatives such as benzyl viologen meet all of the requirements and yet do not possess herbicidal activity. Failure to express phytotoxicity has been attributed to differences associated with uptake of the chemicals *in vivo* and their approach to the site of reduction in the chloroplast. Bulky substituents are considered to hinder a close approach to the site that provides the reducing potential.

IV. PHYSIOLOGICAL STUDIES WITH INTACT PLANTS

A. REQUIREMENT FOR LIGHT

The electron transport inhibitors have been shown to produce phytotoxicity only in the light, and severity of the response is proportional to light intensity. Studies with light quality have suggested that the chlorophylls are the principal absorbing pigments involved in the production of toxicity symptoms (Moreland, 1967). The requirement for light was first reported by Minshall (1957) for monuron, and later by Ashton (1965) for atrazine. Dependence on light for phytotoxic action has also been shown for many other herbicidal Hill inhibitors including the 3,5-dihalogenated 4-hydroxybenzonitriles (Carpenter *et al.* 1964), the substituted 3-phenyl-5-alkyloxazolidine-2,4-diones (Billaz, 1965), pyrazon (Eshel, 1969), solan (Colby and Warren, 1965), and 1-(3-chloro-4-methylphenyl)-3-methyl-2-pyrrolidinone (Eshel, 1967). The bipyridyliums, although they are not Hill inhibitors, are also more active in the light than in the dark.

B. INHIBITION OF PHOTOSYNTHESIS

Insofar as they have been studied, all herbicides that inhibit the Hill reaction of isolated chloroplasts also inhibit photosynthesis of intact plants and photosynthetic micro-organisms. Effects on photosynthesis of intact plants are discussed in detail by van Oorschot in Chapter 10. However, to provide an extrapolation from the *in vitro* observations to *in vivo* events, results will be summarized here in general terms. The technique used by many investigators was developed by van Oorschot and Belksma (1961). In this technique, CO_2 utilization by plants maintained in an enclosed chamber is monitored by infrared gas analysis after application of the herbicide to the nutrient solution in which the plants are growing. Hypothetical examples of results obtained by this technique are shown in Fig. 3.

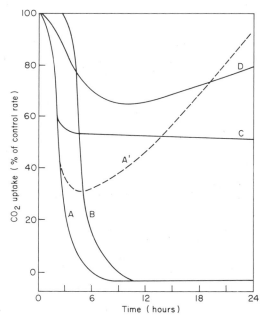

Fig. 3. Diagrammatic presentation of inhibition of photosynthesis by inhibitors of chloroplast electron transport. Photosynthesis is measured by CO_2 fixation in leaves of seedlings during continuous (solid lines) or temporary (dashed line) exposure of roots to herbicides in nutrient solution. The different types of inhibitory responses reflect differences in absorption, transloca-tion, or detoxication of inhibitors. See text for details.

If the herbicide is absorbed readily by the root system and translocated freely in the xylem with the transpiration stream to the foliage, inhibition of photosynthesis can be detected within 1 to 2 h. Complete inhibition is obtained usually within 6 to 24 h. Curve A in Fig. 3 is typical of responses obtained with s-triazine, phenylurea, and uracil herbicides in both sensitive and resistant plants. If the herbicide is not absorbed rapidly, or if it is translocated less readily, a delay may be observed before inhibition of CO_2 fixation can be detected. Curve B represents results obtained with one such herbicide, a pyridazinone. Inhibition did not occur until 3 h after introduction of the pyridazinone in barley (*Hordeum vulgare*) and 24 h in maize (*Zea mays*).

Curves A and B are obtained typically with herbicides classified as electron transport inhibitors, which have a single site of action in the chloroplasts associated with photosystem II. In addition, these herbicides are generally used in pre-emergent treatments and are neither absorbed readily by leaves nor are they mobile in the phloem.

Responses illustrated by Curve C have been obtained with ioxynil and PCP when they were applied to the root environment. There is an initial inhibition of CO_2 uptake, but the inhibition is not complete and rapidly reaches a limiting value. Because these herbicides (inhibitory uncouplers) also interfere with oxidative phosphorylation, the supply of energy required for a continued active uptake of the herbicide by roots is not available, and uptake is arrested. Consequently, the concentration of the herbicides in the foliage may not be sufficient to inhibit photosynthesis completely. However, if the compounds are applied directly to the foliage, complete inhibition of CO_2 fixation results.

Responses represented by Curve D are obtained in tolerant plants usually with low concentrations of inhibitors. The D-type response can be changed to an A-type curve by increasing the concentration of the herbicide. However, some herbicides such as propanil, when applied to the root system even in high concentrations, produce the D-type curve. Propanil is usually applied in the field as a foliar spray and possesses little activity when used as a pre-emergence treatment, in which herbicidal action depends on root absorption. Propanil is hydrolyzed readily in the roots of some plants. Therefore, the possibility exists that the herbicide is inactivated metabolically about as rapidly as it is absorbed by roots. Propanil also inhibits oxidative phosphorylation, hence, uptake might be limited by the unavailability of energy in the root system required for continued absorption.

Selective action of herbicides that inhibit photosynthesis can be detected by removing the source of herbicide when inhibition of CO_2 fixation approaches 50%. Initially, photosynthesis is inhibited in both tolerant and sensitive species of plants. In sensitive species, inhibition of CO_2 fixation continues to increase after replacement of the herbicide-containing solution. Consequently, in sensitive species, inhibition will continue as shown in Curve A (Fig. 3) because the herbicide already in the root continues to move up to the leaves. In tolerant species, removal of the herbicide from the root environment is followed by a recovery of the capacity to fix CO_2, as illustrated by Curve A'. By this method, for example, a basis for tolerance of maize and susceptibility of sugar beets (*Beta vulgaris*) to simazine (van Oorschot, 1965), tolerance of plantain (*Plantago lanceolata*) and susceptibility of maize to monuron (van Oorschot, 1965), tolerance of sugar beets and greater susceptibility of oats (*Avena sativa*) to pyrazon (van Oorschot, 1965), tolerance of peppermint (*Mentha piperita*) and susceptibility of couch grass (*Agropyron repens*) to terbacil (van Oorschot, 1968), and tolerance of sugar beets to phenmedipham (Kötter and Arndt, 1970) have been shown. For the most part, these laboratory observations agree with selective phytotoxicity observed in the field. The selective action has been related to the differential capacity of the tolerant plants to metabolize the herbicides to derivatives that are not inhibitory to photosynthesis.

C. CIRCUMVENTION OF PHYTOTOXICITY

The appearance of toxic symptoms on plants treated with electron transport inhibitors can be prevented if the plants are supplied exogenously with respirable carbohydrates (Fig. 4). These studies have frequently used 10-day-old barley seedlings in the two-leaf stage, when seed reserves were essentially exhausted. Application of an electron transport inhibitor to the roots results in immediate cessation of new leaf growth, if the inhibitor is translocated rapidly

Fig. 4. Glucose protection of barley seedlings during a 10-day growth period against the lethal effect of simazine (Moreland *et al.* 1959). The group of plants at the left (control) received no simazine. Those on the right received 0·75 mg/cup (1 lb/acre) at the 2-leaf stage of growth. Tips of the first two leaves of all plants were clipped and (1) not immersed, (2) immersed in water, and (3) immersed in a 6% glucose solution.

to the leaves without metabolic alteration. New leaves are produced normally when one of the two established leaves is supplied with a solution of carbohydrate through severed veins. Sucrose is the most effective source of carbohydrate, but must be limited to a 5 or 6% solution to avoid plasmolysis of leaf cells. Within another 10 d, new leaves appear on sucrose-treated seedlings, whereas no new growth occurs on the plants that do not receive sucrose. By this technique, carbohydrate circumvention of growth inhibition has been demonstrated for certain phenylurea (Gentner and Hilton, 1960), s-triazine (Moreland *et al.* 1959), uracil (Hilton *et al.* 1964), and pyridazinone (Hilton *et al.* 1969) herbicides. Apparently glycolysis and oxidative phosphorylation can provide sufficient energy to prevent the appearance of phytotoxic symptoms when the photochemical source of energy is inhibited, if respirable substrates are provided. In contrast, carbohydrate circumventions could not be demonstrated with inhibitory uncouplers such as the 3,5-dihalogenated 4-hydroxybenzonitriles and substituted phenols. The latter compounds are known to interfere also with oxidative phosphorylation.

V. EFFECTS PRODUCED ON THE STRUCTURE AND COMPOSITION OF CHLOROPLASTS

Correlation of herbicide-induced alterations of the ultrastructure of plant tissue with results obtained at the biochemical level may provide an insight into events that precede, are associated with, and follow the appearance of external symptoms of phytotoxicity. All techniques have their limitations, and this is true of electron microscopy. For example, the amount of tissue that can be examined is extremely small. Consequently, an adequate sampling of the normal range of variation in the tissue is difficult to achieve, and much is left to the imaginative interpretation of the investigator. Some changes also may be produced in fixing and preparing the tissue for examination.

Herbicides induce changes that are similar in many respects to those that occur in normal senescence, as summarized by Butler and Simon (1971). Degeneration and death of a cell follow a characteristic pattern. In the sequence generalized by Butler and Simon, the first detectable changes include a decrease in the population of ribosomes and the onset of chloroplast breakdown. Unattached ribosomes are the first to disappear, followed by those attached to the endoplasmic reticulum. The chloroplast stroma disappears, thylakoids swell and disintegrate, and there is generally a marked increase in the number and size of the osmiophilic globules. In some instances, the mitochondria also show signs of an early change, a reduction in size, swelling of cristae, and reduction of their number, but in most cases they are more resistant than other cell components and are still present at a late stage. The endoplasmic reticulum swells, vesiculates, and disappears, as do the dictyosomes. The tonoplast breaks down long before the organelles have degenerated completely, but the plasmalemma is usually one of the last recognizable components. The nucleus remains relatively stable, but at a late stage the nuclear membrane becomes irregular, vesiculates, and breaks down with the chromatin shrinking and disappearing. When the tonoplast breaks down, senescence is irreversible. Prior to tonoplast disruption, senescence may be gradual or rapid, but after tonoplast rupture and the release of vacuolar contents into the cytoplasm, death occurs rapidly.

A. ULTRASTRUCTURAL MODIFICATIONS OF CHLOROPLASTS

Alterations of the ultrastructure of chloroplasts produced by herbicides that interfere with their photochemistry have been described by several investigators. A comparative evaluation of the studies is complicated in that investigators have used different test plants, that herbicides were applied differently, and that the age of the tissue examined has varied. One common factor is associated with these various studies, namely, that the plant has to be exposed to light after application of the herbicide before effects become obvious. The severity of the alterations, for the most part, is related directly to the intensity of the light.

Anderson and Thomson (1973) included, in their review, alterations to the fine structure of chloroplasts produced by herbicides. In general, ultrastructural deviations produced by herbicides are similar to those associated with senescence. The effect can be described as aggravated senescence or herbicide-induced senescence. The earliest study of this type was conducted by Ashton et al. (1963) on atrazine-treated bean plants (*Phaseolus vulgaris*), and the observations were confirmed subsequently on barnyard grass (*Echinochloa crusgalli*) by Hill et al. (1968). The first effect observed was a swelling of the intergranal thylakoids. Subsequently, the thylakoids of the grana began to swell, beginning with the outer thylakoids, until the whole lamellar system became disorganized. Initial effects were observed by Hill *et al.* within 2 h after application of atrazine to barnyard grass seedlings. Subsequently, the tonoplast and chloroplast envelopes ruptured, and finally the thylakoid membranes also ruptured. Mixing of plastid and cytoplasmic contents was detected within 4 h by Hill *et al.* External symptoms of injury are usually not apparent until several days after treatment with electron transport inhibitors. Hence, there is complete destruction of the internal morphology in leaves before the appearance of external symptoms.

Results similar to those described above have been reported by different investigators with bromacil, haloxydine, pyrazon, pyriclor, 2,4-D, monuron, diuron, dichlormate, metflurazone, and the bipyridylium herbicides (Anderson and Thomson, 1973). In every test, light was required for the effects to become apparent. Chloroplasts of herbicide-treated plants kept in the dark resembled, in all respects, chloroplasts from the untreated, dark-control plants.

B. CHLOROPLAST DEVELOPMENT AND COMPOSITION

1. *Biochemistry of Development*

In the prior sections of this chapter, effects of herbicides on the photochemical activity of mature chloroplasts were stressed. However, there is evidence that some herbicides affect the development of chloroplasts. These chemicals may inhibit photosynthesis by preventing the formation of photosynthetically capable plastids.

In higher plants, chloroplasts develop from small spherical bodies of uncertain biochemical composition that lack a distinctive internal structure. In darkness, the initial structures develop into proplastids which are rich in protein and lipids, contain small amounts of carotene and protochlorophyllide, and possess distinct internal structures such as 70S ribosomes and the large crystalline lattice-like structure known as the prolamellar body. Upon illumination, proplastids produce large amounts of chlorophyll, synthesize additional carotenoids, lipids, nucleic acids, and proteins, shift fatty acid composition toward more unsaturated lipids and mature into functional chloroplasts. Prolamellar bodies dissociate, and the plastid forms functional photosynthetic lamellar structures. Carotenoid pigment biosynthesis consists of a series of dehydrogenation reactions leading to carotenoids containing fully conjugated double bonds as follows: phytoene (three conjugated double bonds)→ phytofluene (five)→ zeta-carotene (seven)→ neurosporene (nine) → carotenes and xanthophylls (eleven) (Goodwin, 1971). The final light-mediated maturation process required for normal chloroplast development includes large increases in both chlorophyll and the structural galactolipids which may comprise as much as 80% of the chloroplast polar lipids. The polyunsaturated fatty acid, linolenic acid (18:3), is the predominant fatty acid in the galactolipids. Many of the light-induced syntheses appear mutually interdependent. Thus, failure in one process may prevent a series of developments.

2. Interference with Development

Some herbicides selectively inhibit chloroplast development. Foliage formed after treatment with these chemicals is white (bleached or albino) and is distinctly different from the chlorosis that results from a reduction in chlorophyll synthesis or from destruction of existing chlorophyll. The white foliage is observed only in new tissue formed after the chemical treatment. Once formed, the foliage is affected permanently and will not become green even after inhibitory levels of the herbicide have dissipated.

Inhibition of chloroplast development has been reported for a variety of compounds having unrelated chemical structures and widely different physical properties. These include amitrole (Miller and Hall, 1957), fluometuron (Bruinsma, 1965), dichlormate (Herrett and Berthold, 1965), haloxydine (Slater, 1968, Dodge and Lawes, 1972), pyriclor (Burns et al. 1971), o-methylthreonine (Gray and Hendlin, 1962), norflurazon, metflurazone and Sandoz 9774 (4-chloro-5-amino-2-(α,α,α-trifluoro-m-tolyl)-3(2H)-pyridazinone) (Hilton et al. 1969), and a phytotoxin, rhizobitoxin (Owens, 1973). No essential common structural feature has been recognized among these inhibitors. However, minor modifications in their individual structures usually reduce activity drastically. This was particularly true for amitrole activity (Schweizer and Rogers, 1964). The trifluoromethyl substitution on the phenyl ring is essential for the bleaching action of fluometuron and the pyridazinones.

Amitrole was the first herbicide recognized as an inhibitor of chloroplast development. Early literature referred to it as an inhibitor of chlorophyll formation. However, biochemical studies have failed to confirm that interference with porphyrin biosynthesis was involved. Wolf (1960) recognized that both carotenoid and chlorophyll pigments were affected, and concluded that the effect on plastid pigments was secondary to an effect on the plastid itself. Subsequently, Burns et al. (1971) and Bartels and Hyde (1970a) obtained evidence that amitrole, dichlormate, pyriclor, and metflurazone directly inhibited a desaturation step in the carotenoid biosynthetic pathway. Amitrole and pyriclor caused accumulation of colourless phytofluene, phytoene, and yellow zeta-carotene; dichlormate caused accumulation of zeta-carotene (Burns *et al.* 1971); and metflurazone caused accumulation of colourless phytoene and phytofluene (Bartels and McCullough, 1972). Concurrently, Hilton *et al.* (1970) obtained evidence for interference with the desaturation of fatty acids required for galactolipid biosynthesis.

Amitrole, dichlormate, pyriclor, and metflurazone affect carotenoid synthesis in treated seedlings regardless of whether they are grown in darkness, dim light, or high light intensity. In contrast, treated plants grown under low light intensity form chlorophyll which bleaches out on subsequent exposure to higher light intensities. Furthermore, normal greening occurs in leaves that are allowed to mature (form carotenoids) in the dark prior to treatment with amitrole and exposure to light. Additional effects of chemicals plus light, based largely on ultrastructural and ultracentrifugal data for amitrole, dichlormate, and metflurazone, include the loss of normal grana and fret membranes, loss of 70S chloroplastic ribosomes (but not cytoplasmic ribosomes), loss of 18S fraction I proteins, and loss of chloroplast DNA (Bartels *et al.* 1967, 1968, 1970a, 1970b). Dark-grown, herbicide-treated plants form these components. Furthermore, the ultrastructure of etioplasts of dark-grown treated plants is identical to that of etioplasts of dark-grown control plants. On subsequent exposure to light, 70S ribosomes, fraction I protein, and ultrastructure are destroyed (Bartels and Pegelow, 1968; Bartels and Hyde, 1970a). Identical behaviour is reported for mutant plants that lack the genetic capacity for carotenoid production. These observations indicate that the four herbicides affect the competence of plastids for greening, but that light is responsible for most of the damage done to incompetent plastids.

One of the major effects of metflurazone is interference with the light-induced increase in galactolipids and linolenic acid (Hilton *et al.* 1970). Linoleic acid (18:2) accumulates in treated plants at the expense of linolenic. The results are similar to those obtained by maintaining control plants in complete darkness. However, the formation of 18:3, like formation of carotenoids, is inhibited in darkness as well as in light, and therefore differs from the majority of chloroplast aberrations reported. Because biosynthesis of both carotenes and linolenic acid involves a series of desaturation reactions, the herbicide appears to have a general effect on certain types of plastid desaturase systems.

At least three of the bleaching agents are known to inhibit some specific aspect of amino-acid metabolism. O-methylthreonine is an antagonist of isoleucine (Gray and Hendlin, 1962). Amitrole (Hilton, 1969) and rhizobitoxin (Owens *et al.* 1968) inhibit specific enzymes in the biosynthetic pathways for histidine and methionine, respectively. However, there is little justification for the hypothesis that their interference with chloroplast development results from these specific actions. Many of the other chloroplast inhibitors have no direct effect on amino-acid metabolism, but do inhibit additional processes.

VI. Postulated Mechanisms of Action

In mechanism of action studies, there has been a continuing need to identify sites through which herbicides express their action at the cellular and molecular level. In addition, no criteria are available that can be used to separate primary from secondary (or the direct from the indirect) effects of a given chemical. Because of the complex interrelations between cellular metabolism and growth and development, manifestations of chemical effects may occur at sites quite removed from the original site of action (Moreland, 1967).

With the herbicides that inhibit the photochemical activity of isolated chloroplasts, sites of action can be identified within particular regions of the electron transport and phosphorylation pathways, However, the exact components involved in the interaction are not known. Progress here can be anticipated only when additional details of the pathways become available. Herbicides, by inhibiting the photochemical responses, limit the availability to the plant of chemical energy in the form of ATP and reducing power in the form of NADPH. Suppression of these two pivotal components can be measured within seconds or even a fractional part of a second after the herbicides come in contact with isolated chloroplasts or chloroplast fragments. This can be considered as the primary mode of action for this large group of herbicides. Considerable attention is being given to the mechanisms through which suppresssion or unavailability of ATP and NADPH lead to the production of phytotoxicity.

In any hypothesis proposed to account for the mechanism of action of inhibitors of the photoinduced responses in chloroplasts, the following observations must be accommodated: symptoms of phytotoxicity develop only in the light, and severity is proportional to light intensity; CO_2 fixation is inhibited only in the light; toxic effects can be alleviated by the exogenous application of carbohydrates; severe morphological and cytological changes are induced only in the light; and external toxicity symptoms become apparent in higher plants several days after treatment, except for the bipyridylium compounds. Several hypotheses have been proposed that are substantiated more by speculation than by rigorous experimentation.

A. DEATH BY STARVATION

The early reports that inhibitors of electron transport (Hill inhibitors) limited photosynthesis and that starch disappeared from treated plants, prompted some investigators to refer to these compounds as photosynthesis inhibitors. It is true that photosynthesis is inhibited, but this does not imply necessarily how the plants are killed. There is little evidence to suggest that the plants do starve to death. If this were the only process affected, phytotoxic symptoms should resemble those that appear on plants kept in total darkness. Deficiency of photosynthate does limit or prohibit new growth. However, the mechanisms that lead to phytotoxicity appear to be considerably more complex than would be anticipated to result from limiting carbohydrate synthesis by the suppression of CO_2 fixation.

B. FREE RADICAL MECHANISMS

The appearance of phytotoxic symptoms only in the light after treatment of plants with herbicides such as diuron and atrazine prompted some investigators to propose "light-activation" hypotheses. Other authors speculated that toxic substances formed by an interaction between the herbicide and light could be responsible for phytotoxicity. Suggestions have also been made that the toxic substances might be "free radicals". These ideas took various forms. For example, Ashton *et al.* (1963) speculated that a toxic substance, possibly a "free radical", formed by the interaction between atrazine and light could be responsible for the morphological disruptions that were observed. Wain (1964) proposed a free-radical mechanism to account for phytotoxicity of ioxynil. Davis (1966) speculated that leaf injury produced on live oak (*Quercus terbinella*) seedlings by fenuron was caused by toxic components of the inhibited photosynthetic apparatus. Many of the suggestions were based on the fact that interference with photosynthesis and the subsequent inhibition of CO_2 fixation did not explain adequately the many light-dependent phytotoxic responses associated with the inhibition by herbicides of the light-mediated reactions of chloroplasts. In most cases, the speculations were made to account for effects observed. There is little, if any, direct evidence that toxic components are formed from an interaction between a herbicide and light, or that free radicals are actually produced, except for the bipyridyliums.

Evidence supports the conclusion that free radicals are involved in the phytotoxic action of the bipyridyliums. These compounds are activated by light in the presence of oxygen and express their phytotoxicity quite rapidly, usually within 30 to 60 min. Reduction by one electron transfer to stable free radicals has been shown to occur, and a correlation has been made between the ease of reduction of the bipyridylium and phytotoxicity. As discussed earlier in this chapter, these chemicals are considered to intercept electron flow from photosystem I and, hence, essentially shunt the electron transport chain. The oxidation of the free radical back to the parent compound by molecular oxygen results in the formation of reactive intermediates. Oxidized bipyridyl is not consumed in the reaction, hence, only catalytic amounts are needed in the plant system to produce phytotoxicity. When the bipyridylium radical is reoxidized by oxygen, it is theoretically possible for the oxygen to be reduced to water via three intermediates: hyperperoxy radical, hydrogen peroxide, and hydroxy radical. Studies of the reaction between the paraquat radical and oxygen conducted with ESR and flow-cell techniques failed to detect any radical species other than the paraquat radical itself. Consequently, the bipyridyls appear to initiate phytotoxicity by acting as catalysts for the production of hydrogen peroxide within the chloroplast. Hydrogen peroxide is considered to produce an increase in membrane permeability, possibly by reacting with unsaturated lipids, which results in complete disruption of cellular organization, structure, and function (Calderbank, 1972).

C. INTERFERENCE WITH CAROTENOID SYNTHESIS

Amitrole, fluometuron, dichlormate, metflurazone, Sandoz 9774, haloxydine, and pyriclor are known to inhibit carotenoid biosynthesis (Bartels and Hyde, 1970a; Burns *et al.* 1971; Dodge and Lawes, 1972). One of the functions attributed to desaturated carotenoid pigments in photosynthetic systems is that of serving as protective agents against photosensitized oxidations. These occur when light-excited chlorophylls combine with molecular oxygen. This potentially lethal combination would normally be inactivated by a carotenoid such as zeaxanthin which is oxidized to its epoxide derivative, antheraxanthin (Krinsky, 1966). Epoxidation of zeaxanthin takes place only under conditions (light and oxygen) that lead to photosensitization. The reductive de-epoxidation of antheraxanthin to zeaxanthin is a dark enzymatic reaction. These two reactions constitute an epoxide cycle that is considered to perform the protective function. Stanger and Appleby (1972) have postulated also that diuron might induce phytotoxicity by catalyzing lethal photosensitized oxidations in chloroplasts by limiting the availability of the NADPH required for the reduction of antheraxanthin.

Most of the inhibitors of chloroplast development also inhibit noncyclic electron transport in isolated chloroplasts. The relationship, if it should exist, between inhibition of photochemical reactions of isolated chloroplasts and carotenoid synthesis remains to be resolved. Not all inhibitors of chloroplast electron transport produce effects on plants that are associated with the inhibitors of chloroplast development. Amitrole is the only herbicide, of the group considered to act as inhibitors of chloroplast development, that has no measurable effect on the photoinduced electron transport and phosphorylation reactions of isolated chloroplasts.

D. INTERFERENCE WITH ENERGY (ATP) AVAILABILITY

All of the chemicals classified as inhibitors of electron transport, uncouplers, inhibitory uncouplers, and energy transfer inhibitors that interfere with the photochemical activity of isolated chloroplasts limit the production and availability of ATP and NADPH. Action is expressed at different sites on the electron transport and energy generation pathways, but the net result is the same. These actions can be considered as the primary effects of the inhibitors. The interference with ATP production has focused attention on how this action might relate to the production of phytotoxicity (Moreland *et al.* 1970b). These subsequent actions can be considered as secondary effects. ATP exerts an ubiquitous and dominant role in cellular metabolism. This role can be appreciated only if cognizance is extended to the energy requirements of cells, to the regulation of cellular activity and metabolism imposed by ATP, and to what interference with ATP production means to the welfare of a chlorophyllous plant. Only ADP is phosphorylated to ATP in glycolysis, oxidative phosphorylation, and photophosphorylation. Most biosynthetic reactions are driven by energy which comes directly or indirectly from ATP (Green and Baum, 1970; Lehninger, 1971). Biosynthesis includes not only the formation of the biochemical components of cells, but also their assembly into structures such as membranes and the various organelles. All of the chemical components and organelles of living cells are continually being built up and broken down (turnover). ATP energy is also required to maintain the structure of membranes. The functions of membranes such as active transport and osmotic relations, which regulate the volume of cells, are energy dependent. The structural organization, contraction, and orientation of chromosomes and microtubules of the spindle apparatus during mitosis depend on ATP energy. The intracellular concentrations and stoichiometric relations of ATP, ADP, and AMP are considered also to modulate the course of energy metabolism in the cell. Various hypotheses concerning the regulation of metabolism by the adenylic acid system have been proposed (Atkinson, 1969). ATP functions as a substrate for some enzymes, as an allosteric effector of other enzymes, and even participates stoichiometrically in the chemical modification of some enzymes. Without ATP, growth stops, cellular functions are arrested, and the integrity of the cell's structural morphology is lost.

In evaluating the role of ATP in the cellular metabolism of the higher plant, consideration has to be extended to contributions to the ATP pool made by all sources: glycolysis, oxidative phosphorylation, and photophosphorylation. Even though the photosystem II inhibitors block noncyclic photophosphorylation, ATP can still be produced under some conditions *in vivo* by cyclic photophosphorylation. These compounds do not inhibit oxidative phosphorylation or, as far as is known, glycolysis. Hence, ATP can still be produced through the latter systems even though the noncyclic photophosphorylation pathway has been inhibited. The inhibitory uncouplers interfere with both oxidative phosphorylation and photophosphorylation. Effects of inhibitory uncouplers on the glycolytic production of ATP are for the most part unknown.

The observations that phytotoxic symptoms develop only in the light suggest that there is an increased demand for ATP when chlorophyllous organisms are illuminated. Actually, a large number of biosynthetic reactions are light-activated. These include RNA and protein synthesis; the synthesis of the enzyme which forms δ-aminolevulinic acid; other enzymes involved in the synthesis of chlorophyll, chlorophyll precursors, carotenoids, and lipids; and several of the enzymes of the Calvin cycle and the Hatch—Slack pathway (Marcus, 1971; Preiss and Kosuge, 1970; Zucker, 1972). Some of the light-induced enzymes have a high rate of turnover in the light and disappear rapidly in the dark. Turnover of other cellular components is also activated by light. All of the light-activated synthetic activity places a much higher demand upon the plant for energy in the light than in the dark.

Sufficient energy (ATP) can be provided, apparently, through glycolysis and oxidative phosphorylation, if respirable carbohydrates are supplied exogenously to plants treated with the herbicidal inhibitors of electron transport, to satisfy the light-induced demands and prevent the appearance of phytotoxic symptoms. Applications of carbohydrate do not prevent phytotoxicity if the herbicides also interfere with the mitochondrial production of ATP (inhibitory uncouplers).

Chloroplasts both *in vivo* and *in vitro* contract when illuminated, if conditions permit electron and energy transfer to occur, and swell or expand in the dark, or under conditions that are not conducive to electron transport and energy transfer. The light- and electron transport-dependent shrinkage is associated with an uptake of protons and closely linked to an efflux of K^+ and Mg^{2+} ions, which maintains charge neutrality (Dilley, 1971). An efflux of water accompanies the migration of cations to maintain the osmotic balance. Hence, in the presence of electron transport inhibitors, proton gradients that are established across the chloroplast membrane when electron transport is going on do not develop, and chloroplasts are left in an expanded configuration. Without ATP or an energized configuration to regulate ion movement by the chloroplast membrane, monovalent cations can move into the chloroplast by a passive process. The net movement of monovalent cations into the chloroplast interior would be accompanied by an increase in the volume of water, and swelling of the thylakoids and rupture or perforation of the chloroplast membrane would occur. This effect is seen in the electron micrographs of the foliage of plants treated with herbicides that inhibit electron transport. All membranes that are involved in active transport can also be expected to lose their semipermeable properties if ATP availability becomes limiting. The integrity of membranes can be maintained only by the constant input of energy. There is some evidence that continued protein synthesis is also required for the maintenance of membrane function and integrity.

It is possible to attribute to decreased ATP availability, many of the reported deviations in cellular structure and metabolism produced by herbicides that interfere with the photochemical activity of chloroplasts. A decreased ATP supply would essentially arrest or modify all of the energy requiring reactions of the cell. The sequence of events induced by light, traced in the ultrastructural studies after herbicide treatments, which are similar in many respects to those that occur during senescence, can be attributed to interference with ATP availability. Senescence itself is characterized by a decreased production of mitochondrial ATP and decreased photosynthetic activity.

E. SUMMATION OF HYPOTHESES

Results of future research may suggest that none of the above hypotheses are adequate to explain the action of herbicides that interfere with the photochemistry of isolated chloroplasts. Subsequent observations may necessitate the rejection of some of the current postulates or require that they be modified extensively. No single hypothesis may explain adequately the action of all herbicides under all conditions. It may be necessary to define the conditions under which one or another of the hypotheses might explain the sequence of events that are observed after absorption of a herbicide by a plant. With a given herbicide, at a particular concentration, when applied to a certain species or variety of plant, and under given environmental conditions, it may be found that one of the hypotheses will account for the observed phytotoxicity. However, under other conditions or situations, another hypothesis may be more applicable.

Based on the current status of knowledge on the subject, it seems likely that whatever form the final hypothesis may take, it will centre around what happens when the formation of ATP, or NADPH, or both, are inhibited after interference with the photochemical reactions of the chloroplasts. Hopefully, the currently available postulates will serve as models that can be subjected to rigorous and sophisticated experimentation, and can be modified as our knowledge of biochemical control systems in higher plants increases.

REFERENCES

Anderson, J. L. and Thomson, W. W. (1973). *Residue Rev.* **47,** 167–189.
Ashton, F. M. (1965). *Weeds,* **13,** 164–168.
Ashton, F. M., Gifford, E. M., Jr. and Bisalputra, T. (1963). *Bot. Gaz.* **124,** 336–343.
Atkinson, D. E. (1969). *A. Rev. Microbiol.* **23,** 47–68.
Avron, M. (1967). *In* "Current Topics in Bioenergetics" (D. R. Sanadi, ed.) Vol. 2, pp. 1–22. Academic Press, New York and London.
Avron, M. (1971). *In* "Structure and Function of Chloroplasts" (M. Gibbs, ed.) pp. 149–167. Springer-Verlag, Berlin.
Avron, M. and Neumann, J. (1968). *A. Rev. Pl. Physiol.* **19,** 137–166.
Avron, M. and Shavit, N. (1965). *Biochim. biophys. Acta,* **109,** 317–331.
Baldwin, B. C. (1969). *In* "Progress in Photosynthesis Research" (H. Metzner, ed.) Vol. III, pp. 1737–1741. Tübingen.
Bartels, P. G. and Hyde, A. (1970a). *Pl. Physiol., Baltimore,* **45,** 807–810.
Bartels, P. G. and Hyde, A. (1970b). *Pl. Physiol., Baltimore,* **46,** 825–830.
Bartels, P. G., Matisuda, K., Siegel, A. and Weier, T. E. (1967). *Pl. Physiol., Lancaster,* **42,** 736–741.
Bartels, P. G. and McCullough, C. (1972). *Biochem. biophys. Res. Commun.* **48,** 16–22.
Bartels, P. G. and Pegelow, E. J. (1968). *J. Cell. Biol.* **37,** C1–C6.
Barth, A. and Michel, H.-J. (1969). *Pharmazie,* **24,** 11–23.
Baruffini, A., Borgna, P., Calderara, G. and Mazza, M. (1970a). *Farmaco (Ed. Sci.),* **25,** 427–441.
Baruffini, A., Borgna, P., Calderara, G., Mazza, M. and Gialdi, F. (1970b). *Farmaco (Ed. Sci.),* **25,** 10–35.
Bellotti, A., Coghi, E., Baruffini, A., Pagani, G. and Borgna, P. (1968). *Farmaco (Ed. Sci.),* **23,** 591–619.
Billaz, R. (1965). *Fruits,* **20,** 371–381.
Bishop, N. I. (1971). *A. Rev. Biochem.* **40,** 197–226.
Black, C. C., Jr. (1973). *A. Rev. Pl. Physiol.* **24,** 253–286.
Bracha, P., Luwisch, M. and Shavit, N. (1972). *In* "Proceedings Second International IUPAC Congress of Pesticide Chemistry" (A. S. Tahori, ed.) Vol. V, pp. 141–151. Gordon and Breach, New York.
Bruinsma, J. (1965). *Residue Rev.* **10,** 1–39.
Büchel, K. H. (1972). *Pestic. Sci.* **3,** 89–110.
Büchel, K. H. and Draber, W. (1969). *In* "Progress in Photosynthesis Research" (H. Metzner, ed.) Vol. III, pp. 1777–1788. Tübingen.
Büchel, K. H., Draber, W., Trebst, A. and Pistorius, E. (1966). *Z. Naturf.* **21b,** 243–254.
Burns, E. R., Buchanan, G. A. and Carter, M. C. (1971). *Pl. Physiol., Baltimore,* **47,** 144–148.
Butler, R. D. and Simon, E. W. (1971). *Adv. generontol. Res.* **3,** 73–129.
Calderbank, A. (1972). *In* "Proceedings Second International IUPAC Congress of Pesticide Chemistry" (A. S. Tahori, ed.) Vol. V, pp. 29–40. Gordon and Breach, New York.
Carpenter, K., Cottrell, H. J., DeSilva, W. H., Heywood, B. J., Leeds, W. G., Rivett, K. F. and Soundy, M. L. (1964). *Weed Res.* **4,** 175–195.

Cheniae, G. M. (1970). *A. Rev. Pl. Physiol.* **21,** 467–498.
Colby, S. R. and Warren, G. F. (1965). *Weeds,* **13,** 257–263.
Cooke, A. R. (1956). *Weeds,* **4,** 397–398.
Davis, E. A. (1966). *Weeds,* **14,** 10–17.
Dilley, R. A. (1971). *In* "Current Topics in Bioenergetics" (D. R. Sanadi, ed.) Vol. **4,** pp. 237–271. Academic Press, New York and London.
Dodge, J. D. and Lawes, G. B. (1972). *Ann. Bot.* **36,** 315–323.
Döring, G., Renger, R., Vater, T. and Witt, H. T. (1969). *Z. Naturf.* **24b,** 1139–1143.
Draber, W., Büchel, K. H., Dickoré, K., Trebst, A. and Pistorius, E. (1969). *In* "Progress in Photosynthesis Research" (H. Metzner, ed.) Vol. III, pp. 1789–1795. Tübingen.
Draber, W., Büchel, K. H. and Dickoré, K. (1972). *In* "Proceedings Second International IUPAC Congress of Pesticide Chemistry" (A. S. Tahori, ed.), Vol. V, pp. 153–175. Gordon and Breach, New York.
Duysens, L. N. M. and Amesz, J. (1962). *Biochim. biophys. Acta,* **64,** 243–260.
Eshel, Y. (1967). *Weeds,* **15,** 147–149.
Eshel, Y. (1969). *Weed Res.* **9,** 167–172.
Gabbott, P. A. (1968). *Soc. Chem. Ind. Monograph Series No.* **29,** pp. 335–347.
Gabbott, P. A. (1969). *In* "Progress in Photosynthesis Research" (H. Metzner, ed.), Vol. III, pp. 1712–1727. Tübingen.
Gabbott, P. A. and Barnsley, G. E. (1968). *J. Sci. Fd. Agric.* **19,** 16–19.
Gentner, W. A. and Hilton, J. L. (1960). *Weeds,* **8,** 413–417.
Good, N. E. And Izawa, S. (1973). *In* "Metabolic Inhibitors" (R. M. Hochster, M. Kates, and J. H. Quastel, eds.), Vol. IV, pp. 179–214. Academic Press, New York and London.
Good, N. E., Izawa, S. and Hind, G. (1966). *In* "Current Topics in Bioenergetics" (D. R. Sanadi, ed.), Vol. I, pp. 76–112. Academic Press, New York and London.
Goodwin, T. W. (1971). *Biochem. J.* **123,** 293–329.
Gray, R. A. and Hendlin, D. (1962). *Pl. Physiol., Lancaster,* **37,** 223–227.
Green, D. E. and Baum, H. (1970). "Energy and the Mitochondrion". Academic Press, New York and London.
Hansch, C. (1969). *In* "Progress in Photosynthesis Research" (H. Metzner, ed.), Vol. III, pp. 1685–1692. Tübingen.
Hansch, C. and Deutsch, E. W. (1966). *Biochim. biophys. Acta,* **112,** 381–391.
Hatch, M. D. and Slack, C. R. (1970). *A. Rev. Pl. Physiol.* **21,** 141–162.
Herrett, R. A. and Berthold, R. V. (1965). *Science, N.Y.* **149,** 191–193.
Hill, E. R., Putala, E. C. and Vengris, J. (1968). *Weeds,* **16,** 377–380.
Hill, R. (1937). *Nature, Lond.* **139,** 881–882.
Hilton, J. L. (1969). *J. agric. Fd. Chem.* **17,** 182–198.
Hilton, J. L., Monaco, T. J., Moreland, D. E. and Gentner, W. A. (1964). *Weeds,* **12,** 129–131.
Hilton, J. L., Scharen, A. L., St. John, J. B., Moreland, D. E. and Norris, K. H. (1969). *Weed Sci.* **17,** 541–547.
Hilton, J. L., St. John, J. B., Christiansen, M. N. and Norris, K. H. (1970). *Pl. Physiol., Baltimore,* **48,** 171–177.
Hopkins, T. R., Neighbors, R. P. and Phillips, L. V. (1967). *J. agric. Fd. Chem.* **15,** 501–507.
Izawa, S. and Good, N. E. (1972). *In* "Methods of Enzymology" (A. San Pietro, ed.), Vol. **24, B,** pp. 355–377. Academic Press, New York and London.
Kok, B. (1965). *In* "Plant Biochemistry" (J. Bonner and J. E. Varner, eds.), pp. 903–960. Academic Press, New York and London.
Kok, B. and Cheniae, G. M. (1966). *In* "Current Topics in Bioenergetics" (D. R. Sanadi, ed.), Vol. 1, pp. 1–47. Academic Press, New York and London.
Kötter, C. and Arndt, F. (1970). *Z. PflKrankh. PflPath. PflSchutz.* **5,** 81–88.
Krinsky, N. I. (1966). *In* "Biochemistry of Chloroplasts" (T. W. Goodwin, ed.), Vol. I, pp. 423–430. Academic Press, New York and London.
Lehninger, A. L. (1971). "Bioenergetics". W. A. Benjamin, Inc., Menlo Park, California.
Marcus, A. (1971). *A. Rev. Pl. Physiol.* **22,** 313–336.
Meikle, R. W. (1970). *Weed Sci.* **18,** 475–478.
Miller, C. S. and Hall, W. C. (1957). *Weeds,* **5,** 218–226.
Minshall, W. H. (1957). *Weeds,* **5,** 29–33.

Moreland, D. E. (1967). *A. Rev. Pl. Physiol.* **18**, 365–386.
Moreland, D. E. (1969). *In* "Progress in Photosynthesis Research" (H. Metzner, ed.), Vol. III, pp. 1693–1711. Tübingen.
Moreland, D. E. and Blackmon, W. J. (1970). *Weed Sci.* **18**, 419–426.
Moreland, D. E., Blackmon, W. J., Todd, H. G. and Farmer, F. S. (1970a). *Weed Sci.* **18**, 636–642.
Moreland, D. E. and Boots, M. R. (1971). *Pl. Physiol., Baltimore,* **47**, 53–58.
Moreland, D. E., Farmer, F. S. and Hussey, G. G. (1972). *Pestic. Biochem. Physiol.* **2**, 342–353.
Moreland, D. E., Gentner, W. A., Hilton, J. L. and Hill, K. L. (1959). *Pl. Physiol., Lancaster,* **34**, 432–435.
Moreland, D. E., Gruenhagen, R. D. and Blackmon, W. J. (1970b). *Abstr. Weed Soc. Am. Meeting, Montreal, Canada,* pp. 11–12.
Oorschot, J. L. P. van (1965). *Weed Res.* **5**, 84–97.
Oorschot, J. L. P. van (1968). *Proc. 9th Br. Weed Control Conf.* pp. 624–632.
Oorschot, J. L. P. van and Belksma, M. (1961). *Weed Res.* **1**, 245–257.
Owens, L. D. (1973). *Weed Sci.* **21**, 63–66.
Owens, L. D., Guggenheim, S. and Hilton, J. L. (1968). *Biochim. biophys. Acta,* **158**, 219–225.
Park, R. B. (1965). *In* "Plant Biochemistry" (J. Bonner and J. E. Varner, eds.), pp. 124–150. Academic Press, New York and London.
Preiss, J. and Kosuge, T. (1970). *A. Rev. Pl. Physiol.* **21**, 433–466.
Renger, R. (1973). *Biochim. biophys. Acta,* **314**, 113–116.
Rosenberg, J. L., Sahu, S. and Bigat, T. K. (1972). *Biophys. J.* **12**, 839–850.
Schweizer, E. E. and Rogers, B. J. (1964). *Weeds,* **12**, 7–10.
Slater, J. W. (1968). *Weed Res.* **8**, 149–150.
Stanger, C. E., Jr. and Appleby, A. P. (1972). *Weed Sci.* **20**, 357–363.
Templeman, W. G. and Sexton, W. A. (1945). *Nature, Lond.* **156**, 630.
Wain, R. L. (1964). *Proc. 7th Br. Weed Control Conf.* pp. 306–311.
Warburg, O. (1919). *Biochem. Z.* **100**, 230–270.
Warburg, O. (1920). *Biochem. Z.* **103**, 188–217.
Wessels, J. S. C. and van der Veen, R. (1956). *Biochim. biophys. Acta,* **19**, 548–549.
Wolf, F. T. (1960). *Nature, Lond.* **188**, 164–165.
Zucker, M. (1972). *A. Rev. Pl. Physiol.* **23**, 133–156.
Zweig, G., Shavit, N. and Avron, M. (1965). *Biochim. biophys. Acta,* **109**, 332–346.

CHAPTER 17

ACTIONS ON NUCLEIC ACID AND PROTEIN METABOLISM

JOE H. CHERRY

Department of Horticulture, Agricultural Experiment Station,
Purdue University, West Lafayette, Indiana, USA

I. Introduction

In the late 1940s studies were made involving indol-3yl-acetic acid and 2,4-dichlorophenoxyacetic acid (2,4-D) in an attempt to determine the molecular interaction with the plant cell. Most of the research at the time was directed toward the cell wall because it was believed that the hormonal action required a change in the cell wall for cellular expansion. Thus, the effects of the hormone were thought to change either the cell wall deposition or to hydrolyze certain cross linkages in the cell wall making it more elastic. This type of research was carried out for several years without a clear understanding of how auxin controlled such processes in the cell wall. In the 1950s a research team headed by Professor Folke Skoog began a study on the effects of auxin on nucleic acid synthesis in tobacco callus. Silberger and Skoog (1953) were the first to report that the auxin, indol-3yl-acetic acid, remarkably affected RNA and DNA contents in plants. Auxin increased the content of nucleic acids in tobacco tissue cultured on a sucrose agar medium. This increase occurred prior to the auxin-induced growth of the tissue at concentrations of indol-3yl-acetic acid which were optimal for cell enlargement. However, the concentration of auxin favoured DNA synthesis and cell division. Higher concentrations of auxin favoured DNA synthesis and cell division when compared with the optimum concentration for cell enlargement and RNA synthesis. Essentially these studies opened the door for a whole era of research involving chemical control at the genome level. Studies involving nucleic acid synthesis were continued in Professor John Hanson's laboratory at the University of Illinois. His students, particularly West and Key (West *et al.* 1960) showed that 2,4-D produced a wide range of morphological and physiological changes in the hypocotyl of soybean and the mesocotyl of corn. Within 15 to 24 h after 2,4-D treatment enlargement of the nucleus preceded the noted changes in cell division. Accompanying these changes were dramatic increases in RNA, most of which were ribosomal (Key and Hanson, 1961). Chrispeels and Hanson (1962) suggested that 2,4-D acts on the nucleus causing the cells to revert to a meristematic metabolism. The role of the nucleus in such a sequence of events is of obvious importance, since the nuclei had been shown to accumulate RNA in response to auxin. In the early 1960s Professor Joe Key began experiments on auxin regulation of nucleic acid synthesis. Key's students and post-doctoral colleagues at Purdue University classically demonstrated that auxin controls the synthesis of other nucleic acids besides ribosomal RNA. Their data revealed that auxin causes the reproduction of nucleic acid which appears to be of messenger RNA type (Key, 1964 and 1966; Key *et al.* 1967; Key and Ingle, 1964; Key and Shannon, 1964). Experiments carried out in my laboratory (Hardin and Cherry, 1972; Hardin *et al.* 1970 and 1972; O'Brien *et al.* 1968) over the past six or seven years have borne out the observations made by Key and his workers. Our data show that *in vivo* treatment of the soybean plant by 2,4-D increases the capacity of chromatin (O'Brien *et al.* 1968*a, b*) and RNA polymerase solubilized from chromatin (Hardin and Cherry, 1972) to synthesize nucleic acids. Very recently we (Hardin *et al.* 1972) have supported the idea that auxin reacts with a cytoplasmic component, a receptor, which moves into the nucleus where it then controls the activity of RNA polymerase. However, in terms of overall auxin regulation of plant growth and development there are now two major theories. One involves cell wall metabolism while the other involves control of the genome. In this chapter I intend to discuss primarily the research and data involved in the establishment and support of these two theories.

Data involving 2,4-D and IAA will be heavily utilized because of all the herbicides discovered and tested thus far, we know the most about the actions of 2,4-D on nucleic acid and protein synthesis. In the second portion of the chapter I will discuss actions of several other herbicides on either protein or nucleic acid synthesis. For example, EPTC appears to antagonize the action of 2,4-D in sorghum and in corn. Trifluralin leads to radial enlargement of the cortical cells of the roots and shoots in sorghum and corn. It also inhibits RNA and protein synthesis in root tips. Other experiments have showed that the carbamates inhibit the incorporation of ^{14}C-methyl-labelled methionine into protein. In still other experiments, complexes and metabolites of herbicides have also been shown to inhibit a number of physiological responses. Many of these herbicides block nucleic acid and protein synthesis by reducing ATP production.

II. Regulation of Cell-Wall Extensibility

The mode of action of auxin is to promote cell elongation. The phenoxyacetic acids have been widely utilized to study the effect of auxin on cell metabolism because the herbicide auxins are not rapidly metabolized. Furthermore, it is clear that 2,4-D and 2,4,5-T greatly promote stem elongation in a large number of plants. Therefore, a discussion of effects of auxin and auxin herbicides on cell wall properties is essential.

When plant cells elongate an accompanying increase in weight and volume takes place. This increase in size requires that the cell wall increase in mass as well as area. Even though the dry weight of wall material greatly increases due to cellular enlargement, the thickness and density of the wall remains constant. Therefore, it appears that wall synthesis is a fundamental requirement during cell elongation. In addition, an increase in wall mass must come from the deposition of polysaccharides. It was believed for many years that auxin, in this particular case phenoxyacetic acids, increases cell elongation by promoting cell wall synthesis. An alternative to this possibility is that the auxin-herbicide regulates enzymes involved in cell-wall loosening by breaking various cross linkages between the wall microfibrils which then allows wall extension. Plant tissues which respond to auxin by an increased rate of cellular elongation do indeed exhibit an increase in cell wall loosening. However, the kinetics of auxin-induced wall loosening may vary considerably from tissue to tissue. For example, in dark-grown maize coleoptiles wall loosening is induced by concentrations of 2,4-D at 10^{-4} M and proceeds very slowly with time (Morré and Eisinger, 1968). However, the same tissue responds to 2,4-D very dramatically between 2 and 6 h after exposure to auxin when maximal increases in total length occur. It is of interest, therefore, that the rate of cell-wall elongation, as measured by increased section length, is much greater than the increase in the cell wall extensibility (Figs 1 and 2). Therefore, the same

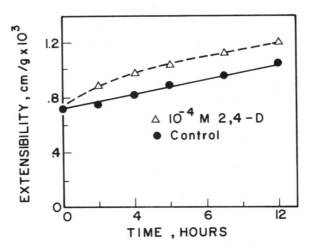

Fig. 1. Time course of extensibility of dark-grown maize coleoptile sections incubated in the presence and absence of 0·1 mM 2,4-D. Courtesy of Professor Morré.

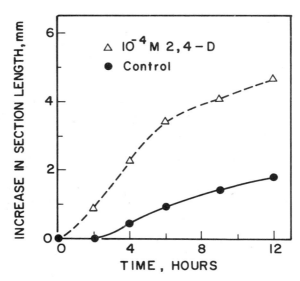

Fig. 2. Time course of growth of dark-grown maize coleoptile sections in the presence and absence of 0·1 mM 2,4-D. Courtesy of Professor Morré.

changes in wall extensibility caused by auxin in a given tissue may affect growth in different ways. In certain tissues, the rate of growth is constant while total extensibility changes. In other tissues the wall extensibility remains constant during the period of changing growth rate. But in general, with a given rate of tissue elongation it is possible to obtain a good correlation between growth rate and wall extensibility. Morré (1965) and Coartney *et al.* (1967) showed that increasing concentrations of auxin over the growth-promoting range also progressively increased wall loosening. Thus, it appears that some aspects of wall extensibility involved in cell elongation are not controlled by wall extensibility as measured by strain-stress methods. The early studies of Heyn (1940) as well as a more recent investigation by Cleland (1958) suggests that auxin affects the plastic components of the total cell wall extensibility more than other components in the cell elongation process. These studies and others indicate that both elastic and plastic components of wall extensibility are enhanced in auxin-treated tissue. Two types of cell-wall extension can be distinguished, reversible, which is elastic and irreversible which is plastic. Over the years there have been considerable disagreements over the relative importance of elasticity and plasticity in determining cell wall extension. Preston and Hepton (1960) suggested that the control of elastic extensibility by auxin during elongation of coleoptiles was consistent with the earlier results of Heyn (1940). However, Masuda (1961) working with oat coleoptiles reported that auxin-induced elasticity remained constant whereas the plastic component increased in proportion to total elongation. In the past it was considered that the breaking of chemical bonds within the cell wall leading to "wall loosening" would occur even if cell expansion was prevented by an external hypertonic solution. Cross links having been broken, the wall could then extend in the absence of auxin when tissue is returned to a normal external solution. However, the idea of wall loosening which could be "stored" and used later in expansion of the wall must now be discarded as a result of the information published by Cleland and Rayle (1972). They observed that auxin-treated oat coleoptiles when placed in an isotonic solution of mannitol would stop growing. However, the tissue would continue a normal rate of growth and have a normal rate of wall extensibility when returned to normal external solutions. In these types of experiments no "burst" in growth rate was noted when the mannitol treated coleoptiles were returned to the normal conditions. Since there was no over-shoot in growth, it appears that there could be no "stored growth" due to auxin treatment.

At the present time it seems that auxin causes a biochemical change in the wall, probably by breaking or modifying the cross links between the wall polysaccharide chains. These changes in the cell wall are then translated under cell turgor pressure into cell elongation. It is now relevant to discuss how auxin affects the cross links in a cell wall, and what the biochemical changes are which take place in a cell wall during cell elongation. In this regard it is necessary to look at the effect of auxin on various enzymes associated with the cell wall in the context of how these enzymes affect cell-wall loosening properties.

III. Regulation of Enzymes by Auxins

Auxin stimulates not only the synthesis of cellulosic polysaccharides (Hall and Ordin, 1968) but also the levels of several glucanases and pectic enzymes (Datko and Maclachlan, 1968). Masuda (1968) and Masuda and Yamamoto (1970) found that the fungal β-1,3-glucanase isolated from cultures of *Sclerotinia libertan* (Ebata and Satomura, 1963) induced rapid elongation of excised oat coleoptile segments. This enzyme was also shown to increase cell-wall extensibility as measured by a stretching method. In another study, Masuda *et al.* (1970) compared the activity of the endo β-1,3-glucanase and its activity on the *Avena* coleoptile cells with those of the exo enzyme. They found that exo glucanase enhanced elongation and extensibility of the cell wall, but the effect was not additive to the effect of indol-3yl-acetic acid. At least 3 h of incubation with exo glucanase was required for enhancement of elongation. The endo glucanase showed no significant effect on elongation and no interaction with the exo enzyme. Later, Masuda and Yamamoto (1970) fractionated the homogenate of the *Avena* coleoptile segment to determine the localization of the types of β-1,3-glucanase. The enzymes associated with the cell wall could be partially released when treated with a detergent. The bound cell-wall glucanase activity was increased when the coleoptile segments were treated with auxin. Furthermore, treatment of the tissue for only 10 min increased the glucanase activity as well as the incorporation of labelled leucine into protein. Auxin-induced activities were completely inhibited by 10 µg/ml of cycloheximide, an inhibitor of protein synthesis. The evidence suggests that regulation of cell wall enzymes plays an important role in auxin-induced cell wall loosening and in the elongation process of plant cells. Thus, in the past few years several laboratories have begun studies to determine the intercellular localization of the endogenous β-1,3-glucanases and the effect of auxin on its activity.

Of particular importance in reviewing the problem are the recent publications of Ray (1973a, b) on the regulation of β-glucan synthetase activity by auxin in pea stems. Some of his experiments illustrate the possibility that auxin control of cell elongation may be due to the regulation of particular enzymes involved in cell-wall formation. As discussed previously, the enzyme β-1,3-glucanase is a likely candidate. For several years, it has been known that GDP-glucose and UDP-glucose participate in the synthesis of cellulose in higher plant systems (Ordin and Hall, 1968). Furthermore, cellobiose, another polysaccharide containing β-1,3- and β-1,4-linkages, is synthesized through the utilization of UDP-glucose and GDP-glucose (Hall and Ordin, 1968). In particular, it appears that the UDP-glucose is most widely used in the synthesis of polysaccharides. In this regard, Ray (1973a) using a particle-bound glucanase has worked out a method to study the activity of the β-1,4-glucanase which requires UDP-glucose as its substrate. The enzyme was isolated by centrifuging a material which sedimented after 45 min at 40,000 g in an angle rotor. The effect of auxin on the β-1,3-glucanase activity was studied by cutting segments from the pea plant and incubating in buffer or other solutions containing auxin. When pea stem segments were incubated in water or in 17 μM indol-3yl-acetic acid at 25°C, he found that the enzyme activity rapidly declined over a period of 2 to 8 h before reaching a plateau. However, the rate of loss in enzyme activity was prevented by incubating the tissue in indol-3yl-acetic acid. Ray suggested that one of the reasons for the decline in the enzyme activity in the segment was that the growing plant provides various non-auxin factors to the tissue to help maintain the glucan-synthetase activity. Therefore, a search was begun to identify these factors. Even though nothing was found to maintain the enzyme activity unchanged at its initial level, sucrose was effective in maintaining a high level of enzyme activity. Tissue sections were incubated with and without auxin in the presence of 50 mM sucrose and the enzyme activity was determined at times ranging from 2 to 20 min. As indicated in Fig. 3 auxin greatly enhanced the enzyme activity within a few minutes. Maximal enzyme activity had been reached virtually within 10 to 12 min. It was found that with the addition of indol-3yl-acetic acid to the reaction mixture, the particles or the homogenate did not affect glucan synthetase activity. To characterize the product, ^{14}C-labelled glucan synthetase material was incubated with *Streptomyces* cellulase. The alkalie-soluble and alkalie-insoluble polymerases were completely degraded by cellulase. The disaccharide produced by this enzyme was completely converted to glucose by the almond β-glucosidase suggesting the original product to be celliobiase.

Tests on the binding of the enzyme to UDP-glucose revealed that in the presence of indol-3yl-acetic acid the enzyme appeared to have a higher affinity

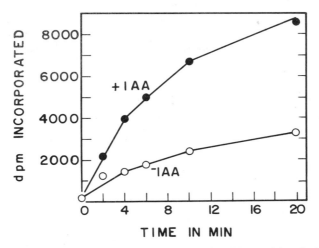

Fig. 3. Time course for glucan synthetase reaction catalyzed by particles obtained from 50 pea segments that had been kept 3 h in moist air at 35°C followed by 2 h at 25°C in 50 mM sucrose plus or minus 17 μM IAA. UDPG (60,000 d/min) was supplied in each assay which represents incorporation by 1/17 of the entire particle preparation. Courtesy of Professor Ray (1973a).

for the substrate than in the absence of auxin. It was shown that the synthetase was produced approximately proportional to the log of the IAA concentration. A response was detected to a concentration as low as 0·1 μM indol-3yl-acetic acid. Tests of various auxin analogues revealed that compounds such as naphth-1yl-acetic acid and 2,4-D were effective in increasing synthetase activity whereas related non-auxin analogues such as 2,6-D, 4-chlorophenoxy-isobutyric acid, phenylacetic acid and TIBA were without effect. In another study Ray (1973b) found that the two- to four-fold rise in the particle-bound β-glucan synthetase activity, which can be induced by indol-3yl-acetic acid in pea stem tissue, is not prevented by concentrations of actinomycin D (an inhibitor of RNA synthesis). The rise in enzyme activity is therefore concluded to be the hormonally induced activation of enzyme which previously existed as deactivated molecules. The activation of β-glucan synthetase activity is not a simple effector action of auxins on the enzyme system. This is shown both by the lack of effect when IAA or 2,4-D is added to the particles and also by the fact that activity does not become elevated if tissue is placed at 0 to 20°C for 10 min after exposure to IAA. The increase in β-glucan synthetase activity is blocked by respiratory inhibitors, by 2-deoxyglucose and by a high osmotic concentration of mannitol. However, the increase in β-glucan synthetase activity is not inhibited by calcium at concentrations that inhibit auxin-induced elongation and prevent promotion of sugar uptake by indol-3yl-acetic acid. The increase in enzyme activity is not inhibited by 2,2-dipyridyl at concentrations which inhibit the formation of 4-hydroxy-proline. Regulation of the

β-glucan synthetase system could be due either to an ATP-dependent activating reaction which effects the enzyme or to changes in the levels of a carbohydrate primer or possibility even to a lipid cofactor. Ray (1973a) concludes that the increase in β-glucan synthetase activity caused within an hour by indol-3yl-acetic acid qualified as a genuine auxin effect. Although the glucan synthetase enhancement is not as large as certain other plant growth regulatory effects on enzyme activity, it occurs considerably more rapidly than most other effects.

Generally, it should be noted that many of the other auxin-regulated effects have a lag time from one to several hours. As evidenced in the data the enhancement of glucan synthetase activity by indol-3yl-acetic acid clearly stands out as one of the most rapid auxin effects on a metabolic function yet demonstrated. However, it is to be remembered that Masuda and Yamamoto (1970) claimed a comparably rapid effect of indol-3yl-acetic acid on β-glucan synthetase of *Avena* coleoptiles. The well-demonstrated enzyme activity and metabolic effects of auxin, however, generally began an hour or more after exposure to indol-3yl-acetic acid. The most spectacular effects require many hours or days to develop. Neumann (1971), however, recently reported a non-enzymic metabolic auxin effect, the production of a glycerolphosphate, in pea tissue which occurred within 5 min.

Only time remains to demonstrate how auxin enhances the activity of β-glucan synthetase. It is possible that auxin merely releases the enzyme from an immobilized form such as that bound to the wall or plasma membrane. If this were so, release of the enzyme would lead to an enhancement of apparent activity without directly affecting the stereo-specificity of the protein in any way. The possibility also exists that the interaction of auxin herbicides with the cell surface may enhance a number of enzymes such as the β-glucan synthetase which would put into play a number of processes leading to the production of cell wall material and synthesis of other macromolecules including protein and nucleic acid. The first step in the enhancement of these many activities may be the attachment of the auxin to a single site which turns on all of these processes (Lembi *et al.* 1971).

IV. The Effects of the Phenoxyacetic Acids on Nucleic Acid and Protein Synthesis

Herbicides may control protein synthesis at many different levels. One of these controls, at the transcriptional level, relates to the specific regulation of DNA-directed RNA synthesis. A control at the genome might be achieved by gene derepression (removal of histones or acidic proteins) or activation of a specific RNA polymerase. A second control, at the level of translation, relates to the read-out of messenger RNA into protein. This type of regulation might be achieved by affecting specific transfer RNAs, ribosomes and various factors involved in protein synthesis.

Of these two possibilities, it is thought that chemical regulation of transcription by the auxin herbicides is more likely. The first indications came from the publication of Silberger and Skoog (1953) who showed that the native auxin, indol-3yl-acetic acid, remarkably effected RNA and DNA synthesis of tobacco callus tissue. Over the years since that time, research has been directed towards determining whether or not auxin increases cell enlargement directly by affecting cell wall structure via cell wall loosening and deposition as mentioned in the previous section. Alternatively, the hormone might affect the cells by changing the expression of the genome which would result in the synthesis of new types of RNA which in turn would be translated into new proteins. These new proteins may be synonymous with the so-called "growth limiting" proteins that are required for the growth of plant cells (Morré and Eisinger, 1968). A decade ago studies on nucleic acid synthesis as controlled by auxin, particularly by the phenoxyacetic acids, was begun in the laboratories of Hanson. West et al. (1960) examined the types of nucleic acids produced in soybean hypocotyl and corn coleoptiles after treatment with 2,4-D (Tables I and II). They found that within a few hours following 2,4-D treatment a large increase in ribosomes was noted. This was a result of a large increase in ribosomal RNA (Table III). At that time Key, working with Hanson, began to study RNA synthesis increased by 2,4-D. Also at about that time Bonner (1958) discussed the concept of "One gene—one microsome—one enzyme" implying that the synthesis of a single protein was controlled by one ribosome. Therefore, it seemed possible at the time that by controlling the synthesis of ribosomes the synthesis of protein would in turn be controlled (Key and Hanson, 1961). However, very soon it became more evident that an increase in ribosomes alone would not allow the hormone specifically to regulate protein synthesis. After the discovery of messenger RNA it became obvious that the regulation of the genome by auxin required a specific control of messenger RNA synthesis. Using a series of inhibitors Key (1966) and Key and Ingle (1964) noted that treatment of soybean hypocotyls with 5-fluorouracil (5-FU) reduced total RNA synthesis by 80% while growth proceeded normally. Further studies on the fractionation of nucleic acid from soybean hypocotyls on a methylated albumin Kieselguhr column showed that

TABLE I

CHANGES IN FRESH WEIGHT, PROTEIN NITROGEN AND RIBONUCLEIC ACID OF LEAVES AND HYPOCOTYL OF CUCUMBER PLANTS FOLLOWING 2,4-D TREATMENT

Treatment	Tissue	Fresh wt. gm/4 plants	Protein N mg/gm fr. wt.	RNA mg/gm fr. wt.
Control	Leaves	3·22	3·31	0·50
	Hypocotyl	1·94	0·39	0·08
2,4-D	Leaves	2·58	3·09	0·58
	Hypocotyl	2·22[1]	0·70[1]	0·40[2]

[1] Interaction significant, 5% level.
[2] Interaction significant, 1% level.
Printed by courtesy of West et al. (1960).

TABLE II

CHANGES IN RIBONUCLEIC ACID CONTENT OF CUCUMBER HYPOCOTYL AND CORN MESOCOTYL
SECTIONS FLOATED ON 2,4-D SOLUTIONS

2,4-D	Percentage increase in fresh wt.	Mg RNA per gm initial fresh weight					
		Experiment					
		1	2	3	5	5	Average
		Cucumber hypocotyl					
Control	17·6	0·586	0·590	0·643	0·613	0·700	0·626
5 parts/10^6	32·6	0·571	0·570	0·667	0·589	0·684	0·616
500–800 parts/10^6	20·0	0·621	0·655	0·717	0·730	0·841	0·713[2]
		Corn mesocotyl					
Control	12·9	0·497	0·537	0·729	—	—	0·588
5 parts/10^6	32·7	0·459	0·507	0·655	—	—	0·540[1]
500–800 parts/10^6	13·5	0·706	0·737	0·905	—	—	0·783[2]

[1] Significant, 5% level.
[2] Significant, 1% level.
Printed by courtesy of West et al. (1960).

TABLE III

CHANGES IN FRESH WEIGHT AND IN THE TOTAL AND PARTICULATE RNA AND PROTEIN OF
CUCUMBER AND CORN SECTIONS INCUBATED IN 2,4-D[1]

2,4-D parts/10^6	% Increase in fresh weight	Mg RNA/gm fresh weight			Mg protein/gm fresh weight		
		Cyto-plasm	Mito-chondria	Micro-somes	Cyto-plasm	Mito-chondria	Micro-somes
		Cucumber hypocotyl					
Initial	—	0·813	0·267	0·373	5·71	2·10	1·36
0	27·3	0·447	0·109	0·237	3·49	1·14	0·88
200	30·7	0·550	0·094	0·340	4·14	1·10	1·07
300	29·1	0·577	0·097	0·368	4·47	1·11	1·13
800	10·2	0·657	0·100	0·445	4·72	1·26	1·23
		Corn mesocotyl					
Initial	—	0·939	0·364	0·423	6·72	2·69	1·63
0	23·1	0·473	0·323	0·045	5·03	2·64	0·60
200	42·5	0·547	0·133	0·150	5·48	2·19	1·25
300	36·1	0·547	0·133	0·180	5·82	2·25	1·30
800	14·8	0·746	0·240	0·240	6·34	2·70	1·36

[1] Sections excised and treated as in Table II. Incubation was in the dark for 15 h at 28°C.
Tissue was homogenized, centrifuged and analyzed as described in methods.
Printed by courtesy of West et al. (1960).

5-FU inhibited the incorporation of labelled adenosine diphosphate into ribosomal RNA but had little effect on the labelling of m-RNA (or D-RNA). Additional experiments with soybeans showed that 5-FU did not affect the labelling of the tenaciously bound (tb) RNA whose synthesis is promoted by 2,4-D. Thus, those experiments were the first indication that the phenoxyacetic acids controlled protein synthesis through nuclear RNA production (apparently the mRNA type).

Sen and his workers (Roychoudury and Sen, 1964; Roychoudury et al. 1965) then showed that nuclei isolated from coconut milk responded to auxin by making more RNA. Many people thereby concluded that the hormone directly affected the nucleus by the regulation of gene expression (e.g. the presence of the hormone led to the exposure of more DNA template). In an attempt to test this hypothesis our first experiments (Cherry, 1967) showed that nuclei did not respond in vitro to concentrations of 2,4-D at 10^{-8} to 10^{-6}M. From many other experiments, Cherry (1967) demonstrated that only a few (approximately 10%) revealed that the nuclei responded to the hormone to make more RNA. In general, we found absolutely no effect of the auxin herbicide on the capacity of the isolated nuclei to produce RNA. On the other hand, it was found that isolated nuclei from 2,4-D-treated soybean seedlings produced twice as much nucleic acid as did control nuclei (Table IV). Furthermore, experiments of O'Brien et al. (1968b) showed that chromatin isolated from soybean hypocotyls did not respond in vivo to 2,4-D. In all cases the tissue needed to be treated with the herbicide for at least 2 h before any effect on chromatin-directed RNA synthesis could be noted (Table V). Subsequently, a progressive increase was noted in chromatin-directed RNA synthesis with time up to 12 h. The optimal 2,4-D concentration for

TABLE IV

INDUCTION OF INCREASED RNA SYNTHESIS BY ISOLATED NUCLEI FROM
SOYBEAN HYPOCOTYLS

Nucleic acid fraction	2,4-D Concentration parts/10^6		
	0	10	100
	Ct/min/fraction $\times 10^{-3}$		
Soluble RNA	34	76	58
DNA	26	63	57
Light ribosomal RNA	56	114	78
Heavy ribosomal RNA	181	385	367

Soybean seedlings (four-day-old) were sprayed with 2,4-D at indicated concentrations. After 24 h the nuclei were isolated and incubated in ^{3}H-adenosine. Labelled nucleic acids were extracted and fractionated on a MAK column. The amount of radioactivity incorporated was then calculated.

Printed with courtesy of O'Brien et al. (1968b).

TABLE V

INDUCTION OF INCREASED CHROMATIN ACTIVITY AFTER 2,4-D
TREATMENT[1]

pmoles ^{3}H-UMP incorporation per 100 μg DNA			
Hours after treatment	Control	2,4-D	Increase over control (%)
2	120	108	0
4	98	143	46
8	93	199	114
12	150	352	135
24	161	483	201

[1] Plants were sprayed with 1,000 μg/ml 2,4-D at the appropriate
time before harvesting.
Printed by courtesy of O'Brien et al. (1968b).

chromatin-directed RNA synthesis is in the range of 10 to 200 parts/10^6. We
suggested that this enhancement was a result either of a more active RNA
polymerase or of gene derepression leading to an increased availability of
DNA template (O'Brien et al. 1968b). A third possibility was that both were
involved. Subsequent experiments employing Escherichia coli RNA
polymerase revealed that chromatin from both control and 2,4-D-treated
plants saturated with the enzyme to the same extent and at the same level.
Even though the total Escherichia coli polymerase activities with chromatin
from 2,4-D-treated plants was slightly higher than that with control chromatin,
it was concluded that the herbicide primarily promoted the endogenous RNA
polymerase.

The rate of progress with chromatin studies was then hampered primarily
because of the failure to isolate soluble RNA polymerase. For a number of
years, we had tried to purify RNA polymerase from soybean hypocotyls by
methods similar to those employed by Stout and Manns (1967) for maize
shoots. However, we could not isolate successfully quantities of RNA
polymerase from soybeans to do the necessary experimentation. However,
within the past three years successful solubilization of RNA polymerase from
soybean hypocotyl chromatin has been achieved (Hardin and Cherry, 1972).
Recently we (Rizzo et al. 1974) learned two important aspects of solubilizing
RNA polymerase from chromatin. First, we found that purification of
chromatin over dense solutions of sucrose tended to eliminate not only a large
amount of the total chromatin DNA, but selectively removed the
nucleoplasmic chromatin, as judged by the small amount of RNA
polymerase II, the nucleoplasmic enzyme responsible for messenger RNA
synthesis, fractionated on a DEAE-cellulose column. Second, we have found
that the solubilized enzyme can be separated into at least three activity peaks
on a DEAE-cellulose column (see Rizzo et al. 1974 for details). The first

enzyme to be eluted from the column at low KCl concentrations is not inhibited by the mushroom toxin, α-amanitin, an inhibitor specific for RNA polymerase II. However, the second enzyme eluted from the column is totally inhibited by α-amanitin. The third fraction is probably composed of several proteins which contain cytoplasmic RNA polymerases, probably those RNA polymerases associated with cytoplasmic organelles such as mitochondria and proplastids. It is also likely that this third RNA polymerase fraction contains a non-nuclear RNA polymerase (probably a cytoplasmic organelle) other than the nucleoplasmic and the nucleolar polymerases. We have found from preliminary experiments that the third RNA polymerase fraction is partially inhibited by rifampicin, an inhibitor which blocks transcription of bacterial RNA polymerase and polymerases isolated from cytoplasmic organelles.

Attempts have been made a number of times to determine whether indol-3yl-acetic acid or 2,4-D directly promotes the activity of solubilized RNA polymerase (Hardin et al. 1972). However, no in vitro stimulation of RNA polymerase by the auxin herbicides has been detected. This fact led us to think of alternative ways in which the hormone may control the nucleic acid-synthesizing machinery. One of these models involved a cytoplasmic component found only in target tissues, a component that the hormone may react with or release by binding to a common cellular component. The released or complexed factor could then move into the nucleus and there control the activity of RNA polymerase. A proposed model involving a cytoplasmic component which modulates RNA synthesis in the presence of auxin will be discussed in detail in the following section.

V. REGULATION OF RNA POLYMERASE ACTIVITIES BY A CYTOPLASMIC FACTOR

A few years ago two groups of researchers, Professor Jensen and Professor Gorski, independently began studies on the animal steroid hormones, particularly estrogen. Their research (Shyamala and Gorski, 1969; Jensen et al. 1968) showed that the hormone entered the cell and formed a complex with a cytoplasmic receptor. This cytoplasmic receptor appeared to change in structure as it was transported into the nucleus where it apparently brought about changes in the capacity of the nucleus to make nucleic acids. Thus, the concept of hormone receptors was initiated and led to a number of other research investigations. At that time Professor Morré and I began to think of possible mechanisms in which the phenoxyacetic acids might control nucleic acid synthesis through the action of such a receptor. We felt that a logical site of a receptor would be on the plasma membrane. In accord with our hypothetical model (Hardin et al. 1970) we felt that the hormone first made contact with the outer surface of the membrane. It was thought that the auxin herbicide either led to a confirmational change in the membrane thereby releasing the receptor or penetrated the membrane to attach directly to the receptor. Either the free receptor or the hormone-receptor complex would then be free to move through the cytoplasm and into the nucleus where it could bring about the change in RNA polymerase activity. Again, utilizing some of the information that had been found in other systems and, in this particular case, systems involving *E. coli* RNA polymerase, we felt that the receptor complex could increase the initiation of transcription by RNA polymerase. An analogy for specific DNA transcription was made with the *Escherichia coli* system where sigma factor binds to DNA and then allows the core enzyme to initiate DNA transcription specifically. We therefore proposed that some factor, perhaps a specific auxin-receptor, of target cells could be released by the hormone. It then would move into the nucleus and bind to unique DNA sites and thereby control the transcription of genetic information. To test the hypothesis presented in the model we had to overcome two technical difficulties. One was to purify the plasma membrane, the supposed site of the factor, and the other to purify RNA polymerase. As has been mentioned previously, RNA polymerase has now been solubilized and partially purified from soybean tissues. Morré's group (Lembi *et al.* 1971) was also able to isolate and purify the plasma membrane. The aim of our two laboratories was to determine whether or not the plasma membrane from plant tissues would stimulate *in vitro* RNA polymerase activity (Hardin *et al.* 1972). The first three experiments (Table VI) show that the greatest stimulation of RNA polymerase

TABLE VI

EFFECT OF SOYBEAN MEMBRANE FRACTIONS ON RNA POLYMERASE ACTIVITY

$[^3H]$ UMP incorporated[1] (pmol/30 min per mg of protein)					
Experiment:	*I*	*II*	*III*	*Average*	*Difference*
RNA polymerase	200	199	280	226	—
+ fraction A	220	204	280	234	8
+ fraction B	240	196	293	243	17
+ fraction C	227	261	360	283	57
+ fraction D	346	256	267	256	30

[1] Experiment II was with membranes from five-day-old hypocotyls. Printed by courtesy of Hardin *et al.* (1972).

occurred from the fraction richest in plasma membranes, since the "C" fraction was previously shown to be composed of 70% plasma membrane. After demonstrating that the plasma membrane fraction stimulated the activity of solubilized RNA polymerase on calf thymus DNA as template, the release of factor from the plasma membrane was tested. According to our hypothesis treatment of the plasma membrane with auxin would release the factor from the membrane. Therefore, we suspended plasma membrane in buffer with or without 0·1 M 2,4-D, incubated and then centrifuged down the membrane fraction. The repelleted plasma membrane and supernatant derived from each treatment were tested with solubilized RNA polymerase and calf thymus DNA. The results (Table VII) show that most factor activity goes with the pelleted membrane unless auxin is included in the pre-incubation media. Thus,

TABLE VII

ENHANCEMENT OF RNA POLYMERASE ACTIVITY BY INCUBATION OF PLASMA MEMBRANES IN 2,4-DICHLOROPHENOXYACETIC ACID

	[^{3}H] UMP incorporated (pmol/30 min per mg of protein)
RNA polymerase	72
+ PM suspended in TGMED[1]	162
+ Repelleted PM suspended in TGMED	116
+ Supernatant after repelleted PM in TGMED	74
+ PM resuspended in TGMED + 0·1 μM 2,4-dichlorophenoxyacetic acid	236
+ Repelleted PM resuspended in TGMED + 0·1 μM 2,4-dichlorophenoxyacetic acid	162
+ Supernatant after PM repelleted in TGMED + 0·1 μM 2,4-dichlorophenoxyacetic acid	106
+ TGMED + 0·1 μM 2,4-dichlorophenoxyacetic acid	84

PM, plasma-membrane fraction.
[1] TGMED is a buffer composed of 0·05 M Tris (pH 7·9), 25% glycerol (v/v), 0·005 M MgCl$_2$, 0·0001 M EDTA and 0·0005 M dithiothreitol.
Printed by courtesy of Hardin *et al.* (1972).

it appears that incubation of membrane with auxin releases the RNA polymerase stimulating factor. To determine whether the release of the factor was specific for auxins we tested the effect of indol-3yl-acetic acid along with the non-auxin, 3,5-dichlorphenoxyacetic acid to compare with 2,4-D. The results (Table VIII) showed that only the auxins, indol-3yl-acetic acid and 2,4-D released the factor from the plasma membranes. Furthermore, first treatment of the membranes with the non-auxin, 3,5-D did not destroy the stimulating activity of the factor as judged by the fact that subsequent

TABLE VIII

COMPARISON OF RNA POLYMERASE ACTIVITY BY INCUBATION OF PLASMA MEMBRANE IN 0·1 μM 2,4-DICHLOROPHENOXYACETIC ACID, INDOL-3YL-ACETIC ACID OR 3,5-DICHLOROPHENOXYACETIC ACID

	[^{3}H] UMP incorporated (pmol/30 min per mg of protein)
RNA polymerase	88
+ PM-indol-3yl-acetic acid supernatant	146
+ PM-2,4-dichlorophenoxyacetic acid supernatant	138
+ PM-3,5-dichlorophenoxyacetic acid supernatant	96

Printed by courtesy of Hardin *et al.* (1972).

treatment of the membranes with 2,4-D released the factor from the membranes (data not shown).

Finally, we tried to determine which of the various RNA polymerases the factor stimulated, by using the inhibitor, α-amanitin which inhibits only RNA polymerase II (Lindell *et al.* 1970). According to Roeder and Rutter (1970), RNA polymerase II is found in the nucleoplasm, and is the enzyme that transcribes unique DNA into messenger RNA. Therefore, in the presence of the toxin, it is possible to determine whether the factor stimulated RNA polymerase I or II. Our results show that the factor released by auxin (either 2,4-D or indol-3yl-acetic acid) stimulates primarily the activity of RNA polymerase II, since the large increase in activity over the control is totally inhibited by α-amanitin. At the moment, we have not completely reconciled these data in terms of how the hormone controls nucleic acid synthesis and in turn how this leads to the production of new proteins. Neither do we know the composition of the factor. We do not know whether or not it is a protein, even though our preliminary data imply that the stimulatory factor is inhibited by protease. Other data (unpublished of Rizzo and Cherry) indicate that the factor is released when the membranes are treated with lipase. We have not been able to show that the factor is released with any other chemical yet tried, including cyclic AMP. Therefore, presently we can only speculate about the factor even though several other factors have been found in plant tissues. Matthysse and Phillips (1969) demonstrated the presence of a factor in tobacco callus tissue that stimulated the activity of *E. coli* RNA polymerase with pea-bud chromatin as template. Venis (1971) using a 2,4-D affinity column isolated a factor that stimulated chromatin-directed RNA synthesis. From soybean cotyledons Hardin *et al.* (1970) detected a factor that stimulated chromatin-directed RNA synthesis. Of particular significance is the finding by Mondal *et al.* (1972) of a number of transcription factors which appear to be involved in various steps of RNA polymerase transcription. These included factors involved in the initiation of transcription as well as the propagation and termination of the newly synthesized RNA chain. It therefore appears that transcription factors are present in many plant tissues, if not all. From these fragmented bits of evidence, and information involving hormones in other systems, particularly the steroids and insulin, we conclude that the phenoxyacetic acids mediate their actions by first binding to the plasma membrane surface. This either leads to a confirmational change in lipoprotein structure thereby releasing the transcription factor or, because the auxin binds to the receptor within the cytoplasm, directly sets off the hormonal response. The released receptor then moves through the cytoplasm and into the nucleus where it controls the transcription process. The mechanism of transcriptional control by the factor is up for debate. The transcription factor could specifically remove or release basic proteins or histones from DNA (Fellenberg, 1969*a, b*). Subsequently, the exposed DNA template would be transcribed by RNA polymerase already existing in the nucleus. Alternatively, we suggested that the factor binds to already exposed DNA sites and allows the core RNA polymerase to initiated RNA synthesis.

VI. ALTERATION OF NUCLEIC ACID AND PROTEIN BIOSYNTHESIS BY OTHER HERBICIDES

By far, more is known about the actions of 2,4-D and the other phenoxyacetic acids on nucleic acid and protein metabolism than any of the other herbicides. However, there are some herbicides that appear to act on plant cells in a manner somewhat analagous to the phenoxyacetic acids. Malhotra and Hanson (1966) stated that the herbicidal activity of picloram was like that of 2,4-D. Picloram enhanced both DNA and RNA synthesis in soybean and cucumber tissues, but no effect was noted in barley and wheat tissues which are tolerant of the herbicides. It is also of interest that EPTC appears to antagonize 2,4-D-stimulated growth (Beste and Schreiber, 1970). 2,4-D appears not to interfere with the uptake or movement of radioactively labelled EPTC within the tissues of sorghum. Respiration of sorghum plants was inhibited by EPTC and this inhibition was overcome by treating the tissue simultaneously with 2,4-D. Chen *et al.* (1972) working with roots of wheat and cucumber seedlings, studied the effects of 2,4-D, 2,4,5-T, dicamba and picloram on nucleic acid and protein changes. Among these herbicides 2,4-D enhanced DNA and protein production of four-day-old seedlings. The greatest difference between the effect of these herbicides on wheat, which is very tolerant, and cucumber, which is very sensitive, was in the RNA levels. As the concentrations of all four herbicides increased progressively, the amount of RNA within wheat tissue decreased. However, 10 and 100 parts/10^6 2,4-D increased the levels of RNA by as much as two-fold in cucumber. When protein levels per unit of RNA in wheat and cucumber were compared, the ratio was higher in the treated cucumber tissue. Chen *et al.* (1972) suggested that differential alterations of the RNA species (assume messenger RNAs to be controlled) produced would interfere with protein synthesis and in this way provide a basis of selectivity. However, it is more likely that the mechanism for controlling RNA and protein synthesis is provided by specific reactions with a receptor site or the binding of the hormone to a site which in turn controls the activity of an enzyme.

Zukel (1963) summarized work which indicated that labelled maleic hydrazide (MH) accumulated within the nucleus. Evans and Scott (1964) found that MH treatment of broad bean seedlings increased the time for mitotic division in the root tip cells. There are other herbicides which also appear to effect either DNA synthesis or chromosomal changes primarily in the roots of plants. Schultz *et al.* (1968) found that germination of corn seed was inhibited by trifluralin. This herbicide leads to radial enlargement of the cortical cells of roots and shoots and inhibits the elongation of *Avena* coleoptile sections at concentrations of 0·1 to 10 μM. Synthesis of DNA, RNA and protein is suppressed in the root tips while no significant effect is noted by these materials in the shoot. As judged by MAK chromatography, trifluralin inhibits ^{32}P incorporation into all fractions within the root cell. However, in shoots an enhancement was noted in the labelling of the nucleic acids. Hacskaylo and Amato (1968) found that trifluralin caused small and dense multinucleated cells and other forms of disorganized cell division within the extreme tip of cotton roots. Immediately behind the root tip region the cells were large, thin walled and aberrant. Treated tissue did not develop in the meristem cell-plate nor cell-wall. In no instance were sequences of mitosis evident. Therefore, it appears that trifluralin inhibits cell division in some fashion. It is not known whether the herbicidal action is via inhibition of DNA synthesis. There are other herbicides noted to inhibit cell division. Canvin and Friesen (1959) while working with the root tips of barley and peas showed that concentrations of the herbicides allidochlor and propham inhibited cell division. Propham at a low concentration (1 part/10^6) completely inhibited cell division in barley roots. At higher concentrations it caused endopolyploidy, binucleate cells and other abnormalities. In these experiments pea roots were only slightly effected by the herbicides.

It is of interest that other herbicides, particularly the phenylureas appear also to affect DNA synthesis. However, they may act in a fashion somewhat analogous to cytokinins. Cytokinins are known to regulate cell division. Interestingly, Mann et al. (1967) showed that the herbicidal derivative of phenylurethanes (phenyl-NH-CO-O-alkyl) and other synthetic cytokinins produced essentially the same result in a variety of assays. In one of these assays the induction of α-amylase by gibberellic acid in barley endosperm is inhibited by both the phenylurethanes and cytokinins. In another assay the herbicides prevented the loss of chlorophyll from excised barley leaves.

Amitrole has been found to inhibit the incorporation of labelled precursors into DNA, RNA and acid-soluble nucleotides of wheat seedlings (Bartels and Wolf, 1965). This herbicide causes a slight decrease in the ribosomal RNA content in light-grown plants. Schweizer and Rogers (1964) found that amitrole decreased the content of acid-soluble nucleotides in maize roots and in etiolated coleoptiles. Mann et al. (1965) found that protein synthesis was inhibited by concentrations of propham ranging from 1 to 50 parts/10^6. Chlorpropham and other carbamates as well inhibited the incorporation of ^{14}C-methyl labelled methionine into proteins of several plants. Resistance to inhibition by the carbamates varied with plant species and was in accord with previously observed field responses in terms of tolerance or resistance to the herbicides. Mann et al. (1965) used 23 herbicides in tests on the incorporation of leucine into protein (Table IX) and found that allidochlor, chlorpropham, endothal, ioxynil and PCP had pronounced effects. Barban inhibited ^{32}P incorporation into both RNA and DNA by about 75% in excised sesbania roots. Shokraii and Moreland (1966) found that propanil, chlorpropham and dinoseb inhibited leucine incorporation into protein and ATP incorporation into nuclear, mitochondrial, ribosomal and soluble RNA of excised soybean hypocotyls. Observed uncoupling of oxidative phosphorylation by ioxynil, dinoseb and PCP probably reduced ATP production rather than directly inhibiting either nucleic acid generation or protein synthesis.

Even though many workers have looked for effects of various herbicides (other than the phenoxyacetic acids) on protein and nucleic acid metabolism

TABLE IX

INHIBITION OF LEUCINE-1-C[14] INCORPORATION INTO PROTEIN BY VARIOUS HERBICIDES

Herbicide	Barley			Sesbania		
	2 parts/10⁶	5 parts/10⁶	Importance of inhibition	2 parts/10⁶	5 parts/10⁶	Importance of inhibition
None	0[1]	0[1]	—	0[2]	0[2]	—
Ethanol	−1	8	—	−6	23	—
Amiben	−20	—	—	4	22	—
Amitrole	−13	−4	—	19	33	—
Atrazine	7	11	—	37	32	—
Allidochlor	51	70[3]	+	58	87[3]	+
CDEC	25	29	—	7	0	—
Chlorpropham	26	84	+	32	98	+
Dacthal	18	34	—	15	21	—
Dalapon	7	8	—	4	8	—
Diallate	34	18[3]	—	22	8[3]	—
2,4-D	15	—	—	4	—	—
Dichlobenil	26	0	—	24	14	—
Diphenamid	21	27	—	17	22	—
Endothal	21	24	—	63	72	+
EPTC	38	22	—	14	11	—
Hadacidin[4]	12	15	—	18	1	—
Ioxynil	44	70	+	82	88	+
Maleic hydrazide	6	−1	—	0	−3	—
Monuron	32	19	—	24	24	—
Naptalam	−20	—	—	20	12	—
PCP	13	62	+	42	65	+
Propanil	—	8	—	−7	14	—
Pyrazon	25	29	—	35	26	—
Trifluralin	8	3	—	13	29	—

[1] Control values of protein synthesis by barley ranged in six experiments from 4,405 to 5,785 ct/min; with a mean of 4,850 ct/min; total uptake of leucine was approximately four-fold greater than incorporation into protein.

[2] Control of protein synthesis by Sesbania ranged in five experiments from 17,785 to 24,150 ct/min; with a mean of 19,980 ct/min, compared with total leucine uptake of 80,000.

[3] For reasons of solubility, 4 ct/min rather than 5 parts/10⁶ used.

[4] N-hydroxyl-N-formyl sodium glycinate.

Printed by courtesy of Mann et al. (1965).

few effects have been noted. In many cases the effects noted on protein and nucleic acid synthesis may be due indirectly to limitation of photosynthetic energy supply such as caused by diuron, a known inhibitor of the Hill reaction. In the cases of ioxynil, dinoseb and PCP it is likely that the major effect of the herbicides in limiting macromolecular synthesis is due to the reduction in ATP synthesis. It is known that approximately 90% of all the ATP generated in the cell is used to produce macromolecules. Therefore, a serious limitation of ATP production would drastically inhibit the synthesis of macromolecules including nucleic acids and protein. Along this line of thinking (Moreland *et al.* 1969) studied the effect of 22 different herbicides on RNA and protein synthesis and ATP generation in maize mesocotyl and soybean hypocotyl sections. The assays measured ATP and orotate incorporation into RNA, leucine incorporation into protein and gibberellin induction of α-amylase. Average results of the four assays suggested that 14 of the herbicides inhibited RNA and protein synthesis *in vivo*. The most inhibitory compounds were ioxynil, dinoseb, propanil, pyriclor, and chlorpropham. Isocil, allidochlor and picloram inhibited only the α-amylase assay while atrazine inhibited ATP incorporation. Trifluralin and chloramben had no measurable effects in any of the assays; dichlobenil was also inactive.

Jones and Foy (1971) studied the inhibition of α-amylase activity in barley half seeds as induced by endogenous gibberellic acids. They found that the induction of α-amylase was almost completely inhibited by 5×10^{-4} M fenac, bromoxynil and endothal. At the same concentration, paraquat inhibited α-amylase induction by 72%. Another herbicide, dalapon inhibited α-amylase induction by 42%. Schultz *et al.* (1968) also found that trifluralin inhibited α-amylase induction as well as the incorporation of ^{32}P into all RNA fractions.

Even though interference with nucleic acid and protein biosynthesis by herbicides implies disorder of a phytotoxic nature, it is not clear that all alterations in the synthesis of these macromolecules is a casual effect. Usually, past research has not shown direct killing by the herbicide. We know that both chloroplasts and mitochondria contain DNA differing in base composition from the nuclear DNA. DNA dependent synthesis of RNA and protein synthesis are also thought to take place in those organelles. Many of the herbicides which inhibit oxygen uptake by mitochondria and oxygen evolution by chloroplasts are also reported to interfere with nucleic acid metabolism and protein synthesis. Other sites of reaction by these herbicides, not found in the short-term oxygen-measurement studies, may well contribute to the overall biochemical responses to the herbicide. Therefore, regulation of RNA and protein synthesis, suppression or stimulation of DNase and RNase activity, and uncoupling of oxidative phosphorylation may be associated with the activity of many herbicides.

Many herbicides form complexes with amino acids, carbohydrates and cellular constituents such as conjugates. In this way, the herbicides may affect cell metabolism and reduce energy production which in turn would reduce protein and nucleic acid biosynthesis.

VII. Summary

The group of phenoxyacetic acids probably are most widely researched in terms of their mechanism of action. These synthetic auxins appear to control nucleic acid biosynthesis as well as controlling some aspect of cell wall loosening, deposition and relaxation. The net result of treating target cells of sensitive plants with the auxin type phenoxyacetic acids results in a large increase in ribonucleic acid production including messenger RNA. These herbicides then act in the cell as do indol-3yl-acetic acid by bringing about a large enhancement in RNA polymerase activity. The resulting synthesis of RNA and protein caused by the auxin-herbicides are accompanied by massive cell proliferation. In case of the very sensitive plants, low concentrations of herbicide lead to such a large proliferation of growth, swelling and appearance of gall-like areas on the roots and leaves that the vascular tissues are crushed causing death.

Finally, it can be stated that herbicides generally affect several complex areas of cell metabolism. Herbicides may uncouple oxidative phosphorylation or block photosynthetic phosphorylation thereby reducing ATP and energy production which in turn leads to cell destruction and degeneration. Herbicides may complex with reactants of cell metabolism and therefore, prevent them from participating in metabolism. Herbicides may directly effect enzyme activity as noted by the activation of β-glucan synthetase by auxin. Also, *in vivo* treatment of soybean hypocotyls with phenoxyacetic acids result in altered DNA-directed RNA synthesis. We believe that these herbicides indirectly alter genome transcription by controlling RNA polymerase activity.

References

Bartels, P. G. and Wolf, F. T. (1965). *Physiologia Pl.* **18**, 805–812.
Beste, C. E. and Schreiber, M. M. (1970). *Weed Sci.* **18**, 484–488.
Bonner, J. (1958). *Engng. Sci. Monthly* (California Institute of Technology) October.
Canvin, D. T. and Friesen, G. (1959). *Weeds*, **7**, 153–156.
Chen, L. G., Switzer, C. M. and Fletcher, R. A. (1972). *Weed Sci.* **20**, 53–55.
Cherry, J. H. (1967). *Ann. N.Y. Acad. Sci.* **144**, 154–168.
Chrispeels, M. J. and Hanson, J. B. (1962). *Weeds*, **10**, 123–125.
Cleland, R. (1958). *Physiologia Pl.* **11**, 599–609.
Cleland, R. and Rayle, D. L. (1972). *Planta*, **106**, 61–71.
Coartney, J. S., Morré, D. J. and Key, J. L. (1967). *Pl. Physiol., Lancaster*, **42**, 434–439.
Datko, A. H. and Maclachlan, G. A. (1968). *Pl. Physiol., Lancaster*, **43**, 735–742.
Ebata, J. and Satomura, Y. (1963). *Agric. biol. Chem.* **27**, 478–483.
Evans, H. J. and Scott, D. (1964). *Genetics*, **49**, 17–38.
Fellenberg, G. (1969a). *Planta*, **84**, 195–198.
Fellenberg, G. (1969b). *Planta*, **84**, 324–338.
Hacskaylo, J. and Amato, V. A. (1968). *Weed Sci.* **16**, 513–515.
Hall, M. A. and Ordin, L. (1968). *In* "Biochemistry and Physiology of Plant Growth Substances" (F. Wightman and G. Setterfield, eds), pp. 659–671. Runge Press, Ottawa.
Hardin, J. W. and Cherry, J. H. (1972). *Biochem. biophys. Res. Commun.* **48**, 299–306.

Hardin, J. W., Cherry, J. H., Morrë, D. J. and Lembi, C. A. (1972). *Proc. natn. Acad. Sci. U.S.A.* **69,** 3146–3150.
Hardin, J. W., O'Brien, T. J. and Cherry, J. H. (1970). *Biochim. biophys. Acta,* **224,** 667–670.
Heyn, A. N. J. (1940). *Bot. Rev.* **6,** 515–574.
Jensen, E. V., Susuki, T., Kawashima, T., Stumpf, W. E., Jungblut, P. W. and DeSombre, E. R. (1968). *Proc. natn. Acad. Sci. U.S.A.* **50,** 632–638.
Jones, D. W. and Foy, C. L. (1971). *Weed Sci.* **19,** 595–597.
Key, J. L. (1964). *Pl. Physiol., Lancaster,* **39,** 365–370.
Key, J. L. (1966). *Pl. Physiol., Lancaster,* **41,** 1257–1264.
Key, J. L., Barnett, N. M. and Lin, C. Y. (1967). *Ann. N.Y. Acad. Sci.* **144,** 49–62.
Key, J. L. and Hanson, J. B. (1961). *Pl. Physiol., Lancaster,* **36,** 145–152.
Key, J. L. and Ingle, J. (1964). *Proc. natn. Acad. Sci. U.S.A.* **52,** 1382–1388.
Key, J. L. and Shannon, J. C. (1964). *Pl. Physiol., Lancaster,* **39,** 360–364.
Lembi, C. A., Morrë, D. J., Thomson, K. and Hertel, R. (1971). *Planta,* **99,** 37–45.
Lindell, T. J., Weinberg, F., Morris, P. W., Roeder, R. G. and Rutter, W. J. (1970). *Science, N.Y.* **170,** 447–449.
Malhotra, S. S. and Hanson, J. B. (1966). *Pl. Physiol., Lancaster,* **41,** Abstr. vi.
Mann, J. D., Jordan, L. S. and Day, B. E. (1965). *Pl. Physiol., Lancaster,* **40,** 840–843.
Mann, J. D., Yung, K. H., Storey, W. B., Pu, M. and Haid, H. (1967). *Pl. Cell Physiol., Tokyo,* **8,** 613–622.
Masuda, Y. (1961). *Pl. Cell Physiol., Tokyo,* **2,** 129–138.
Masuda, Y. (1968). *Planta,* **83,** 171–184.
Masuda, Y., Oi, S. and Satomura, Y. (1970). *Pl. Cell Physiol., Tokyo,* **11,** 631–638.
Masuda, Y. and Yamamoto, R. (1970). *Development, Growth and Differentiation,* **11,** 287–296.
Matthysee, A. G. and Phillips, C. (1969). *Proc. natn. Acad. Sci. U.S.A.* **63,** 897–903.
Mondal, H., Mandal, R. K. and Biswa, B. B. (1972). *Eur. J. Biochem.* **25,** 463–470.
Moreland, D. E., Malhotra, S. S., Gruenhagen, R. D. and Shokraii, E. H. (1969). *Weed Sci.* **17,** 556–563.
Morrë, D. J. (1965). *Pl. Physiol., Lancaster,* **40,** 615–619.
Morrë, D. J. and Eisinger, W. R. (1968). *In* "Biochemistry and Physiology of Plant Growth Substances" (F. Wightman and G. Setterfield, eds), pp. 625–645. Runge Press, Ottawa.
Neumann, P. M. (1971). *Planta,* **99,** 56–62.
O'Brien, T. J., Jarvis, B. C., Cherry, J. H. and Hanson, J. B. (1968a). *In* "Biochemistry and Physiology of Plant Growth Substances" (F. Wightman and G. Setterfield, eds), pp. 747–759. Runge Press, Ottawa.
O'Brien, T. J., Jarvis, B. C., Cherry, J. H. and Hanson, J. B. (1968b). *Biochim. biophys. Acta,* **169,** 35–43.
Ordin, L. and Hall, M. A. (1968). *Pl. Physiol., Lancaster,* **43,** 473–476.
Preston, R. D. and Hepton, J. (1960). *J. exp. Bot.* **11,** 13–27.
Ray, P. M. (1973a). *Pl. Physiol., Baltimore,* **51,** 601–608.
Ray, P. M. (1973b). *Pl. Physiol., Baltimore,* **51,** 609–614.
Rizzo, P. J., Cherry, J. H., Pedersen, K. and Dunham, V. L. (1974). *Pl. Physiol., Baltimore,* **54,** 349–355.
Roeder, R. G. and Rutter, W. J. (1970). *Proc. natn. Acad. Sci. U.S.A.* **65,** 675–682.
Roychoudhury, R., Datta, A. and Sen, S. P. (1965). *Biochim. biophys. Acta,* **107,** 346–351.
Roychoudhury, R. and Sen, S. P. (1964). *Physiologia Pl.* **17,** 352–362.
Schultz, D. P., Funderburk, H. H. Jr. and Negi, N. S. (1968). *Pl. Physiol., Lancaster,* **43,** 265–273.
Schweizer, E. E. and Rogers, B. J. (1964). *Weeds,* **12,** 310–311.
Shokraii, E. H. and Moreland, D. E. (1966). *Abstr. Weed Sci. Soc. Am.* pp. 46–47.
Shyamala, G. and Gorski, J. (1969). *J. biol. Chem.* **224,** 1097–1103.
Silberger, J. and Skoog, F. (1953). *Science, N.Y.* **118,** 443–444.
Stout, E. R. and Mans, R. J. (1967). *Biochim. biophys. Acta,* **134,** 327–336.
Venis, M. A. (1971). *Proc. natn. Acad. Sci. U.S.A.* **68,** 1824–1827.
West, S. H., Hanson, J. B. and Key, J. L. (1960). *Weeds,* **8,** 333–340.
Zukel, J. W. (1963). "A Literature Summary on Maleic Hydrazide", p. 111. U.S. Rubber Co., Naugatuck, Conn.

AUTHOR INDEX

Numbers in *italic* type indicate pages in the References
at the end of each chapter

A

Aamisepp, A., 173, *187*
Abe, S., 285, *301*
Abeles, A. L., 272, *277*
Abeles, F. B., 43, *52*, 196, 202, 207, *214, 215*,
  256, 257, 258, 259, 260, 261, 262, 264,
  265, 268, 269, 270, 271, 272, 273, *277*,
  278, 279, 280, 430, 432, *441*
Åberg, B., 42, 44, *50*, 70, *78*, 478, *486*
Åberg, E., 173, *187*
Abraham, P. D., 258, *277*
Abu-Shakra, S., 168, *187*
Adachi, M., 414, 415, *422*
Adam, N. K., 342, 355, *360*
Adams, R. S., 243, 247, *250*
Adamson, D., 61, *78*
Addicott, F. T., 134, *161*, 192, 193, 194, 195,
  197, 198, 199, 200, 201, 202, 205, 207,
  210, 214, *214, 215, 217*
Aebi, H., 327, *330*, 369, 372, 390, *394*
Agamalian, H., 146, *163*
Ahlgren, G. H., 481, *491*
Ahmad, S., 173, *188*
Ahrens, J. F., 171, *188*
Akhavein, A. A., 182, *187*
Akita, S., 310, 318, *329*
Akobundu, I. O., 221, 224, *250*, 457, *486*
Alexander, A. G., 224, *250*
Ali, A., 482, *487*
Alimova, F. R., 228, *250*
Allard, R. W., 76, *78*
Allaway, M., 311, *332*
Allaway, W. G., 293, 295, *301*, 325, *329*
Allen, W. S., 320, *329*, 467, *487, 490*
Alley, H. P., 67, *80*, 100, *123*
Amato, V. A., 56, 57, 58, *80*, 95, *123*, 541,
  545
Amchem Products, 455, *487*
Amelunxen, F., 339, *360*
Amer, S., 119, *121*
Ames, B. N., 238, *251*
Amesz, J., 501, *522*
Andersen, R. N., 234, *250*
Andersen, S., 154, *161*

Anderson, D. B., 337, *360*
Anderson, H. L. 60, 73, *79*
Anderson, I. C., 206, *214*
Anderson, J. D., 435, *442*
Anderson, J. L., 59, 76, *81*, 91, 99, 108, 109,
  121, 122, 124, 320, *329*, 513, *521*
Anderson, O. R., 121, *121*
Anderson, R. N., 374, 375, *393*, 410, *424*,
  478, 479, *487*
Anderson, W. P., 78, *78*, 292, *301*
Andreae, W. A., 270, 271, *277*, 398, 400, *422*
Anon, 296, *301*
Antognini, J., 10, *50*
Antoszewski, R., 319, *332*
Apelbaum, A., 62, 63, 69, *78*, 94, *121*, 258,
  261, 262, 267, 268, 269, 270, 271, *277*
Appleby, A. P., 46, *54*, 76, *81*, 320, *333*, 518,
  523
Arai, M., 169, 170, *187*
Arch, P. D., 192, *217*
Ard, J. S., 44, *52*, 478, *488*
Arditti, J., 296, *303*
Arglebe, C., 111, *124*
Arlt, K., 129, 141, 145, 146, 148, 159, 161,
  161
Armstrong, D. J., 339, 349, *364*
Arndt, F., 9, *51*, 316, 322, *330*, 510, *522*
Arnold, W. E., 291, *302*, 405, *422*
Arnold, W. R., 148, *161*
Arntzen, C. J., 60, *78*, 95, *121*, 135, *161*, 300,
  303
Artz, T., 337, *360*
Ashton, F. M., 39, 43, 46, 47, 48, 49, 50, *51*,
  59, 73, 75, *78, 79*, 84, 95, 101, 105, 107,
  108, 110, *121, 122*, 129, 135, 137, 148,
  159, *161, 162*, 209, *214*, 225, 235, 236,
  239, 240, 243, 245, 248, *250, 252, 253*,
  285, 286, 290, 291, 293, 295, *302, 303*,
  311, 314, 320, 324, 327, *330, 331, 332*,
  365, 378, 389, *393, 395*, 398, 405, 419,
  420, *422*, 465, 466, 475, 477, 480, 482,
  483, 484, *487, 491*, 508, 513, 517, *521*
Åslander, A., 4, 40, 42, *51*
Assche, Ch. J., van, 134, *161*

547

SUBJECT INDEX

Types of reference are indicated as follows:
   Simple page number   Text reference
   Page number followed in parentheses by:
       (I)   Line drawing, graph or plate
       (F)   Structural formula of compound
       (T)   Data in tabular form

A

ABA, 134, 175, 202–203, 206–207, 307, 318
   abscission, 200
      promotion, 210
   dormancy induction by, 184
   ethylene synthesis and, 264, 272
   IAA interaction with, 434
   moisture stress and, 195, 196(I)
   stomatal closure by, 319(T), 325
   synthesis, 264
Abnormalities
   flowers, 149
   fruits, 156
   leaves, 144
   roots, 134
   stems, 138, 140
Abscisic acid (see ABA)
Abscisins, 429–430, 436
Abscission, 46, 110–111, 200, 207, 268
   anomalous, 213–214
   auxin levels and, 194, 200
   correlation, 195
   cotton, ethylene effects, 201(I)
   ecological conditions and, 192
   flower, 203
   fruit, 204
   hormone levels and, 196
   hormone metabolism and, 192, 195
   hydrolysis patterns and, 193
   hydrolytic enzymes, synthesis and, 194
   mature fruit, 208–209
   mineral nutrients and, 194
   minor element deficiency and, 195
   natural history of, 191
   photosynthate supply and, 194
   physiology of, 193, 197(I)
   prevention of, 202, 204(I)
   promotion of, 209–211
   respiration and, 194
   stem and branch, 211
   twig, 212(T)
   water relations and, 195

Abscission zone, 111, 192–193(I)
   ultrastructure, 115(I)
Absorption of herbicides, 231, 375
   effect of previous environment, 349
   environmental factors and, 347
   foliar, 351
   humidity and, 348
   by stomata, 349, 355
   stomatal opening and, 349
   temperature and, 347
   water relations and, 347
Abutilon theophrasti, 168, 172
Acacia, 213
Acacia farnesiana, 158, 181, 386–387
Accumulation of herbicides
   apical tissues, 384
   lysigenous glands, 390–391
Acer (*see also* Maple), 74
Acer macrophyllum, 382, 384, 399
Acer pseudoplatanus, 349
Acer rubrum, 87, 92, 387
Acer saccharinum, 421
Acetate metabolism, 474, 484
Acetyl-CoA, 476(I)
cis-Aconitic acid, 470
Acrolein, 412
   membrane integrity, 285
Actinomycin D, 118, 531
Action potentials, 298
Activation by metabolism, 43
Activation energy of enzymes, 428
Acylanilides, 313, 315, 503
   as electron transport inhibitors, 504
   structure-activity relations, 505
ADP, 445, 447, 449
Adventitious root development, 78, 141
African blood lily, 116–117
Agathis, 212
Agropyron desertorum, 148, 159
Agropyron repens, 7, 10, 90, 98, 143, 167,
      172, 175, 176(I), 177, 178(I),
      179–180, 182, 184, 186, 228, 236,
      275, 340(I), 374, 375, 384, 392, 510

573